*Towards an Understanding of*

# the Mechanism

# of Heredity

## H. L. K. WHITEHOUSE

*Sc.D.*
*Fellow of Darwin College, Cambridge*
*and Reader in Genetic Recombination,*
*University of Cambridge*

Foreword by
Professor G. PONTECORVO
*Dr. Agr., Ph.D., F.R.S.*
*Visiting Professor, University College, London*

EDWARD ARNOLD

© H. L. K. Whitehouse 1973

*First published 1965*
by Edward Arnold (Publishers) Limited,
25 Hill Street,
London, W1X 8LL

*Reprinted 1967*

*Second edition 1969*

*Reprinted 1970*

*Third edition 1973*

*Reprinted 1975*

Boards edition ISBN 0 7131 2411 3
Paper edition ISBN 0 7131 2412 1

To my MOTHER and FATHER
who first aroused my interest in the
scientific study of plants and animals

*Made and Printed in Great Britain by William Clowes & Sons Limited, London, Beccles and Colchester.*

# *Foreword*

Genetics—the study of heredity and variation—is going through a phase of obvious success. It is inevitable that a proportion of the young and more able artificers of this success should have come to think that what matters in genetics dates, say, from 1953. That was the time of the remarkable elucidation by Watson and Crick of the structure of DNA with its all embracing implications. In the last two years quite a number of books, some of them admirable, have been written with that conscious or unconscious attitude.

Dr. Whitehouse's book is refreshingly different. It traces the basic ideas of Genetics through their development and describes concisely, yet without omitting any essential detail, the critical experiments which led to the establishment of each of those ideas. This historical approach, invaluable for teachers and no doubt appealing to the more thoughtful students, is not at the expense of keeping abreast with the advancing frontiers. Suffice it to say that in one chapter Dr. Whitehouse deals with the latest genetical and biochemical approaches to recombination—a stagnant field since about 1930—which gives hope that this basic biological process may soon become understood.

To the honour of having been asked to write this foreword I can add the pleasure of having been compelled to read a most valuable book.

Glasgow  
September, 1965 G. Pontecorvo

# Preface to the First Edition

This publication is intended as an account of the primary evidence for current beliefs concerning the mechanism of heredity. Other aspects of genetics are referred to only if their study has illuminated fundamental features of heredity. Thus, the book deals only with the central core of genetics, omitting the more peripheral components. There is no discussion of how the hereditary mechanism may have evolved, nor of how evolutionary divergence of populations and species may have come about, and discussion of the many applications of genetics, such as to medicine, and to crop and livestock improvement, is omitted. The aspects of genetics which are included occupy such a central position in biology that in varying degrees all the rest of biology may be said to be related to them.

The theme of the book has been to trace the development of ideas about heredity from the earliest times. The approach, however, is not strictly historical. Each of the 16 chapters deals with a particular hypothesis concerning a fundamental feature of the hereditary mechanism, and sets out to show how experimental evidence has either confirmed the theory, or led to its modification or its abandonment. The classic experiments which gave rise to major advances are described, together with the inferences drawn from them. Particular trouble has been taken to distinguish observation from deduction. In this way it is hoped that the nature of the scientific method will be clearly demonstrated. The chemical and physical evidence for the structure of the components of DNA (Chapter 11), and the evidence from biochemical studies for the steps in protein synthesis (Chapter 14), are not given. On the other hand, some idea of the nature of the evidence for all the conclusions reached about the hereditary mechanism is included, with the exception of some of the inferences about genetic recombination in bacteria. Some features of the study of the hereditary mechanism are rather complex, but no attempt has been made to omit them on that account. In consequence, certain parts of the book are not easy reading, for instance, §§ 4.8, 8.6 and 9.8.

The book is intended for any interested person who has some background knowledge of biology and chemistry. No attempt has been made to describe the structure or life-cycle of the organisms mentioned in the text, but a systematic list of them is included (Appendix 1). It is hoped that the book will be useful to Biology students at Universities, including research students. Much of the book is based on the courses of lectures in genetics which I have given to first-, second- and third-year students reading Botany at Cambridge University. A glossary of genetical terms is included (Appendix

2) for the benefit of readers unfamiliar with the subject, and an extensive bibliography of original papers for those wishing to obtain more detailed information about the topics discussed.

The book has been written in the hope that an account of the remarkable progress that has been made (beginning with Mendel just a century ago) in understanding the mechanism of heredity, will give some readers the pleasure of intellectual stimulation. The immense advances of the last 20 years have greatly illuminated many of the earlier discoveries by revealing their molecular basis, and I hope readers will be able to recapture something of the intellectual excitement that this has caused. Much of the basic fabric of living organisms now stands revealed, with its central enigma of how the extraordinary interlocked system of nucleic acids and proteins (both of which, in different ways, are necessary for the synthesis of each) first evolved.

Cambridge
February, 1965                                          H. L. K. Whitehouse

# Preface to the Second Edition

In revising the book to take account of the remarkable developments in the last three years, I have largely rewritten the last three chapters, dividing the final one into three numbered 15, 17 and 18, the former Chapter 15 being renumbered 16. Some changes have also been made in Chapter 12 and slight alterations to several others. Over 250 references have been added to the bibliography.

I am grateful to Dr T. Butterfass of the Max-Planck-Institut für Pflanzengenetik at Ladenburg, Dr J. A. Hunt of the Department of Genetics, University of Hawaii, and Dr N. W. Simmonds, Director of the Scottish Plant Breeding Station, for drawing my attention to errors in the previous edition. I thank Dr S. A. Henderson of the Department of Genetics, University of Cambridge, for helpful discussions about subchromatid exchanges. Some of the revision of the book was carried out while I was a member of the Faculty of the Division of Biology at the Southwest Center for Advanced Studies at Dallas, Texas, and I would like to thank Dr Carsten Bresch and other colleagues for the opportunity to discuss many questions in genetics.

Cambridge
August, 1968

H. L. K. Whitehouse

# *Preface to Third Edition*

In revising the book for a new edition I have taken the opportunity to bring up-to-date the discussion of the transcription of genetic information from DNA to RNA and its translation into protein, and to consider in more detail than hitherto the diverse mechanisms which these processes offer for controlling gene activity. Considerable revision has also been made of the chapters on DNA replication, on the operon hypothesis, and on recombination. In the final chapter new sections have been added on satellite DNA, on antibody diversity, the source of which poses a major riddle for geneticists, and on chloroplast and mitochondrial DNA.

I am grateful to Dr. Ann P. Wylie of the University of Otago, New Zealand, for drawing my attention to a mistake in Chapter 10; to Professor P. R. Bell of University College, London, for telling me of Knapp and Schreiber's early evidence pointing to DNA as the genetic material; and to Dr. C. Milstein of the M.R.C. Laboratory for Molecular Biology, Cambridge, for discussion of antibody structure. I thank Dr. R. W. Davis of Stanford University Medical Center, Dr. O. L. Miller, Jr., of Oak Ridge National Laboratory, U.S.A., and Professor D. von Wettstein of the University of Copenhagen for their generosity over photographs.

Cambridge
August, 1972                                           H. L. K. Whitehouse

# Contents

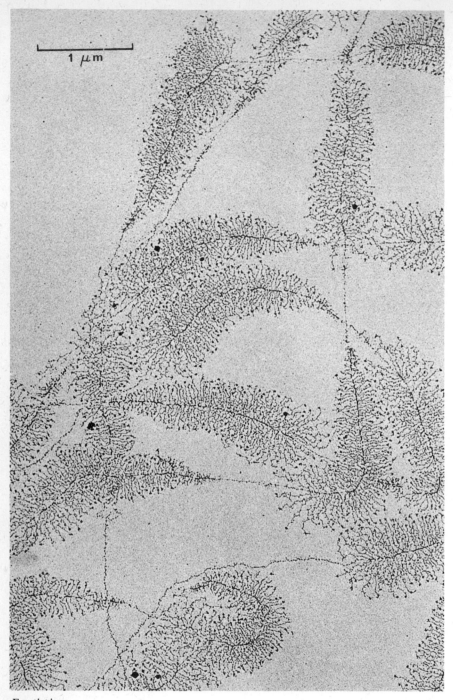

*Frontispiece*

Nucleolar genes isolated from an oocyte of the spotted newt, *Triturus viridescens*. For explanation see Plate 18.1.

(*Courtesy of O. L. Miller, Jr. and Barbara R. Beatty*, Biology Division, Oak Ridge National Laboratory, U.S.A.)

# 1. *The classical theory of direct inheritance of characters*

## § 1.1 The ideas of Hippocrates

Among the earliest known writings on the subject of heredity are those of Hippocrates (ca. 400 B.C.). He believed that the reproductive material came from all parts of the individual's body and hence that characters were directly handed down to the progeny. As evidence in support of this hypothesis he referred to a race of mankind called the Macrocephali, who, immediately after a child was born, fashioned its head by hand to give it an elongated shape. He also said that, at a later period, the elongated head was formed naturally, without the necessity of moulding it after birth. Hippocrates asked why, since such characters as baldness and blue eyes are inherited, should not the long-headed character be similarly handed down. Since he believed that the reproductive material coming from unhealthy parts of the body was correspondingly unhealthy, he was willing to accept the notion that mutilations, such as the moulding of the head, would in the course of one or a few generations become inherited. Hippocrates must have relied on hearsay for supporting his theory of heredity, as he could hardly have made direct observations of a change in a human population over a number of generations.

## § 1.2 The views of Aristotle

Aristotle (ca. 350 B.C.) questioned the Hippocratic view and pointed out the difficulties, even impossibilities, which it presented. He referred to the inheritance of such characters as the voice, nails, hair, and way of moving, which he believed could not contribute to the reproductive material, since they were intangible or concerned dead tissues, and he also referred to characters such as a beard or grey hair which may not be present at the time of reproduction. Further, Aristotle pointed out that children may resemble a grandparent more closely than they resemble their parents. He also drew attention to the dilemma which the Hippocratic theory presents when applied to plants, where, for various reasons, a part may be absent at the times of reproduction and yet be inherited. He was puzzled, too, as to how the two parents could each contribute something from all their parts, and yet their progeny have but one and not two of each part.

Aristotle argued that the reproductive material, instead of being derived

1

from all the parts of the individual as Hippocrates thought, should be regarded as made up of nutrient substances which, while on their way to the various parts, had been diverted to the reproductive path. These nutrient substances would differ from one another depending on the part of the organism for which they were originally destined. He also favoured the idea that the contributions to the progeny from the two sexes were not the same, the female contributing the material and the male something to define the form that the embryo is to assume, just as the carpenter's contribution to the making of a chair is merely its shape and form.

The ideas of Hippocrates and of Aristotle concerning heredity differ, but share the notion that inheritance is direct: that is, that substances derived from or intended for specific parts of the individual are handed down to the progeny. Hypotheses of this kind represent the simplest theory of heredity which one can postulate.

## § 1.3   Darwin's theory

Over the centuries many variations of the theories of Hippocrates and Aristotle have been proposed, but the basic idea that characters are in some way transmitted directly from parent to offspring was never effectively challenged until the year 1883 when Weismann proposed his theory of the continuity of the germ-plasm. Mendel's conclusions (1866) implied the discarding of the classical view of heredity, but this was not appreciated at the time, as his contribution was largely overlooked until 1900. Thus, Darwin (1868, Chapter 27) suggested that all the cells and tissues of an organism threw off minute granules, both during development and when the individual had reached maturity. He further supposed that these granules circulated through the plant or animal, multiplied, and were passed to the reproductive cells, which thus contained a multitude of components thrown off from each individual part of the organism. In this way, hereditary characters were thought to be transmitted to the progeny. In the offspring the granules were regarded as responsible for the development of cells or tissues corresponding to those from which they had been derived in the parent. The granules were supposed to be transmitted occasionally in a dormant state for several generations before developing to reveal an ancestral character. As Darwin (1875, Chapter 27, footnote 42) himself remarked, this theory of heredity resembles that of Hippocrates. It differs chiefly in specifying that granules were responsible for the hereditary transmission.

## § 1.4   Experimental results obtained by Knight and by Goss

Experimental evidence which appeared to establish the existence of a hereditary mechanism of the kind which Hippocrates described has been obtained on a number of occasions. One of the most recent concerns the effect of fertilizer treatment on *Linum usitatissium* (Flax) (Durrant, 1962; see also Mather and Jinks, 1971). Rigorous attention to many points of detail is necessary in order to exclude the possibility of selection, and no claim for the existence of a direct adaptive influence of the environment on inheritance has gained general acceptance. In view of this lack of confirmation, it is remarkable that Hippocrates's hypothesis remained unchallenged for 23 centuries. Numerous experiments on the crossing of different species or varieties of plants and animals were performed during the 18th and 19th centuries, but although these studies failed to provide support for the classical theory, they did not lead to any alternative hypothesis. That is, of course, with the notable exception of Mendel's work. Some of the most promising earlier experimental studies were made by Knight and by Goss with *Pisum sativum* (Edible Pea) and since this was the plant which Mendel subsequently used so successfully, their work will be described.

Knight (1799) lived near Ludlow, England, and his primary intention was to obtain new and improved varieties of fruits and vegetables. He chose the Edible Pea as his initial experimental material, because of its short generation time, the numerous varieties available and the self-fertilizing habit, which made the protection of the flowers from insects carrying pollen unnecessary.

He crossed two varieties which differed in colour. One was unpigmented, having green stems, white flowers and colourless (white) seed-coats, while the other had purple stems and flowers and grey seed-coats. He found that when an unpigmented plant was pollinated by a pigmented one, only pigmented progeny were obtained the next year (the first filial generation), but these on self-pollination, or on pollination by unpigmented plants, produced some pigmented and some unpigmented plants in the succeeding year. This diversity was manifest even when the peas from within one pod were sown. However, he did not record the numbers of the two kinds, either in the progeny of individual plants, or in total, and so failed to discover the underlying mechanism of heredity and was able merely to deduce that there was a 'stronger tendency' to produce coloured than colourless plants.

Knight also established that reciprocal crosses gave identical results. Thus a tall variety pollinated by a dwarf one, and *vice versa*, gave plants which were the same as regards their vigour of growth, the size of the seeds and the season of maturity. Unknown to Knight, the identity of the progeny from reciprocal crosses had already been discovered by Kölreuter (1763) using species of *Nicotiana* and other flowering plants.

Goss (1824) made similar discoveries to those of Knight, but took the analysis a little further. He lived at Hatherleigh in Devonshire, and had also been attempting to produce new varieties of vegetables. He had removed the stamens from a normally green (blue) seeded pea, pollinated it from a yellow (white) seeded variety, and had been surprised to find that the seeds set (the first filial generation) were yellow like the male parent (Fig. 1.1). He sowed these the following year and was equally surprised when, after self-pollination, he found some pods with all green, some with all yellow, and many with both green and yellow peas in the same pod (the second filial generation) (Fig. 1.2). On sowing these the next year, he found on further 'selfing' that whereas the green peas bred true and gave only green progeny, the yellow yielded some pods with all yellow and some with both green and

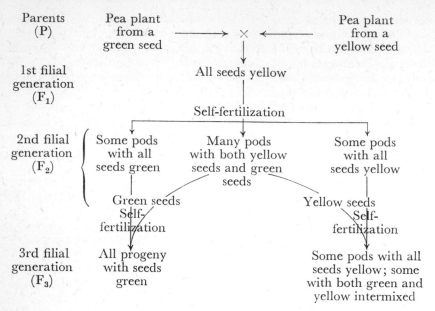

| Parents (P) | Pea plant from a green seed | → × ← | Pea plant from a yellow seed |

FIGURE 1.1   Goss's results from crossing green and yellow peas (*Pisum sativum*).

yellow peas intermixed (the third filial generation) (Fig. 1.1). Forty-two years later, Mendel (1866) reported similar results. However, Goss, like Knight, did not count the numbers of the two kinds of peas (or if he did count them, he failed to see the significance of the figures and did not publish them), and so he failed to discover the hereditary mechanism which Mendel found.

The studies by Knight and by Goss with peas illustrate the kind of result that was obtained in the 18th and 19th centuries with many organisms. The experimental work at this time appears to have been designed, either to obtain new and improved varieties of animals and plants for use in agriculture or

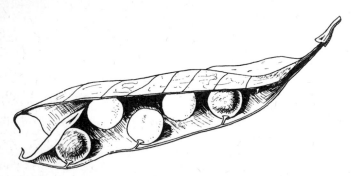

FIGURE 1.2   Drawing of a pea pod with seeds of two colours, similar to that used to illustrate Goss's paper.

horticulture, or with the object of understanding the nature of species by means of artificial hybridizations, rather than with the specific intention of obtaining information about the mechanism of heredity. Probably this is the reason that the experimental results were not recorded with sufficient detail and precision for anything about heredity to emerge except 'an inscrutable medley of contradictions' (Bateson 1909, page 2). To take one example, Goss states that his second generation yellow peas produced, on selfing, some pods with a mixture of green peas and yellow peas, and some with all the peas yellow. If he had studied the plants more carefully he would have found no doubt, as Mendel did, that one can substitute 'plants' for 'pods' in the previous sentence; in other words some of the second generation yellow peas breed true. He would then have been much nearer to understanding the underlying mechanism of inheritance.

# 2. *The theory of indirect inheritance through particulate determinants*

## § 2.1 Mendel's experimental results

Mendel (1866) had made crossing experiments with *Pisum sativum* similar to those of Knight, Goss and others but, unlike them, he recorded the numbers of progeny of each kind. This was the rewarding advance which enabled him, by a stroke of genius, to detect the underlying mechanism and so to put forward an entirely new hypothesis concerning heredity.

One of the character-differences which Mendel studied was the same as Knight had used about 60 years earlier, namely the presence or absence of pigmentation in the plant (stems, flowers and seed-coats). Like Knight, Mendel found that the progeny from a cross between the two forms were pigmented (the first filial generation), and that, on self-pollination of these, both pigmented and unpigmented plants were obtained (the second filial generation). But Mendel went further and counted the numbers of each kind. He found that among 929 plants, 705 were pigmented and 224 were not, and he observed that these frequencies closely approximate to $\frac{3}{4}$ and $\frac{1}{4}$ of the total (696·75 and 232·25), respectively. The character—presence of pigment, in this instance—which is manifest in all the immediate progeny of the cross, and in $\frac{3}{4}$ of the following generation, Mendel called the *dominant* character, the other being the *recessive*. He also confirmed Kölreuter's and Knight's observations that reciprocal crosses ($A$ female $\times$ $B$ male, and $B$ female $\times$ $A$ male) give similar results.

Mendel also worked with the character-difference which Goss (1824) had used, confirming that yellow seed (embryo) is dominant to green, and that both kinds appear in the second generation after crossing. Unlike Goss, Mendel counted his peas, and recorded that out of 8023, 6022 were Yellow* and 2001 green, or again a very close approximation to $\frac{3}{4}$ ($= 6017\cdot25$) and $\frac{1}{4}$ ($= 2005\cdot75$). Mendel confirmed Goss's third-generation observations, namely, that the green peas bred true, whereas the Yellow often did not. But Mendel went much further than this. Of 519 such Yellow seeds, he found that 166 bred true, while 353 did not (but gave Yellow and green seeds in the proportion of 3 to 1 like the previous generation). Mendel observed that the frequencies 166 and 353 are a close fit with $\frac{1}{3}$ ($= 173$) and

* A capital initial letter is used for dominant characters, lower case for recessive ones.

$\frac{2}{3}$ ($= 346$) of the total, respectively. This implied that the second-generation ratio of 3 dominants to 1 recessive was really a ratio of 1 pure-breeding dominant to 2 impure dominant to 1 recessive (always pure-breeding). Mendel pursued his inheritance studies through 5 or 6 generations and showed that the pure-breeding types in the second filial generation remained pure-breeding, and the others gave a 1:2:1 ratio in each generation.

Essentially similar results to those just described were obtained by Mendel for a total of 7 character-differences in *P. sativum*. In the first generation from a cross, one character was always dominant to its alternative, and in the second generation showed a 3 to 1 ratio, and this, on further analysis, was shown to be a disguised 1 to 2 to 1 ratio. What was the significance of these simple numerical ratios?

## § 2.2   Mendel's theory

In building an hypothesis to account for his results, Mendel introduced symbols, *A* for the dominant and *a* for the recessive character. This use of

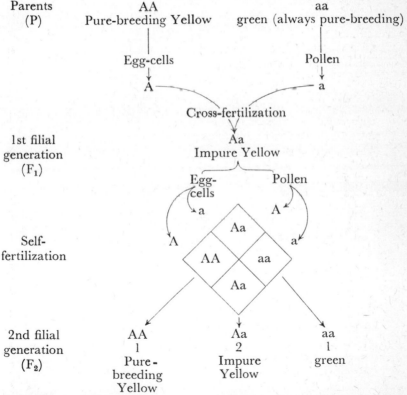

FIGURE 2.1   Representation of one of Mendel's experiments, using symbols (*A* and *a*) for the hypothetical hereditary determinants (*A* for Yellow, *a* for green seed-colour).

symbols is most significant, as it implies the existence of something which 'stands for' the character. Although Mendel used 'Charakter' or 'Merkmal' almost throughout his paper, it is clear that he was thinking in terms of 'factors' or 'determinants' responsible for the manifestation of the character. This is the crucial feature of his theory of heredity, that characters are not 'transmitted' directly from generation to generation as the classical theory supposed, but that there exist discrete particles responsible for the appearance of particular characters. Furthermore, each individual receives one particle from each of its two parents in respect of a particular character-difference; and these particles do not influence one another in any way but separate uncontaminated at the time of formation of the reproductive cells. If *A* denotes the particle which determines Yellow seeds and *a* that for green seeds, then on this hypothesis, Mendel's experiment described above can be represented as in Fig. 2.1. A further corollary implicit in the hypothesis was that egg-cells carrying *A* were equally likely to be fertilized by pollen cells carrying *A* or those carrying *a*, and similarly with egg-cells carrying *a*. In other words, fertilization was at random between egg-cells and pollen-cells, irrespective of what factors they carried.

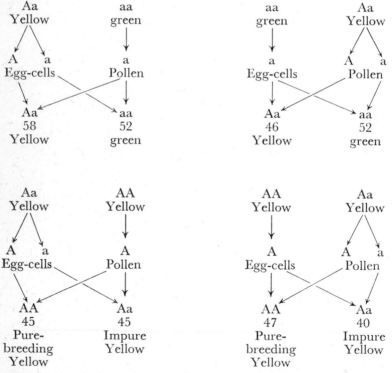

FIGURE 2.2   Representation of a set of back-cross experiments used by Mendel to test the validity of his theory of heredity. *A* = determinant for Yellow seed-colour, *a* for green.

In order to test the validity of this theory of heredity, Mendel devised a series of experiments. The Yellow peas obtained in the first generation from a cross between pure-breeding Yellow and green (necessarily pure-breeding) were germinated and the flowers on the resulting plants, instead of being allowed to self-pollinate, were crossed with the parental forms. On Mendel's hypothesis, from such a 'back-cross' to the green parent, a 1 to 1 ratio of Yellow to green peas is expected and this prediction was confirmed experimentally: the first generation Yellows as female parent crossed with green as male parent gave 58 Yellow and 52 green peas as progeny, and the reciprocal cross (green as female parent, first generation Yellow as male parent) gave 46 Yellow and 52 green. Comparable results were obtained on back-crossing the first generation Yellow peas to the Yellow parental type, but here it was necessary to grow the progeny and allow them to self-pollinate in order to reveal the 1:1 ratio of pure-breeding and impure Yellows. These experiments confirming the hypothesis are shown diagrammatically in Fig. 2.2. Mendel made similar tests with all the 7 character-differences in peas with which he worked, and in every case he obtained the expected 1:1 ratio.

## § 2.3   The rediscovery of Mendel's work

Mendel's work made no impression on contemporary thought and indeed was almost entirely overlooked for 34 years. It was 'rediscovered' independently by De Vries (1900), Correns (1900) and von Tschermak (1900). Each of these authors had made experiments similar to those of Mendel and obtained comparable results. Thus, De Vries had found that various character-differences in a whole range of species of flowering plants showed dominance in the first generation from a cross and 3:1 ratios in the second generation. The species and characters are listed in Table 2.1. Correns had worked with *Zea mays* (Maize) and both he and von Tschermak with *Pisum sativum*. The characters in *P. sativum* which they studied included those used by Knight and by Goss and which, unknown to them, had also been used by Mendel. Correns introduced the expression 'Mendels Regel' (Mendel's law) for the basic principle, namely, the separation or segregation at the time of formation of the reproductive cells of the discrete particulate determinants of alternative characters, and their reassociation at fertilization.

So, when Mendel's work was rediscovered, a number of his experiments had already been repeated and his results confirmed independently by two observers. The truth of Mendel's work was thus immediately established. Moreover, its wide applicability was demonstrated by De Vries' experiments. Further evidence for this came from the work of Bateson and Saunders (1902)*, who established that Mendel's law applied to *Gallus*

---

* Bateson and Saunders introduced several terms which have since been generally adopted: *allelomorph* (now often abbreviated as *allele*) for each of the alternative factors responsible for a particular character-difference; *heterozygote* for the product of fertilization of gametes differing by an allelomorph; and *homozygote* for the product of fertilization of gametes carrying the same allelomorph. They also proposed the symbols $P$ for the parental generation, $F_1$ for the first filial generation, $F_2$ for the second filial generation, and so on.

TABLE 2.1  Character-differences found by De Vries (1900) to show Mendelian inheritance as indicated by 3:1 ratios in the second filial generation.

| Name of Plant | Characters | |
|---|---|---|
| | Dominant | Recessive |
| *Agrostemma githago* (Corn Cockle) | Normal reddish-purple petals | Petals pale-coloured |
| *Aster tripolium* (Sea Aster) | Normal blue flower-colour | White flower-colour |
| *Chelidonium majus* (Greater Celandine) | Normal pinnately cut leaves and petals | Laciniate leaves and petals |
| *Chrysanthemum roxburghi* | Normal yellow flower-colour | White flower-colour |
| *Coreopsis tinctoria* | Normal yellow flower-colour | Brown flower-colour |
| *Datura tatula* and *D. stramonium* (Thorn Apple) | Purple stem- and flower-colour | Green stem- and white flower-colour |
| ——— | Thorny fruit | Unarmed fruit |
| *Hyoscyamus niger* (Henbane) | Normal purple-veined corolla | Corolla pale-coloured |
| *Oenothera lamarckiana* De Vries (Evening Primrose) | Normal style | Short style |
| *Papaver somniferum* (Opium Poppy) | Dark spot at base of petals | Petals without basal dark spot |
| ——— | Normal single flower | Double flower |
| *Silene alba* (White Campion) | Normal hairy shoots | Glabrous shoots |
| *Silene dioica* (Red Campion) and *S. alba* | Red petal colour | White petal colour |
| *Solanum nigrum* (Black Nightshade) | Normal black fruit | Green fruit |
| *Trifolium pratense* (Red Clover) | Normal red flowers | White flowers |
| ——— | Quinquefoliate leaves | Normal trifoliate leaves |
| *Veronica longifolia* | Normal blue flower-colour | White flower-colour |
| *Viola cornuta* | Normal blue flower-colour | White flower-colour |
| *Zea mays* (Maize) | Normal starchy endosperm | Sugary endosperm |

*domesticus* (Domestic Fowl), as well as to further plants. In *G. domesticus* they found an extra toe was dominant to the normal foot; pea comb was dominant to single comb; rose comb was also dominant to single comb; and white shanks and bill were dominant to yellow. Each of these character-differences appeared in a 3:1 ratio of dominants to recessives in $F_2$. Simultaneously and independently, Cuénot (1902) also established that Mendelian inheritance occurred in animals, when he showed that the normal grey coat-colour in *Mus musculus* (House Mouse) was dominant to albino in $F_1$ from a cross, and segregated to give 198 grey and 72 albino in $F_2$.

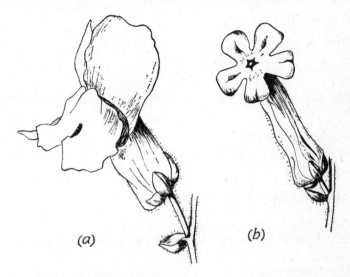

(a)                                           (b)

FIGURE 2.3   Drawings of (a) normal, and (b) peloric *Antirrhinum majus* (Snapdragon).

Bateson and Saunders (1902) drew attention to the fact that many studies on inheritance dating from before 1900, previously inconclusive, were now explicable in terms of Mendelian inheritance. They quoted, for instance, the work of Darwin (1868, vol. 2, chapter 14) with peloric *Antirrhinum majus* (Snapdragon). This form differs from the normal in having a regular radially-symmetrical flower instead of the normal irregular two-lipped structure (Fig. 2.3). Darwin's results are shown in Fig. 2.4. They were interpreted by him as indicating that the tendency to produce normal flowers prevailed in the first filial generation from the cross between the two forms, whilst 'the tendency to pelorism appeared to gain strength by the intermission of a generation', and prevailed to a large extent in the second filial generation. But on Mendel's theory of heredity it is merely necessary to postulate that the factor for pelorism is recessive to the normal. Darwin's $F_2$ frequencies, omitting the two intermediate plants, are in good agreement with the 3:1 ratio expected on the Mendelian hypothesis. Later studies have confirmed the Mendelian explanation.

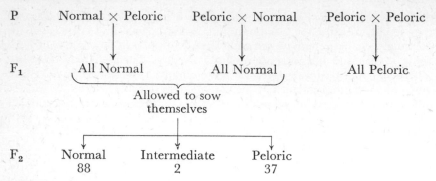

FIGURE 2.4   Darwin's results from studies of the inheritance of peloric *Antirrhinum majus* (Snapdragon).

## § 2.4   The difference between the classical and Mendelian theories

The example from Darwin's work provides a good illustration of the difference between the classical and Mendelian theories of heredity. On the classical theory, something is transmitted directly from each part of the organism, for example, the flower of the *Antirrhinum*, to the corresponding part in the progeny. On the Mendelian hypothesis, on the other hand, inheritance is indirect through the agency of particulate determinants. It is these determinants which are transmitted from generation to generation, and the character of the flowers which develop in any individual is merely the fortuitous consequence of the set of determinants which it happens to receive at fertilization. Thus, the classical theory deals with the tendencies of characters to gain strength or lose strength, while Mendel's is concerned with chance combinations in pairs of dominant and recessive particulate determinants of characters. The fundamental difference is that on the classical theory inheritance is direct and hence the extent of transmission of a particular character to the progeny may be influenced by the character itself, whereas on the Mendelian theory the determinant of a character is considered to be in no way modified by its presence in an organism, whether it possesses that character or not, and hence the inheritance of a character is thought to be unaffected by the character itself. Thus, on the Mendelian theory, the inheritance of mutilations, such as Hippocrates quoted in support of the classical theory, is impossible.

## § 2.5   The diversity of Mendelian characters

Bateson, Saunders and Punnett (1905) established one of the first instances in Mendelian inheritance of the occurrence of incomplete dominance, that is, where the heterozygote is intermediate between the two homozygotes. They found that the so-called 'blue' Andalusian variety of *Gallus domesticus* (really a grey) is the heterozygote from a cross between a black and a white variety. Many further instances of incomplete dominance were established soon afterwards, both in animals and in plants. Dominance and recessiveness were thus seen not to be an essential part of Mendel's theory of heredity, the central notion of which is the coming together at fertilization, and the separation at gamete-formation, of pairs of particulate determinants.

Within a few years of the rediscovery of Mendel's work, innumerable examples of Mendelian inheritance had been established. Bateson (1909), without any attempt at completeness, listed about 100 examples in plants and a similar number in animals. Both the organisms and the characters involved were highly diverse, including susceptibility to *Puccinia glumarum* (Yellow Rust) in *Triticum aestivum* (Wheat); eye-colour in *Homo sapiens* (Man); waltzing behaviour in *Mus musculus*; and shell colouring in *Helix hortensis* and *H. nemoralis* (Snails).

## § 2.6   The Hardy-Weinberg equilibrium

Confirmation of Mendelian inheritance for particular character-differences can sometimes be obtained from study of the frequencies in a population of individuals of different genetic constitutions. Hardy (1908) and Weinberg (1908) showed independently that, in the absence of mutation* or selection, the frequencies of the two homozygotes and the heterozygote for a pair of alleles will remain constant from generation to generation, provided the population is large enough for chance fluctuations in their proportions to be neglected, and provided crossing is at random with respect to the alleles. If the pair of alleles is denoted by $A$ and $a$, and their frequencies in the population by $p$ and $q$, respectively, where $p + q = 1$, then with random crossing the frequencies of $AA$, $Aa$ and $aa$ individuals in the population will be given by the terms of the expansion of $(p + q)^2$, that is, by $p^2$, $2pq$ and $q^2$, respectively. The frequencies of all the possible crosses, such as $AA \times Aa$, $Aa \times Aa$, and so on, can then be expressed in terms of $p$ and $q$, and likewise with their progeny. When the frequencies of $AA$, $Aa$ and $aa$ progeny from the various crosses are summed, it is found that they are given by the expressions $p^2$, $2pq$ and $q^2$, respectively. In other words, the population is in equilibrium, and the relative proportions of individuals with the different genetic constitutions are the same as in the previous generation.

Snyder (1931) discovered that ability to taste phenyl-thiocarbamide was due to a single dominant Mendelian gene, and that in 440 individuals from 100 American families, 301, or 68.5%, could taste it and 139, or 31.5%, could not. There were 9 families in which both parents were taste-deficient, and their 17 children were all taste-deficient also, as expected if this is a recessive character. Assuming that the Hardy-Weinberg equilibrium conditions apply (that is, large population size, random mating with respect to tasting, no differential viabilities or fertilities of the heterozygote or homozygotes, and negligibly low mutation rates), the frequency of non-tasters ($q^2$) is 0.315. It follows that $q = 0.561$, $p = 1 - q = 0.439$, $p^2 = 0.192$ and $2pq = 0.493$; that is, 19.2% of the total population are expected to be

---

* Mutation is an abrupt change from one allele to another: see § 9.2.

homozygous and 49·3% heterozygous for the dominant gene for tasting ability. This means that $\frac{0·192}{0·685}$ or 28% of the tasters are homozygous and 72% heterozygous. It is evident that in 72% of families where one parent is a taster and the other not, there will be equal numbers of tasters and non-tasters among the children, while in the other 28% of such families, the children will be expected all to be tasters. Snyder tested 117 children from 51 such families and found that 37 were taste-deficient. The expected number is half of 72% of 117 = 42·1, in agreement with the observed number. In 40 families where both parents were tasters, he found 16 taste-deficient children out of a total of 106. The expected number is $\frac{1}{4} \times \left(\frac{72}{100}\right)^2 \times 106 = 13·7$. Thus, the numbers of tasters and non-tasters among the children in both these classes of families are in agreement with the Hardy-Weinberg predictions. This confirms that the taste deficiency is due to one recessive gene, and this confirmation has been obtained without the need to test tasting ability over several generations.

One of the conditions for Hardy-Weinberg equilibrium is that there shall be no selection favouring one or other allele. In practice, the maintenance of a pair of alleles in a population, if not due to recurrent mutation, is likely to be due to *balanced polymorphism*, that is, the possession by each allele of certain advantages over the other, depending on the conditions. It is a simple possibility if the heterozygote has an advantage over either homozygote. An example of this is provided by heterozygotes for a pair of alleles, $A$ and $S$, affecting human haemoglobin. Allele $A$ gives normal adult haemoglobin, and $S$ gives sickle-cell haemoglobin (see Chapter 13). Allison (1954) showed that the heterozygote, $AS$, has increased resistance to falciparum malaria, caused by *Plasmodium falciparum*, compared with normal $AA$ individuals (see Livingstone, 1971). On the other hand, sickling homozygotes, $SS$, suffer from a severe anaemia which is usually lethal in childhood. As expected, the $S$ allele occurs in human populations only where malaria is endemic, such as in West Africa, or in populations recently derived from such areas.

## § 2.7   Non-Mendelian inheritance

Although Mendelian inheritance appears to be so widespread, it is not universal. One of the first exceptions to be established was through the work of Correns (1909). He had been studying the inheritance of a number of pale-leaved or variegated forms in various flowering plants. A strain of *Urtica pilulifera* (Roman Nettle) with yellowish-green leaves, and a variegated strain of *Lunaria annua* (Honesty) with white-margined leaves were found to be inherited as Mendelian recessives. In *Mirabilis jalapa* (Marvel of Peru), a strain called *chlorina* with leaves wholly yellowish-green, another called *variegata* with leaves variegated with yellowish-green, and a third called *striata* with striped flowers, were all found to show Mendelian inheritance. However, a fourth strain of this plant, called *albomaculata*, with leaves variegated with yellowish-white, was found to show a different kind of inheritance. Correns observed ,that, in addition to the variegated ones, occasional shoots on the *albomaculata* plants were normal, being wholly green, while others were wholly white. He found that flowers on green shoots gave only green progeny, irrespective of whether the pollen came from green, white or variegated shoots. Similarly, whatever the source of pollen, flowers on white shoots gave only white progeny, which necessarily died soon after seed-germination owing to the lack of chlorophyll for photo-synthesis. Flowers on variegated shoots, again irrespective of the type of shoot from which the pollen was obtained, gave three kinds of progeny: wholly green; wholly white; and variegated; and moreover, the proportions of the three kinds were highly variable, never consistently fitting any particular ratio. Plants or branches with only traces of white gave a higher proportion of wholly green progeny than did strongly variegated individuals or shoots.

Here were results differing markedly from those expected on Mendel's hypothesis, notably in the maternal inheritance, with divergent results from reciprocal crosses, and in the lack of constant proportions of the different kinds in segregating progenies. Hypotheses which have been put forward to explain these results will be discussed in Chapter 5, when other examples of non-Mendelian inheritance will be described. However, it has been evident ever since this work by Correns that Mendelian inheritance, although widespread and abundant, is not the only method of inheritance in living organisms. The relative importance of the different methods of inheritance is discussed in Chapter 18.

# 3. *The theory of independent inheritance of determinants for different characters*

## § 3.1 Mendel's results for two character-differences

Mendel (1866) in his classic paper on inheritance in *Pisum sativum* did not confine himself merely to the study of the inheritance of various character-differences considered in isolation. He made experiments in which the parents differed in two or three characters. Thus, a pure-breeding strain with round (that is, smooth) yellow seeds was crossed with a pure-breeding strain with wrinkled green seeds. The $F_1$ seeds were all Round Yellow, these being the dominant characters. In $F_2$ he obtained 556 seeds, which comprised 315 Round Yellow, 101 wrinkled Yellow, 108 Round green and 32 wrinkled green. These figures are in good agreement with a ratio of $9:3:3:1$. Similar results were obtained with all the combinations of character-differences in peas which Mendel chose to test, apart from such features as presence or absence of pigment in the seed-coat and flower, where Mendel confirmed Knight's observation that these are merely two aspects of one character-difference: the presence or absence of pigment in the plant as a whole.

## § 3.2 Mendel's hypothesis

As an hypothesis to account for his results, Mendel postulated that when two character-differences are present each is inherited independently. The $9:3:3:1$ ratio is then predicted, for it is the consequence of superimposing two $3:1$ ratios, one for each character-difference. Its derivation is shown in Table 3.1, where $R$ represents the determinant for Round seeds, $r$ for wrinkled, $Y$ for Yellow and $y$ for green. Mendel assumed that at pollen formation in the $F_1$ plants, pollen-grains with the four possible combinations of the two pairs of factors, that is, $RY$, $Ry$, $rY$, $ry$, arise with equal frequency, and similarly at egg-cell formation. Mendel also assumed that fertilization is at random between the different kinds of pollen and egg-cells. It then follows, as indicated in Table 3.1, that on the average $\frac{9}{16}$ of the combinations will carry both $R$ and $Y$, $\frac{3}{16}$ will carry $R$ but not $Y$, $\frac{3}{16}$ $Y$ but not $R$, and $\frac{1}{16}$ neither $R$ nor $Y$.

16

TABLE 3.1   The proportions of progeny of different kinds expected in $F_2$ from a cross between pure-breeding strains differing in two characters, on Mendel's hypothesis that in the double heterozygote the four possible combinations of two pairs of factors arise with equal frequency, both in the formation of the pollen and of the egg-cells, and that fertilization is at random. $R$ = factor for Round, $r$ for wrinkled, $Y$ for Yellow, $y$ for green seeds in *Pisum sativum*. (*a*) Individual frequencies. (*b*) Totals.

(*a*)

| Reproducive Cells from $F_1$ Heterozygote | | | Pollen | | | |
|---|---|---|---|---|---|---|
| | | | *RY* 1 | *Ry* 1 | *ry* 1 | *rY* 1 |
| Egg-cells | *RY* 1 | | *RRYY* 1 | *RRYy* 1 | *RrYy* 1 | *RrYY* 1 |
| | *Ry* 1 | | *RRYy* 1 | *RRyy* 1 | *Rryy* 1 | *RrYy* 1 |
| | *ry* 1 | | *RrYy* 1 | *Rryy* 1 | *rryy* 1 | *rrYy* 1 |
| | *rY* 1 | | *RrYY* 1 | *RrYy* 1 | *rrYy* 1 | *rrYY* 1 |

(*b*)

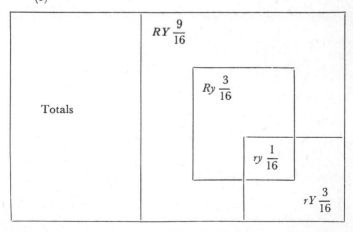

| Totals | | | |
|---|---|---|---|
| | $RY \dfrac{9}{16}$ | | |
| | | $Ry \dfrac{3}{16}$ | |
| | | | $ry \dfrac{1}{16}$ |
| | | | $rY \dfrac{3}{16}$ |

Mendel confirmed this hypothesis in two ways. Firstly, he bred from his $F_2$ plants, and showed that the Round Yellow, Round green and wrinkled Yellow peas consisted of the appropriate mixture of pure-breeding and impure types and moreover in the appropriate proportions, while the wrinkled green peas bred true. Secondly he backcrossed the $F_1$ plants to the parental forms, and demonstrated that such crosses gave a $1:1:1:1$ ratio in the progeny. Thus Round Yellow $F_1$ crossed with wrinkled green gave a progeny of 55 Round Yellow, 51 Round green, 49 wrinkled Yellow, and 53 wrinkled green seeds.

When Mendel's work was rediscovered, confirmation of the independence of inheritance of different character-differences was immediately available. Thus, De Vries (1900) reported an approximation to a $9:3:3:1$ ratio in $F_2$ from a cross between a white trifoliate and a red quinquefoliate strain of *Trifolium pratense* (Red Clover), the latter two characters being dominant; and von Tschermak had unknowingly confirmed Mendel's results for Round Yellow and wrinkled green Peas. Bateson and Saunders (1902) established $9:3:3:1$ ratios in $F_2$ in respect of flower-colour and fruit prickliness in *Datura* (Thorn Apple), and in respect of type of comb and number of toes, and other pairs of character-differences, in *Gallus domesticus*.

## § 3.3   Bateson, Saunders and Punnett's discovery of partial linkage

Bateson, Saunders and Punnett (1905), working with *Lathyrus odoratus* (Sweet Pea) were the first to discover an exception to the law of independence. In an $F_2$ progeny 381 plants with coloured petals comprised 305 with a purple pigment and 76 with a red pigment. These figures are a rather poor fit with a $3:1$ ratio, but later work established this ratio, indicating that purple differs from red by a single dominant factor. The plants also showed diversity in respect of another character-difference, namely, shape of pollen-grain, whether elongated (long) or disc-shaped (round). This character-difference had been chosen for study in the hope (unavailing as it transpired) that it might reveal more precisely when segregation of character-differences took place, for Mendel had postulated its occurrence at the time of formation of the pollen and egg-cells. However, all the pollen-grains on any individual plant were found always to be alike.* By a coincidence, the 381 plants showed the same proportion with the two different shapes of pollen (305 long, 76 round) as for the two different flower colours. Although the fit with a $3:1$ ratio is poor, later work confirmed its truth, and hence established that long-pollened plants differ from round-pollened by a single dominant factor.

When the 381 plants were classified simultaneously for pigment colour and pollen shape, a startling discovery was made. The relative frequencies of the four classes were quite different from those expected on the basis of all

---

* Pollen characters do not always behave like this—see § 5.4.

previous work. The results obtained were:

$$284 \ (74 \cdot 6 \%) \ \text{with Purple flowers and Long pollen}$$
$$21 \ ( \ 5 \cdot 5 \%) \ \text{with Purple flowers and round pollen}$$
$$21 \ ( \ 5 \cdot 5 \%) \ \text{with red flowers and Long pollen}$$
$$\text{and} \quad 55 \ (14 \cdot 4 \%) \ \text{with red flowers and round pollen.}$$

More precise frequencies, based on 6952 $F_2$ plants derived from the same cross (Purple Long × red round) gave:

$$4831 \ (69 \cdot 5 \%) \ \text{Purple Long}$$
$$390 \ ( \ 5 \cdot 6 \%) \ \text{Purple round}$$
$$393 \ ( \ 5 \cdot 6 \%) \ \text{red Long}$$
$$\text{and} \quad 1338 \ (19 \cdot 3 \%) \ \text{red round} \qquad \text{(Punnett, 1917).}$$

This *coupling* of 'Purple' with 'Long' (the two dominant characters), as Bateson, Saunders and Punnett called it, implies that at the time of pollen

TABLE 3.2   The proportions of progeny of different kinds expected in $F_2$ from a cross between pure-breeding strains of *Lathyrus odoratus* with purple flowers and long pollen and with red flowers and round pollen on the assumption that reproductive cells of constitution *PL* and *pl* each have a frequency of 44%, and *Pl* and *pL* of 6%, and with fertilization at random. $P$ = factor for Purple, $p$ for red flowers; $L$ for Long, $l$ for round pollen. (*a*) Individual frequencies. (*b*) Totals.

(*a*)

| Reproductive Cells from $F_1$ Heterozygote | | Pollen | | | |
|---|---|---|---|---|---|
| | | *PL* 4·4 | *Pl* 0·6 | *pl* 4·4 | *pL* 0·6 |
| Egg-cells | *PL* 4·4 | 19·36% | 2·64% | 19·36% | 2·64% |
| | *Pl* 0·6 | 2·64% | 0·36% | 2·64% | 0·36% |
| | *pl* 4·4 | 19·36% | 2·64% | 19·36% | 2·64% |
| | *pL* 0·6 | 2·64% | 0·36% | 2·64% | 0·36% |

(*b*)

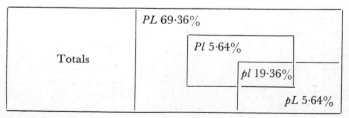

| Totals | PL 69·36% | | |
|---|---|---|---|
| | | Pl 5·64% | |
| | | | pl 19·36% |
| | | | pL 5·64% |

and egg-cell formation in the $F_1$ plants, the four different combinations of the factors $P$ (Purple) and $p$ (red flower-colour), and $L$ (long) and $l$ (round pollen) do not arise with equal frequency, but that $PL$ and $pl$ are of appreciably higher frequency than the other two combinations ($Pl$ and $pL$). Frequencies of 44% for each of $PL$ and $pl$, and of 6% for $Pl$ and $pL$, instead of the 25% for each which Mendel would have expected, were found (Haldane, 1919) to give a close fit with the data from the 6952 plants (see Table 3.2).

As Bateson, Saunders and Punnett continued their breeding work with *Lathyrus odoratus*, several other examples of such *partial linkage* between different character-differences were soon found. The cause of partial linkage will be discussed in Chapter 7.

## § 3.4   Doncaster and Raynor's discovery of sex-linkage

Linkage of another kind, in which two of the four possible ways of associating two character-differences appeared not to arise at all, was discovered

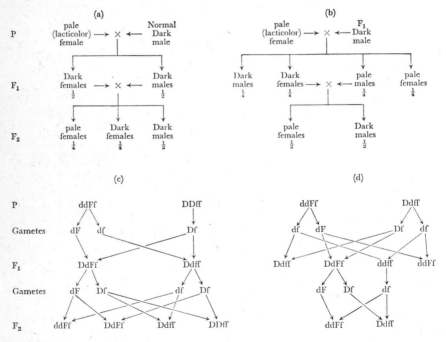

FIGURE 3.1   Doncaster and Raynor's results ((*a*) and (*b*)) from study of the inheritance of the pale (*lacticolor*) form of *Abraxas grossulariata* (Magpie Moth), and Punnett and Bateson's hypothesis ((*c*) and (*d*)) to explain them. $D$ = dominant factor for normal Dark form of the moth, $d$ = recessive allele for pale form, $F$ = dominant factor for Femaleness, $f$ = recessive allele for maleness. The hypothesis was later modified: see § 6.11.

by Doncaster and Raynor (1906) working with *Abraxas grossulariata* (Magpie Moth). They studied the inheritance of a pale form of this moth called *lacticolor*, which first appeared in a female individual. They established that the pale character behaved as a normal Mendelian recessive, disappearing in $F_1$ and reappearing in one quarter of the $F_2$ and in one half of the back-cross progeny, that is, the progeny from crossing the $F_1$ males (normal) with their mother (pale) (see Fig. 3.1(*a*) and (*b*)). However, the following results were unexpected:

(i) The quarter of the $F_2$ which were pale were all female, and hence resembled their grandmother (Fig. 3.1(*a*)).

(ii) Pale males from the backcross progeny, when crossed with their normal sisters, gave only pale females and normal males (the grandparental forms), instead of equal numbers of each form in each sex (Fig. 3.1(*b*)).

Earlier work by Cuénot with albino *Mus* and by Bateson and Punnett with comb characters in *Gallus* had always shown character-differences to be equally distributed between the sexes, so this *sex-linkage* in *Abraxas* was unexpected.

For linkage between two pairs of characters (factors $A/a$ and $B/b$) to be detected it is necessary to have an individual heterozygous for both character-differences, in order to provide the four possible kinds of gamete ($AB$, $ab$, $Ab$, $aB$). Hence, to explain Doncaster and Raynor's results, in addition to accepting that the pale form is recessive, Punnett and Bateson (1908) made two postulates:

(1) Sex in *Abraxas* is a Mendelian character, with femaleness dominant to maleness, such that all Females are heterozygous for the sex factors ($Ff$), and all males are the homozygous recessive ($ff$).

The 1 : 1 ratio of the sexes in each generation was thereby explained, and the normal females, which provide one parent for both the aberrant progenies (i) and (ii) above, would be the double heterozygote, $DdFf$, where $D$ is the dominant factor for the normal Dark insect and $d$ the recessive factor for paleness.

(2) The double heterozygote ($DdFf$) produces eggs of only two kinds, bearing the factors for Dark male ($Df$) and for pale Female ($dF$), respectively, the other two combinations ($DF$ and $df$) not being produced, for some unknown reason.

Punnett and Bateson spoke of this kind of linkage as *repulsion* since the dominant factors stayed apart, and as complete linkage unlike the partial linkage of the *Lathyrus* example discussed earlier. Their hypothesis was found to explain satisfactorily the sex linkage data in Doncaster and Raynor's results (see Fig. 3.1(*c*) and (*d*)).

However, shortly afterwards, Doncaster (1908) made a remarkable discovery. He found that whenever a pale male was crossed with a normal Dark female, all the male offspring were normal and all the females pale.

This was true even when the normal female parent came from a part of the country where the rare pale form had never been seen. This result was totally unexpected as it implied, on Punnett and Bateson's hypothesis, that all female Magpie Moths were heterozygous for the pale character! As will be shown in § 6.11, their hypothesis was later modified so as to make this unlikely deduction unnecessary.

Morgan (1910) reported results with *Drosophila melanogaster* (Fruit Fly) essentially similar to those of Doncaster and Raynor with *Abraxas*, except in one important respect. Morgan studied the inheritance of white eye-colour in *Drosophila* and found this to differ from the normal dull red eyes by a single recessive factor. The white-eyed condition first appeared in a male individual, unlike *Abraxas* where, as already mentioned, the pale form first appeared in a female. This difference was reflected throughout the inheritance studies on the white-eyed condition in *Drosophila*, where the role of the sexes appeared to be exactly reversed compared with *Abraxas*. To explain these results, Morgan put forward an hypothesis identical to that of Punnett and Bateson for *Abraxas*, except that the position of the sexes was reversed, that is, maleness was regarded as dominant to femaleness and the male *Drosophila* was postulated to be heterozygous for the sex factors, instead of the female *Abraxas*. Furthermore, Morgan found that when a white-eyed female *Drosophila* was crossed with a normal Red-eyed male, even if the latter came from an unrelated stock with no known ancestry of white eyes, the progeny always consisted of Red-eyed females and white-eyed males. This was the exact counterpart of Doncaster's discovery with *Abraxas* and implied that all normal red-eyed *Drosophila* males were heterozygous for white eyes! As with *Abraxas*, the hypothesis was subsequently modified to eliminate this unlikely deduction (see § 6.11).

## § 3.5   Conclusion

In spite of a number of exceptions such as those just described, Mendel's hypothesis of independent inheritance of the determinants for different characters was found to hold quite widely, and ultimately was christened 'Mendel's Second Law' by Morgan (1919), the 'First Law' being the fundamental concept of the coming together of the determinants for a particular character-difference at fertilization, and their separating again unchanged at gamete formation.

# 4. The theory of Mendelian inheritance for characters showing a continuous range of variation

## § 4.1 The analysis of quantitative data

The method which Mendel used so successfully to obtain information about the mechanism of heredity was to find a well-marked character-difference in the chosen organism and to cross pure-breeding strains of each kind. The method is clearly dependent on the existence of sharply defined character-differences. However, many characters are not suitable because they do not reveal clear-cut discontinuities. The most familiar

TABLE 4.1  Johannsen's data for the weight-class of 5494 $F_2$ beans (seeds of *Phaseolus vulgaris*), obtained by self-pollination for two successive generations of 19 beans of diverse origin.

| Weight-class in Centigrams | 5–15 | 15–25 | 25–35 | 35–45 | 45–55 | 55–65 | 65–75 | 75–85 | 85–95 |
|---|---|---|---|---|---|---|---|---|---|
| Number of Beans | 5 | 38 | 370 | 1676 | 2255 | 928 | 187 | 33 | 2 |

examples of characters which in most organisms commonly show a continuous range of variation are size measurements such as height or weight. In order to study the inheritance of such characters statistical methods are required. Before describing some of the results obtained from study of the inheritance of characters showing a continuous range of variation, it is necessary to consider the techniques for analysing quantitative data. Only a superficial account is given below of a few statistical methods. For a sound treatment of these and other methods, reference should be made to standard works such as Fisher (1963), Bailey (1959) or Snedecor (1956).

Data of Johannsen (1909) for the weight of dwarf bean (*Phaseolus vulgaris*) seeds will be used as an illustration. He weighed 5494 beans and classified them into 9 categories depending on their weight (see Table 4.1). Such data can be plotted as a frequency histogram, as illustrated in Fig. 4.1. Such histograms for biological data commonly approximate to a smooth curve of a

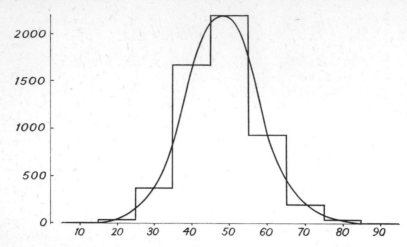

FIGURE 4.1    Frequency histogram for bean weight in *Phaseolus vulgaris* from the data of Johannsen. The number of beans is plotted as ordinate and the weight in centigrams as abscissa. A normal curve has been fitted to the data.

particular kind called the *normal* curve. Non-biological data also frequently fit a normal curve, for example, the frequencies of different numbers of 'heads' (or 'tails') after repeatedly tossing a number of coins simultaneously. With a small number of coins, a stepped graph (histogram) is obtained approximating to the 'binomial distribution' (Fig. 4.2(a)), but as the number of coins tossed simultaneously is increased the graph of number of 'heads' (or 'tails') in the throw plotted against the frequency of occurrence of that number approaches the smooth curve of the 'normal distribution' (Fig. 4.2(b) and (c)). With the coins, whether a 'head' or a 'tail' shall turn up depends on a number of causes which in the long run act with equal frequency in two opposite directions. Similarly, bean weight may be supposed to depend on many factors, such as the position of the seed in the pod, the fertility of the soil and the hereditary constitution of the plant, which again may be supposed on the average to act equally in two directions, some favouring an increased and some a decreased weight.

A set of data which show an approximately normal distribution, such as bean weight in Johannsen's experiment, may be specified by two quantities: firstly, the *mean*, which is the sum of the measurements divided by the total number; and secondly, the *variance* (or its square root, the *standard deviation*) which are measures of the extent to which the various measurements differ from the mean value. The variance is calculated by finding the deviation of each individual measurement from the mean value, squaring each deviation, then summing these squared values, and dividing by the total number of measurements (or strictly speaking, by one less than this total number). On the normal curve, the standard deviation is represented by the horizontal distance from the position of the mean to the point of steepest slope of the curve on either side. Thus a wide curve, implying considerable variability in the quantity measured, will have a correspondingly large standard deviation. However, the area under the curve measured out as far as the standard deviation on each side is always the same fraction of the total area under the curve, namely 68·26% or approximately $\frac{2}{3}$. This means that $\frac{2}{3}$ of the observations will differ from the mean by less than the standard deviation, and the remaining $\frac{1}{3}$ will exceed it.

A further quantity of use in the analysis of quantitative data is the *co-variance*. Suppose pairs of related measurements are available, such as heights of a number of men and of one adult daughter of each of them. Data on human stature usually fit a normal distribution. The covariance of fathers' and daughters' height is obtained by the same method as for variance except that instead of squaring deviations, the deviation of each father's height from the fathers' mean is multiplied by the deviation of the corresponding daughter's height from the daughters' mean. A positive value for the covariance means that on the average when the fathers are tall so are the daughters, and when the fathers are short so are the daughters. A negative value would mean that tall fathers tend to have short daughters, and *vice versa*. If the covariance is zero, a tall father is equally likely to have a tall or short daughter. The co-variance of fathers' and daughters' height divided by the variance of the fathers is called the *regression coefficient* of daughters on fathers; if the covariance is divided by the variance of the daughters, this is the *regression coefficient* of fathers on daughters; and if the covariance is divided by the geometric mean (that is, the square root of the product) of the

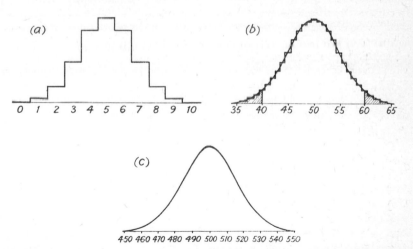

FIGURE 4.2   Frequency histograms or curves for numbers of 'heads' (or 'tails') when (*a*) 10, (*b*) 100, and (*c*) 1000 coins are tossed simultaneously. In (*b*) the shaded areas indicate the probability of obtaining 60 heads and 40 tails or a larger deviation from equality.

two variances (the fathers' and the daughters'), this gives the *correlation coefficient* of the fathers' and daughters' heights. These three coefficients are all measures of the degree of association between the two variables. A value of zero for the correlation coefficient means no association, of $+1$ means complete positive association, and $-1$ complete negative association. Intermediate values between $-1$ and $0$ and between $0$ and $+1$ would indicate the appropriate degree of association between the fathers' and daughters' heights.

## § 4.2   The significance of experimental data

The fact that data from the tossing of coins (in sufficient numbers) fit a normal distribution illustrates how the effects of chance can be assessed in testing the significance of experimental data in biology (and elsewhere). Suppose 100 coins are tossed simultaneously. Assuming the coins are not biassed, 50 heads and 50 tails is the most likely result, but there are so many other possibilities such as 49 heads and 51 tails, or 52 heads and 48 tails, that are individually only slightly less likely to occur, that in fact one would be surprised if in a single throw exact equality were achieved. Indeed, on the average 50 heads and 50 tails is expected only about once in every seven or eight throws of 100 coins. The likelihood of each alternative turning up can be precisely calculated. Now suppose at the first toss 60 heads and 40 tails are found. Is this a reasonable result, or should one suspect that the coins are biassed? The probability of this particular ratio turning up by chance is obviously quite low, but this has no significance because, as just indicated, there are so many possible alternative ratios that the probabilities are low for all of them. In order to judge the significance of the 60:40 ratio it is necessary to group together the alternatives, and consider the probability of getting by chance not only 60 heads and 40 tails, but all ratios showing a deviation from 50:50 as great or greater than the one observed, such as 61 heads and 39 tails, 62 heads and 38 tails, 40 heads and 60 tails, and so on. To find this probability is equivalent to finding the area under the tails of the normal curve (that is, the sum of the two shaded regions in Fig. 4.2($b$)) compared with the total area under the curve, since as already indicated coin data as extensive as these tend to fit a normal curve.

Statisticians have derived formulae and tables to enable one to calculate the approximate area under the tails of the normal curve in specific instances with the minimum of trouble. The most generally useful method is that developed by Pearson (1900). It involves calculating a variance ratio called *khi squared* ($\chi^2$). To calculate $\chi^2$, take each observed frequency, find its deviation from expectation, square this deviation and divide by the expected value, and then sum the figures so obtained. In the present instance, the observed frequencies are 60 and 40, the expected 50 and 50, so the deviations are 10 and 10, and $\dfrac{(\text{deviation})^2}{\text{expected}}$ is $\dfrac{100}{50} = 2$ in each instance, and $\chi^2 = 4$.

One must then refer to a table showing the values of $\chi^2$ associated with particular areas under the tails of the normal curve. From the way in which

$\chi^2$ is calculated, it is evident that the larger the difference between observed and expected values the larger will $\chi^2$ be, and hence the smaller the area in question. Moreover, since $\chi^2$ is obtained by adding a series of numbers one for each class, in general, the larger the number of classes the larger the value of $\chi^2$. Evidently the number of classes must be taken into account in finding the area under the tails. In the table of $\chi^2$ values (Table 4.2), the quantity $n$ in the first column is one less than the number of classes (technically $n$ is the number of degrees of freedom, or the number of classes which can be filled up arbitrarily). In the present example there were two classes (heads and tails) and hence $n = 1$. The value 4 for $\chi^2$ in the $n = 1$ row of the table corresponds to $P = 0.05$ approximately. In other words, the two

TABLE 4.2

Table of $\chi^2$

(based on Fisher (1963) by permission of Oliver and Boyd)

| $n$ | $P$ | | | | | | | | |
|---|---|---|---|---|---|---|---|---|---|
| | 0·99 | 0·98 | 0·95 | 0·90 | 0·50 | 0·10 | 0·05 | 0·02 | 0·01 |
| 1 | 0·00016 | 0·00063 | 0·0039 | 0·016 | 0·46 | 2·7 | 3·8 | 5·4 | 6·6 |
| 2 | 0·020 | 0·040 | 0·10 | 0·21 | 1·4 | 4·6 | 6·0 | 7·8 | 9·2 |
| 3 | 0·12 | 0·19 | 0·35 | 0·58 | 2·4 | 6·3 | 7·8 | 9·8 | 11·3 |

shaded areas under the tails of the normal curve in Fig. 4.2(*b*) represent in total approximately $\frac{1}{20}$ ($\frac{1}{40}$ each) of the total area under the curve. Thus once in 20 times of throwing 100 coins one can expect to obtain a deviation from equality at least as large as 60:40. It is usual to say that a 1 in 20 probability such as this indicates a suspiciously large departure from expectation and to describe the deviation as significant. However, if one made the deduction that the coins were biassed there is a distinct possibility that one would be wrong since with unbiassed coins this deviation is equalled or exceeded once in every 20 times purely by chance. Thus a probability of 0·05 suggests further experimenting is needed, before reaching a conclusion as to whether the hypothesis of 1:1 for heads:tails is true or not. A deviation giving a $P$ value of 0·05 is equal to approximately twice the standard deviation.

## § 4.3   The significance of Mendel's data

In Chapter 2 Mendel's results were quoted from a back-cross of heterozygous Yellow peas with green peas. The expected ratio of Yellow:green peas in the progeny is 1:1. The situation is therefore strictly comparable to the example given above of tossing coins, and the effects of chance can be assessed in the same way. Thus, in one of the experiments quoted, Mendel found 58 Yellow and 52 green peas in the progeny. It was taken for granted in Chapter 2 that this was a satisfactory fit with a 1:1 ratio, but strictly speaking the $\chi^2$ test ought to be applied in order to confirm this. The expected frequency in each class is 55, the deviation is 3 and so

$$\chi^2 = \frac{9}{55} \times 2 = 0.33.$$

Reference to Table 4.2 shows that for $n = 1$, the probability of getting a deviation as large or larger than the observed one is between 0·5 and 0·9. In other words, more often than not one would expect a bigger difference from 1:1 purely by chance, and the data agree very well with the hypothesis.

The $\chi^2$ test can also be applied when the expected frequencies of the classes are unequal. Thus, Mendel's $F_2$ data discussed in Chapter 2 are expected on his hypothesis to fit a 3:1 ratio. The observed frequencies for the seed colours were 6022 Yellow and 2001 green. The expected frequencies are 6017·25 and 2005·75 respectively. The deviation is 4·75 for each class and

$$\chi^2 = \frac{(4.75)^2}{6017.25} + \frac{(4.75)^2}{2005.75} = 0.015.$$

Referring again to Table 4.2, for $n = 1$ the likelihood of getting a deviation as large or larger than that observed is approximately 0·9. Thus only once in 10 times would one expect to get so good a fit with the expected ratio.

The $\chi^2$ method can also be used when the number of classes is more than two. Thus, Mendel's $F_2$ data quoted in Chapter 3 for the Round/wrinkled and Yellow/green character-differences were 315, 101, 108 and 32 in the four classes, where the expected frequencies on the hypothesis of a 9:3:3:1 ratio are 312·75, 104·25, 104·25 and 34·75 respectively. The deviations of the observed from the expected values are 2·25, 3·25, 3·75 and 2·75, respectively, and

$$\chi^2 = \frac{(2.25)^2}{312.75} + \frac{(3.25)^2}{104.25} + \frac{(3.75)^2}{104.25} + \frac{(2.75)^2}{34.75} = 0.47.$$

Referring to Table 4.2, when $n = 3$, $P = 0.90 - 0.95$, implying a close fit with the hypothesis. Indeed 14 times out of 15 a larger deviation is expected by chance.

This finding points to a remarkable feature of Mendel's data discovered by Fisher (1936), namely, that Mendel's results taken as a whole fit the expected ratios better than they should! When the $\chi^2$ test is applied to the results of his individual experiments, the probability values obtained usually lie between 0·5 and 0·95 as in the three examples just discussed. Individually, these are acceptable, but when one considers the data in the aggregate one's suspicions are aroused. For, with a series of experiments, the $P$ values must on the average fall below 0·5 as often as above, assuming the data fit the hypothesis. From each experiment one reaches the conclusion that more often than not one would expect a bigger difference than that observed. Mendel's work was conducted on such a massive scale that the cumulative effect of the many experiments, each individually fitting the theory slightly better than expected on the average, is to produce an aggregate set of data which is exceedingly unlikely to be valid. Thus, Fisher found that Mendel's total data give a $P$ value of 0·99993, or in other words only once in some 14,300 times would one expect to get so good a fit with expectation purely by chance!

This extraordinary discovery naturally raises the question of whether Mendel in fact carried out the experiments he describes! Fisher found that many points of detail were entirely consistent with the belief that Mendel did do what he says he did. Thus, the number of plants he would have to have grown in each year was reasonable for the size of the monastery garden at Brno (Brünn); and the scoring of embryo characters in the seed, such as yellow and green, covered an extra generation compared with mature characters, like tall and dwarf, as expected since the former but not the latter character-difference can be observed without germinating the seeds.

Accepting that Mendel conducted the experiments which he describes, why did he apparently adjust the frequencies of progeny a little nearer to the theoretical values? Here a second puzzling feature of Mendel's paper may provide a clue. Fisher has pointed out that although Mendel's first experiments were with strains of peas differing by single characters (as described in Chapter 2), yet commercial varieties of peas, with which he started, always differ by many characters. The published experiments covered five years (1859–1863 inclusive), although Mendel states that the paper is the result of eight seasons' work. It is evident that the earlier work (1856–1858 inclusive), during which the strains differing by single characters were obtained, was never published. Fisher suggests that it was on the basis of these unpublished experiments, estimated to have involved the growing of nearly 7000 pea plants, that Mendel must have framed the entire theory of heredity. At the latest this would have been when he reviewed his 1858 results. The gigantic series of experiments upon which he then embarked, involving over the succeeding five years the cultivation, in Fisher's estimation, of over 26,500 plants and forming the basis of his classic paper, is then seen as a carefully planned demonstration of the conclusions which he had already reached before the outset. The slight adjusting of frequencies would then presumably have been done in order to make the conclusions a little more obvious. In this he was singularly unsuccessful, for no-one in his lifetime saw the significance of his epoch-making discovery. Even if the adjusted figures in his paper now make the recognition of the underlying mechanism look simple, yet Mendel himself would clearly not have had this advantage when he inferred that inheritance was particulate, with segregation at gamete-formation! His brilliant discovery, probably made in the very year that Darwin's 'The Origin of Species' was published, apparently remained unknown to Darwin and was not confirmed by anyone for 40 years.

## § 4.4   Further uses of the $\chi^2$ test

Mendel's results have been taken as an example of data which give an exceptionally high probability value with the $\chi^2$ test. In general, $P$ values within the range 0·1 to 0·9 cause no suspicion and indicate that the observations fit the hypothesis under test, although they may of course also fit other hypotheses. Values of 0·05 or less are conventionally said to indicate a real discrepancy, although this is quite an arbitrary line of demarcation and one would less often be misled if one took lower figures such as 0·02 or 0·01 as the significance level. Conversely values of 0.95 or more indicate an abnormally close fit with hypothesis, while probabilities as high as 0·98 or 0·99 would certainly arouse suspicion that, either consciously or unconsciously, the data had become biassed in favour of the hypothesis.

The $\chi^2$ test can still be used when one does not know the expected proportions of the classes, but wishes to know whether the observed frequencies are in the same proportion following different treatments. Where there are two classes $a$ and $b$ and two treatments 1 and 2, the four observed frequencies may be denoted by $a_1$, $b_1$, $a_2$ and $b_2$ respectively. Then $\chi^2$ is given by the expression

$$\frac{(a_1 + b_1 + a_2 + b_2)(a_1 b_2 - a_2 b_1)^2}{(a_1 + b_1)(a_2 + b_2)(a_1 + a_2)(b_1 + b_2)} \quad \text{and} \quad n = 1.$$

The application of this formula is called the *contingency* $\chi^2$ test, and examples of its use are given in § 7.5 and § 8.6.

If the observed frequency in any class is less than 5, the $\chi^2$ test becomes so inaccurate as a means of estimating the area under the tails of the normal curve that it should not be used. However, Yates (1934) pointed out that when there is only one degree of freedom ($n = 1$), reasonable accuracy is restored if one adds 0·5 to the smaller observed value and subtracts 0·5 from the larger one before applying the test. With the contingency test there will be two small values and 0·5 should be added to each, and similarly there will be two large values from each of which 0·5 should be subtracted. An example showing the application of Yates's correction is given in § 4.8.

## § 4.5  Galton's and Pearson's studies on the inheritance of characters showing continuous variation

One of the first applications of statistical methods to the study of the inheritance of characters showing continuous variation was made by Galton (1889). His method was to obtain data on the attributes of relatives, notably in *Homo sapiens*, and then calculate regression and correlation coefficients. He obtained the data in several ways, but chiefly by offering prizes to anyone who would send in detailed particulars of their parents, grandparents, great-grandparents, brothers, sisters, uncles, aunts, great-uncles and great-aunts. The information asked for included stature, eye-colour, artistic aptitude, health and temperament. Data were received for about 150 families and concerned over 1000 individuals.

From the data on stature he obtained evidence of positive association between relatives. As one would expect, the parental and fraternal relationships gave the highest values for the coefficients, and cousins the lowest. He thus obtained the first sound evidence that human stature is to some extent an inherited character. Moreover, the coefficients for the two sexes did not differ significantly, indicating that their contributions to heredity were equal. From the values of the correlation coefficients he deduced that on the average a child receives $\frac{1}{4}$ of its heritage of stature from its two parents (total $\frac{1}{2}$), $\frac{1}{16}$ from each of its four grandparents (total $\frac{1}{4}$), and so on. This geometric scale continued indefinitely backwards would account for the total heritage. He found evidence that the other attributes for which he had data were inherited similarly. Later he called this geometric series of contributions the 'law of ancestral heredity'.

Pearson (1897) extended Galton's work in a number of ways, using his data. Thus, he showed that in assessing the significance of the correlation coefficients for human stature, allowance must be made for assortive mating, for it was evident from the data that tall husbands tend to have tall wives, and short husbands short wives. For some characters the marital correlation is remarkably high: for instance, Penrose (1963) has shown that for intelligence it may be as high as 0·5. Pearson and Lee (1903) collected further data on human stature in several hundred families, by the same means as Galton had done. They also obtained data on arm-span and elbow to finger-tip distances. For stature, the parental and fraternal correlation coefficients were 0·51 and 0·54 respectively; for arm-span they were 0·45 and 0·54 respectively; and for the forearm measurements they were 0·42 and 0·46 respectively*. It will be noted that in each instance the fraternal correlation is greater than the parental. Pearson and Lee's figures for both the parental and fraternal correlation coefficients for stature are appreciably larger than those which Galton had obtained.

Galton and Pearson also made statistical studies on the inheritance of characters such as coat colour in *Canis familiaris* (Dog) and *Equus caballus* (Horse), using the information in pedigree records. They found that the correlation coefficients were similar to those in man.

## § 4.6   The multiple factor hypothesis

With the rediscovery of Mendel's work, Bateson and Saunders (1902) raised the possibility that characters showing a continuous range of variation, such as human stature, might be determined by a large number of Mendelian factors. Pearson (1904) analysed this possibility mathematically, assuming that all the postulated Mendelian factors had equal and additive effects, and that there was complete dominance of each over its recessive allele when heterozygous, and that the two alleles of a pair were equally frequent. He found that, however many pairs of alleles were postulated, the expected parental correlation was 0·33 and the fraternal 0·42. Since these values did not agree with his observed values, he concluded that the Mendelian explanation was incorrect. Yule (1907) pointed out that the assumption of complete dominance was not necessarily justified, and that if the heterozygotes were intermediate between the homozygous dominants and the recessives, higher values of the correlation coefficients would be expected, such as had been found. However, much confusion of thought on the applicability of Mendelian heredity to continuous variation persisted until Johannsen (1909)

---

* The weighted means are given, as calculated by Fisher (1918).

published the results of his experiments with *Phaseolus vulgaris* (Dwarf Bean) and other plants, and drew for the first time a clear distinction between the hereditary determinant, or *gene*, as he called it, and its effect.

## § 4.7   Johannsen's experiments with *Phaseolus vulgaris*

Johannsen (1903) introduced the term 'pure line' for the progeny of single individuals in organisms which are regularly self-fertilised. Examples of such organisms are *Hordeum vulgare* (Barley), *Lathyrus odoratus* (Sweet Pea) and *Phaseolus vulgaris*. He found that the weights of the seeds of the members of a pure line of *P. vulgaris* fitted a normal curve, but whether he selected heavy, medium or light beans, the progeny of each by self-fertilization in the succeeding year again fitted a normal curve, and moreover they all had the same mean value. In other words, within a pure line selection for heavy or light seeds had no effect: the mean weight remained constant. Thus, his pure line no. 2 gave him in the year 1901 beans ranging in weight from about 40 to 70 centigrams. He classified them into four groups according to their weight (35–45, 45–55, 55–65, and 65–75 cg) and the following year he sowed each group separately. Those in the 35–45 cg class produced 86 seeds in 1902, and they showed a normal distribution about a mean weight of 57·2 cg. Those in the 45–55 cg class (1901 seed) gave 195 seeds in 1902 with a mean of 54·9 cg. Similarly the 55–65 cg class gave 120 seeds of mean weight 56·5 cg and the 65–75 cg class produced 74 seeds with mean weight 55·5 cg. The mean weights of the seeds produced by the various classes clearly do not differ significantly. He concluded that the variation in bean weight within a pure line was entirely due to environmental factors such as the position of the seed in the pod.

Johannsen studied 19 different pure lines of *Phaseolus vulgaris*. Each had been obtained from a different source, and had a characteristic mean weight of seed ranging from 64·2 cg in no. 1 down to 35·1 cg in no. 19. Within each pure line he obtained results similar to those described above for no. 2. Each gave a normal distribution of bean weight about its characteristic mean value. Within a pure line selection for heavy seeds had no effect on succeeding generations.

The total population of bean seeds under observation also gave a normal distribution, with a mean of 47·9 cg (Table 4.1 and Fig. 4.1). This population was a mixture of the 19 pure lines. Selection for heavy beans within this population would produce a response, because one would tend to pick out seeds from the heavier pure lines and hence put up the average weight the following year. Selection would continue to have an effect until all the seeds belonged to the heaviest pure line, but thereafter it would be ineffective in stepping up the mean weight further.

Johannsen's work demonstrated the distinction between variation due to hereditary factors and that due to environmental factors. He coined the term *genotype* for the hereditary constitution of an individual, and *phenotype* for the appearance of the individual as a consequence of the interaction of genotype and environment. Thus a bean seed might be heavy, say of 60 cg,

as a consequence of favourable environmental conditions during development, even though belonging to, say, pure line no. 15 with a mean weight
of only 45·0 cg. Conversely, a bean of the same weight might belong to pure
line no. 1 grown under rather unfavourable conditions. Both beans would be
said to have the same phenotype, but the genotypic and environmental
contributions to this end-result would have acted in opposite directions in the
two examples. It is evident that the two kinds of variation, genetic and
environmental, cannot be distinguished by observation but only by experiment. Thus, it is necessary to study the progeny of two beans before one
can determine whether they belong to the same pure line or not. Differences
in genotype were responsible for differences in mean seed weight of the order
of 30 cg between the most extreme pure lines. This difference is of about
the same magnitude as the variation in weight within pure lines due to
environmental factors. It is evident that in a mixed population of a number
of pure lines the variation due to the environment will wholly obscure the
discontinuities in mean weight of the different lines, and give rise to a
continuous range of variation.

Johannsen was able to demonstrate the respective contributions of heredity
and environment to variation by choosing an organism in which, owing to
the regular self-fertilization, many individuals have the same genetic constitution. In most organisms, where cross-fertilization is the rule, it is more
difficult to distinguish the two kinds of variation. Nevertheless there is every
reason to suppose that the concepts of genotype and phenotype developed
by Johannsen with *Phaseolus vulgaris* are of general application. They are of
particular importance in plant and animal breeding, since his work showed
that selection for favourable characters in an organism will be unsuccessful
unless the character-difference in question has a genetic basis and is not
merely a consequence of favourable environmental conditions. Thus a first
essential in crop and livestock improvement is to have available sources of
hereditary variability such as strains of the organism from different regions
of the world, or allied species which will give fertile hybrids with that which
is being bred.

## § 4.8  Confirmation of the multiple factor hypothesis

From studies on crop plants, Nilsson-Ehle (1909) and East (1910) obtained
evidence in support of the multiple factor hypothesis to explain the inheritance of quantitative characters. Nilsson-Ehle had been studying the
inheritance of glume and ligule characters in *Avena sativa* (Oats) and of ear
and grain pigmentation in *Triticum aestivum* (Wheat). In *Avena* he found that
black glumes usually behaved as a simple Mendelian dominant to white
glumes, when strains with these characters were crossed, but in one instance
he obtained 630 plants with black and 40 with greyish-white glumes in $F_2$
from such a cross. He pointed out that this ratio approximated to 15:1
(expected frequencies 625·3 and 44·7, deviations 4·7, $\chi^2 = 0.53$, $n = 1$,
$P = 0.5$ approximately), and so could be regarded as a 9:3:3:1 ratio in

which the first three classes looked alike. Hence he postulated that the two strains differed by two independently-inherited genes, such that each alone or both together produced a black pigment in the glumes. Similarly, with ear pigmentation in *Triticum*, a brown variety crossed with white gave brown in $F_1$, and usually a 3:1 ratio of brown to white in $F_2$, but with winter wheats he obtained 15:1 ratios in $F_2$.

Nilsson-Ehle also crossed strains of *Triticum* having a red pericarp (which is maternal tissue surrounding the seed) with others having a white pericarp. All the crosses gave a red $F_1$ and some gave 3:1 ratios of red:white in $F_2$. However, a cross between an old red variety called Swedish Velvet-wheat and a white variety, produced only red plants in both $F_1$ and $F_2$. This unexpected result led him to obtain an $F_3$ generation by self-fertilization of the 78 $F_2$ plants. He found that 8 of the 78 gave a 3:1 ratio in $F_3$ (aggregate frequencies 307 red, 97 white); 15 gave a 15:1 ratio (total progeny 727 red, 53 white); 5 gave a 63:1 ratio (total progeny 324 red, 6 white); and the remaining 50 $F_2$ plants bred true, giving in total 2317 red progeny. The 63:1 ratio is what is expected if 3 independently inherited pairs of alleles are segregating simultaneously and each of the dominants *A*, *B* and *C*, has the same effect. Nilsson-Ehle suggested that there were three such dominant genes responsible for the red pericarp colour in Velvet-wheat and that the white variety with which it was crossed carried the recessive alleles *a*, *b* and *c*. One would then expect a 63:1 ratio in $F_2$, with which the 78:0 ratio observed is in satisfactory agreement. (For the $\chi^2$ test it is necessary to apply Yates's correction, since one of the observed frequencies is less than 5. The 'corrected' observed frequencies are 77·5 and 0·5, the expected frequencies on the 63:1 hypothesis are 76·8 and 1·2 respectively. Hence the deviations are 0·7 each, $\chi^2 = 0·41$, $n = 1$ and $P = 0·5$ approximately.) Moreover, Nilsson-Ehle showed that on his three gene hypothesis the expectations in $F_3$ agree satisfactorily with what he had found. On the average, out of every 63 red $F_2$ plants, 6 are expected to give 3 reds:1 white in $F_3$, 12 to give a 15:1 ratio, 8 to give a 63:1 ratio and the remainder to breed true.

Nilsson-Ehle suggested that the hereditary basis of continuous variation might be similar to what he had found in *Avena* and *Triticum* for discontinuous variation. In other words, if there were a number of genes affecting the same character, such that the discontinuities of phenotype were sufficiently small to be obscured by environmental variation, then continuous variation would be manifest.

East (1910) had studied the inheritance of various grain characters and of the number of rows of grains on the ear in *Zea mays*. Some of the grain characters gave 15:1 ratios in $F_2$ suggesting control by two independently-inherited pairs of alleles, but the number of rows of grains on the ear showed continuous variation. East found that there was always an even number of rows, and that the number of pairs of rows varied from 4 to 15. He made a number of crosses between strains differing in mean number of pairs of rows and studied the progeny. For example, he crossed a strain of *Zea* in which 28 ears had a mean of 4·14 pairs of rows and standard deviation 0·45 with one in which 107 ears had a mean of 6·59 pairs and standard deviation

0·79\*. He obtained 21 $F_1$ ears with a mean of 5·62 pairs of rows and standard deviation 0·51. In other words, the $F_1$ was intermediate between the two parents and relatively uniform. He selected an $F_1$ ear with 6 pairs of rows and obtained 77 $F_2$ ears by self-fertilization. These had a mean of 5·44 pairs of rows and standard deviation 0·82. Thus, although the $F_2$ mean was similar to that of the $F_1$, the $F_2$ plants were much more variable. East pointed out that this tendency to recover in $F_2$ some individuals that approached the parental forms in appearance is what would be expected if segregation of several pairs of alleles controlling row number was taking place, and hence he favoured the multiple factor hypothesis of inheritance of characters showing a continuous range of variation.

TABLE 4.3  Emerson and East's data for the frequencies of maize ears of various lengths from a cross between a short-eared and a long-eared strain.

| Generation | No. of Ears Measd. | Ear Length to Nearest Centimetre | | | | | | | | | | | | | | | | | Mean Lgth. cm | Standard Deviation cm |
|---|---|---|---|---|---|---|---|---|---|---|---|---|---|---|---|---|---|---|---|---|
| | | 5 | 6 | 7 | 8 | 9 | 10 | 11 | 12 | 13 | 14 | 15 | 16 | 17 | 18 | 19 | 20 | 21 | | |
| P | 57 | 4 | 21 | 24 | 8 | | | | | | | | | | | | | | 6·6 | 0·8 |
| P | 101 | | | | | | | | | 3 | 11 | 12 | 15 | 26 | 15 | 10 | 7 | 2 | 16·8 | 1·9 |
| $F_1$ | 69 | | | | | 1 | 12 | 12 | 14 | 17 | 9 | 4 | | | | | | | 12·1 | 1·5 |
| $F_2$ | 401 | | | 1 | 10 | 19 | 26 | 47 | 73 | 68 | 68 | 39 | 25 | 15 | 9 | 1 | | | 12·9 | 2·3 |

An essential feature of this multiple factor hypothesis is that the effects of the different genes should be cumulative. If they merely duplicated one another, discontinuous variation would result, with $F_2$ ratios of 15:1, 63:1, and so on, as in the examples of these ratios found by Nilsson-Ehle and East. However, if the phenotype depends on the dosage of a series of different genes, then even with quite a small number of them, a considerable number of alternative phenotypes becomes possible, leading to greatly increased variability in the $F_2$ generation from a cross between two pure-breeding lines.

This increased variability in $F_2$, which Kölreuter (1763) was the first to discover, and which East (1910) had found, has since been amply confirmed for numerous characters showing continuous variation and in a great diversity of organisms. One of the clearest examples was from the work of Emerson and East (1913). They had studied the inheritance of numerous quantitative characters in *Zea mays*, such as lengths and diameter of ears, weight of grains, height of plant, number of nodes, internode length and duration of growth. From a cross between two strains differing markedly in ear length, the results shown in Table 4.3 and Fig. 4.3 were obtained. The

\* From such data as these, it is possible to calculate the *standard error* of the mean, which is a measure of the precision of the estimate of the mean. The standard error of the mean is obtained by taking the standard deviation of the sample and dividing it by the square root of the number of observations. Thus the first parental strain has a mean of 4·14 ± 0·09 pairs of rows, since its standard error is $\frac{0·45}{\sqrt{28}} = 0·09$, and similarly the second parental strain had a mean of 6·59 ± 0·08 pairs of rows, since its standard error is $\frac{0·79}{\sqrt{107}} = 0·08$. The first mean does not differ significantly from 4, since once in twenty times one would expect a deviation of plus or minus twice the standard error or more. By a similar argument one would deduce that the second mean differs significantly from both 6 and 7.

short-eared parent was relatively uniform with a mean ear-length of 6·6 cm
and standard deviation 0·8 cm. The long-eared parent was more variable
(mean length 16·8 cm, standard deviation 1·9 cm). The $F_1$ plants were
intermediate and relatively uniform (mean length 12.1 cm, standard deviation 1·5 cm) while the $F_2$ plants were much more variable (mean length
12·9 cm, standard deviation 2·3 cm).

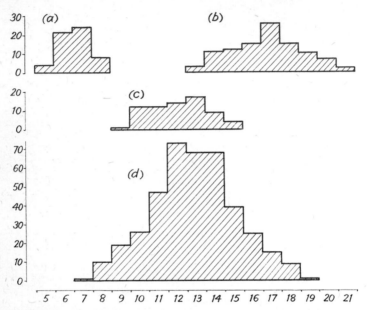

FIGURE 4.3   Frequency histograms for ear length in *Zea mays*
from the data of Emerson and East. The number of
ears is plotted as ordinate and the ear length in
centimetres as abscissa. (*a*) and (*b*) represent the
two parental strains, (*c*) the $F_1$ generation, and (*d*)
the $F_2$ generation.

Fisher (1918) took up the problem of testing the agreement between the
observed values of the correlation coefficients between relatives and those
expected on the multiple factor hypothesis, without making the arbitrary
simplifications which had misled Pearson. Fisher made no assumptions about
equality of effect of the various factors, nor about dominance or equality of
frequency between alleles. He began by pointing out that when there are two
or more independent causes of variability in a population, their contributions
to the variance are additive. This necessarily implies that their contributions to the standard deviation are not additive. Hence, in considering the
causes of variation in a population it is desirable to use the variance, rather
than the standard deviation, as a measure of the dispersion about the mean
value. He interpreted the fraternal correlation coefficient for human stature
of 0·54, which Pearson and Lee had obtained, by saying that 54% of the
variance of brothers is accounted for by ancestry. On the hypothesis of a

large number of Mendelian factors determining stature, the greater part of the remaining 46% of the variance would be due to the segregation of factors for which the parents were heterozygous, and a small part to environmental differences during the brothers' childhood.

Fisher confirmed Yule's finding that the effect of dominance between the alleles is to reduce the parental correlation coefficient by a certain amount. But Fisher took the analysis much further and showed that the effect of dominance is to reduce the fraternal correlation to only half the extent to which the parental correlation is reduced. This enables the effects of dominance to be distinguished from the effects of environmental factors, and moreover explains why the fraternal correlation coefficient exceeds the parental. The observation of this excess by Pearson and Lee is thus in itself evidence in favour of the theory of cumulative Mendelian factors, and moreover with a considerable amount of dominance between the alleles. Fisher estimated that not more than 5% of the fraternal variance was due to environmental factors. His estimates for the correlation coefficients between relatives on the basis of the multiple factor hypothesis were in excellent agreement with the observations of Pearson and Lee, and others. Moreover, on making certain assumptions about assortative mating Fisher was able to deduce Galton's 'law of ancestral heredity'. Thus the controversial question of how far the inheritance of continuous variation could be attributed to Mendelian factors was finally resolved.

Much further support for the multiple factor theory has been obtained subsequently. Fisher's idea of partitioning the variance between the various causes of variability has been greatly extended, and has led to a much fuller understanding of the genetics of quantitative characters. Thus, Mather and Jinks (1971) have demonstrated that the genes responsible for the inheritance of continuously variable characters are not all inherited independently, but may show linkage, comparable to that known for certain genes controlling discontinuous characters. The method of establishing linkage is based on the analysis of variance. Consider two inbred lines differing in a quantitative character such as abdominal bristle number (*Drosophila*), or height or length of parts (plants), and assume that this character is determined by numerous genes. The character is counted or measured in a number of individuals of each line, and the means and variances calculated. Assuming that the two strains have each been inbred for a number of generations, the individuals of each will be almost uniform genetically, like Johannsen's pure lines of *Phaseolus*. In consequence, the observed variance can be attributed almost entirely to environmental factors. The two lines are then crossed and a number of individuals of the $F_1$ measured. The mean value for the character is expected to be intermediate in the $F_1$ between the two parents, as East (1910) and Emerson and East (1913) had found. Moreover, the variance of the $F_1$ will be due wholly to environmental factors, since all the individuals will have the same genotype, being heterozygous for the alleles by which the parents differ. An $F_2$ generation is raised by self-pollination (plants) or inter-crossing of $F_1$ individuals, and again the mean and variance are calculated from measurements of a number of individuals. The variance will now have both genetic and environmental components, since segregation of alleles will have taken place. As a first approximation it may be assumed that the environmental variance is constant in the parental, $F_1$ and $F_2$ generations, if they have been grown under similar conditions, and hence the genetic contribution to the $F_2$ variance can be estimated. However, recent work has indicated that the $F_1$ generation commonly has a smaller variance than the initial inbred lines, indicating that the development of the latter is more readily modified by the environment. Some of the $F_1$ individuals are

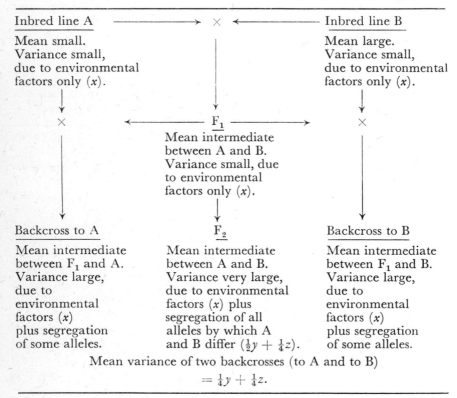

Inbred line A      ———————→   ×   ←———————   Inbred line B

Mean small.
Variance small,
due to environmental
factors only ($x$).

Mean large.
Variance small,
due to environmental
factors only ($x$).

×

$F_1$
Mean intermediate
between A and B.
Variance small, due
to environmental
factors only ($x$).

×

Backcross to A

Mean intermediate
between $F_1$ and A.
Variance large,
due to
environmental
factors ($x$)
plus segregation
of some alleles.

$F_2$
Mean intermediate
between A and B.
Variance very large,
due to environmental
factors ($x$) plus
segregation of all
alleles by which A
and B differ ($\frac{1}{2}y + \frac{1}{4}z$).

Backcross to B

Mean intermediate
between $F_1$ and B.
Variance large,
due to
environmental
factors ($x$)
plus segregation
of some alleles.

Mean variance of two backcrosses (to A and to B)
$$= \tfrac{1}{4}y + \tfrac{1}{4}z.$$

FIGURE 4.4   Diagram showing the means and variances expected when two inbred lines are crossed and the progeny inbred and also backcrossed. It is assumed that the environmental conditions are constant, that each population responds similarly to environmental effects (not necessarily true—see text), and that the character under study shows continuous variation determined by a number of independent Mendelian factors.

$x$ = environmental contribution to variance.

$y$ = measure of difference in phenotype between corresponding homozygotes, such as $AA$ and $aa$.

$z$ = measure of dominance, that is, difference in phenotype between $AA$ and $Aa$ and between $Aa$ and $aa$, and similarly with the other factors.

back-crossed to each parent, and the means and variances of the progeny determined. As with the $F_2$, the back-cross progeny will have genetic and environmental contributions to the variance, since the alleles by which the inbred lines differed will also segregate in the back-crosses.

The hereditary component of variance can be considered to consist of two parts. One is concerned with the magnitude of the effect of the various genes, that is, the difference of phenotype between $AA$ and $aa$, and similarly for all the other pairs of alleles affecting the character. The other part is concerned with dominance, that is, how far the phenotype of $Aa$ is inter-mediate between $AA$ and $aa$, and similarly for the other pairs of alleles. If one compares the average variance of the progenies from back-crossing to each parent with the variance of the $F_2$, the environmental effects may be considered as a first approximation to contribute equally to each, and similarly with the dominance effects ($Aa$ *versus* $AA$ and $aa$). On the other hand, the homozygote differences ($AA$ *versus* $aa$) will contribute much less to the average variance of the backcrosses than to the $F_2$ variance, because in the back-crosses one cannot have both homozygotes for a pair of alleles (for example, $AA$ and $aa$) present at once. This explains why the variance of the back-crosses is less than that of the $F_2$ (see Fig. 4.4). From a comparison of the average variance of the backcrosses with the $F_2$ variance, the respective contributions of homozygote differences and of dominance can be estimated. In most experiments the former considerably exceeds the latter.

From the analysis of variance in $F_3$, further estimates can be obtained of the size of the various components of the variance. From studies of the variance for plant height in inbred lines of *Antirrhinum majus* and *A. glutinosum* and of the $F_1$, $F_2$, $F_3$ and back-cross progenies, Mather (1949) found internal dis-crepancies in the various estimates for the components of the variance. However, by making allowance for the possibility of linkage between some of the genes, an excellent agreement with observation was obtained. He deduced that linkage was present. Confirmation of linkage between genes responsible for continuously variable characters has been obtained, for example in *Drosophila*, by studying quantitative characters in conjunction with one or more discontinuous characters.

## § 4.9  Conclusion

The hypothesis of multiple Mendelian factors, each with additive effects, has been found satisfactorily to explain data on the inheritance of many quantitative characters. The phenomena of dominance and of linkage, which are manifest in Mendelian inheritance of discontinuous characters, have also been demonstrated with characters showing a continuous range of variation. There is no reason, therefore, to suppose that the inheritance of quantitative characters differs in any fundamental way from that of quali-tative characters which show Mendelian inheritance. There is evidence, however, that at least some of the genes affecting quantitative characters may occur in special regions of the hereditary material (see Chapter 18).

# 5. The theory that Mendelian inheritance is determined by the cell-nucleus and non-Mendelian by the cytoplasm

## § 5.1 Weismann's theory of the continuity of the germ-plasm

The term *nucleus* was applied by Brown (1833) to the more or less spherical structure which usually occupies a central position in cells. He observed the nucleus in the cells of both the vegetative and reproductive parts of many flowering plants. He made particular mention of its occurrence in the leaf-cells of orchids and in the stigma and pollen of *Tradescantia*. The fate of the nucleus in cell-multiplication was a source of debate and speculation until Strasburger (1879) reached the conclusion that nuclei arise only from pre-existing nuclei. He had made extensive studies of the cells of the re-productive organs of numerous seed-plants. Meanwhile, Hertwig (1875) with *Paracentrotus lividus* (a species of Sea Urchin) and van Beneden (1875) with *Oryctolagus cuniculus* (Rabbit) had independently established that at fertilization there occurs not merely a fusion of cells but also of nuclei, and moreover these are derived from the egg and the sperm, respectively. Strasburger's discovery that the cell-nucleus has itself the property of heredity, in that it never arises afresh but is always handed down, led naturally to the conclusion that the nucleus is the vehicle of heredity.

It was against this cytological background that Weismann (1883, 1885) put forward his theory of the continuity of the germ-plasm. The essence of this theory was 'that heredity is brought about by the transference from one generation to another, of a substance with a definite chemical, and above all, molecular constitution' (1885, p. 168). He called this substance 'germ-plasm', and he supposed that in the development of the body or 'soma' of an individual from the fertilized egg-cell, this substance, or part of it, was 'reserved unchanged for the formation of the germ-cells of the following generation' (1885, p. 168). This notion had a considerable influence on contemporary thought. Weismann (1883) was the first person to challenge the existence of the inheritance of mutilations, or 'acquired characters', as they were called by Lamarck (1809), an idea which, as indicated in Chapter 1, had been widely accepted ever since the time of Hippocrates. For it was fundamental to Weismann's theory that this chemical (the germ-plasm) could not be

40

modified by the environment during the life of an individual, but was handed down unaltered. Weismann accepted the opinions of Hertwig and Strasburger that 'the nature of heredity is based upon the transmission of nuclear substance with a specific molecular constitution' (Weismann 1885, p. 180) and went on to equate this substance with his postulated germ-plasm.

## § 5.2   Boveri's experiments with echinoderms

Strong support for the view that the nucleus is the primary carrier of hereditary material in the cell was obtained by Boveri (1889) from studies on echinoderms. Working at the Zoological Station at Naples, he added sea urchin sperm to eggs which had been fragmented by shaking, and he found that even pieces of egg lacking a nucleus could sometimes be fertilized. The resulting larvae were quite normal in appearance, except that they were only about one-quarter of the size of those developed from unfragmented eggs or from fragmented eggs with a nucleus. He deduced from this experiment that

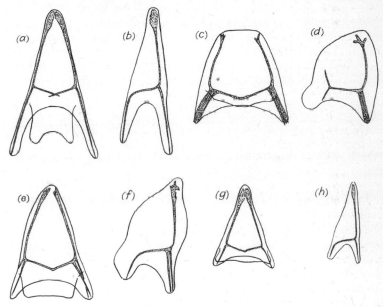

FIGURE 5.1   Posterior and side views (×100) of the calcereous skele-
tons of the larvae of sea urchins (from Boveri, 1889).
(a) and (b) *Psammechinus microtuberculatus*
                 (a) Posterior view          (b) Side view
(c) and (d) *Sphaerechinus granularis*
                 (c) Posterior view          (d) Side view
(e) and (f) *S. granularis* × *P. microtuberculatus*
                 (e) Posterior view          (f) Side view
(g) and (h) *S. granularis* (enucleated) × *P. microtubercu-
                 latus*
                 (g) Posterior view          (h) Side view

the sperm nucleus possessed all the properties necessary to function like a fertilized egg-nucleus. This was an important conclusion as it implied that the egg-nucleus and the sperm-nucleus were essentially alike in their hereditary contributions.

Boveri made further experiments with even more striking results. He hybridized two species of sea urchin, *Psammechinus microtuberculatus* and *Sphaerechinus granularis*. The calcareous skeleton of the *Psammechinus* larva when seen in posterior view is shaped like a capital A of rather narrow outline (Fig. 5.1(*a*)) and the body is similarly shaped. In side view, the appearance is even narrower (Fig. 5.1(*b*)). On the other hand, the *Sphaerechinus* larva is much broader, the skeleton in posterior view almost resembling a capital H (Fig. 5.1(*c*)), but with the uprights slightly tilted towards the A shape. The body is correspondingly broad, and is almost as wide in side view (Fig. 5.1(*d*)). Boveri found that when normal *Sphaerechinus* eggs were fertilized by *Psamm-echinus* sperm, the resulting hybrid larvae were almost exactly intermediate in appearance between those of the two parents. The skeleton in posterior view was like a capital A of broad outline (Fig. 5.1(*e*)), and the body both in front view and in side view was of intermediate breadth (Fig. 5.1(*f*)). When the hybridization was repeated using fragmented *Sphaerechinus* eggs, the enucleated fragments when fertilized with *Psammechinus* sperm developed into larvae exactly resembling those derived from *Psammechinus* egg and sperm, though dwarf in size (Fig. 5.1(*g*) and (*h*)). Boveri concluded that the maternal cytoplasm has no influence on the hereditary characters, and that these are determined by the nucleus alone.

## § 5.3  Hämmerling's experiments with *Acetabularia*

Another series of experiments which demonstrated the importance of the nucleus in determining heredity were those of Hämmerling (reviews 1953, 1963) with species of *Acetabularia*, small green plants of warm seas. These algae, in their vegetative condition, consist essentially of a single giant uninucleate cell. This cell may be 6 cm or more in length and is differentiated into a basal lobed rhizoid containing the nucleus, an erect photosynthetic stalk, and, in mature plants, a terminal cap up to 1 cm in diameter. Thus the full-grown plant has something of the appearance of a small green toadstool. In *A. mediterranea* which, as its name implies, is a native of the Mediterranean region, the cap has an average of about 81 rays (80·8 $\pm$ 3·2), and these have rounded tips (Fig. 5·2(*a*)). In *A. crenulata* from the W. Indies there are only, on average, 31 rays (30·7 $\pm$ 2·8) per cap, and the tips of the rays are pointed (Fig. 5.2(*b*)).

Hämmerling took immature plants of each species and grafted them together by cutting off the tops of the stalks (the caps had not yet developed) and bringing the cut surfaces into contact. A new stalk soon developed at the point of grafting, and this gave rise to a cap which was intermediate in appearance between the two parents. A number of such caps had an average of about 42 rays (42·3 $\pm$ 4·8) and hence approached *A. crenulata*, but with tips usually rounded like *A. mediterranea*. The intermediate character of this graft-hybrid is to be expected, since it contains two nuclei, one from each species. Control experiments in which two plants of *A. mediterranea* were grafted together gave caps resembling this species. Corresponding results were obtained with the other species.

Further grafts were achieved using the stalk of an immature plant of one

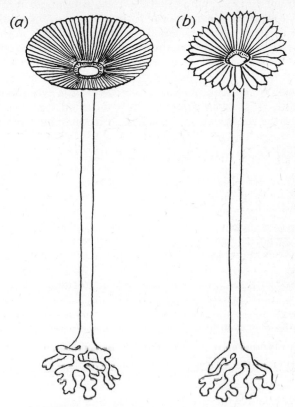

FIGURE 5.2   Drawings to show the appearance
(×2) of (*a*) *Acetabularia mediterranea*, and (*b*) *A. crenulata*.

species and the rhizoid system (and hence nucleus) of the other. Such grafts contained only one nucleus. The type of cap which developed was found to depend on the length of stalk grafted. If it was about 1 to 2 cm long, the cap was often of intermediate appearance between that of the two species, whereas if the piece of the stalk that was grafted on was about 0·5 cm long or less, the cap usually resembled that of the rhizoid (and hence nuclear) parent. It was noteworthy that, however long the stalk, the cap never wholly resembled the species from which the stalk was obtained. Hämmerling deduced that the nucleus in the rhizoid had an over-riding control over the form of the cap, but that the stalk also had some influence.

This result is not so clear cut as that obtained by Boveri with echinoderms. However, Hämmerling found that if in these graft-hybrids of the rhizoid of one species of *Acetabularia* with the stalk of the other, he cut off the first cap to form, in due course a second would develop and this invariably resembled the cap of the species which provided the rhizoid (and hence the nucleus). Hämmerling concluded that although substances in the cytoplasm are the immediate determinants of the form of the cap, these substances are ultimately controlled by the nucleus. It is to be noted that in these grafting experiments one cannot rule out the possibility that the cytoplasm in the rhizoid is playing a vital part. It seems more probable, however, that it is the nucleus which is responsible, just as in Boveri's hybrid echinoderms from enucleated eggs. In the latter instance, there is so little cytoplasm in a sperm that one is fully justified in drawing the conclusion that the nucleus was responsible for the resemblance of the hybrid to the male parent.

## § 5.4   Delayed Mendelian inheritance

It will be recalled that one of the characteristic features of Mendelian inheritance is that reciprocal crosses give identical results. Such identity gives strong support to the hypothesis that the cell nucleus is the bearer of the Mendelian factors, since the nuclear contributions of sperm and egg appear to be the same, whereas their cytoplasmic contributions clearly differ. The over-riding importance in heredity of the nucleus compared with the cytoplasm, as demonstrated by Boveri's and Hämmerling's experiments, appears to be matched by the over-riding importance of Mendelian in-heritance in living organisms compared with non-Mendelian. However, as will be discussed in § 18.11, the possibility that non-Mendelian inheritance does play a major part in heredity must not be overlooked.

Before discussing non-Mendelian heredity, some experiments will be described which demonstrate Mendelian inheritance, but with delayed action such that the effect of the nuclei presumed to be responsible is carried over in the cytoplasm into cells which may have a different nuclear con-stitution. Such an effect may resemble non-Mendelian inheritance, and needs to be carefully distinguished from it.

One example of such delayed Mendelian heredity has already been described in Chapter 3, namely, the inheritance of pollen shape in *Lathyrus odoratus* studied by Bateson, Saunders and Punnett (1905). Although segregation of factors must occur before the formation of the pollen grains, as first postulated by Mendel, yet when a pure-breeding strain of *Lathyrus odoratus* with long pollen-grains (Fig. 5.3(*a*)) is crossed with a pure-breeding strain with round grains (Fig. 5.3(*b*)), the $F_1$ plants as already mentioned have pollen grains which are all long in shape, and in $F_2$ three-quarters of the plants have long pollen and the remaining one-quarter have round pollen. It is evident from these results that long pollen-shape is determined by a dominant factor ($L$) and round pollen-shape by the recessive allele ($l$). It follows that the $F_1$ plants have the genetic constitution $Ll$ and half their pollen grains must contain the $L$ factor and the other half the $l$ factor.

Nevertheless, all the grains from these plants have the L shape. It is evident that pollen shape must be controlled by the genetic constitution of the anther and not by that of the pollen grain itself. The simplest hypothesis is that some substance determining pollen shape is produced by the nuclei of the pollen

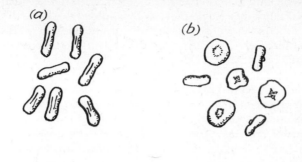

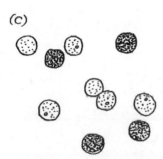

FIGURE 5.3   (a) and (b) Pollen (×175) of *Lath-yrus odoratus* in the dry state (from Bateson, 1909).

(a) The normal Long (cylindrical) pollen-grains (usually with three germ-pores)

(b) The mutant round (disc-shaped) pollen-grains (usually two-pored)

(c) Pollen (×125) of *Oryza sativa* segregating for starchy (shown dark) and waxy (shown light) contents (from Parnell, 1921).

mother cells (or other anther cells) and carried over to the developing pollen-grains in the cytoplasm. This delayed manifestation of a nuclear effect is comparable to that found by Hämmerling in *Acetabularia* species grafts, where the second cap to form resembles that of the parent contributing the nucleus, whereas the first cap is more often intermediate. It is known that other characters of pollen in flowering plants may not show this delayed inheritance. Thus, Parnell (1921) discovered that in *Oryza sativa* (Rice) if a pure-breeding strain with normal starchy pollen-grains (staining blue with iodine) is crossed with a pure-breeding strain with waxy pollen-grains containing erythrodextrin (staining reddish-brown with iodine), the $F_1$ plants have half their pollen-grains starchy and the other half waxy (Fig. 5.3(c)). Evidently a single pair of alleles determines this difference, and the factors function after the formation of the pollen grains. A similar situation was discovered in *Zea mays* independently by Demerec (1924) and by Brink and MacGillivray (1924).

A second example of a delayed nuclear effect is provided by the work of Gossop, Yuill, and Yuill (1940) with *Aspergillus niger*. This asexual fungus normally has chains of black conidia (asexual spores), but a strain is known with cinnamon-coloured conidia and another with brown conidia. If the cinnamon strain and the brown strain are allowed to grow together, the hyphae may fuse so that the mycelium has within its cells nuclei from both strains (there is no nuclear fusion). The occurrence of nuclei of different genotype within one cell is called *heterokaryosis*. When such a heterokaryon produces conidia, the erect aerial conidiophores are composed of multi-nucleate cells, but the conidia produced from these are uninucleate. Hence

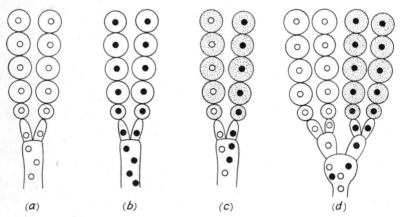

(a)  (b)  (c)  (d)

FIGURE 5.4  Conidia (×2000) of *Penicillium* and *Aspergillus* (from Pontecorvo, 1947).

(a–c) *Penicillium notatum*. Small open circles indicate nuclei carrying the gene for white conidia, solid circles those carrying the gene for yellow conidia, and dots in the cytoplasm indicate synthesis of the normal green conidial pigment. (a) White mutant. (b) Yellow mutant. (c) White-yellow heterokaryon with green conidia.

(d) *Aspergillus nidulans*. Small open circles indicate nuclei carrying the gene for white conidia, solid circles those carrying the normal allele for green conidia, and dots in the cytoplasm indicate synthesis of the green pigment. The white-green heterokaryon has some chains of spores white and others green.

in the formation of conidia, the heterokaryon breaks down and each conidium receives either a 'cinnamon' or a 'brown' nucleus. Nevertheless, all the conidia are black. Evidently the black pigment, or a precursor of it, is manufactured in the heterokaryotic conidiophore, where both types of nuclei can co-operate in its production, and is then transmitted in the cytoplasm to the conidia. The colonies derived from these conidia, if isolated, will be either cinnamon or brown in conidial colour according to the nuclei they have received. This is comparable to the long and round pollen shapes in *Lathyrus odoratus*. An even more striking example was described by Pontecorvo and Gemmell (1944), where white-spored and yellow-spored mutants of the normally green-spored *Penicillium notatum* were found to produce a heterokaryon which had green conidia, although these are uninucleate. When isolated, these conidia produced strains with either white or yellow conidia (Fig. 5.4(a)–(c)).

Other characters in fungi may not show this delay in the manifestation of the effects of nuclei. Thus, Gossop, Yuill, and Yuill found that in *Aspergillus nidulans* a heterokaryon between a white-spored mutant and the normal green-spored fungus produced mixed heads in which single chains of conidia were either green or white (Fig. 5.4(d)). This is analogous to starchy and waxy pollen in *Oryza* or *Zea*.

## § 5.5   Boycott and Diver's experiments with *Limnaea*

One of the best-known examples of delayed Mendelian inheritance concerns the direction of coiling in the fresh-water snail *Limnaea peregra* studied by Boycott and Diver (1923). The normal snail has the body and shell coiled in a right-handed spiral but in a pond near Leeds, Yorkshire,

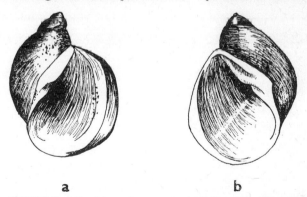

a                                            b

FIGURE 5.5   *Limnaea peregra* (×2·5). (*a*) Normal dextral snail shell. (*b*) Sinistral snail shell.

along with the normal form another was found which was an exact mirror image of it (Fig. 5.5). Not only were the body and shell coiled in a left-handed spiral, but the positions of the heart and kidney were complementary to those in the normal animal, and the alimentary and reproductive openings were on the left instead of the right side of the neck. Earlier work by other authors had indicated that the difference between dextral and sinistral snails becomes apparent at the earliest stage of development, since the plane of the first cell-division in the fertilized egg is different in the two forms.

*Limnaea peregra* is hermaphrodite, and although cross-fertilization is usual, self-fertilization can also occur. Individuals which had not yet reached sexual maturity were isolated and allowed to fertilize themselves. Broods of several hundred snails were obtained in this way from each of 144 individuals, all of related ancestry. It was found that, apart from rare exceptions which were possibly due to environmental factors, all the snails in a brood were alike as regards their direction of coiling, but either type of parent (dextral or sinistral) could produce either type of brood!

Boycott and Diver explained these remarkable results by postulating that 'the appearance of the individual is the result of, and is determined by, the product of the parental nuclear composition . . .' (1923, p. 211). Sturtevant (1923), on the basis of their results, was more specific and suggested that, in view of the very early development of the difference between dextral and sinistral forms, it was likely to be the nuclear composition of the mother which determined the direction of coiling. Two further observations made by Boycott and Diver, namely, that sinistral snails obtained by self-fertilization bred true, whereas dextral snails similarly obtained often did not, led Sturtevant to postulate that the dextral condition was dominant to sinistral.

Later work by Boycott, Diver, Garstang and Turner (1930) confirmed both of Sturtevant's postulates. In crosses between pure-breeding dextral and pure-breeding sinistral forms, the appearance of the $F_1$ progeny was the same as their mother, the $F_2$ snails were all dextral, and in $F_3$ 1192 dextral and 401 sinistral broods were obtained. These results are intelligible as simple Mendelian inheritance, provided the direction of a snail's coiling is always regarded as a property of its mother. Evidently the mother's nuclear constitution leads to the production of some substance in the cytoplasm of the cells of the ovary or primary oocyte, and this substance reaches the next generation by way of the cytoplasm of the egg, where, after fertilization, it determines the plane of the first cell-division, and hence the dextrality or sinistrality of the offspring.

## § 5.6  Kühn's experiments with *Ephestia*

A rather similar example of delayed Mendelian inheritance concerns eye-colour in the moth *Ephestia kuhniella*. The eyes are normally black but a variant with red eyes is inherited as a Mendelian recessive. Kühn (1937) discovered that if red-eyed females (*aa*) are crossed with heterozygous black-eyed males (*Aa*), the black-eyed (*Aa*) and red-eyed (*aa*) progeny can be distinguished even in the young larvae. However, when the reciprocal cross is made, all the young larvae have black eyes, but in the individuals of constitution *aa* the colour gradually changes to red as they develop. Evidently a precursor of the pigment is carried via the cytoplasm of the egg to the progeny, but its effect does not persist to adult life, except in those progeny

(*Aa*) which carry the black-eye factor and hence can renew production of the precursor and pigment. Kühn confirmed this hypothesis by grafting testes from a black-eyed (*Aa*) larva into a red-eyed (*aa*) female one. It then developed a darker eye-colour which also appeared in the young *aa* progeny larvae. The substance which was shown by these experiments to be carried over in the cytoplasm of the egg to the next generation has been identified as kynurenine, which is also a precursor of *Drosophila* eye-pigment.

## § 5.7   Spurway's experiments with grandchildless *Drosophila*

A more extreme example of delayed Mendelian inheritance is provided by the mutant called *grandchildless* described by Spurway (1948) in *Drosophila subobscura*. During breeding work with this species of fly, all the males in a particular family were found to have abnormally small internal reproductive organs. These organs have an orange pigment and can be seen in living flies through the body wall. On dissection no testes or only the minutest rudiments were found. The same condition appeared in the progeny of sister cultures. Moreover, it was found that females were similarly affected, the ovaries remaining vestigial.

It was established that this sterility of the progeny was associated with particular female parents and was independent of the male parent. In order to determine the hereditary basis of the condition, brothers of grandchildless females were crossed with unrelated females and the progeny inbred. After two generations the sterile families reappeared, thus showing that their cause could be carried by males. Crosses were also made between brothers and sisters of 'sterile-progeny' females, and some the the female progeny were found to show the condition by having sterile offspring. In this way the sterility was followed through more than twenty generations of flies. Grandchildless females formed a minority of the females in each generation.

On the evidence of the crossing experiments, it was postulated that the sterility was due to a Mendelian recessive gene called *gs*. The sisters with fertile offspring, by which the *gs* stock was maintained, were assumed to be heterozygous normals ($+/gs$)\*, but it was impossible to determine whether the brothers with which these were crossed in the inbreeding experiment were heterozygous ($+/gs$) or homozygous ($gs/gs$). On the hypothesis just mentioned of a Mendelian recessive factor determining the grandchildless condition, in each generation following brother-sister mating, either a 3:1 or a 1:1 ratio of 'fertile-progeny' to 'sterile-progeny' females is expected, but the individual families were too small to distinguish these ratios. Taking Spurway's total data for fifteen consecutive generations ($F_8$ to $F_{22}$ inclusive) comprising 23 brother-sister matings, the female progeny consisted of 309 with fertile offspring, 90 with sterile offspring and 112 themselves sterile. Since grandchildless females are less fertile than normal, it is probable that the 112 sterile females include many *gs/gs* females. With any reasonable distributions of the 112 between the other two classes, the ratio of the latter lies between 3:1 and 1:1, and hence is in agreement with the hypothesis.

\* The $+$ sign is used for the normal allele.

It thus appears that a Mendelian recessive factor, when homozygous, leads to the production of some substance in the ovary of female flies. This substance, or some other effect of the gene, may lower their fertility somewhat, but its main effect is on the progeny. It is presumably carried via the cytoplasm of the egg to the next generation, where it leads to a failure of development of the ovaries and testes. The abnormality cannot be detected before the last instar in male larvae, or the mid-pupal stage in females, and is most readily observed in the mature insects. This represents one of the most extreme instances known of the delayed effect of a Mendelian gene.

## § 5.8   Rhoades's experiments with iojap *Zea*

An instructive comparison can be drawn between the inheritance of grandchildless *Drosophila subobscura* and that of a particular kind of leaf-striping in *Zea mays* called 'iojap'. The name 'iojap' is abbreviated from 'Iowa', as the source of the maize strain, and 'japonica' as the name of a similar striped variety. Jenkins (1924) described experiments which established that the striped leaves found in iojap plants are due to a recessive Mendelian gene. Maize is usually cross-fertilized. Certain normal green plants when self-pollinated were found to give progenies segregating for iojap striping. Omitting one family where the stripes were very narrow and hard to recognize, Jenkins's total data for these segregating progenies were 2498 green and 782 iojap (including 12 white plants) (Fig. 5.6(*a*)) (Expected frequencies on $3:1$ hypothesis: 2460 and 820, deviations 38, $\chi^2 = 2\cdot3$, $n = 1$, $P > 0\cdot1$.) Furthermore, 26 of these 2498 green plants, on self-pollination, were found to consist of 9 which bred true and 17 which again

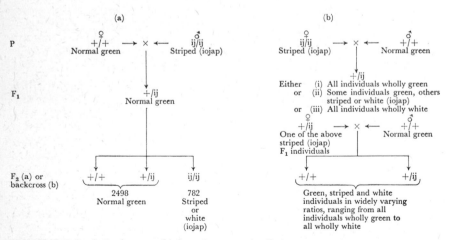

FIGURE 5.6   Inheritance of iojap character in *Zea mays*.
(*a*) Mendelian inheritance when iojap plants used as male parents (from Jenkins, 1924).
(*b*) Non-Mendelian inheritance when iojap plants used as female parents (from Rhoades, 1943).

segregated for iojap, in good agreement with the 1:2 ratio of homozygotes to heterozygotes which Mendel would have expected. However, an un-expected result was obtained when female inflorescences of iojap plants (obtained as above) were pollinated from a pure-breeding normal green plant. The resulting $F_1$ plants often had striped leaves of typical iojap appearance. Rhoades (1943) showed that so long as iojap plants are used as female parents, the inheritance of the condition is similar to that of the *albomaculata* variety of *Mirabilis jalapa* described by Correns (1909) and discussed in § 2.7. Thus, widely varying ratios of green, white and striped progeny were obtained in this way from iojap plants, and this result was found irrespective of the male parent used (Fig. 5.6(*b*)).

Rhoades proposed the following hypothesis to account for these remark-able results. The nuclear gene, *ij*, when homozygous (*ij*/*ij*) causes striping of the leaves, but this striped condition, once it has been initiated, is then inherited through the cytoplasm. Cytoplasm is handed down only through the egg-cell and not through the pollen, and so the iojap condition shows maternal inheritance. Rhoades confirmed this hypothesis by crossing the $F_1$ striped plants as female parent with an unrelated normal green plant as male parent (see Fig. 5.6(*b*)). The $F_1$ striped plants would be heterozygous for the iojap gene (+/*ij*), the other parent would be homozygous for the normal allele (+/+), and so the progeny from this cross would consist of equal numbers of individuals homozygous for the normal allele (+/+) and heterozygous (+/*ij*). Nevertheless, for some ears the progeny consisted entirely of white seedlings, half of which by the above reasoning must not have possessed the iojap gene at all. From other ears, there were green, variegated and white progeny in diverse ratios, and here again many of the iojap plants must have lacked the iojap gene.

The essential difference between the inheritance of grandchildless *Drosophila* and iojap *Zea* is that in the former the cytoplasmic modification initiated by the nuclear gene is passed only to the immediate progeny and no further, whereas in the latter it appears to be capable of indefinite passage from generation to generation. This distinction constitutes the essential difference between Mendelian and non-Mendelian inheritance. It can of course be argued that in grandchildless *Drosophila* there is no further opportunity for cytoplasmic passage beyond the immediate progeny, on account of the sterility induced. The initiation of iojap *Zea* shows Mendelian heredity, but once it has been initiated, the inheritance is non-Mendelian. The ability to be handed down indefinitely, independently of the nuclear constitution, implies that whatever is responsible for non-Mendelian inheritance must possess the ability to reproduce itself in the cytoplasm. The nature of such reproducing structures is discussed later in this chapter and in § 18.11.

## § 5.9  Baur's experiments with *Pelargonium*

Almost simultaneously with Correns's (1909) discovery of the non-Mendelian inheritance of the *albomaculata* character in *Mirabilis jalapa* (see Chapter 2), Baur (1909) reported an example in *Pelargonium zonale*

(Geranium) which differed in an important respect from that in *Mirabilis*. Baur's example also concerned leaf-variegation, but he found that in *Pelargonium* reciprocal crosses gave similar results. It will be recalled that in 'albomaculata' *Mirabilis* (and also in 'iojap' *Zea*) such crosses give different results. Plants of a strain of *Pelargonium* with white-margined leaves when self-pollinated or inter-crossed gave wholly white progeny, which necessarily died young. Baur found that the sub-epidermal layer in white-margined plants was white. Since this is the layer which gives rise to the reproductive cells, the wholly white progeny could be explained. However, when the white-margined plants were crossed with normal green plants, the progeny consisted of green plants, white-margined plants and wholly white plants in widely-varying ratios, and this was true which ever direction the cross was made. Baur concluded that in *Pelargonium* cytoplasm is inherited from the male as well as the female parent. Thus the difference between reciprocal crosses, which in the *Mirabilis* and *Zea* examples was quoted as one of the distinctions from Mendelian heredity, cannot be maintained as an invariable feature of non-Mendelian heredity. The most striking feature of the latter kind of inheritance is the failure of the character-difference to segregate in a regular and predictable way at the time of formation of the reproductive cells.

## § 5.10   Renner's experiments with *Oenothera*

Further examples of non-Mendelian heredity of the *Pelargonium* type were found by Renner (1924) in *Oenothera* (Evening Primrose) hybrids. Thus, he found that when *O. hookeri* was crossed as female parent with the not very closely related *O. muricata* as male parent, the hybrids had yellow leaves and died. Plants of the same nuclear constitution obtained from the reciprocal cross* had green leaves with yellow flecks. He postulated that these yellow flecks represented cytoplasm contributed by the male parent (*O. hookeri*) and that in the presence of the hybrid nucleus, the *O. hookeri* cytoplasm became defective, giving the yellow colour. From self-pollination of flowers on wholly yellow shoots on the flecked plants, Renner was able to show that normal green progeny were recovered whenever the cytoplasm was restored to an *O. hookeri* nucleus. This result confirmed his hypothesis that the yellow colour was a consequence of lack of harmony between *O. hookeri* cytoplasm and hybrid nucleus, and showed that the change was reversible.

## § 5.11   Hypotheses to explain non-Mendelian inheritance of plastid characters

From the discussion of non-Mendelian heredity, both in this chapter and in Chapter 2, it is apparent that this type of inheritance is best explained as due to the transmission of some material in the cytoplasm. All the examples mentioned have referred to abnormalities of the green pigmentation of the

* To obtain such plants is more difficult than might be supposed, owing to the occurrence in *O. muricata* of an abnormal hereditary mechanism (see § 9.2).

leaves. The chlorophyll occurs in particles in the cytoplasm called plastids. The simplest hypothesis to account for the non-Mendelian inheritance of chlorophyll-pigmentation defects is therefore to postulate that plastids are autonomous self-replicating structures and that they may occur in abnormal yellow or colourless forms. This hypothesis was first proposed by Baur (1909) to account for his results with *Pelargonium*, and is supported by several pieces of evidence, notably that the white areas in variegated plants are found to have minute colourless plastids visibly different from the normal green ones characteristic of the green areas. However, it does not follow from this that the plastids are necessarily themselves the hereditary determinants. Indeed, the following observations made by Correns (1909) with 'albomaculata' *Mirabilis jalapa* suggest that cytoplasmic constituents other than the plastids are responsible. First, at the boundary between green and white areas in variegated plants, a gradation in plastid colour was observed from cell to cell, suggesting diffusion from the white areas of some substance which interferes with normal plastid development. Secondly, cells were rarely found to contain a mixture of two different kinds of plastids such as on Baur's hypothesis would be expected to occur frequently, particularly in the boundary zone between green and white areas. Thirdly, there are insufficient cell-divisions during development for the chance segregation of two types of plastids into different cells to occur often, yet green and white areas are constantly being initiated. Similar observations to these have been made in a number of other instances of non-Mendelian inheritance of variegation, for example by Rhoades (1947) with iojap *Zea*.

If the cytoplasmic hereditary determinants are not the plastids, what is their nature? In attempting to answer this question, an observation by Anderson (1923) is important. He was the first to discover non-Mendelian inheritance of leaf-striping in *Zea mays*, and in studying the progeny of such a striped plant used as female parent, he took care to record the position of each grain on the ear. He found that all the grains in one area of the ear produced green plants, and all those in another area produced white plants that died, while striped plants tended to come from grains forming a transitional zone between the 'green' and 'white' areas (see Fig. 5.7). These areas extended longitudinally up the ear and corresponded with the cell-multiplication during development, just as in the vegetative parts the striping reflects the way in which the leaf develops. It is evident that, in sexual reproduction as well as in leaf growth, green cells tend to give green, and white to give white. This behaviour suggests that the hereditary determinants do not readily diffuse from one cell to another, or in other words are particles rather than a fluid. It should be noted that the gradation in plastid colour at the boundary of green areas, referred to in the last paragraph, need not imply that the hereditary determinant is fluid but merely that its effects can diffuse from one cell to another. Similar gradations in plastid colour often occur when variegation shows Mendelian inheritance, and then one concludes that the hereditary determinant resides in the cell nucleus, and hence cannot be directly responsible for the gradation in colour in the cytoplasm.

FIGURE 5.7   Reconstruction ($\times 0.5$) of an
ear on a striped plant of *Zea
mays* studied by Anderson
(1923). White (unshaded)
grains gave albino seed-
lings, cross-hatched grains
gave striped seedlings, and
black grains gave normal
green seedlings. All the
grains on the ear looked
alike. Almost all those on
the rear side of the ear gave
green seedlings.

This hypothesis of particulate cytoplasmic hereditary determinants,
capable of replication and indefinite passage from generation to generation,
appears to be consistent with the data on non-Mendelian inheritance of
variations in the green pigmentation of plants. However, these hypothetical
particles have not yet been identified with specific cytoplasmic structures.
Certain of the data, notably those of Baur for *Pelargonium* and of Renner for
*Oenothera*, are consistent with the equating of these particles with the plastids,
but this equation lacks confirmation.

## § 5.12   Further examples of cytoplasmic inheritance

Cytoplasmic inheritance in green plants is not confined to plastid characters. Michaelis (1954) studied cytoplasmic inheritance in *Epilobium*
(Willow-herb) by making reciprocal crosses between different species and

varieties, and then pollinating the progeny with the male parent, and repeating this in each successive generation. On the assumption that no cytoplasm is inherited from the pollen, plants will be obtained within a few generations which have essentially the nuclear constitution of one parent (that used as male in each generation), while the cytoplasm has all been derived from the other parent. Michaelis found that *E. luteum* as female crossed for 24 successive generations with *E. hirsutum* as male gave male-sterile plants resembling *E. hirsutum*. The reciprocal cross could not be continued for more than 3 generations of back-crossing because of sterility or inviability. Within the one species, *E. hirsutum*, similar back-crossing between strains from different parts of Europe also gave evidence of persistent cytoplasmic differences: such characters as male-sterility, stunted growth, or deformed flowers distinguished the back-cross progeny from their male parent. It is evident that some features of *Epilobium* cytoplasm are autonomous and persist, apparently indefinitely, in the presence of an alien nucleus.

Some cytoplasmically-inherited variants, for example, the killer character in *Paramecium aurelia*, and carbon dioxide sensitivity in *Drosophila melanogaster*, appear to be due to infection by a virus (cf. review in Jinks, 1964).

## § 5.13   The heterokaryon test

Many examples of non-Mendelian inheritance are known in Fungi. For this group of organisms, Jinks (1954) has devised a test for the occurrence of cytoplasmic variants, based on the phenomenon of heterokaryosis described in § 5.4. A mutant such as the white conidial character in *Aspergillus nidulans* is known to differ from the normal green conidial form by a nuclear gene, and a heterokaryon of the two produces conidia of each colour. These conidia are uninucleate, and so the colonies derived from them will be homokaryotic. If another character-difference also distinguishes the strains which give rise to the heterokaryon, these characters may reappear only in the parental combinations in the daughter homokaryons. It is then inferred that the second character-difference is also determined by the nucleus. However, it has been found that when a white sexually-fertile and a green sexually-sterile strain form a heterokaryon, the daughter homokaryons may all be sexually fertile whether white or green. It is then inferred that the sexual sterility of the green parent is of cytoplasmic origin. In a particular strain of *A. nidulans* which had been treated with ultraviolet light, a variant with red mycelium continually reappears in a proportion of the colonies derived from the conidia, even when these conidia are obtained from colonies of normal appearance. This is comparable to the origin of cytoplasmically inherited petite yeast described in § 5.15. Arlett, Grindle and Jinks (1962) isolated single conidia from a heterokaryon between this strain and a strain with white conidia. They found that recombination had occurred and the red variant now arose in some isolates of the white-conidial strain, as well as in some of those with normal green conidia. It was evident that the red variant was cytoplasmically inherited, since there is normally no recombination between nuclear factors in a heterokaryon.

## § 5.14   Specific agents which induce cytoplasmic variants

One of the characteristic features of cytoplasmic inheritance is the specific induction of variants by certain agents. This is in contrast to the production of Mendelian mutants, where mutagens produce a whole range of kinds of mutants, not one specific one (see Chapter 9). Agents known to cause specific cytoplasmic variants include acridine dyes, streptomycin, and high temperatures during growth. Treatment with streptomycin (Provasoli, Hutner and Schatz, 1948) or growth for 5 or 6 days at 34°C (Pringsheim and Pringsheim, 1952) can cause a permanent loss of the plastids in *Euglena gracilis*, which then becomes a colourless saprophytic organism, formerly regarded as a distinct species and called *Astasia longa*. Sager (1962) discovered that streptomycin caused cytoplasmic mutation in *Chlamydomonas reinhardi*. Such mutants are transmitted usually by one parent only, exceptionally by both. From study of the progeny of such exceptional zygotes, Sager and Ramanis (1963) found that segregation of two cytoplasmic differences (streptomycin dependence ($sd$) and sensitivity ($ss$), and acetate requirement ($ac^-$) and independence ($ac^+$)) occurred independently, the acetate character usually at an earlier stage of development than the streptomycin character.

## § 5.15   Ephrussi's experiments with yeast

One of the clearest examples of non-Mendelian heredity in Fungi concerns 'petite colonie' in *Saccharomyces cerevisiae* (Baker's Yeast) studied by Ephrussi and colleagues (see Ephrussi, 1953). When individual yeast cells are spread on culture medium, they soon grow to form small circular colonies containing many cells. Ephrussi observed that one or two out of every thousand such colonies were only about one-third or one-half the diameter of the remainder. Cells from one of the normal large colonies, when again spread on culture medium produced a further small proportion of 'petite' colonies, and this happened time after time. On the other hand, the cells from the small colonies bred true and gave only petites. Biochemical studies established that the slow growth of the petite strain of yeast was due to the loss of the ability to respire aerobically. In the absence of oxygen, the growth rate of large colonies was slowed to that of petites. Furthermore, the lack of aerobic respiration in the petite strain was found to be associated with the absence of a number of respiratory enzymes. These enzymes occur in particles in the cytoplasm called mitochondria. Ephrussi and associates discovered that in the presence of acridine dyes, such as acriflavine or euflavine, all the newly-formed daughter-cells in a yeast colony, if isolated, gave rise to petite colonies. This change from normal to petite, with its associated loss of respiratory enzymes, persisted indefinitely after the removal of the dye. When the petite strain, whether arisen spontaneously or induced by acridines, was crossed with a normal strain, the progeny did not show the petite character, and it did not reappear in subsequent generations apart from the low percentage of petite colonies which were always arising. Other character differences, when present, showed normal Mendelian inheritance, with precise ratios.

A number of similar variants with deficiencies in several respiratory enzymes are also known in *Neurospora crassa* (see Fincham and Day, 1971, and Jinks, 1964).

To account for his results, Ephrussi put forward the hypothesis that the petite strain differs from the normal in the loss of cytoplasmic particles postulated to be necessary for the synthesis of the respiratory enzymes. If these hypothetical particles are distributed at random between the parent cell and the daughter-cell budded off from it, and if it is further assumed that there are normally only about 10 of the particles per cell, then about one per thousand daughter-cells will contain none of the particles and will give rise to a petite colony. This would be in agreement with observation. The acridine dyes are assumed to destroy the particles or to prevent them from multiplying. When a petite strain is crossed with a normal one, the petite character is lost because some of the cytoplasm of the normal parent is passed to all the progeny, which therefore receive the postulated particles. Thus it can be seen that this hypothesis will account satisfactorily for all the observations.

It will be noticed that this hypothesis resembles that proposed to explain non-Mendelian inheritance of green pigment variants in flowering plants discussed above. Both theories invoke particulate cytoplasmic determinants. The resemblance may extend further, for the white areas in variegated plants may be composed of cells which have lost the particles necessary for the development of normal plastids, just as the petite yeast cells are thought to have lost particles necessary for the development of normal mitochondria. Moreover, just as plastid variants may show Mendelian inheritance (for example, yellow *versus* green seeds in *Pisum*), Ephrussi found a dwarf strain of yeast which showed Mendelian inheritance. This strain lacked the ability to respire aerobically, just like the non-Mendelian petite strain already described. It appears that particulate hereditary factors responsible for fundamental cell phenomena such as photosynthesis and respiration occur in both nucleus and cytoplasm.

## § 5.16  Conclusion

Evidence has been brought together in this chapter for the belief that Mendelian heredity, with its regular segregation of character-differences to give simple ratios of different kinds of progeny, and with its identical results from reciprocal crosses, is brought about by particulate determinants in the cell-nucleus, although their effects may sometimes be delayed. Conversely, non-Mendelian heredity, characterized by a lack of regular segregation of character-differences, by variable progeny ratios, and usually by maternal inheritance, appears to be determined by cytoplasmic particles. Whereas the identity of the nuclear particles is well-established and will be discussed in subsequent chapters, that of the cytoplasmic particles is less certain (see § 18.15). The relative importance of nucleus and cytoplasm in heredity is discussed in Chapter 18.

# 6. The chromosome theory of Mendelian inheritance

## § 6.1 Chromosomes

Thread-like structures which readily stain with many dyes had long been noticed at nuclear divisions before careful and detailed observations, notably by Flemming (1882) with *Salamandra maculosa* (Spotted Salamander), Strasburger (1882) with various plants, and van Beneden (1883) with *Parascaris equorum) (Ascaris megalocephala*, Horse Thread-worm), established the essential features of the division process. Flemming (1879) introduced the term *chromatin* for the deeply staining material of nuclei, and (1882) *mitosis* for the process of division. Various nouns such as 'threads', 'loops', 'segments' or 'elements', qualified by such adjectives as 'primary', 'nuclear' or 'chromatic' had been used for the structures seen during mitosis, until Waldeyer (1888), in reviewing what was known about nuclear division, proposed the name *chromosome* for them, because 'they are so important', adding that if this name 'is practically applicable it will become familiar, otherwise it will soon sink into oblivion' (English translation, p. 181).

## § 6.2 Mitosis

The key observations on mitosis were made by Flemming (1882) and van Beneden (1883). Flemming discovered in the salamander larva that the chromosomes divided longitudinally during mitosis, and van Beneden observed in *Parascaris equorum* that the two daughter-chromosomes derived from one original chromosome were exactly alike down to the smallest detail, and moreover that they separated and passed one to each daughter-nucleus. It thus became apparent that mitosis was essentially a mechanism for distributing the halves of longitudinally-split chromosomes equally to two daughter-nuclei. The longitudinal splitting was seen to be of such paramount importance that Strasburger (1884) coined the terms *prophases* and *metaphases* to refer to the stages of mitosis before (*pro-*) and after (*meta-*) the chromosomes had separated into daughter-halves, with *anaphases* for the stages when the daughter groups of chromosomes were going back (*ana-*) to the appearance of the original parent-nuclei. Subsequently, more careful observation revealed that the chromosomes divide earlier than was originally supposed, and indeed it is now generally accepted that the threads are double from the earliest stage of mitosis at which they can be seen. The term *telophase* was added by Heidenhain (1894) for the final stages of mitosis, with particular reference to human leucocytes, and about 1905 Strasburger began to use his terms for earlier stages of mitosis than he had originally. These revised usages are those now universally adopted.

The so-called resting-nucleus, prior to the beginning of mitosis, is a more or less spherical structure, bounded by the nuclear membrane and containing the nuclear sap, one or more *nucleoli*, and the chromatin. The nuclear membrane has been shown from electron micrographs to be a double membrane with pores about 84 nm in diameter (Roberts and North-cote, 1970). When a cell is fixed and treated with dyes, the nucleoli and the chromatin usually stain deeply whereas the nuclear sap remains unstained. The nucleoli are small spherical structures. Each is attached to a *nucleolar organizer* on a particular chromosome. A possible function for the nucleoli is discussed in §18.7. The chromatin gives the appearance of a network of threads filling the nucleus, but even at the time of Flemming and Strasburger's pioneer observations, the existence of a physical reticulum was questioned, and it is now generally agreed that the chromosomes retain their identity through the resting-stage, although invisible. The fact that there appeared to be characteristic chromosome numbers for each species argued in favour of the continuity of the chromosomes. Moreover, Boveri (1888) showed for the first few mitoses in the fertilized egg of *Parascaris equorum* that the chromosomes re-appeared at prophase in the position they had occupied at the preceding telophase. In recent years this has been clearly shown by phase-contrast photography of living nuclei, such as Bajer's ciné films of mitosis in the endosperm of *Haemanthus katherinae* and *Clivia cytranthiflora* (see Bajer, 1957–58). The reticular appearance of nuclei in the resting-stage is thought to be an optical effect associated with the existence of the chromosomes in the form of threads so fine as individually to be beyond the resolving-power of the microscope.

Mitosis is initiated (*prophase*) by the chromosomes gradually becoming more distinct. It was formerly thought that all the chromosomes might be joined end-to-end at this stage but here also it is now generally agreed that they are not joined. The early studies were made by fixing the plant or animal material, embedding it in wax, and then cutting a series of thin sections with a microtome. The thickness of these sections was commonly of the same order as the diameter of the nucleus, so that nuclei were frequently cut open. If free ends of chromosomes were seen, it was often thought that these were artefacts due to the cutting. Nowadays, squash techniques rather than sections are usually employed.

The chromosomes become thicker and shorter during prophase through spiral contraction. The direction of coiling is variable. As already indicated, the prophase chromosomes are seen to be double structures, each having divided longitudinally into two identical halves called *chromatids*. The two halves, however, remain closely associated throughout their length (see Fig. 6.1(*b*) and Pl. 6.1(*a*)), and are twisted round one another. In late pro-phase this relational coiling disappears. The uncoiling is well seen in Bajer's film. The nucleoli diminish in size and ultimately disappear, and shortly

PLATE 6.1   Stages of mitosis seen in acetic-orcein squash preparations of root-tip
cells of *Crocus balansae* ($2n = 6$). Magnification ca. ×750.

*(Courtesy of S. A. Henderson)*

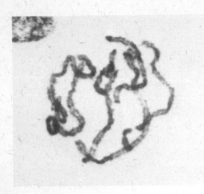

(*a*)  *above* Mid-prophase                        (*b*)  *above* Metaphase

(*c*)  Early anaphase                              (*d*)  Late anaphase

afterwards the nuclear membrane breaks down, thereby releasing the
chromosomes in the cytoplasm of the cell. It is customary nowadays to
identify this relatively abrupt breakdown of the membrane with the end of
prophase.

At the beginning of the succeeding stage, *metaphase*, a series of faint lines,
which do not stain with dyes, become visible forming a spindle-shaped
structure in the cytoplasm of the cell. In animals the spindle has well-defined
poles—star-like structures from which the faint lines radiate. In plants, the
lines fade away before the poles are reached. The chromosomes, now relatively short and thick, each with its two chromatids still closely paired
(Pl. 6.1(*b*)), become arranged in the equatorial plane across the middle of the
spindle. Each chromosome is usually bent at a sharp angle at one point to give
a shape like the letter *V* as seen from the poles of the spindle. The angles
of the *V*'s all point towards the centre of the equatorial plate. The tip of the

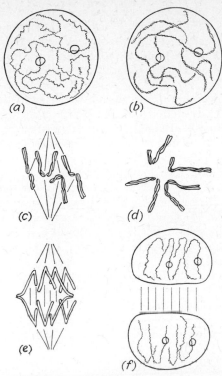

FIGURE 6.1   Stages of mitosis in an organism with two pairs of chromosomes.
(*a*) Early prophase. The small circles represent nucleoli.
(*b*) Late prophase.
(*c*) Side view of metaphase.
(*d*) Polar view of metaphase.
(*e*) Side view of anaphase.
(*f*) Telophase.

angle in each chromosome coincides with a non-staining constriction called the *centromere*. It is the centromeres which appear to be responsible for the chromosome movement on to the metaphase plate. The spindle-fibres, as the fine non-staining lines in the cytoplasm are called, pass through individual centromeres. The chromosome arms, if long, usually point towards one or other pole (see Fig. 6.1(*c*)). In a few organisms, for example, species of *Luzula* (Wood-rush) and certain hemipterous insects, chromosome movement appears to be determined through a generalized property of the chromosomes rather than through a discrete entity, the centromere. Such organisms are said to have diffuse centromeres. Their chromosomes do not show the characteristic *V* or *L* shape of normal chromosomes at metaphase.

*Anaphase* is said to begin when the daughter-centromeres start to separate, each moving along a spindle-fibre in opposite directions towards the poles. The daughter-chromosomes thus separate first at the angle between the two arms (Pl. 6.1(c)). The last point of separation is the tip of the longer arm (Fig. 6.1(e)). Mazia (1955) has isolated the spindle from fertilized eggs of the echinoderm *Strongylocentrotus purpuratus* undergoing mitosis, by removal of the rest of the cytoplasm with the natural steroid detergent digitonin. Chemical analysis showed that the spindle fibres are composed largely of protein, and the amino-acid composition of this protein was found to resemble that of the protein of animal muscles. It was therefore suggested that the shortening of the spindle fibres, which occurs in the anaphase movement, has features in common with muscular contraction. A second factor in the anaphase separation of the daughter-chromosomes is the elongation of the middle region or body of the spindle lying between the two groups of chromosomes. Ris (1949) showed that this elongation is inhibited by chloral hydrate, whereas the spindle fibre contraction is not.

*Telophase* is said to be initiated when nuclear membranes form round each group of chromosomes (Fig. 6.1(f)). Nucleoli re-appear within each daughter-nucleus, and the chromosomes, each now composed of a single chromatid, gradually become longer and thinner until the resting-stage or *interphase* is reached, when they can no longer be clearly seen.

From this brief outline of mitosis it is evident that the duplication of the chromosomes takes place during the resting-stage. This has been confirmed from isotope labelling and other experiments (see Chapter 11).

## § 6.3   Meiosis

An observation of great significance was made by van Beneden (1883) in his studies of *Parascaris equorum*. He noticed that the egg and sperm nuclei each contained two chromosomes, whereas at the mitoses in the fertilized egg four chromosomes were visible. He thus established that the gametes each contribute the *haploid* number of chromosomes to the zygote, which thus contains the *diploid* number in its nucleus. Weismann (1887) predicted that there must be a special nuclear division, repeated in every generation, at which the chromosome number is reduced again to half the number contained in the parent nucleus. It was already known that the nuclear divisions just before the maturation of sperm and eggs in animals, and before the formation of pollen-grains and embryo-sacs in flowering plants, were abnormal. Two nuclear divisions were found to occur in quick succession. Flemming (1887) called them *heterotype* and *homotype*, respectively, since the first was the more abnormal of the two. Strasburger (1888) observed in the pollen-mother-cells and in the embryo-sac-mother-cells of various flowering plants that at late prophase of the first of these divisions, when the chromosomes first become clearly visible, the haploid number is present. Whether these were individual chromosomes, or pairs of chromosomes, and if the latter, whether the chromosomes of each pair were associated laterally or terminally, were long disputed, but it soon became evident that Weismann's predicted reduction was occurring during these divisions, since at the preceding mitosis the diploid number of chromosomes was visible. Farmer and Moore (1905) gave the name *meiosis*, meaning reduction, to the whole process of two successive nuclear divisions. Grégoire (1904) had already suggested that the two divisions should be distinguished by the Roman numerals I and II.

## § 6.4   Von Winiwarter's observations on meiosis in rabbits

As already indicated, much difference of opinion persisted for many years about the precise sequence of events during the early stages of meiosis I. This was partly owing to the difficulty of distinguishing younger and older cells in fixed preparations. However, von Winiwarter (1900) discovered that the ovary of *Oryctolagus cuniculus* (Rabbit) provided ideal material, since meiosis was found to extend over a long period of time. By examining the stage reached in young rabbits of known age, an unequivocal seriation was achieved.

The majority of the nuclei in the ovaries of a half-day old animal were found to contain chromosomes in the form of exceedingly fine single threads (Fig. 6.2(*a*) and Pl. 6.2(*b*)). Cells at this stage he described as having *leptotene* nuclei, meaning that the chromosomes were in the form of a 'slender ribbon'. Later work showed that the leptotene chromosomes have a highly characteristic appearance resembling a string of beads of unequal size unequally strung. The granules are called *chromomeres*.

Nearer the centre of the ovaries of the half-day old rabbit, von Winiwarter found a few nuclei in which the chromosomes had associated in pairs side by side for part of their length, but remained independent elsewhere (Fig. 6.2(*b*) and Pl. 6.2(*c*)). These he called 'synaptene' nuclei, meaning 'uniting ribbons', but owing to confusion over the meaning of the term 'synapsis', the synonym *zygotene* was substituted by Grégoire (1907). In rabbits aged $1\frac{1}{2}$ and $2\frac{1}{2}$ days from birth, von Winiwarter found abundant zygotene nuclei, with leptotene nuclei also present but chiefly nearer the periphery of the ovaries.

In the inner parts of the ovaries of 4- and 5-day old rabbits he found nuclei in which the chromosomes were thicker than at the earlier stages, and were associated in pairs throughout their length. To these he gave the name of nuclei with 'thick ribbons', or *pachytene* nuclei (see Fig. 6.2(*c*) and Pl. 6.2(*d*)). Outside these cells there was an extensive zone of cells with zygotene nuclei, and near the periphery of the ovary a few still at the leptotene stage. Wenrich (1916) observed in the grasshopper *Phrynotettix tschivavensis* (= *P. magnus*) that the chromosome pairing at the zygotene and pachytene stages was highly specific, with corresponding chromomeres closely associated with one another. The mechanism of this specific pairing of homologous regions of homologous chromosomes is not understood. Belling (1929) and Darlington (1929*b*) discovered that when three editions of each chromosome are present, as in some horticultural varieties of *Hyacinthus orientalis*, all three associate, but in any one region of the chromosome only two are in actual contact (see §8.14). Riley and Chapman (1958) and Sears and

PLATE 6.2   The course of meiosis as seen in acetic-orcein squash preparations of spermatocytes of the locust *Schistocerca gregaria* (2n = 22 + X in the male) (continued opposite). Magnification ca. ×1000. *X* = X-chromosome. In (*a*) and (*f*), which are autoradiographs (4 days' exposure), the tissue was treated for 1 hour at 30°C with either 100 $\mu$C/0·2 ml of ³H-thymidine (*a*) or 100 $\mu$C/0·1 ml of ³H-uridine (*f*) before fixation. The grains reveal where uptake of the radioactive label has taken place, and show that preleptotene DNA synthesis (*a*) has occurred in the X-chromosome and the autosomal region of the nucleus (cf. p. 178), and that RNA synthesis at early diplotene (*f*) has taken place in the autosomal bivalents, but not in the X-chromosome which is evidently inactive (see p. 422).                    (*Courtesy of S. A. Henderson*)

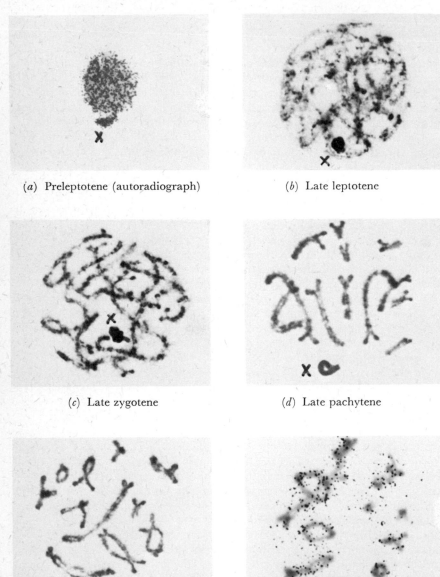

(*a*)  Preleptotene (autoradiograph)          (*b*)  Late leptotene

(*c*)  Late zygotene                          (*d*)  Late pachytene

(*e*)  Early diplotene          (*f*)  Early diplotene (autoradiograph)

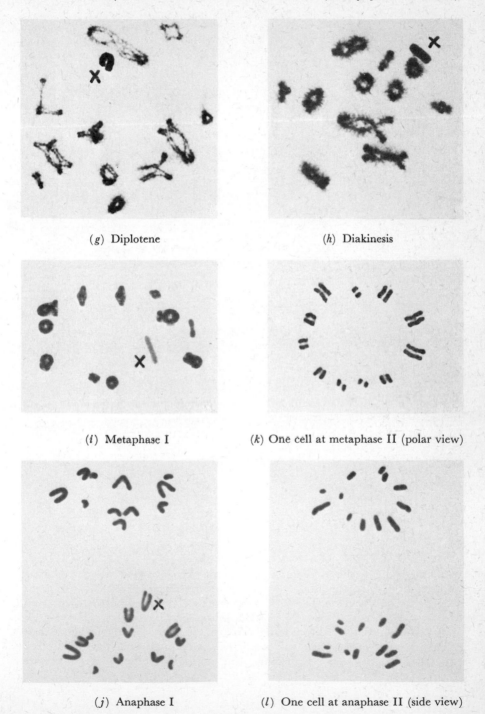

PLATE 6.2 (continued) Meiosis in acetic-orcein squash preparations of male *Schistocerca gregaria* (2n = 22 + X). Magnification ca. ×1000. X = X-chromosome. In (g) and (h) the autosomal bivalents (but not the X-chromosome) have a lampbrush appearance (see Chapters 17 and 18). *(Courtesy of S. A. Henderson)*

(g) Diplotene

(h) Diakinesis

(*i*) Metaphase I

(k) One cell at metaphase II (polar view)

(j) Anaphase I

(l) One cell at anaphase II (side view)

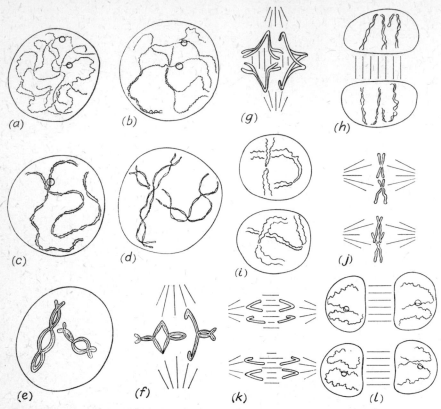

FIGURE 6.2   Stages of meiosis in an organism with two pairs of chromosomes.
   (*a*) Leptotene stage. The small circles represent nucleoli.
   (*b*) Zygotene stage.
   (*c*) Pachytene stage.
   (*d*) Diplotene stage.
   (*e*) Diakinesis.
   (*f*) Metaphase I (side view).
   (*g*) Anaphase I (side view).
   (*h*) Telophase I.
   (*i*) Prophase II.
   (*j*) Metaphase II.
   (*k*) Anaphase II.
   (*l*) Telophase II.

Okamoto (1958) discovered that in *Triticum aestivum* (Wheat) the pairing behaviour of the chromosomes is under genetic control. This plant is thought to have arisen by hybridization of ancestral species, each with a similar set of chromosomes, and followed by doubling of chromosome number. In the absence of a particular chromosome (no. 5 of the B set), it was found that each chromosome associated at meiosis with its counterparts from the other ancestral species, whereas normally the pairing is confined to the strictly

TABLE 6.1   Appearance of nuclei in ovaries of rabbits (von Winiwarter, 1900).

| Age from Birth in Days | Tissue of Ovary | | | |
|---|---|---|---|---|
| | Extreme Outer | Outer | Middle | Inner |
| 0·5 | Resting | Resting | Leptotene | Zygotene |
| 1·5 and 2·5 | Resting | Leptotene | Zygotene | Zygotene |
| 4 and 5 | Resting | Leptotene | Zygotene | Pachytene |
| 10, 11 and 12 | Leptotene | Zygotene | Pachytene | Diplotene |
| 18 | Zygotene | Pachytene | Diplotene | Diplotene |
| 28 | Pachytene | Diplotene | Diplotene | Diplotene |

homologous chromosomes within the contribution from each ancestor. It was inferred that a specific gene in chromosome no. 5B determined the meiotic behaviour of all the chromosomes and this has now been confirmed (Wall, Riley and Gale, 1971).

From rabbits aged 10, 11 and 12 days von Winiwarter found that the nuclei in the central parts of the ovaries contained chromosomes which were even shorter and thicker than the pachytene ones, and moreover had a very evident duality of structure, forming characteristic loops (see Fig. (6.2(*d*) and Pl. 6.2(*e*)). These nuclei he described as *diplotene*. The greater part of the ovaries of this age contained pachytene nuclei, with a peripheral zone at the zygotene and a few at the leptotene stage. An 18-day old rabbit had ovaries containing a thick zone of diplotene nuclei with many pachytene and a few zygotene nuclei near the outside, while at 28 days from birth there were no zygotene and only a few pachytene nuclei and the majority of the ovary cells showed nuclei at the diplotene stage. These results are summarized in Table 6.1, where the sequence leptotene–zygotene–pachytene–diplotene is clearly brought out. In the ovaries of vertebrate animals, meiosis commonly stops at the diplotene stage for a long period of time, so von Winiwarter was not able to follow meiosis beyond this stage in the young female rabbit. He observed the same stages as he had seen in the rabbit in the ovaries of a seven-month old human foetus.

Dispute continued for a further 30 years after von Winiwarter's pioneer studies, before general agreement was reached that his observations were correct and applied throughout the plant and animal kingdoms wherever meiosis took place. The reasons for this long drawn out controversy will be discussed in Chapter 8.

## § 6.5   Diakinesis

Prior to von Winiwarter's studies, Häcker (1897) had subdivided the prophase of the first division of meiosis into two parts. For the second of these, when the chromosomes are relatively short and thick and show characteristic loops or cross-shaped configurations, he coined the name *diakinesis*, meaning 'moving apart', since he saw that the chromosomes were well-separated from one another at this stage, whereas in the first part of the prophase he thought, like many others at the time, that they were joined end-to-end. He illustrated the diakinetic stage from both plants and animals, using the published diagrams of others. Since it is now established that the various chromosomes are not joined end-to-end in early prophase I, the term diakinesis is really a misnomer. It is also out of keeping with von Winiwarter's terms (with the suffix *-tene*) for the earlier stages. In the diakinetic chromosomes the successive loops lie at right angles to one another, whereas at the diplotene stage they are all in one plane. Otherwise the chromosomes do not differ in any important respect from the diplotene ones, apart from being shorter and thicker (Fig. 6.2(*e*) and Pl. 6.2(*h*)), and sometimes with fewer loops (see § 8.18).

Observation of diplotene or diakinetic nuclei in favourable material soon established that the structures within them are not merely double but quadruple (Pl. 6.2(*g*)). In other words, each chromosome, in addition to being associated with a counterpart, is itself divided into two chromatids. How the four strands which make up the diplotene and diakinetic chromosome pairs have originated has been a source of prolonged debate and will be discussed in Chapter 8.

## § 6.6    Later stages of meiosis

With the disappearance of the nucleoli and of the nuclear membrane, prophase I gives way to *metaphase I*. Each chromosome pair takes up a position on the equatorial plate of the spindle such that the centromeres of the two component chromosomes lie one on each side of the mid-plane (Fig. 6.2(*f*) and Pl. 6.2(*i*)). Thus, whereas in metaphase of mitosis the centromeres of the individual chromosomes lie on the equator of the spindle, at metaphase I of meiosis the centromeres of a chromosome-pair lie one in the tropic of Cancer and the other of Capricorn and both on the same line of longitude. At this stage, the chromosomes are exceptionally short and thick.

*Anaphase I* is said to begin when the two whole chromosomes forming each pair start to move apart towards opposite poles of the spindle. At this stage the two chromatids of which each chromosome is composed are readily seen (see Fig. 6.2(*g*)). Nuclear membranes develop round each group of chromosomes at *telophase I* (Fig. 6.2(*h*)). After a brief *interphase*, when the chromosomes become difficult to observe, *prophase II* follows during which they become shorter and thicker again (Fig. 6.2(*i*)). *At metaphase II* the nuclear membranes have broken down again and two new spindles have formed (Fig. 6.2(*j*)). The second division of meiosis resembles mitosis in the centromeres of the chromosomes taking up a position on the equatorial plate, but differs in the two chromatids, of which each chromosome has been composed ever since prophase I, diverging somewhat from one another instead of being closely associated (Pl. 6.2(*k*)). At *anaphase II* the daughter-chromosomes separate and move towards opposite poles on each spindle with the centromeres as usual leading the way (Fig. 6.2($k$)), and at *telophase II* nuclear membranes have formed round each of the four haploid nuclei which result from meiosis (Fig. 6.2($l$)).

The essential features of meiosis which distinguish it from mitosis are, first, that whole chromosomes associate in pairs at prophase *I* and separate to opposite poles at anaphase *I*; and secondly, that there is no duplication of the chromosomes in the brief interphase between the two nuclear divisions, so that the daughter-chromosomes which separate at anaphase II were those first seen at prophase I. Thus, instead of the single nuclear division without chromosome division which Weismann (1887) had anticipated, when the details of meiosis were discovered, it was found that the process was less simple: as indicated above, there are two divisions of the nucleus associated with one of the chromosomes.

## § 6.7   Evidence that paternal and maternal chromosomes associate

Montgomery (1901), from studies of the chromosomes at meiosis in some 42 species of hemipterous insects, came to the conclusion that the chromosomes which associate in pairs at prophase I are of paternal and maternal origin respectively. His evidence for this important deduction was two-fold. First, whenever two chromosomes could be distinguished by their greater size, as for instance in *Protenor belfragei*, these two especially large ones were always found paired with one another at prophase I. Secondly, in *Euschistus variolarius* with 14 chromosomes visible at the somatic mitoses, 7 bivalent associations were always found at prophase I of meiosis. He argued that if chromosome association was paternal with paternal this would give 3 bivalent and 1 univalent association, and likewise with the chromosomes of maternal origin. The regular formation of pairs even when, as here, the haploid chromosome number is odd, led Montgomery to the conclusion that the bivalent associations at prophase I are always made up of one paternal and one corresponding maternal chromosome.

Sutton (1902) extended Montgomery's observations when he showed that in the grasshopper *Brachystola magna* all the chromosomes can be individually recognized. The somatic mitosis in the male showed 23 chromosomes which could be classified by size into 11 pairs and one odd one (the *X*-chromosome— see § 6.10). The largest chromosomes were 5 to 6 times as long as the smallest. At prophase I of meiosis, apart from the *X*-chromosome, 11 double chromosomes were visible in each nucleus and they showed the same size differences as at mitosis. Moreover, though the absolute lengths of the chromosomes varied greatly at different stages of meiosis, their relative lengths were approximately constant. It was evident that each chromosome was distinct and maintained its morphological individuality throughout successive cell-divisions, and that the meiotic pairing was strictly between paternal and maternal homologues, as Montgomery had suggested.

## § 6.8   Evidence of qualitative differences between chromosomes

Boveri (1902) found evidence that the individual chromosomes in *Paracentrotus lividus* (a Sea Urchin) possess different qualities. This remarkable conclusion was based on studies of abnormal progeny resulting from fertilization of eggs by two sperms. Boveri showed that the first mitosis in such *triploid* eggs, that is, with 3 sets of chromosomes, was abnormal, and gave rise to an irregular distribution of the chromosomes to the daughter-nuclei. He was able to show that for normal development, a particular combination of chromosomes was more important than a particular number. It will be recalled that he had shown earlier (see Chapter 5) that haploid individuals (from enucleated eggs) were normal although dwarf. He now established that they were normal because they had the normal combination of chromosomes. This important principle has subsequently been confirmed in many other organisms. Thus, Blakeslee (1922) showed that in *Datura stramonium* (Thorn Apple) triploid and *tetraploid* individuals, having respectively 3 and 4 complete sets of 12 chromosomes in their somatic cells, only differed rather slightly in appearance from the normal diploid, whereas individuals which he described as *trisomic*, having one of the chromosomes represented three times but the remaining 11 represented twice, were abnormal. The discovery of individuals with each of the 12 different chromosomes in turn present in triplicate revealed that each had a different morphology. Thus, three doses of one chromosome gave rise to a variety called Globe with broader leaves than normal, a less toothed leaf-margin, and a spherical instead of ovate capsule with shorter spines than normal (Fig. 6.3(*a*) and (*b*)).

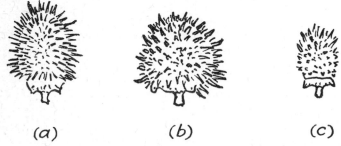

(a)                         (b)                         (c)

FIGURE 6.3   Fruits (×0·5) of normal and trisomic plants of *Datura stramonium* (from Blakeslee, 1922). (*a*) Normal. (*b*) Globe. (*c*) Cocklebur.

Individuals trisomic for a different chromosome gave a variety called Cocklebur with weak stems, narrow twisted leaves and a narrow short-spined capsule (Fig. 6.3(*c*)). Each trisomic variety differed from the normal in a whole range of characters affecting all parts of the plant.

Knowledge of the abnormality of trisomic individuals has recently been extended to man. From squash preparations of mitosis in cultures of lung tissue from 4 embryos of unknown sex after legal abortion, Tjio and Levan (1956) showed that the human somatic chromosome number was 46.

PLATE 6.3    Metaphase of mitosis in a leucocyte from peripheral blood of a human male (2n = 44 + X + Y). The material was treated with colchicine before fixation, and is a squash preparation stained with acetic-orcein. Magnification ca. ×400. X = probable X-chromosome. Y = probable Y-chromosome.          *(Courtesy of D. Briggs)*

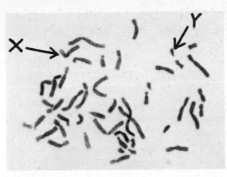

Confirmation of this number was obtained by Ford and Hamerton (1956) from study of meiosis in fresh operative specimens of testis tissue from 3 individuals. Human chromosomes are readily classified by size into 7 classes, and within these classes the individual chromosomes can be recognized by the relative lengths of their arms (Pl. 6.3). By international agreement, the largest of the 23 pairs is called no. 1, and the others have been allocated identification numbers in order of decreasing size. Lejeune, Gautier and Turpin (1959) discovered that Down's syndrome (mongolism) is associated with the presence of an additional small chromosome. It appears that those with this disorder have three editions of chromosome no. 21 in their somatic cells, instead of the normal two.

## § 6.9   The chromosome theory of Mendelian heredity

Sutton (1903), on the basis of his studies on *Brachystola magna* and with the support of Montgomery's and Boveri's findings, gave the first clear formulation of the chromosome theory of Mendelian heredity. He drew attention to the resemblance between the separation of homologous chromosomes at meiosis and Mendel's postulated separation of character-differences at gamete-formation. Moreover, Sutton pointed out that in all probability each homologous pair of chromosomes is orientated on the metaphase I spindle independently, so that any pair of chromosomes may lie with maternal or paternal chromosome indifferently towards either pole irrespective of the orientation of other pairs. He noted that this would lead to a large number of different possible combinations of maternal and paternal chromosomes in the gamete nuclei, if the chromosome number were at all large: indeed $2^n$ combinations, where $n$ is the haploid chromosome number. He quoted as an example the Sea Urchin *Lytechinus variegatus* with 36 chromosomes in the somatic cells and hence $2^{18}$ or 262,144 possible gametic combinations of paternal and maternal chromosomes, so that the number of types of offspring from a single pair would be $2^{36}$ or approximately $6 \cdot 87 \times 10^{10}$! The possible number of different assortments of parental chromosomes will vary greatly with the species, because chromosome numbers show such a big range. The extremes recorded for the haploid number are 2 in a few arthropods and in the flowering plant *Haplopappus gracilis* (Jackson, 1957), and approximately 510 in the fern *Ophioglossum petiolatum* (Manton and Sledge, 1955). Sutton drew attention to the parallel between the presumed independent orientation of homologous chromosome-pairs at meiosis, and Mendel's observation of independent inheritance of different character-differences.

Confirmation of this random orientation of different chromosome-pairs on the first-division spindle was obtained by Carothers (1913). She examined 300 cells showing metaphase I or anaphase I of meiosis from males of *Brachystola magna* heterozygous for an inequality in the size of one of the chromosomes. She found that this size difference always segregated at the first division of meiosis. The single $X$-chromosome moves in its entirety to one pole at this division. In 146 of the cells, the smaller component of the unequal pair went to the same pole as the $X$-chromosome, and in the remaining 154 cells it was the larger component which segregated with the

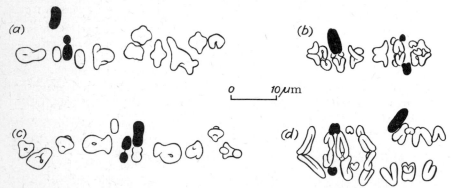

FIGURE 6.4    Metaphase I (*a* and *c*) and anaphase I (*b* and *d*) of meiosis in males of *Brachystola magna* heterozygous for an inequality in the size of one of the chromosomes (from Carothers). The unequal chromosome-pair and the $X$-chromosome are shown black. (*a* and *b*) Smaller component orientated to move to the same pole as the $X$-chromosome. (*c* and *d*) Larger component so orientated.

$X$-chromosome (see Fig. 6.4). Similar results were obtained with another grasshopper, *Arphia simplex*. Some exceptions to this random orientation of different chromosome-pairs on the meiotic first-division spindle are discussed in §18.3.

## § 6.10   Sex chromosomes

The first definite evidence for a connection between the chromosomes and a character-difference showing Mendelian inheritance came from work on sex-determination. Henking (1891) in studies on meiosis at sperm-formation

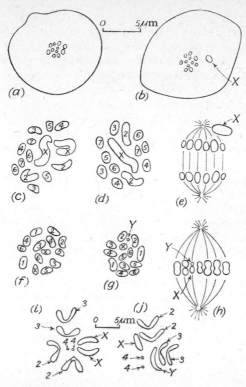

FIGURE 6.5   Sex chromosomes in various insects. $X = X$-chromosome. $Y =$ $Y$-chromosome. Numbers refer to the autosomes in order of diminishing size

(*a* and *b*) *Pyrrhocoris apterus.* Polar views of the two groups of chromosomes on one of the spindles at anaphase II of meiosis in the male (from Henking, 1891). The single $X$-chromosome moves to one pole at anaphase II.

(*c–e*) *Protenor belfragei* (from Wilson).

(*c*) Female. Late prophase of mitosis in an oogonium. Two $X$-chromosomes present.

(*d*) Male. Late prophase of mitosis in a spermatogonium. A single $X$-chromosome present.

(*e*) Male. One spindle at anaphase II of meiosis. The single $X$-chromosome moves to one pole.

(*f–h*) *Lygaeus turcicus* (from Wilson).

(*f*) Female. Late prophase of mitosis in an oogonium. Two $X$-chromosomes present.

(*g*) Male. Late prophase of mitosis in a spermatogonium. An $X$- and a $Y$-chromosome present.

(*h*) Male. One spindle at metaphase II of meiosis. The $X$- and $Y$-chromosomes separate to opposite poles at anaphase II.

(*i* and *j*) *Drosophila melanogaster* (from Stevens).

(*i*) Female. Polar view of metaphase of mitosis in an oogonium. Two $X$-chromosomes present.

(*j*) Male. Polar view of metaphase of mitosis in a spermatogonium. An $X$- and a $Y$-chromosome present.

in the hemipterous insect *Pyrrhocoris apterus* observed a deeply-staining chromatin-element, which passed to one pole on each anaphase II spindle, so that half the sperms received it and half did not (see Fig. 6.5(*a*) and (*b*)). He labelled it *X* as he was uncertain whether it was a chromosome or not. Later workers on insect cytology established its chromosomal nature, and after observing a similar unpaired chromosome in various grasshoppers, McClung (1902) suggested that it was concerned with sex-determination. This was established by Wilson (1905) who discovered that the single *X-chromosome* (as he called it, Wilson 1909) of the male *Protenor belfragei* has two counterparts in the female, so that the somatic chromosome numbers of male and female are 13 and 14 respectively (Fig. 6.5(*c*)–(*e*)). He argued from these observations that the sperms which receive an *X*-chromosome will give rise to females and those that do not to males. Similar observations were made on several other hemipterous insects, while in a second group of these insects, exempified by *Lygaeus turcicus*, both sexes were found to have the same chromosome number ($2n = 14$ in this instance), but one chromosome in the male was much smaller than the corresponding one in the female (Fig. 6.5(*f*)–(*h*)). Again Wilson argued that there must be a causal connection between these chromosomes and sex.

Stevens (1905) independently made similar observations on other insects, for instance the beetle *Tenebrio molitor* (Meal worm) where she found that the male has 19 large and 1 small chromosome in its somatic cells, while the female has 20 large chromosomes. Later she showed (1908) that *Drosophila melanogaster* has four pairs of chromosomes in its somatic cells and that in the male the two components of one pair are of unequal size (Fig. 6.5(*i*) and (*j*)). The term *Y-chromosome* was proposed by Wilson (1909) for the odd chromosome when present, so *Lygaeus*, *Tenebrio* and *Drosophila* could be described as female *XX* and male *XY*, and *Protenor* as female *XX* and male *X*. Wilson (1909) suggested that in both these categories sex was determined by the number of *X*-chromosomes. He argued that since a *Y*-chromosome was not always present it played no part in sex-determination.

## § 6.11   The chromosome theory of sex-linkage

On the basis of Wilson's theory of sex-determination, Morgan (1911*a*) postulated that the gene for white eyes in *Drosophila melanogaster* is carried by the *X*-chromosome. This hypothesis to explain the sex-linkage which he had found (see Chapter 3) represented an important advance over his earlier theory (based on that of Punnett and Bateson for *Abraxas*), since it was no longer necessary to assume that all male Drosophilas were heterozygous for white eye-colour, for, with only one *X*-chromosome in the male, heterozygosity was impossible for sex-linked characters. Moreover the criss-cross inheritance of the X-chromosomes was exactly paralleled by that of white eye-colour (see Figs 6.6 and 6.7).

The sex-linkage of the pale form of *Abraxas grossulariata* was similarly explained when it was established that in Lepidoptera it is the female which has either an *X*- and a *Y*-chromosome, or an unpaired *X*-chromosome, while

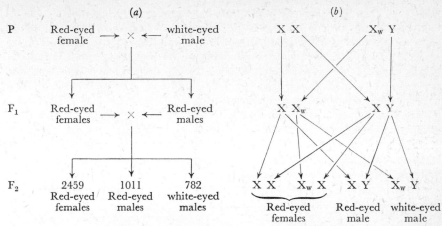

FIGURE 6.6   (a) Morgan's data (1910) for the inheritance of white eye-colour in the normally red-eyed fly *Drosophila melanogaster*.

(b) The hypothesis which he proposed (1911a) to account for these results, namely, that the gene for white eyes is carried by the X-chromosome. Explanation of symbols:

$X$ = X-chromosome with dominant normal allelomorph of white-eye gene, and hence causing red eyes.

$X_w$ = X-chromosome with recessive gene for white eyes.

$Y$ = Y-chromosome (plays no part in inheritance).

Two X-chromosomes give a female, one X-chromosome a male. The shortage of males in $F_2$, particularly those with white eyes, compared with expectation, can probably be accounted for by lowered viabilities.

the male has 2 X-chromosomes, that is, the converse of the situation in *Drosophila*.

Red-green colour blindness in man was found to show criss-cross inheritance similar to white eye-colour in *Drosophila*. That is to say, colour-blindness is usually manifest only in males, whose daughters act as carriers of it to

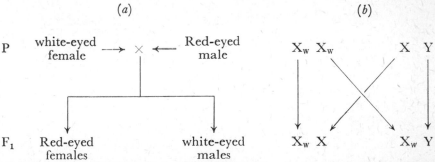

FIGURE 6.7   (a) Further data from Morgan (1910) concerning the inheritance of white eye-colour in *Drosophila melanogaster*, and

(b) his chromosome hypothesis (1911a) to account for them. Symbols as in Fig. 6.6.

half their sons. The female $XX$, male $XY$ sex-determining mechanism in man was confirmed subsequently (cf. Pl. 6.3). By contrast, the dominant character barred feather-colouring of the Barred Plymouth Rock and its recessive allele the black feather-colouring of the Black Langshan variety of *Gallus domesticus* were found to show sex-linkage of the *Abraxas* type, and the cytology confirmed that birds, like Lepidoptera, have the female $XY$, male $XX$ sex-determining mechanism.

Baur (1912) obtained the first example of sex-linkage in plants. He discovered a form of *Silene alba* (White Campion) with narrow grass-like leaves. It was found to be a staminate plant. Pollen from it was used to fertilize a female plant with normal broad leaves. All the progeny were normal, and when intercrossed they gave 167 plants with broad leaves and 60 with narrow leaves, in agreement with the Mendelian expectation of a $3:1$ ratio (expected frequencies 170·25 and 56·75, deviations 3·25, $\chi^2 = 0·24$, $n = 1$, $P = 0·9–0·5$). When 52 of the narrow-leaved plants were grown to maturity all were found to be male, while 23 of the broad-leaved ones comprised 15 female and 8 male, in good agreement with the $2:1$ ratio of the sexes expected in $F_2$ individuals with the dominant phenotype, if *Silene alba* is like *Drosophila* in its sex-linkage (compare Fig. 6.6), that is, female $XX$, male $XY$, with the gene for narrow leaves carried by the $X$-chromosome. Subsequently, cytological studies confirmed that *Silene* has this chromosome formula.

## § 6.12   XXY Drosophila

The chromosome theory of Mendelian heredity may be said to have become established with the publication of the results of combined cytological and genetical studies by Bridges (1914, 1916) with an abnormal strain of *Drosophila melanogaster* having two $X$-chromosomes and a $Y$ chromosome.

Morgan (1910) had reported that a white-eyed female *Drosophila* crossed with a normal red-eyed male always gave equal numbers of the converse types in $F_1$, that is, red-eyed females and white-eyed males. On the hypothesis that the gene for white eyes is carried by the $X$-chromosome these results were readily explained (Morgan 1911a) by the criss-cross inheritance of the $X$-chromosomes, as indicated in Fig. 6.7. However, Morgan and Bridges discovered a strain of white-eyed females which gave some unexpected progeny from such a cross. These unexpected individuals comprised about 4·3% of the progeny, and consisted of approximately equal numbers of the two parental forms (white-eyed females and red-eyed males). The occurrence of these *matriclinous* daughters (that is, resembling their mother) and *patriclinous* sons (resembling their father) appeared to contradict the chromosome hypothesis.

Bridges studied the progeny in the succeeding generation, with the following results. The matriclinous daughters (white eyed; class (i) in Table 6.2) behaved like their mother and again gave about 4.3% of parental forms when crossed with normal red-eyed males. The patriclinous sons (red-eyed; class (ii) in Table 6.2) when crossed with unrelated females gave no abnormal progeny in the next or succeeding generations. Half of

TABLE 6.2   (*a*) Bridges's results from crossing an abnormal white-eyed female *Drosophila melanogaster* with a normal Red-eyed male, (*b*) his total data (1916) from this and similar crosses using various sex-linked mutant characters, and (*c*) his hypothesis to explain these results. Symbols as in Fig. 6.6.

(*a*)

| Parents | Red-eyed (normal) male | |
|---|---|---|
| White-eyed female from abnormal stock | | 2·1% of white-eyed females (i), all with some exceptional progeny. |
| | 2·2% of Red-eyed males (ii), all with normal progeny. | |
| | 49·0% of Red-eyed females, half (iii) with normal progeny, and half (iv) with some exceptional progeny. | 46·7% of white-eyed males, half (v) with normal progeny, and half (vi) with some exceptional progeny. |

(*b*)

| Parents | Normal males | |
|---|---|---|
| Females from abnormal stock, with various sex-linked mutant characters | | 1169 females of mutant appearance (i). |
| | 1235 males of normal appearance (ii). | |
| | 27,679 females of normal appearance (iii and iv). | 26,391 males of mutant appearance (v and vi). |

(*c*)

| Parents | | | | $XY$ (Red-eyed male) | | | |
|---|---|---|---|---|---|---|---|
| | | | | Sperm | | | |
| | | | | $X$ | | $Y$ | |
| $X_wX_wY$ (white-eyed female from abnormal stock) | Eggs | 8·2% | $X_wX_w$ | $XX_wX_w$ | Dies | (i) $X_wX_wY$ | |
| | | | $Y$ | (ii) $XY$ | | $YY$ | Dies |
| | | 91·8% | $X_w$ | (iii) $XX_w$ | | (v) $X_wY$ | |
| | | | $X_wY$ | (iv) $XX_wY$ | | (vi) $X_wYY$ | |

the ordinary patriclinous daughters (red-eyed; class (iii)) gave normal progeny indefinitely, and the other half (class (iv)) inherited the property of giving about 4·3% of unexpected progeny. Half of the ordinary matriclinous sons (white eyed; class (v)) gave normal progeny indefinitely and the other half (class (vi)) transmitted the property of producing exceptional progeny to some of their daughters, although not themselves having this property. These results are summarized in Table 6.2(*a*).

To account for these results, Bridges postulated that the original white-eyed female was abnormal in possessing a $Y$-chromosome in addition to two $X$-chromosomes. Furthermore he postulated that at the first division of meiosis in the maturation of the eggs there was variability of behaviour such that in about 92% of meioses one $X$-chromosome and the $Y$-chromosome went to one pole and the remaining $X$-chromosome to the other pole, while in the other 8% the two $X$-chromosomes separated from the $Y$. This hypothesis will account perfectly for all the observations, as indicated in Table 6.2(*c*). Individuals with 0 or 3 $X$-chromosomes were thought not to survive.

Confirmation of this $XXY$ hypothesis was obtained from cytological studies: Bridges found that the white-eyed females and half the red-eyed ones had two $X$-chromosomes and a $Y$-chromosome, exactly as the theory predicts. It is presumed that the original $XXY$ female arose through fertilization by a $Y$ sperm of an abnormal egg containing two $X$-chromosomes. This egg might have been the consequence of a meiotic abnormality whereby both $X$-chromosomes passed to the same pole of the spindle. Bridges (1916) reported several instances of this failure of the $X$-chromosomes to separate at meiosis even in the absence of a $Y$-chromosome. Altogether he bred 56,474 $F_1$ progeny from $XXY$ females carrying various sex-linked characters crossed to normal males, with the results shown in Table 6.2(*b*). The percentages in Table 6.2(*a*) are based on these results.

## § 6.13   Sex determination

In addition to establishing the truth of the chromosome theory of Mendelian inheritance, Bridges' experiments with $XXY$ flies also confirmed Wilson's theory of sex-determination, for it was evident that an $XXY$ fly was female and an $XYY$ fly male, as expected if sex is determined by the number of $X$-chromosomes irrespective of the number of $Y$-chromosomes. Bridges (1922) later showed that sex is determined by the proportion of $X$-chromosomes to other chromosomes (*autosomes*). If $A$ stands for a haploid set of autosomes and $X$ for an $X$-chromosome, $\frac{X}{A}$ is $\frac{1}{2}$ in the normal male and $\frac{2}{2}$ in the normal female. With three $X$-chromosomes $\left(\frac{X}{A}=\frac{3}{2}\right)$ abnormal sterile females of low viability and delayed development were obtained. Bridges also obtained triploid flies with three sets of autosomes and found that $\frac{X}{A}=\frac{1}{3}$ gave an abnormal sterile male of slow development, $\frac{X}{A}=\frac{2}{3}$ gave a sterile fly of appearance intermediate between the two sexes. while $\dfrac{X}{A} = \dfrac{3}{3}$ gave a female of almost normal appearance, although sterile. Thus his results were consistent with the view that sex was determined by the proportion of $X$-chromosomes and autosomes, which were presumed to carry genes for female and male characters, respectively. Later work has confirmed that, irrespective of the number of $Y$-chromosomes, $\dfrac{X}{A}$ of $\dfrac{1}{2}$ or less gives males, 1 or more gives females, and between $\frac{1}{2}$ and 1 gives intermediates.

On the other hand, in *Silene alba*, *Homo sapiens* and *Mus musculus* a different mechanism appears to operate. Westergaard (1958) in reviewing his own and others' work with *Silene* showed that the presence of the $Y$-chromosome gives a male and its absence a female, irrespective of the number of $X$-chromosomes and sets of autosomes. The only exceptions are tetraploid plants with 4 $X$-chromosomes but only 1 $Y$-chromosome, which are usually hermaphrodite. It appears that the male-determining genes reside on the $Y$-chromosome and the female-determining on the $X$-chromosome, and that a ratio of $4X : 1Y$ is required even to produce an hermaphrodite. Work with fragmented $Y$-chromosomes suggests that this chromosome contains genes which lead to the development of male characters and others which suppress the development of female characters.

In *Homo sapiens* and *Mus musculus* normal females have two $X$-chromosomes and normal males an $X$- and a $Y$-chromosome. Compared with the other human chromosomes, the $X$-chromosome is in the third class for size, while the $Y$- is the seventh or smallest class (Pl. 6.3). In both *Homo* and *Mus* an abnormal condition occurs occasionally where there is a single $X$-chromosome and no $Y$-chromosome in the somatic nuclei. Such individuals are female, and fertile in *Mus* (Welshons and Russell, 1959), but usually sterile and with impaired development in *Homo* (Turner's syndrome). Individuals with two $X$-chromosomes and a $Y$-chromosome are sterile males, both in *Mus* (Russell and Chu, 1961) and *Homo* (Jacobs and Strong, 1959). In man, such individuals have certain slight female characteristics (Klinefelter's syndrome). It thus appears that, as in *Silene*, sex in these mammals is determined by the proportion of $X$- and $Y$-chromosomes.

## § 6.14   Conclusion

Although Bridges's work on *XXY Drosophila* may be said finally to have established the truth of the chromosome theory of Mendelian heredity, nevertheless the theory still met opposition. This was primarily because it was found necessary to postulate that the homologous chromosomes regularly exchanged segments at meiosis, a notion which some biologists thought conflicted with the concept of the individuality and permanence of the chromosomes. This will be discussed in the next chapter.

# 7. The theory of chromosomal crossing-over

## § 7.1 De Vries's theory of factor exchange

In his formulation of the chromosome theory of Mendelian heredity, Sutton (1903) pointed out that the number of distinct characters in an organism must exceed the number of chromosomes. He argued from this that the basis of an allele was only a part of a chromosome, and that if the chromosomes permanently retain their individuality all the factors represented by one chromosome must be inherited together. However, within a short time, four apparently independently-inherited character-differences were known in *Pisum sativum* in addition to the seven studied by Mendel, and yet the haploid chromosome number was known to be only seven. Similarly, the number of apparently freely-inherited character-differences in *Antirrhinum majus* was found to exceed the haploid chromosome number of 8. These findings appeared to be a fatal objection to the chromosome theory. De Vries (1903), however, had offered a solution to this dilemma.

De Vries accepted Sutton's view that the chromosomes formed the material basis of Mendelian heredity, and that many factors must co-exist in each chromosome, but to account for the apparently independent inheritance of each character-difference he supposed that an exchange of material took place between the maternal and paternal homologous chromosomes while they were closely associated in pairs during prophase I of meiosis. He argued that if two like factors lay opposite each other, a simple exchange of factors could occur. He assumed that in any given instance it was a matter of chance whether particular factors were exchanged or not. He argued further that a factor could be exchanged only for a like one, that is, for one which represented the corresponding hereditary character, otherwise each chromosome would not retain the entire series of factors. He regarded it as essential that the progeny should receive from their parents the sum total of these factors, as representing all the characters of the species. He pointed out that after these hypothetical exchanges had occurred, each chromosome would contain some paternal and some maternal units. Furthermore, since the two chromosomes of a pair pass into different gamete-nuclei, this redistribution of factors between the homologous chromosome-pairs would lead to the formation of gametes with all possible combinations of the parental character-differences, just as Mendelian theory required.

## § 7.2   Morgan's evidence for crossing-over

De Vries's theory of exchanges between homologous chromosomes gained little support at the time. In the first place there was no evidence that such exchanges occurred, and secondly the idea of exchanges was thought by many to be contrary to the concept of the permanence and individuality of the chromosomes. For the same reasons, when Bateson, Saunders and Punnett (1905) discovered partial linkage between two different characters in *Lathyrus odoratus* (see Chapter 3), they did not favour the chromosome theory as an explanation. But when Morgan (1911*a*), working with *Drosophila*

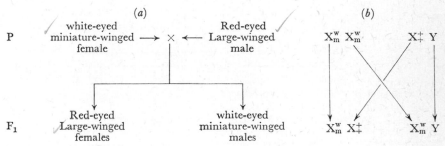

FIGURE 7.1   Morgan's results from crossing a white-eyed miniature-winged female with a normal male *Drosophila melanogaster*.
(*a*) $F_1$ results.
(*b*) Hypothesis to explain (*a*).
$X_m^w$ = X-chromosome with recessive genes for white eyes (*w*) and miniature wings (*m*).
$X_+^+$ = X-chromosome with dominant normal allelomorphs of *w* and *m*.
$Y$ = Y-chromosome.

*melanogaster*, turned up a similar example of partial linkage, he had no alternative but to postulate a De Vriesian exchange of material between homologous chromosomes, because the characters concerned, white eye-colour and miniature wings, both showed sex-linkage. If the genes for each of these characters were carried by the X-chromosome, as the sex-linkage data suggested, then the only way to explain the occurrence of new combinations of them was to postulate exchanges between the two X-chromosomes (in the female insect, since the male has only one X-chromosome). Thus, when he crossed a white-eyed miniature-winged female with a normal (Red-eyed Large-winged) male, the $F_1$ progeny were similar except that the sexes were reversed, that is, the females were all normal and the males all white-eyed and miniature-winged (see Fig. 7.1). This result is what is expected (see Chapter 6) if the genes for white eyes and miniature wings are both carried by the X-chromosomes (see Fig. 6.7). The $F_1$ flies were allowed to interbreed, and 2441 progeny ($F_2$) obtained (Morgan 1911*b*). The numbers of each kind are shown in Table 7.1(*a*). It will be noticed that flies with the grand-parental character-combinations (white-eyed miniature-winged and Red-eyed Large-winged) are the most frequent in both sexes (classes (i) to (iv)).

The surprising discovery, however, was the occurrence of appreciable numbers of flies with the other two combinations of the eye and wing characters (classes (v) to (viii)). Altogether there were just 900 of these out of the 2441 or 36·9%. Morgan's hypothesis to explain these results is shown in

TABLE 7.1   Morgan's results from crossing a white-eyed miniature-winged female with a normal male *Drosophila melanogaster*.

        (*a*) $F_2$ results.
        (*b*) Hypothesis to explain (*a*).
        Symbols as in Fig. 7.1.

(*a*)

| Parents($F_1$) | White-eyed Miniature-winged Males | | | | |
|---|---|---|---|---|---|
| | Eyes | Wings | Females | Males | Totals |
| Red-eyed Large-winged females | white | miniature | (i) 359 | (ii) 391 | 750 ⎱ 1541 |
| | Red | Large | (iii) 439 | (iv) 352 | 791 ⎰ |
| | white | Large | (v) 218 | (vi) 237 | 455 ⎱ 900 |
| | Red | miniature | (vii) 235 | (viii) 210 | 445 ⎰ |
| | Totals | | 1251 | 1190 | 2441 |

(*b*)

| Parents ($F_1$) and their Gametes | | | | $X_m^w Y$ | |
|---|---|---|---|---|---|
| | | | | Sperm | |
| | | | | $X_m^w$ | $Y$ |
| $X_m^w X_+^+$ | Eggs | 63·1% | $\begin{cases} X_m^w \\ X_+^+ \end{cases}$ | (i) $X_m^w X_m^w$ <br> (iii) $X_+^+ X_m^w$ | (ii) $X_m^w Y$ <br> (iv) $X_+^+ Y$ |
| | | 36.9% | $\begin{cases} X_+^w \\ X_m^+ \end{cases}$ | (v) $X_+^w X_m^w$ <br> (vii) $X_m^+ X_m^w$ | (vi) $X_+^w Y$ <br> (vii) $X_m^+ Y$ |

Table 7.1(*b*). He assumed that in 36·9% of the eggs an exchange of factors had occurred, just as De Vries had anticipated, between the two *X*-chromosomes.

Another Mendelian character in *Drosophila* which Morgan found to show sex-linkage was yellow body-colour. When he crossed a white-eyed yellow-bodied female with a normal (Red-eyed Grey-bodied) male, the $F_1$ progeny comprised normal females and white-eyed yellow-bodied males, as expected with sex-linkage (Fig. 7.2). When the $F_1$ flies were intercrossed, 2205 $F_2$ flies were obtained. The numbers of each kind are shown in Table 7.2. Again it is flies with the grandparental character-combinations (white-eyed yellow-bodied and Red-eyed Grey-bodied) which were most frequent, but

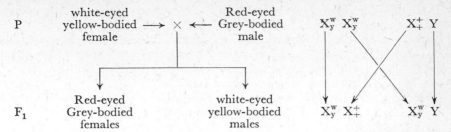

FIGURE 7.2   Morgan's results from crossing a white-eyed yellow-bodied female with a normal male *Drosophila melanogaster*.

(*a*) $F_1$ results.

(*b*) Hypothesis to explain (*a*).

$X_y^w$ = *X*-chromosome with recessive genes for white eyes (*w*) and yellow body (*y*).

$X_+^+$ = *X*-chromosome with dominant normal allelomorphs of *w* and *y*.

$Y$ = *Y*-chromosome.

TABLE 7.2   Morgan's results from crossing a white-eyed yellow-bodied female with a normal male *Drosophila melanogaster*.

(*a*) $F_2$ results.

(*b*) Hypothesis to explain (*a*).

Symbols as in Fig. 7.2.

(*a*)

| Parents ($F_1$) | White-eyed Yellow-bodied Males | | | | |
|---|---|---|---|---|---|
| | Eye-colour | Body-colour | Females | Males | Totals |
| Red-eyed Grey-bodied females | white | yellow | (i) 543 | (ii) 474 | 1017⎫ 2176 |
| | Red | Grey | (iii) 647 | (iv) 512 | 1159⎭ |
| | white | Grey | (v) 6 | (vi) 11 | 17⎫ 29 |
| | Red | yellow | (vii) 7 | (viii) 5 | 12⎭ |
| | | Totals | 1203 | 1002 | 2205 |

(*b*)

| Parents ($F_1$) and their Gametes | | | $X_y^w Y$ | |
|---|---|---|---|---|
| | | | Sperm | |
| | | | $X_y^w$ | $Y$ |
| $X_y^w X_+^+$ | Eggs | 98·7% $\begin{cases} X_y^w \\ X_+^+ \end{cases}$ | (i) $X_y^w X_y^w$ (iii) $X_+^+ X_y^w$ | (ii) $X_y^w Y$ (iv) $X_+^+ Y$ |
| | | 1·3% $\begin{cases} X_+^w \\ X_y^+ \end{cases}$ | (v) $X_+^w X_y^w$ (vii) $X_y^+ X_y^w$ | (vi) $X_+^w Y$ (viii) $X_y^+ Y$ |

this time the other two combinations of these character-differences appeared only infrequently: 29 out of 2205, or 1·3%. Morgan concluded that the factors responsible for these character-differences tended to remain together much more often than those in the previous experiment. This principle of partial linkage he called the *third law of heredity* (Morgan 1919), the first two laws being those of Mendel (see Chapter 3).

Morgan (1911*b*) reported results for over 9000 $F_2$ flies from crosses involving these characters (white eyes, miniature wings and yellow body) in different combinations. It was evident that the grandparental combinations were always the most frequent. Thus, when a homozygous Red-eyed yellow-bodied female was crossed with a white-eyed Grey-bodied male, the male $F_2$ progeny comprised 349 with Red eyes and yellow bodies and 374 with white eyes and Grey bodies (the grandparental combinations), while there were just two of each of the other combinations. It was the latter which were so frequent in the $F_2$ progeny from the cross described in the last paragraph, where these were the grandparental combinations. Morgan concluded that during segregation of character-differences at meiosis certain genes tend to remain together, not because of any attraction between them but because they lie near together in the same chromosomes. Furthermore, he argued that characters remain associated together to a greater extent the nearer the corresponding genes are together in the chromosome. The following year Morgan and Cattell (1912) introduced the term *crossing-over* for the process of interchange by which new combinations of linked factors arise.

## § 7.3  Linkage maps

Sturtevant (1913) took Morgan's argument a stage further. He pointed out that the proportion of cross-overs, that is, progeny derived from gametes in which an exchange had occurred, could be used as an index of the distance between any two genes. Morgan had already shown that this proportion was approximately constant for any particular genes, irrespective of their initial combinations. Thus, he had found that white *versus* Red eyes showed about 1% of cross-overs with yellow *versus* Grey body, whether initially the factors were in *coupling* (dominants in one parent, recessives in the other) or in *repulsion* (one dominant and one recessive in each parent). Sturtevant suggested that the unit of distance between two factor-pairs should be taken to be of such length that, on the average, one cross-over will occur in it out of every 100 gametes formed. In other words, he proposed that the percentage of cross-overs should be used as the index of distance. This concept of distance led him to an important discovery, namely, that the distances between a series of linked factors were additive, implying that the factors could be represented by a linear map. Thus, Morgan's $F_2$ data (1911*b*) from a cross between a yellow-bodied white-eyed miniature-winged female and a normal male gave 1·3% of cross-overs for the body-colour and eye-colour characters, 32·6% for eye-colour and wing-size, and 33·8% for body-colour and wing-size (see Table 7.3). The latter percentage is very

TABLE 7.3   Data of Morgan (1911*b*) for the $F_2$ progeny of a cross between a yellow-bodied white-eyed miniature-winged female and a normal male *Drosophila*. The shortage of yellow-bodied flies compared with the normal Grey in each complementary pair is probably due to a lower viability.

| Characters | | | | Cross-overs | | |
|---|---|---|---|---|---|---|
| Body colour | Eye colour | Wing size | Number | Body colour and eye colour | Eye colour and wing size | Body colour and wing size |
| Grey | Red | Large | 758 | — | — | — |
| yellow | white | miniature | 700 | — | — | — |
| Grey | Red | miniature | 401 | — | 401 | 401 |
| yellow | white | Large | 317 | — | 317 | 317 |
| Grey | white | miniature | 16 | 16 | — | 16 |
| yellow | Red | Large | 12 | 12 | — | 12 |
| Grey | white | Large | 1 | 1 | 1 | — |
| yellow | Red | miniature | 0 | 0 | 0 | — |
| | | Totals | 2205 | 29 | 719 | 746 |
| | | Percentage | 100 | 1·3 | 32·6 | 33·8 |

close to the sum of the other two and implies that the eye-colour factors lie between the others. The linkage map, using the percentages as distances, is shown in Fig. 7.3(*a*). Sturtevant pointed out that there was no means of knowing whether or not the distances as drawn represent the actual spatial distances, because some parts of the chromosome might be more liable to exchange than others. However, the ability to represent the factors by a linear map provided a new argument in support of the chromosome theory, since the chromosome was itself of thread-like proportions. The principle of the linear order of genes was called by Morgan (1919) the *fourth law of heredity*.

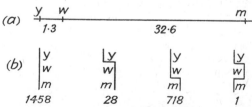

FIGURE 7.3   (*a*) Linkage map of the factors for yellow body (*y*), white eyes (*w*), and miniature wings (*m*) in *Drosophila melanogaster* derived from Morgan's data (1911*b*). (*b*) Summary of the data in Table 7.3 on the basis of the linkage map in (*a*). The lines indicate the position of the exchanges.

## § 7.4 Double crossing-over

Sturtevant found that the additive property of cross-over frequencies did not hold if all three frequencies were large. He attributed this discrepancy to double cross-overs. The data in Table 7.3 illustrate the effect of such double exchanges. There were 29 exchanges between the factors for body-colour and eye-colour and 719 between those for eye-colour and wing-size, making a total of 748, but between body-colour and wing-size there appeared to be only 746. The discrepancy is due to the single Grey-bodied white-eyed Large-winged fly. In the production of the egg from which this fly developed two exchanges must have occurred between the X-chromosomes, with the result that the body-colour and wing-size factors ('Grey' and 'Large', respectively) appeared together as in the grandparent, although the intervening factor (white eyes) was derived from the other grandparent (see Fig. 7.3(*b*)). It is evident that large cross-over frequencies give an inaccurate map, since, unless intervening factors are available, there will be no means of detecting double cross-overs. Various empirical formulae have been devised to allow for the effects of double crossing-over by 'correcting' recombination frequencies to give 'map units'.

In the light of these findings by Morgan and Sturtevant, earlier observations by others took on a new significance. Thus, characters due to factors placed far apart in a chromosome would evidently assort almost at random in the gametes, owing to the frequent occurrence of crossing-over between them. Here was evidently the explanation of how the number of apparently freely-inherited character-differences could exceed the haploid chromosome number.

## § 7.5 Test-crosses

The characters in *Lathyrus odoratus* found by Bateson, Saunders and Punnett (1905) to show 12% of cross-over gametes (Purple *versus* red flower colour and Long *versus* round pollen-shape—see Chapter 3) could now be interpreted as due to factors at a distance apart of 12 units (or slightly more allowing for double cross-overs) on one of the chromosomes. Although *L. odoratus* was thus the first plant (or indeed organism of any kind) for which linkage map data became available, it is not a particularly favourable organism for linkage studies. This is because self-fertilization ($F_2$) data are not easily interpreted (see Table 3.2). In studying linkage it is preferable to cross the heterozygous $F_1$ individuals with the homozygous recessive, since the relative frequencies of the different kinds of gametes from the $F_1$ heterozygote are then reflected directly in the frequencies of different kinds of progeny. Such a cross is called a *test-cross*. In *L. odoratus*, cross-fertilization is tedious and gives few seeds per pollination. An ideal plant for linkage studies was found in *Zea mays*. In this grass the male and female flowers are widely separated at the apex and near the base of the plant, respectively, with the result that controlled cross-pollinations are readily made and moreover, several hundred seeds are normally obtained from a single pollination.

The first example of linkage in *Zea* was found by Collins (1912) between two endosperm characters: the colour of the aleurone layer and the horny (starchy) or waxy texture of the endosperm tissue. An aleurone layer with a purplish pigment was found to be dominant to a colourless aleurone, and

TABLE 7.4  Data of Bregger (1918) for coupling and repulsion test-crosses with Coloured *versus* colourless aleurone (C/c) and Starchy *versus* waxy endosperm (*Wx/wx*) in *Zea mays*.
> (*a*) Primary data.
> (*b*) Totals for contingency $\chi^2$ test (see text).

(*a*)

| Testcross | Coloured Starchy | Colourless Waxy | Coloured Waxy | Colourless Starchy | Total |
|---|---|---|---|---|---|
| Coupling | 147 | 133 | 65 | 58 | 403 |
| Repulsion | 46 | 32 | 103 | 111 | 292 |

(*b*)

| Testcross | Parental | Recombinant | Total |
|---|---|---|---|
| Coupling | 280 | 123 | 403 |
| Repulsion | 214 | 78 | 292 |
| Total | 494 | 201 | 695 |

the normal horny (starchy) endosperm giving a translucent grain was dominant to the opaque waxy endosperm, which had been found to occur in a maize variety from China. Several genes affect the aleurone pigmentation, and the first straightforward test-cross data for the waxy character and the colour factor showing linkage to it were obtained by Bregger (1918). He crossed a pure-breeding strain having Coloured aleurone and Starchy endosperm (*C C Wx Wx*) with one having colourless aleurone and waxy endosperm (*c c wx wx*). The doubly heterozygous $F_1$ plant so obtained had the factors in coupling $\left(\dfrac{C\ Wx}{c\ wx}\right)$. It was test-crossed with the strain having colourless aleurone and waxy endosperm (*c c wx wx*), and gave an ear with 403 grains, details of which are given in Table 7.4(*a*). There were 280 grains with the grandparental character combinations but only 123 with the other two combinations. The cross-over frequency is therefore 30·5%.

Bregger also crossed a pure-breeding strain having Coloured aleurone and waxy endosperm (*C C wx wx*) with one having colourless aleurone and Starchy endosperm (*c c Wx Wx*), and test-crossed the doubly heterozygous

$F_1$ progeny, which had the factors in repulsion $\left(\dfrac{C\ wx}{c\ Wx}\right)$, with the homozygous recessive strain ($c\ c\ wx\ wx$). From this cross he obtained an ear with 292 grains, the details of which are also given in Table 7.4($a$). Again the grand-parental forms were recovered with higher frequency than the others, and the cross-over frequency is 26·7%. The results of the two test-crosses are set out in Table 7.4($b$). Application of the contingency $\chi^2$ test (see Chapter 4) gives

$$\chi^2 = \frac{695[(280 \times 78) - (214 \times 123)]^2}{403 \times 292 \times 494 \times 201} = 1·2, \qquad n = 1,$$

and from Table 4.2 (p. 27), $P = 0·5\text{--}0·1$. Hence, the two estimates of the cross-over frequency do not differ significantly. It might be thought that the best estimate of the recombination frequency would be obtained simply from the total data, that is, $\dfrac{201}{695}$ or 28·92%. However, Muller (1916) pointed out that when coupling and repulsion linkage data are available, the effects of differential viabilities can be minimized by calculating the geometric mean of the ratios of recombinant to parental progeny from each cross. In the present instance, these ratios are $\dfrac{123}{280}$ and $\dfrac{78}{214}$, and their geometric mean is

$$\sqrt{\frac{123}{280} \times \frac{78}{214}} = 0·4001.$$

If there are 40·01 recombinants to every 100 parental genotypes, the recombination frequency is $\dfrac{40·01}{140·01} = 28·58\%$.

A third character of the endosperm of *Zea mays*, namely, a shrunken condition such as gives a concave surface to the apex or side of the mature grain, instead of the normal convex surface, was found to be determined by a recessive gene, *sh*, which showed linkage to $C/c$ and $Wx/wx$. Stadler (1926) made extensive studies using these three linked character-differences. In one of these experiments the triple heterozygote (with all three factors in coupling) was crossed as female parent with a homozygous recessive plant as male parent, thereby forming a 'three-point test cross'. The 45,832 grains on 63 progeny plants were classified in respect of the three pairs of characters and the results are given in Fig. 7.4($a$). From the data, the frequency ($p$) of recombination between $C/c$ and $Sh/sh$ is $\dfrac{1033 + 32}{45832} = 2·32\%$. Similarly, the frequency ($q$) of recombination between $Sh/sh$ and $Wx/wx$ is $\dfrac{9109 + 32}{45832} = 19·94\%$, and the frequency ($r$) of recombination between $C/c$ and $Wx/wx$ is $\dfrac{1033 + 9109}{45832} = 22·13\%$. On the basis of these frequencies, it is evident that the locus of the gene for shrunken endosperm lies between the other two. The linkage map is shown in Fig. 7.4($b$). As in the *Drosophila*

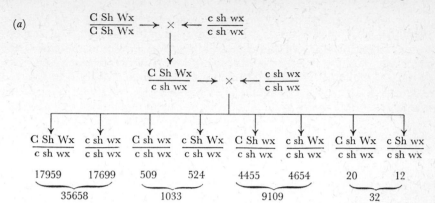

FIGURE 7.4 (*a*) Numbers of progeny of various genotypes from a *Zea* test-cross
(Stadler 1926).

     *C* = Dominant gene for Coloured aleurone layer of the endo-
       sperm.
     *c* = Recessive allele for colourless aleurone.
    *Sh* = Dominant gene for normal Non-shrunked endosperm.
    *sh* = Recessive allele for shrunken endosperm.
    *Wx* = Dominant gene for Starchy (translucent) endosperm.
    *wx* = Recessive allele for waxy (opaque) endosperm.
  (*b*) Linkage map based on the data in (*a*).

example discussed in § 7.4, the slight discrepancy in the figures is due to
double cross-overs, since

$$p + q = r + \frac{64}{45832}.$$

If another gene *X/x* were available between *Sh/sh* and *Wx/wx*, the recombination frequency of *Sh/sh* and *X/x* plus the recombination frequency of
*X/x* and *Wx/wx* would no doubt exceed 19·94%.

  The frequency of recombination between *C/c* and *Wx/wx* in Stadler's
experiment (22·1%) was significantly less than that found by Bregger
(28·6%). Such variation in recombination frequencies is of frequent occurrence, and may be due to genetic or to environmental factors affecting the
frequency of crossing-over. It is evident that distances on linkage maps
have no absolute values, but are likely to differ in different experiments.

## § 7.6 Evidence that crossing-over is associated
## with chromosomal exchange

  Creighton and McClintock (1931) were the first to obtain convincing
evidence that genetic crossing-over of linked characters is associated with
exchange of parts between homologous chromosomes. They bred a strain

of *Zea mays* heterozygous for two peculiarities of chromosome 9 and for two hereditary characters known to be carried by this chromosome. The chromosomal characters were the presence or absence of a knob at the end of the short arm, and the presence or absence of a part of chromosome 8 in place of part of the long arm. The genetic characters were two of those referred to in § 7.5, namely, a Coloured (*C*) or colourless (*c*) aleurone layer to the endosperm of the maize grain, known to be due to a gene difference located near the knob, and Starchy (*Wx*) or waxy (*wx*) endosperm, due to a factor difference located near the point of attachment of the portion of chromosome 8. The combinations of these characters in the heterozygote were as follows:

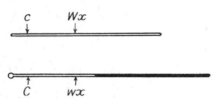

FIGURE 7.5   The pair of no. 9 chromosomes in Creighton and McClintock's strain of *Zea mays* heterozygous for two structural characters of the chromosome and for two endosperm characters. The structural characters were the presence and absence at one end of the chromosome of a knob, and at the other end of an interchange of a segment with a segment (shown black) of chromosome no. 8. *C* = Gene for Coloured, *c* = allele for colourless aleurone layer to the endosperm. *Wx* = Gene for Normal Starchy endosperm, *wx* = allele for waxy endosperm.

On one edition of chromosome 9 there was the knob, the dominant gene *C* for Coloured aleurone, the recessive gene *wx* for waxy endosperm, and the portion of chromosome 8. On the other edition of chromosome 9, the other alternative of each of these characters was present (Fig. 7.5). This quadruple heterozygote was crossed with a plant having no knobs and no parts of chromosome 8 on either of its 9th chromosomes, and also homozygous for the recessive gene *c* for colourless aleurone, but heterozygous for the Starchy-waxy character-difference. There were 27 progeny, of which 13 could be fully scored and interpreted (see Table 7.5). Since both parents were heterozygous for the Starchy-waxy character-difference, any progeny similarly heterozygous have been omitted from the table since it cannot be established from which parent the *Wx* and *wx* factors were derived. (As indicated in Chapter 5, the *Wx Wx* and *Wx wx* genotypes can be distinguished by examination of the pollen after staining with iodine.) All 13 progeny showed complete linkage between the aleurone colour factors and the knobbed or knobless condition, and 11 of them also showed complete linkage between the quality of the endosperm and presence or absence of the interchange. When the genetical characters crossed-over, so did the cytological ones. In two of the progeny there was evidently a cross-over between the *Wx/wx* locus and that of the point of attachment of the part of chromosome 8. The data form striking evidence that genetical crossing-over is associated with exchange of parts between homologous chromosomes. Whether this comes about by breakage and rejoining, or in the process of replication of the hereditary material, is discussed in Chapter 16.

TABLE 7.5 Numbers of progeny found by Creighton and McClintock (1931) from a *Zea* cross, using as one parent a plant heterozygous both for two abnormalities in chromosome 9 and for two characters of the endosperm (aleurone colour and food reserve) known to be due to genes borne by this chromosome. A normal chromosome 9 was present in addition to the one described in the table. Individuals heterozygous for the food reserve characters have been omitted.

| Chromosome 9 Characters \ Endosperm Characters | | Coloured | Colourless | Coloured | Colourless |
|---|---|---|---|---|---|
| | | Waxy | Starchy | Starchy | Waxy |
| Knob present | Interchange with chromosome 8 present | 3 | 0 | 0 | 0 |
| No knob | No interchange | 0 | 5 | 0 | 0 |
| Knob present | No interchange | 0 | 0 | 1 | 0 |
| No knob | Interchange with chromosome 8 present | 0 | 2 | 0 | 2 |

A few weeks later Stern (1931) reported similar results for *Drosophila melanogaster*. He had obtained four different strains of female flies, each heterozygous for two abnormalities in the $X$-chromosome and for two sex-linked characters. Over 27,000 progeny from these females were scored for the latter characters, and 364 of them were also examined cytologically and scored for the chromosomal characters. All four experiments gave similar results, but only one will be described.

The female parents in this experiment carried the factor for the recessive character carnation-coloured eyes on one of their $X$-chromosomes, which was structurally normal. On the other $X$-chromosome was the normal allele of carnation eye and the dominant gene for Bar eye. This chromosome was also structurally abnormal in two respects: firstly, it had part of a $Y$-chromosome attached at one end just beyond the carnation locus; secondly, it was broken into two parts, the break being just beyond the Bar locus. This means that the proximal part carried both the Bar gene and the normal allele of carnation. The distal part was attached to the small fourth chromosome (Fig. 7.6). The male parent was structurally normal, but carried the gene for carnation eyes on its $X$-chromosome. Details of the 8231 progeny of this cross are given in Table 7.6. Out of this number 107 were examined cytologically, including 54 in which crossing-over had occurred between the loci for carnation and Bar. Without exception, these 54 flies also showed crossing-over for the chromosomal characters. Bar eye-shape was always associated

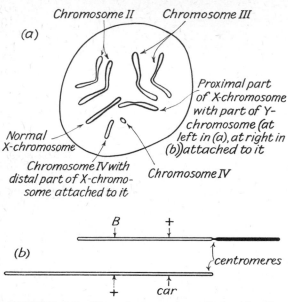

FIGURE 7.6   The chromosomes of the female parents in one of Stern's experiments with *Drosophila melanogaster* which demonstrate that genetical crossing-over is associated with physical exchange between the corresponding homologous chromosomes.
(*a*) Polar view of mitotic metaphase.
(*b*) The *X*-chromosomes. *B* = Dominant gene for Bar eye-shape, and the plus sign (+) at the corresponding position = recessive normal allele; *car* = recessive gene for carnation eye-colour, and the plus sign at the corresponding position = dominant normal allele. The attached part of the *Y*-chromosome is shown black.

with the broken *X*-chromosome to which it was closely linked, and normal Red eye-colour was associated with the attached piece of *Y*-chromosome. Indeed, as indicated in the table, there was only one fly out of the 107 which showed breakdown of these linkages: a cross-over had apparently occurred between the carnation locus and the point of attachment of the *Y*-chromosome segment.

TABLE 7.6  Numbers of progeny found by Stern (1931) from a *Drosophila* cross, using females heterozygous both for two abnormalities in the *X*-chromosome and for two sex-linked characters, carnation eye-colour and Bar eye-shape. A normal *X*-chromosome (in females) or *Y*-chromosome (in males) was present in addition to the *X*-chromosome described in the table. The shortage of normal progeny (Red round eyes) is thought to be due to their failure to survive owing to loss of one of their small fourth chromosomes.

| *X*-chromosome characters / Sex-linked characters | | Normal Red Bar | Carnation Normal round | Carnation Bar | Normal Red Normal round |
|---|---|---|---|---|---|
| Part of *Y*-chromosome attached | Broken | 26 | 0 | 0 | 0 |
| Without *Y*-chromosome | Unbroken | 0 | 26 | 0 | 0 |
| Without *Y*-chromosome | Broken | 1 | 0 | 45 | 0 |
| Part of *Y*-chromosome attached | Unbroken | 0 | 0 | 0 | 9 |
| Total progeny (including those in which chromosomes were not examined) | | 4001 | 4157 | 61 | 12 |

Stern's data constitute further evidence that genetical crossing-over involves exchange of material between the appropriate homologous chromosomes. Thus, just 20 years after Morgan's crucial discovery of new combinations of two sex-linked character-differences in the progeny of a *Drosophila* cross, which led him to postulate interchange of parts between homologous chromosomes, Stern, using both genetical and cytological characters in the identical chromosome, obtained overwhelming evidence of the truth of Morgan's postulate. The mechanism of crossing-over is discussed in Chapter 16.

# 8. The chiasmatype theory

## § 8.1 Chiasmata

As indicated in Chapter 6, the chromosome pairs seen at late prophase I of meiosis usually present the appearance either of a cross or of one or more loops. Janssens (1909) gave the name *chiasma* (meaning a cross) to the nodes between the arms or loops in these diplotene or diakinetic configurations. He also adopted in the same publication a particular hypothesis (described below) to account for the origin of these nodes. The use of the term chiasma was at first intimately bound up with this particular interpretation of them. However it is customary nowadays to use the term for these nodes, irrespective of the theory adopted to account for them. It is highly inconvenient to have a name for an observed structure the use of which is dependent on a theory to explain it. Hence, throughout this book chiasma will be used as Janssens first defined it, for a node in the paired chromosomes at the first division of meiosis. It should not be taken to imply any particular interpretation of its origin.

## § 8.2 Janssens's theory

At the diplotene stage the chromosomes have already divided into two daughter-chromatids, with the result that four threads are present in each chromosome-pair. Janssens had studied meiosis in several organisms, but particularly at sperm-formation in the amphibian *Batrachoseps attenuatus*. At the chiasmata at the diplotene stage, whenever the fixation and staining had been particularly good, he observed that of the four filaments, two cross each other and two do not (Fig. 8.1(*a*) and Pl. 6.2(*g*)). At anaphase I, when the chiasmata break down, the two chromatids which pass to one pole of the spindle were seen to lie together on one side of each chiasma and to lie apart on the other side (Fig. 8.1(*c*) and (*d*)). It was evident that two of the four chromatids crossed at each chiasma. These observations have been amply confirmed in many organisms.

Janssens believed that prior to the formation of the chiasmata, the paternal and maternal chromosomes in the early diplotene stage were loosely coiled round each other, as in Fig. 8.2(*b*). In order to derive chiasmata from such a loose coil, Janssens postulated that the paternal and maternal chromatids made contact at intervals, and that one of each penetrated the other until they broke, whereupon they rejoined in new ways paternal to maternal, and *vice versa*. The other two chromatids (one paternal and one maternal) were considered to remain intact, so giving rise to the typical chiasma with two

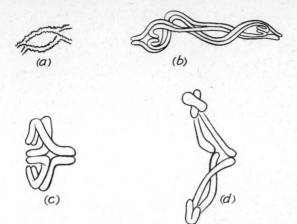

FIGURE 8.1    Chromosome-pairs showing chiasmata
at meiosis in males of *Batrachoseps
attenuatus* (from Janssens, 1909). The
number and position of the chiasmata
varies from cell to cell. (*a*) Diplotene
stage (2 chiasmata). (*b*) Diakinesis
(3 chiasmata). (*c*) Metaphase I (1
chiasma). (*d*) Anaphase I.

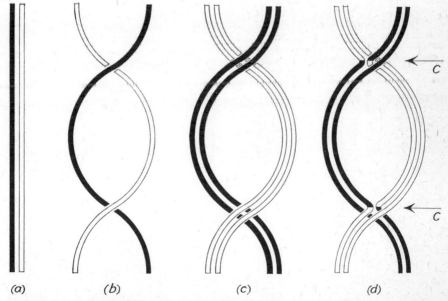

FIGURE 8.2    Diagrams to show the structure of a pair of homologous chromosomes at (*a*) the pachytene stage and (*b–d*) successives early diplotene
stages, according to Janssens's hypothesis of the origin of chiasmata.
*C* = Chiasma.  Black = paternal and white = maternal origin.

chromatids crossing each other and two not (Fig. 8.2(*d*)). This remarkable hypothesis has been a centre of controversy ever since it was first put forward.

Janssens also used the teleological argument that the occurrence of his hypothetical exchanges between one paternal and one maternal chromatid at each chiasma would ensure that the four nuclei, to which the four chromatids of each diplotene chromsome-pair were distributed, had different combinations of paternal and maternal segments of each chromosome, and hence would account for the universal existence of two nuclear divisions in meiosis and four products. He argued that meiosis was a mechanism for the production of four haploid nuclei all of different genotypes. He pointed out that there would then be segregation of some factor differences at each division of meiosis (Fig. 8.3(*b*)) in contrast to the widely held view that segregation of paternal from maternal chromosomes occurred at the heterotype or first division of meiosis (Fig. 8.3(*a*)), with the identical daughter-chromosomes

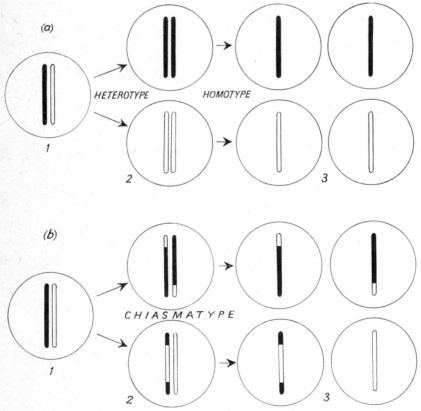

FIGURE 8.3    Diagrams of (1) the pachytene stage of prophase I, (2) telophase I, and (3) telophase II, according to (*a*) the hypothesis that meiotic segregation occurs only at the first division, and (*b*) Janssens's hypothesis that it occurs partly at each division. Black = paternal and white = maternal origin.

composing each of them separating from one another at the homotype or second division (Fig. 8.3(*a*)). To emphasise this difference from the heterotype-homotype theory of meiotic segregation, he called his idea the *chiasmatype* theory.

Janssens also pointed out that his theory provided an escape from the dilemma of the existence of a larger number of freely assorting allelomorphic characters than there are pairs of chromosomes.

## § 8.3   The classical theory

The weak link in Janssens's argument was his postulation that loose coils preceded the formation of chiasmata. Since he was observing fixed and stained material, it was difficult to test this postulate by observation. It was based on the assumption that division of each chromosome into two chromatids did not occur until after the paternal and maternal chromosomes had begun to separate in the early diplotene stage (Fig. 8.2(*b*) and (*c*)). It is true that at the preceding pachytene stage the chromosomes appear to be single threads (Fig. 8.2(*a*)). However, the possibility remained that division of the chromosomes into daughter-chromatids occurs before the early diplotene separation is initiated (Fig. 8.4(*b*)). An alternative explanation of the origin

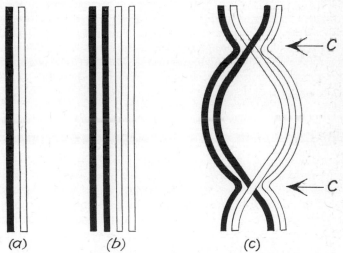

(*a*)                    (*b*)                    (*c*)

FIGURE 8.4   Diagrams to show (*a*) early pachytene, (*b*) late pachytene, and (*c*) diplotene chromosomes according to the classical hypothesis of the origin of chiasmata.   C = Chiasma.   Black = paternal and white = maternal origin.

of chiasmata is then possible, namely that at a chiasma there is a change of pairing partner. At one region of the chromosome two paternal chromatids would be associated and likewise the two maternal ones, but on the other side of a chiasma each pair of chromatids would consist of one paternal and one maternal thread (Fig. 8.4(*c*)). This *classical hypothesis* to explain chiasmata was preferred by most cytologists because it did not require breakage and rejoining of chromosomes, with consequent threat to their permanence and individuality.

## § 8.4   Genetic evidence in support of Janssens's theory

When Morgan (1911*a*) first discovered that new combinations of sex-linked characters were appearing in the $F_2$ generation of a *Drosophila* cross, he favoured De Vries's theory of factor exchange. However, shortly after-wards he realized that the sex-linked factors were not assorting at random, as De Vries's theory demanded, but (as indicated in the last chapter) were showing partial linkage of various intensities. He then favoured Janssens's chiasmatype theory as an explanation.

## § 8.5   An objection to Janssens's theory

Muller (1916) pointed out a difficulty of Janssens's theory, namely that the exchanges were supposed to occur at the diplotene stage when the chromo-somes are relatively short and thick, and yet crossing-over must be between precisely homologous points of paternal and maternal threads if the cross-over chromosomes were not to have duplications or deficiencies of the linearly-arranged genes. Muller argued that factors must be set very close together in the chromosomes, and hence that crossing-over required a high degree of precision, because in *Drosophila* mutations in new 'loci' (positions in the chromosomes) were still turning up regularly. He therefore suggested that the exchanges perhaps took place earlier in meiosis when the chromosomes were still in the form of slender threads, as for example at the zygotene stage. This suggestion raised a fundamental question.

## § 8.6   Crossing-over in relation to chromosome duplication

An essential feature of Janssens's theory was that crossing-over should occur between only two of the four chromatids, otherwise the presence of two threads crossing one another and two not at each chiasma would not be explained. But at the zygotene stage the chromosomes were thought to be undivided. Hence crossing-over at this stage, as postulated by Muller, would presumably involve whole chromosomes and so would be contrary to the chiasmatype theory. To test the truth of Janssens's theory it was therefore essential to know whether crossing-over occurred before or after the chromo-somes had duplicated. If exchange involved only two of the four chromatids, as Janssens supposed, this would be conclusive evidence that crossing-over did not occur before chromosome duplication (Fig. 8.5(*a*)). On the other hand, if crossing-over involved all four chromatids at the same locus, there were two possible explanations: either exchange occurred while the chromo-somes were still single threads and the subsequent duplication produced two editions of each (Fig. 8.5(*b*)); or all four chromatids were broken at corresponding positions and were subsequently joined paternal to maternal, and *vice versa* (Fig. 8.5(*c*)). Janssens himself thought he had cytological evidence that the latter alternative occurred occasionally.

Bridges (1916) was the first to obtain information on this fundamental question of whether crossing-over takes place before or after chromosome duplication. During the course of his work with the *XXY* strain of *Drosophila*

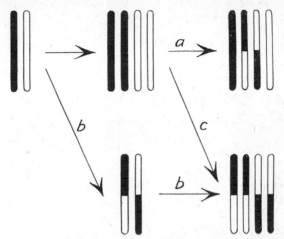

FIGURE 8.5    Possible ways in which crossing-over might
occur.
(*a*) Janssens's chiasmatype hypothesis:
crossing-over after chromosome duplication and involving only two of the
four chromatids.
(*b*) Crossing-over at the two-strand stage
before chromosome duplication.
(*c*) Crossing-over after chromosome duplication but involving all four chromatids at the same locus (thought by
Janssens to occur occasionally).
Black = paternal and white = maternal
origin.

*melanogaster* (see Chapter 6), he obtained *XXY* females which were hetero-zygous for the sex-linked characters vermilion eye-colour and eosin eye-colour. These characters are both recessive to the normal Red eyes and are due to genes showing about 30% recombination with one another. The doubly-heterozygous *XXY* females, some with vermilion and eosin in coupling and some with them in repulsion, were crossed with Bar-eyed but otherwise normal males. The dominant gene Bar is sex-linked and leads to the production of a narrow eye instead of the normal round eye. Any progeny from this cross which receive an *X*-chromosome from their father will have Bar eyes. Hence any female progeny with eyes of the normal round shape must have received both their *X*-chromosomes from their mother. They will correspond to the flies in class (i) of Table 6.2, which represent about 2% of the progeny. From the coupling crosses, 422 non-Bar females

TABLE 8.1 Numbers of different genotypes of female *Drosophila melanogaster* when the two *X*-chromosomes of each individual have both been derived from their mother. Bridges's (1916) data refer to *XXY* flies, Anderson's (1925) to attached *XX*, and Bridges & Anderson's (1925) to triploid *XXX*. A dash (—) as distinct from zero (0) indicates flies of normal phenotype, the genotype of which was not determined. There were 414 of these in the coupling crosses (column 2) and 168 in the repulsion crosses (column 3).

| The Ten Possible Genotypes (Female Parent $\frac{AB}{ab}$) | *XXY* Bridges (1916) | | Attached *XX* Anderson (1925) | | | Triploid *XXX* Bridges & Anderson (1925) | | | |
|---|---|---|---|---|---|---|---|---|---|
| | Vermilion (a) and eosin (b) | Vermilion (a) and eosin (B) | Forked (a) and garnet (B) | Garnet (A) & tan (b) | Tan (a) and cut (B) | Bar or forked and dusky or miniature | Dusky or miniature and lozenge or tan | Lozenge or tan and ruby or bifid | Ruby or bifid and scute or yellow |
| (i) $\frac{AB}{AB}$ | — | 0 | 8 | 15 | 21 | 2 (i+ii) | 3 | 3 | 5 |
| (ii) $\frac{ab}{ab}$ | 2 | 0 | 11 | 25 | 36 | | | | |
| (iii) $\frac{AB}{ab}$ | — | — | 126 | 103 | 104 | 153 | 167 | 140 | 126 |
| (iv) $\frac{Ab}{aB}$ | — | — | 11 | 9 | 8 | 3 | 1 | 2 | 0 |
| (v) $\frac{Ab}{Ab}$ | 0 | — | 0 | 0 | 0 | 0 | 0 | 0 | 0 |
| (vi) $\frac{aB}{aB}$ | 0 | 0 | 0 | 0 | 0 | 0 | 0 | 0 | 0 |
| (vii) $\frac{AB}{Ab}$ | — | — | 2 | 3 | 7 | 0 (vii+viii) | 0 | 2 | 5 |
| (viii) $\frac{aB}{ab}$ | 3 | 1 | 1 | 5 | 4 | | | | |
| (ix) $\frac{AB}{aB}$ | — | 4 | 10 | 13 | 3 | 1 (ix+x) | 2 | 7 | 16 |
| (x) $\frac{Ab}{ab}$ | 3 | — | 19 | 15 | 5 | | | | |

were obtained, of which 414 had Red eyes like their mother and 8 had vermilion, eosin or vermilion-eosin eyes. Similarly, from the repulsion crosses there were 168 non-Bar females with Red eyes and 5 with vermilion or eosin eyes. The genotypes of the vermilion- and eosin-eyed females were established by further breeding and the results are given in Table 8.1. All

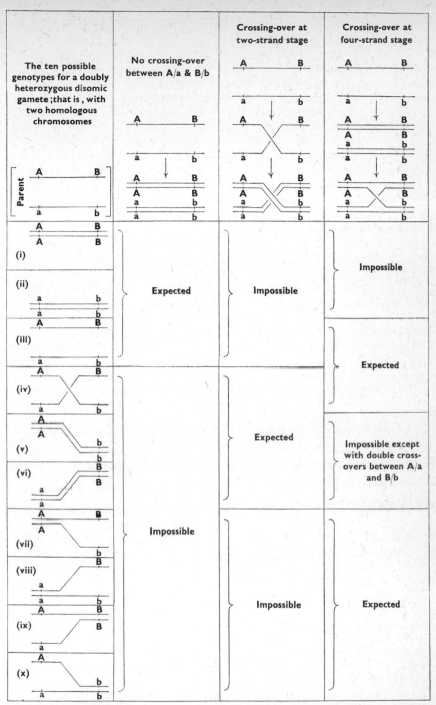

FIGURE 8.6   The genotypes expected for the gametes from a double heterozygote $\frac{AB}{ab}$ if two of the four meiotic chromatids pass into each gamete. The expectations with no crossing-over, with crossing-over at the two-strand stage, and with crossing-over at the four-strand stage, are shown in separate columns.

those that were homozygous for eosin were found to be heterozygous for vermilion, and *vice versa*. To obtain flies of these constitutions requires crossing-over between the two X-chromosomes of the mother during matura-tion of her eggs, and furthermore such crossing-over must take place between individual chromatids and involve only two of them (one derived from each X-chromosome—see Fig. 8.6). With no crossing-over, or with crossing-over before chromosome duplication, any two of the four chromatids will always be either identical or complementary, that is of genotypes (i) to (vi) in Fig. 8.6. The occurrence of genotypes (viii) to (x) was proof of crossing-over after chromosome duplication. Altogether, Bridges obtained information about 11 cross-overs between the loci of vermilion and eosin and every one of them was between two chromatids.

Anderson (1925) obtained more evidence that crossing-over occurred only at the four-strand stage. A female *Drosophila* was discovered after treatment with X-rays, in which the two X-chromosomes were joined together at one end. This union was demonstrated cytologically, for a single V-shaped chromosome replaced the two rod-shaped X-chromosomes. It was also shown by the fact that on crossing with a normal male, sex-linked characters which she possessed were manifest in her daughters and not in her sons. This was in direct contrast to the usual behaviour of sex-linked characters, which show criss-cross inheritance (see Chapter 6). The daughters of the attached-X female will inherit her two joined X-chromosomes, and moreover there will be opportunity at meiosis during maturation of her eggs for crossing-over to occur between them. The parent attached-X female was heterozygous for the recessive sex-linked characters forked bristles, garnet eye-colour, tan body-colour, and cut wing-tips. If no crossing-over occurred the daughters would have a double X-chromosome identical with that of their mother. On the other hand, crossing-over in certain intervals of the chromosome could lead to one or more of the recessive characters becoming homozygous and hence manifest, just as in Bridges's earlier work.

Out of 4344 daughters of attached-X females heterozygous for these genes, Anderson found totals of 226 (5·2%) with forked bristles, 414 (9·5%) with garnet-coloured eyes, 698 (16·1%) with tan-coloured bodies and 672 (15·5%) with cut wings. The occurrence of these individuals, homozygous for one or more recessive characters for which the mother was heterozygous, was evidence of crossing-over between the joined X-chromosomes, and in order to obtain more information, a random sample of 188 of the daughters was subjected to detailed study. The genetic constitution of each X-chromosome of each individual was established from the frequencies of different pheno-types in their progeny from suitable crosses. The results are set out in Table 8.1 with the four pairs of alleles taken two at a time. In each instance, of the 10 genotypes theoretically possible, 8 were represented. The missing ones were always those with two identical cross-over chromosomes (genotypes v and vi) and they can only be obtained by postulating that crossing-over occurs at the two-strand stage (unless double cross-overs are invoked—see Fig. 8.6.). Conversely there were present all the genotypes with one parental and one cross-over chromosome (nos. vii to x) which can be obtained only by assuming that crossing-over occurs between two chromatids, that is, at the four-strand stage. The data demonstrate that every one of 87 cross-overs, about which information was obtained, took place after chromosome replication and involved only two of the four chromatids at any one place.

Bridges and Anderson (1925) obtained further evidence that crossing-over occurs at the four-strand stage from studies of triploid *Drosophila* with sex-linked characters. The evidence was obtained from 182 diploid female progeny with 2 $X$-chromosomes both derived from the mother and between which crossing-over had occurred. The 3 $X$-chromosomes in the triploid flies were each uniquely identified at five different positions by the genes they carried, and this enabled the source of the 2 $X$-chromosomes in the daughters to be determined. The results are given in Table 8.1 for the four consecutive regions of the chromosome delimited by the genes. The totals in each column are less than 182 because progeny containing genes derived from all three maternal $X$-chromosomes have been omitted. The results are in complete agreement with those of Bridges (1916) and Anderson (1925). The total data from the three publications indicate no cross-overs at the two-strand stage (classes v and vi) and 131 at the four-strand stage and involving only two of the four strands (classes vii to x).

These experiments which demonstrate that crossing-over occurs exclusively at the four-strand stage are open to the criticism that they are concerned with crossing-over under abnormal conditions; $XXY$, attached $XX$ or triploid *Drosophila*. The ideal way in which to obtain information about the stage at which crossing-over occurs is to recover the four products of a single meiosis, in isolation from those of neighbouring meioses. This tetrad analysis is possible in many plants.

Pascher (1918) crossed two species of the haploid unicellular green alga *Chlamydomonas* differing in several characters and isolated a number of the resulting zygotes. Meiosis in *Chlamydomonas* occurs on germination of the zygote and leads to the production of four motile cells. Pascher examined the characters of these cells and found that in each tetrad the parental characters always segregated two and two. Thus one parent had pear-shaped cells and a laterally-placed chloroplast, while the other had spherical cells and a basal chloroplast. The four cells from each tetrad always comprised two with pear-shaped and two with spherical cells, and similarly two had lateral and two basal chloroplasts. Pascher found that in some tetrads there were two cells like one parent and two like the other, while in others all four cells differed from one another, the tetrad having, for example, one cell with each of the four combinations of the shape and chloroplast characters. The existence of such 'tetratype' tetrads showed that segregation of character-differences could occur at either division of meiosis and was in agreement with the hypothesis that crossing-over occurred at the four-strand stage, although convincing evidence of this would require linked factors.

Soon afterwards tetrad analysis was achieved in bryophytes and in basidiomycete and zygomycete fungi, but it was not until Lindegren (1933) found linked characters in the ascomycete fungus *Neurospora crassa* that direct evidence was obtained by tetrad analysis that crossing-over occurs at the four-strand stage and involves only two of the four strands at any one place. *Neurospora crassa* is a haploid fungus existing in two mating-types. Meiosis occurs immediately after fusion of the nuclei of opposite mating-type. Wilcox (1928) had shown that in the closely allied species *N. sitophila* the four haploid nuclei resulting from meiosis are linearly arranged within an elongated tube-like cell, the ascus, such that the two second-division spindles do not overlap (see Fig. 8.7). Subsequently mitosis takes place in each

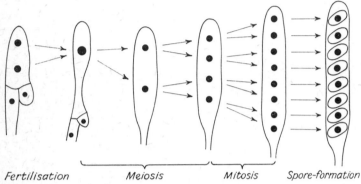

| Fertilisation | Meiosis | Mitosis | Spore-formation |

FIGURE 8.7    Stages in the development of the ascospores of *Neurospora sitophila*.

nucleus, and again the spindles do not overlap. Spore walls are then laid down round each nucleus, and the mature ascus, which is about 150 $\mu$m long and 15 $\mu$m wide, contains a linear sequence of 8 black elliptical spores, each measuring about 25 $\mu$m in length and 15 $\mu$m in width. She dissected a number of asci using a fine needle and isolated the spores in sequence. When the resulting cultures were tested for mating-type four were always of one mating-type and four of the other. Moreover, spores 1 and 2 numbering from the top of the ascus were always alike, and so were 3 and 4, and so on. In other words, the spores were in identical pairs, as would be expected if the third nuclear division in the young ascus was mitotic (cf. Pl. 8.1).

Dodge (1929) made similar ascus dissections and found evidence that the mating-type factors sometimes segregate at the first division of meiosis and sometimes at the second division. This was confirmed by Lindegren (1932) who dissected numerous asci of *N. crassa* and tested the spores for mating-type. His results are shown in Table 8.2. Eleven out of 273 asci showed one or more pairs of non-identical spores, numbering from the end of the ascus, in contrast to what was expected if the cytology was similar to that of *N. sitophila*. However in every instance exchanging the position of two adjacent spores would suffice to explain these anomalies, and he presumed that either third-division spindle overlap or spore displacement during dissection was responsible. The 273 asci can then be classified into six

PLATE 8.1    Asci of the fungus *Sordaria brevicollis* from a cross between a mutant with buff-coloured ascospores and another with yellow ascospores. The wild-type spores are black and the double-mutant spores are white (colourless). The buff and yellow mutants show linkage with about 9% recombination. Magnification ca. ×350.

(*a*) Ascus with 4 buff (to left) and 4 yellow spores, implying no crossing-over between the gene loci.

(*b*) Ascus with 2 yellow, 2 white, 2 black and 2 buff spores (1 cross-over between the loci).

(*c*) Ascus with 2 pairs of black and 2 pairs of white spores (a four-strand double cross-over between the loci: see Fig. 8.12, p. 113).

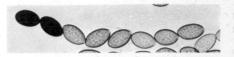

(*d*) Ascus with the spore sequence: 2 black, 2 yellow, 4 buff, indicating conversion from yellow to wild-type (see § 16.2).

classes numbered I to VI in the table. Classes I and II correspond to segregation of *A* from *a* at the first division of meiosis, if the nuclear behaviour is like *N. sitophila*. The numbers of asci in these two classes were not significantly different (observed frequencies 105 and 129, expected 117, deviations 12, $\chi^2 = 2.5$, which with 1 degree of freedom, is associated with a probability of just over 0·1 of obtaining a deviation as large or larger by chance). Classes III to VI correspond to segregation of *A* from *a* at the second division of meiosis, and again the numbers in the classes do not differ significantly (observed frequencies 9, 5, 11 and 14, expected 9·75, deviations 0·75, 4·75, 1·25 and 4·25, $\chi^2 = 4.5$, which for three degrees of freedom implies a probability of between 0·5 and 0·1 of by chance getting a deviation as large or larger than that observed). However the frequencies of classes I and II clearly differ significantly from those of the other four.

Lindegren (1933) reported further data from ascus dissection in *N. crassa*. He crossed a strain of the fungus with pale aerial filaments and conidia (asexual spores) with one lacking conidia, which he described as fluffy from the appearance of the aerial growth. The normal fungus has orange-coloured conidia. Each of the characters (pale and fluffy) appeared to differ from normal by a single gene, for the spores from each ascus produced four pale and four orange cultures, and they could also be classified for the presence or absence of conidia and again there were four progeny per ascus in each class. Lindegren dissected 109 asci showing segregation for these characters. The numbers of asci with first- and second-division segregation which he obtained are shown in Table 8.3. The frequency of second-division segregation for mating-type in this set of data (11·0%) is slightly lower than

TABLE 8.2   Lindegren's data (1932) for the segregation of the mating-type factors in asci of *Neurospora crassa*. The letters *A* and *a* denote the two mating-types.

| Class | Position of spore in ascus | | | | | | | | Number |
|---|---|---|---|---|---|---|---|---|---|
|  | 1 | 2 | 3 | 4 | 5 | 6 | 7 | 8 |  |
| I | *A* | *A* | *A* | *A* | *a* | *a* | *a* | *a* | 102⎫ 105 |
|  | *A* | *A* | *A* | *a* | *A* | *a* | *a* | *a* | 3⎭ |
| II | *a* | *a* | *a* | *a* | *A* | *A* | *A* | *A* | 123⎫ 129 |
|  | *a* | *a* | *a* | *A* | *a* | *A* | *A* | *A* | 6⎭ |
| III | *A* | *A* | *a* | *a* | *A* | *A* | *a* | *a* | 8⎫ 9 |
|  | *A* | *a* | *A* | *a* | *A* | *A* | *a* | *a* | 1⎭ |
| IV | *a* | *a* | *A* | *A* | *a* | *a* | *A* | *A* | 5 |
| V | *A* | *A* | *a* | *a* | *a* | *a* | *A* | *A* | 10⎫ 11 |
|  | *A* | *a* | *A* | *a* | *a* | *a* | *A* | *A* | 1⎭ |
| VI | *a* | *a* | *A* | *A* | *A* | *A* | *a* | *a* | 14 |
|  |  |  |  |  |  |  |  | Total | 273 |

that found previously (14·3%), but the difference is not significant. (Contingency $\chi^2 = \dfrac{382[(234 \times 12) - (39 \times 97)]^2}{273 \times 109 \times 331 \times 51} = 0.74$, which, with one degree of freedom means that $P$ is just under 0·5, or in other words, a difference as large or larger than that found may be expected to occur by chance nearly every other time on the average.) On the other hand the second-division segregation frequencies of the pale and fluffy characters (33·0% and 61·4% respectively) clearly differ significantly from each other and from that for mating-type.

Lindegren's hypothesis to explain these results was that crossing-over occurred at the four-strand stage and involved only two of the four chromatids, and furthermore that the paternal and maternal centromeres of the chromosome pair segregated from one another at the first division of meiosis. Second-division segregation of a pair of alleles would then be the consequence of crossing-over between their locus and the centromere of the chromosome, while first-division segregation would mean that the factors had separated along with the centromeres, without crossing-over in between (Fig. 8.8). Owing to the variability in the position of cross-overs a given pair of alleles would sometimes separate at the first division of meiosis and sometimes at the second division. Lindegren pointed out that it followed from this that the frequency of second-division segregation for a particular character-difference was a measure of the frequency of crossing-over between the gene concerned and its centromere, and hence was a measure of their distance apart on the linkage map.

TABLE 8.3   Lindegren's data (1933) for first- and second-division segregation frequencies for various character-differences in *Neurospora crassa*.

| Character difference | | Numbers of asci showing segregation | | | Percentage of asci showing 2nd division segregation | Distance of gene from centromere |
|---|---|---|---|---|---|---|
| | | First division | Second division | Total | | |
| Mating-type | Data of Table 8.2 | 234 | 39 | 273 | 14·3 | 7·1 |
| | New data | 97 | 12 | 109 | 11·0 | 5·5 |
| | Total | 331 | 51 | 382 | 13·4 | 6·7 |
| Pale/normal orange conidial colour | | 73 | 36 | 109 | 33·0 | 16·5 |
| Fluffy/normal growth-form | | 42 | 67 | 109 | 61·4 | 30·7 |

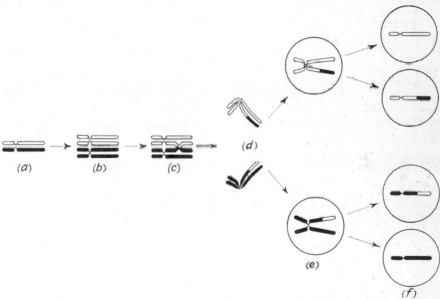

(a)        (b)        (c)

(d)

(e)

(f)

FIGURE 8.8   Diagrams of a pair of homologous chromosomes at various stages of meiosis to illustrate Lindegren's hypothesis of centromere segregation at the first division of meiosis, and crossing-over at the four-strand stage. This hypothesis was put forward to account for the characteristic frequency of second-division segregation for each gene. (a) Pachytene stage. (b) End of pachytene stage. (c) Diplotene stage. (d) Anaphase I. (e) Telophase I. (f) Telophase II.
The constrictions indicate the centromeres, and the shading the parental origin.

It will be recalled that Sturtevant (1913) had proposed that the unit of distance between two gene loci on a linkage map should be such as to give rise to 1% of cross-over gametes (see § 7.3). In a haploid organism such as *Neurospora* meiosis leads directly to the next generation, so the percentage frequency of progeny with non-parental combinations of two character-differences is the measure of the distance apart on the linkage map of the factors concerned. However, if crossing-over occurs at the four-strand stage, one exchange will lead to two cross-over and two non-cross-over progeny (compare Fig. 8.8). Hence every time a pair of allelomorphs segregate from one another at the second division of meiosis, the four products of meiosis will comprise two cross-overs and two non-cross-overs for the interval between the locus of the alleles and the centromere. Thus 1% of cross-over progeny, which is the unit on the map, will be produced by second-division segregation in 2% of meioses, or in other words frequencies of second-division segregation expressed as percentages can be converted into gene-centromere map distances simply by halving them. The centromere distances obtained in this way for the factors studied by Lindegren are given in the right-hand column of Table 8.3.

Lindegren's theory of first-division segregation of centromeres was supported by the recombination data from his 109 asci (see Table 8.4). The numbers of parental and non-parental associations of mating-type and fluffy growth-form (222 parental and 214 non-parental) do not differ significantly, and nor do those for the pale and fluffy characters (224 parental and 212 non-parental). On the other hand, the mating-type and pale characters showed partial linkage, with 22·5% recombination. It follows that the genes responsible for these character-differences are located 22·5 units apart on the same linkage map. This figure agrees well with the sum of their centromere distances (22·0) (see Fig. 8.9). Lindegren concluded that the genes for mating-type and pale conidia are located in opposite arms of the chromosome, and that his hypothesis of first-division segregation of centromeres, together with crossing-over at the four-strand stage and involving only two of the four strands, was correct (cf. Pl. 8.1(*b*)).

TABLE 8.4   Lindegren's data (1933) for the numbers of parental and non-parental combinations of character-differences in the 436 pairs of spores from 109 asci of *Neurospora crassa*.

| Pair of Character-differences | Parental | Non-parental | Percentage Recombination |
|---|---|---|---|
| Mating-type and pale/normal (orange) conidial colour | 338 | 98 | 22·5 |
| Mating-type and fluffy/normal growth form | 222 | 214 | 49·0 |
| Pale/normal (orange) conidial colour and fluffy/normal growth-form | 224 | 212 | 48·6 |

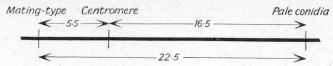

FIGURE 8.9    Linkage map of the mating-type chromosome of *Neurospora crassa*, showing the similarity between the distance apart of the
mating-type and 'pale' gene loci based
on second-division segregation frequencies
(figures above the line) and that based on
recombination frequency (figure below
the line). The data are from the analysis of
109 asci by Lindegren (1933).

The hypothesis of first-division segregation of the centromeres, which is
required by the chiasmatype theory, had been proposed earlier by Bridges
and Anderson (1925), but their evidence was indirect and less convincing.
An examination of their data and those of Anderson (1925) in Table 8.1
shows that there are progressive changes in the frequencies of the various
genotypes as one proceeds from column to column across the table, or in
other words, as one proceeds along the chromosome from region to region
(see Fig. 8.10). By extrapolation, it appeared that at a point beyond the gene
for forked bristles, which in both experiments marked one end of the part of
the chromosome under study, the genotypes homozygous for the $A$ or the $a$
allele (that is, classes i, ii, vii and viii) would diminish to zero. It was suggested that at this hypothetical point alleles (such as $A$ and $a$) always
segregated from one another at the first division of meiosis and sister genes
(such as $A$ and $A$) always passed to opposite poles at the second meiotic
division. Since the $X$-chromosome in *Drosophila melanogaster* is rod-shaped, the
centromere must lie at or near one end, and it was natural to suggest that this
corresponded with the hypothetical point of first-division segregation.

With the attached-$X$ flies this point of zero frequency of homozygotes
would be the place where the two $X$-chromosomes joined, but this was
thought to coincide with the centromere since, as already indicated, the

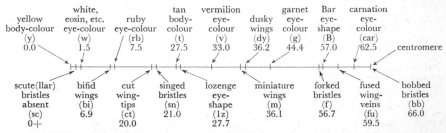

FIGURE 8.10    Linkage map of the $X$-chromosome of *Drosophila melanogaster* to
show the relative positions of genes mentioned in the text. The
figures show the distance from the distal end in map units. (Data
chiefly from Morgan, 1926.)

chromosome when fixed and stained appeared V-shaped with two equal arms. Anderson's data allowed this point of junction to be mapped. From Table 8.1, out of the 188 flies studied, 22 (11·7%) were homozygous at the 'forked' locus (classes i, ii, vii and viii in the 'forked and garnet' column) and 48 (25·5%) were homozygous at the 'garnet' locus (classes i, ii, ix and x in the same column). Since homozygosis is thought to be due to crossing-over between gene and attachment-point, these frequencies are a measure of the distances of the gene loci from this point on the linkage map. The two *X*-chromosomes in an attached-*X* fly are the equivalent of two spores from opposite ends of a *Neurospora* ascus. Homozygous flies (*A/A* or *a/a*) are equivalent to second-division segregation asci, but only half the occurrences of second-division segregation are detected. This is evident from examination of Table 8.2: there are 25 asci (classes V and VI) in which spores 1 and 8 are identical, compared with a total of 39 in which second-division segregation has occurred. Since frequencies of second-division segregation must be halved for use as map distances, homozygosis frequencies, which are equivalent to the half values, can be used directly as map distances between gene loci and the point of junction of the two *X*-chromosomes. This has been done in Fig. 8.11.

Out of the 376 *X*-chromosomes in Anderson's 188 flies, 54 (14·4%) were cross-overs between 'forked and garnet' (classes vii to x, plus twice class iv in Table 8.1). This frequency is close to the difference between the other two (see Fig. 8.11). Thus the recombination and homozygosis data agree,

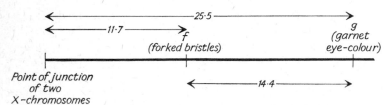

FIGURE 8.11    Linkage map of part of the *X*-chromosome of *Drosophila melanogaster* showing the similarity between the distance apart of the 'forked' and 'garnet' gene loci based on homozygosis frequencies (figures above the line) and that based on recombination frequency (figure below the line). The data are from the analysis of 188 attached-*X* females by Anderson (1925).

and suggest that the point of junction of the two *X*-chromosomes is about 12 units beyond the 'forked' locus. This corresponds well with the position of the terminal gene known (bobbed bristles) in the linkage group (see Fig. 8.10).

Support for the hypothesis of first-division segregation of centromeres could also be obtained from the observations of Carothers (1913) at spermatogenesis in the grasshopper *Brachystola magna*. As mentioned in § 6.9, she found an unequal chromosome-pair, and the inequality always segregated at the first division of meiosis. Carothers concluded that the spindle-fibres became attached at the unequal end at metaphase I (see Fig. 6.4).

Direct evidence for first-division segregation of centromere regions has now been obtained from isotope labelling experiments (see § 16.5).

The discovery that crossing-over takes place, not between paternal and maternal chromosomes, but between individual daughter chromatids, one of paternal and one of maternal origin, constituted strong evidence in support of Janssens's chiasmatype theory. However, the data which led to this discovery provided no information about crossing-over in relation to chiasmata. The next important question, therefore, in testing the truth of Janssens's theory, was to establish whether or not the frequencies and distribution of cross-overs in groups of linked genes were in agreement with the frequencies and distributions of chiasmata in the corresponding chromosomes.

## § 8.7  Genetical interference

Muller (1916) pointed out that the occurrence of one crossing-over in *Drosophila* interferes with the coincident occurrence of another in the same pair of chromosomes, and termed the phenomenon *interference*. For example, Morgan's data for 2205 flies in Table 7.3 indicate 29 cross-overs (1·3%) between the loci for yellow body-colour and white eyes, and 719 (32·6%) between the loci for white eyes and miniature wings. The expected frequency of double cross-overs, if crossing-over occurs at random, is given by the product of these frequencies, or $\dfrac{29}{2205} \times \dfrac{719}{2205} = 0\cdot43\%$. The observed frequency of double cross-overs was $\dfrac{1}{2205}$ or $0\cdot045\%$. Muller suggested that the most useful way to measure the intensity of interference was to divide the observed double cross-over frequency by the expected value (assuming no interference), and so obtain a fraction showing what proportion of the coincidences which would have happened on pure chance really took place. In the present example this ratio is

$$\frac{2205 \times 1}{29 \times 719} = 0\cdot106.$$

He called this quantity the *coincidence*, pointing out that a low value such as this means much interference. He showed that in the *X*-chromosome of *Drosophila* the intensity of interference is very great over short intervals, but falls off with distance, such that for regions 40 units or more apart in the linkage group the coincidence has risen to near 1. Morgan (1919) called the phenomenon of interference the *fifth law of heredity*. Genetical interference has since been found to occur in most organisms that have been investigated. Thus, in *Zea*, from Stadler's data in Fig. 7.4(*a*), the frequency of double cross-overs expected if there is no interference is

$$pq = \frac{1065}{45,832} \times \frac{9141}{45,832},$$

or 212·4 out of the 45,832 progeny. The observed frequency was only 32. Hence, the coincidence is $\dfrac{32}{212\cdot4} = 0\cdot150$.

Stevens (1936) pointed out that for non-adjacent regions of a chromosome, the coincidence had frequently been wrongly estimated. He showed that the best estimate, whether the regions under study were adjacent or not, was given by the standard expression

$$\frac{n \times n_{pq}}{n_p \times n_q}$$

where $n$ is the total number of individuals, and $n_p$, $n_q$, and $n_{pq}$ are, respectively, the numbers of individuals showing a recombination in segment $p$, in segment $q$, and simultaneously in segments $p$ and $q$, but with the important corollary that these numbers must be irrespective of the occurrence of recombination elsewhere. Stevens calculated coincidence values from published recombination data for *Drosophila melanogaster*. Some of his results are given Table 8.5. It is evident that within the one arm of the $X$-chromosome there is strong interference over short intervals, declining to no interference (coincidence 1) at about 40 units separation, as Muller had found. On the other hand, between the two arms of chromosome II he found there was no interference.

TABLE 8.5  Coincidence values obtained by Stevens (1936) from published recombination data for two linkage groups in *Drosophila melanogaster*. The six intervals between gene loci in linkage-group I averaged 9·4 units in length. The interval spanning the centromere in linkage-group II was 3 units long and the other six averaged 17·3 units. The coincidence values in the lower part of the table, and for three and four intervening segments in the upper part, do not differ significantly from unity.

| Region | Number of Intervening Segments | | | | | |
|---|---|---|---|---|---|---|
| | 0 | 1 | 2 | 3 | 4 | 5 |
| Within one arm of linkage-group I (*X*-chromosome) | 0·03 | 0·13 | 0·55 | 1·00 | 1·01 | — |
| | 0·06 | 0·37 | 0·87 | 1·06 | — | — |
| | 0·12 | 0·64 | 0·82 | — | — | — |
| | 0·24 | 0·71 | — | — | — | — |
| | 0·19 | — | — | — | — | — |
| Across the centromere of linkage-group II | — | 1·01 | 1·05 | 0·97 | 1·11 | 0·89 |
| | — | — | 1·00 | 0·97 | 0·97 | — |
| | — | — | — | 0·98 | — | — |

## § 8.8  Chromatid interference

With the discovery that crossing-over is between daughter chromatids and not parent chromosomes, ideas about interference had to be revised. It could no longer be implicitly assumed that interference meant that the position of one exchange between homologous chromosomes affected the position of a neighbouring exchange. Haldane (1931) pointed out that if adjacent cross-overs were usually between different pairs of chromatids, forming a four-strand double cross-over as in Fig. 8.12, a moderate degree of genetical interference was compatible with a random distribution of the points of exchange. This is possible because four-strand double cross-overs lead only to single cross-over progeny, since no chromatid is broken more than once. Mather (1933a) gave the name of *chromatid interference* to the phenomenon of non-random distribution between the chromatids involved in successive cross-overs. There are four possible relationships between the strands concerned in two cross-overs, and they are illustrated in Fig. 8.12. There are two kinds of three-strand double cross-overs, depending on which parent contributes the unbroken strand. If crossing-over is at random as regards the strands involved, the four types will occur with equal frequency, or in other words, the ratio of two-strand:three-strand:four-strand double cross-overs will be as 1:2:1.

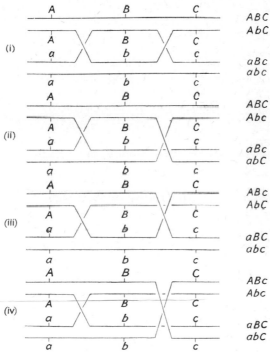

FIGURE 8.12   The four kinds of double cross-overs and the corresponding tetrad genotypes from the cross *ABC* × *abc*. (i)   Two-strand, (ii) and (iii) three-strand, and (iv) four-strand relationship (cf. Pl. 8.1(*c*)).

Emerson and Beadle (1933) were the first to obtain appreciable data concerning the strand-relationship of adjacent cross-overs. They used attached-$X$ *Drosophila melanogaster* heterozygous for several sex-linked characters, and like Anderson (1925) determined the frequencies of the various genotypes in the progeny of a cross. With three or more pairs of alleles segregating simultaneously, it is possible to distinguish between two classes of double cross-overs: ($A$) two-strand and one kind of three-strand, and ($B$) four-strand and the other kind of three-strand. Anderson (1925) found 3 progeny of type $A$ and 4 of type $B$. Emerson and Beadle, from their own data together with unpublished data of Sturtevant, brought these totals to 73 $A$, 62 $B$. The frequencies of the two types do not differ significantly, and the inference was that, as far as the evidence went, there was no chromatid interference in this chromosome.

Tetrad analysis with three or more linked factors provides full information on the strand-relationship of cross-overs. This is evident from the fact that the four tetrad genotypes in Fig. 8.12 are all different. The first data of this kind were obtained by Lindegren with *Neurospora crassa*. Extensive results have been obtained subsequently with this organism by a number of authors, and Bole-Gowda, Perkins and Strickland (1962) quote totals of 423 two-strand, 759 three-strand and 329 four-strand double cross-overs as the pooled results of a number of recent studies. This represents a significant excess of two-strand double cross-overs over four-strand. Data obtained by E. Knapp, E. Möller and W. O. Abel with *Sphaerocarpos donnellii* have been summarized by Abel (1967*b*). When one cross-over occurred in one chromosome arm and one in the other, there were 60 two-strand, 43 three-strand and 15 four-strand double cross-overs, implying a significant excess of two-strand over four-strand and a significant shortage of three-strand relationships. Within chromosome arms, there was a significant excess of four-strand double cross-overs at 10°–20°C (860 two-strand, 1785 three-strand, 995 four-strand), whereas at 22°–28°C two-stranded events were in excess (905 two-strand, 1722 three-strand, 824 four-strand). Evidently, a slight interaction can sometimes occur between one cross-over and another as regards the chromatids involved.

The strands involved in adjacent cross-overs can sometimes be determined cytologically. Some of the most extensive information of this kind was obtained by Brown and Zohary (1955) with *Lilium formosanum*, and is described in § 8.15. It indicated no chromatid interference within a chromosome arm.

## § 8.9  Cross-over position interference

If, as would thus appear, there is usually only slight chromatid interference, genetical interference between cross-overs must often be attributed to the position of one point of exchange interfering with the position of a neighbouring exchange, as indeed was originally supposed. To this phenomenon, Mather (1933*a*) gave the name 'chiasma interference'. Carter and Robertson (1952) suggested that a more explicit title would be 'chiasma position interference'. Both these names are open to the objection that they use a cytological term, chiasma, to describe a genetical phenomenon. In other words, by using these terms one assumes that the chiasmatype theory is true. A more appropriate title would be 'cross-over position interference' or simply *position interference*.

## § 8.10   Chiasma position interference

If chiasmata correspond to cross-overs, as the chiasmatype theory demands, and if cross-overs show position interference, then it is to be expected that chiasmata will also show position interference. Haldane (1931) demonstrated this, using data of Maeda (1930) for chiasma frequencies in the pollen-mother-cells of *Vicia faba*. The 12 somatic chromosomes of this plant comprise five pairs of about equal size and with sub-terminal centromeres, and one pair about twice as long and with the centromere near the middle. Maeda recorded the numbers of chiasmata at metaphase I in 1100 of the five smaller chromosome-pairs (which were not individually distinguishable) and in 1057 of the large one. The mean chiasma frequencies were 3·5 in the short chromosomes and 8·1 in the long one. Haldane drew attention to the distribution of chiasma frequencies about these mean values. In the short chromosomes, a majority had three or four chiasmata and the range was from 1 to 6, while in the long one there was a great concentration at 7 to 9 chiasmata and the total range was from 3 to 13. He pointed out that if the distribution of chiasmata was at random along the chromosomes, the frequencies of chromosomes with different numbers of chiasmata would fit a Poisson distribution, in which the variance is equal to the mean. Analysis of Maeda's data showed that for both the short chromosomes and the long one there was much less variation in chiasma frequency than would be expected on the Poisson distribution. Thus, for the short chromosomes the variance was only about one quarter of the mean and for the long one about one third. These differences from random chiasma distribution were highly significant and indicated that the occurrence of one chiasma was greatly reducing the likelihood of another in its vicinity. Haldane obtained similar results from the analysis of less extensive data for several other flowering plants.

Mather (1933*a*) made the converse analysis. He took published recombination data for linkage groups II and III of *Drosophila melanogaster*, assumed there was no chromatid interference and that chiasmata corresponded to points of crossing-over between two chromatids, and determined the chiasma distribution that would result. He found that the *Drosophila* genetical data would lead to a chiasma distribution in which the variance was $\frac{1}{5}$ to $\frac{1}{4}$ of the mean, in good agreement with the cytological data from Angiosperms.

## § 8.11   Chiasma frequencies and cross-over frequencies

The pattern of distribution of chiasmata in chromosomes thus appears to be in agreement with the hypothesis that chiasmata correspond to the points of crossing-over. However, if the chiasmatype theory is true, one would also expect to find correspondence between chiasma frequencies and cross-over frequencies in individual chromosomes. Morgan, Sturtevant, Muller and Bridges (1915) reported that mutant genes in *Drosophila melanogaster* fell into four linkage groups. Within groups there were various intensities of partial linkage, while between groups parental and non-parental combinations were statistically equal in frequency. These findings agreed with the observation that the haploid chromosome number was 4. Morgan (1919) called this equality between numbers of linkage groups and of chromosomes the *sixth law of heredity*. Although at the time established only for *Drosophila*, later work has confirmed the truth of this law for many organisms including for example *Zea mays* with 10 linkage groups and chromosomes, and *Neurospora crassa* with seven of each. Morgan *et al.* (1915) also found that the number of mutants in each *Drosophila* linkage group was roughly proportional to the physical size of the corresponding chromosome.

Darlington (1934*b*) showed that there was agreement between chiasma and cross-over frequencies for each chromosome of *Zea mays*. He obtained a mean total chiasma frequency of 27·1 for the ten chromosome-pairs in the pollen-mother-cells. Distributing these chiasmata amongst the chromosomes in proportion to their lengths, using more recent estimates of the latter from Rhoades (1950), has given the figures in the third column of Table 8.6. However, Darlington found evidence that the shorter chromosomes had relatively more chiasmata in relation to their length, as indicated in the caption to the table. He compared the estimated chiasma frequencies with

TABLE 8.6   Comparison of chiasma and cross-over frequencies in *Zea mays*. The data in columns 2 and 4 are from Rhoades (1950), and the total chiasma frequency is from Darlington (1934*b*), who found evidence that the short chromosomes have slightly more chiasmata in relation to their length. His estimates of the chiasma frequencies of chromosomes 1, 2, 9 and 10 would approximate to 3·8, 3·2, 2·2 and 2·0, respectively.

| Chromosome Number | Length in $\mu$m at Mid-Pachytene | Estimated Mean Chiasma Frequency if Proportional to Length | Mean Cross-over Frequency of Mapped Portion |
|:---:|:---:|:---:|:---:|
| 1 | 82 | 4·0 | 3·1 |
| 2 | 67 | 3·3 | 2·6 |
| 3 | 62 | 3·0 | 2·4 |
| 4 | 59 | 2·9 | 2·2 |
| 5 | 60 | 2·9 | 1·4 |
| 6 | 49 | 2·4 | 1·3 |
| 7 | 47 | 2·3 | 1·9 |
| 8 | 47 | 2·3 | 0·6 |
| 9 | 43 | 2·1 | 1·4 |
| 10 | 37 | 1·8 | 1·1 |
| Totals | 553 | 27·0 | 18·0 |

the cross-over frequencies derived from published linkage data. More extensive data from Rhoades (1950) are now available and are given in the fourth column of the table. With crossing-over occurring between only two of the four chromatids, one exchange will produce 50% of cross-over gametes and 50% with the parental gene combinations. Hence 50 units on the linkage map is equivalent to an average of one exchange per meiosis in that interval of the chromosome. The figures in column 4 have therefore been obtained by taking the total length of each linkage map and dividing it by 50. In every instance the frequencies of crossing-over are slightly less than the chiasma frequencies (on average two thirds of them), but this is to be expected since it is unlikely that genes have been mapped at the extreme ends of the chromosomes. This agreement with expectation gives further support to the view that chiasmata correspond to the points of crossing-over, and hence supports the chiasmatype hypothesis.

The probability that a newly-obtained mutant will show linkage to a previously known mutant will depend on the chromosome number and the cross-over frequency for that organism. Carter (1955) has shown how total cross-over frequencies per meiosis can be estimated from such linkage data, and has applied the method to data for *Mus musculus* (Mouse). Of 33 mutants, 9 were due to genes situated within 25 units of a previously discovered mutant, from which the total cross-over frequency per meiosis was estimated to be approximately 34. This estimate did not differ significantly from the observed total chiasma frequency for the 20 chromosome pairs, which was approximately 38.

The evidence presented so far in this chapter has established that crossing-over occurs between two of the four chromatids; that these chromatids appear to be chosen more or less at random at each cross-over; that paternal and maternal centromeres segregate from one another at the first division of meiosis; and that chiasmata and points of crossing-over show similar frequencies of occurrence and corresponding intensities of interference. All these findings are in agreement with Janssens's chiasmatype theory, but nevertheless do not constitute convincing evidence of the truth of the theory.

## § 8.12   Belling's modification of Janssens's theory

Janssens's theory, although favoured almost at once by geneticists, received no support from other cytologists for nearly 20 years. They criticized it on the grounds that they could find no evidence for the breakage and rejoining of threads at the early diplotene stage, such as he had claimed to have seen. However, Belling (1928a), after studying the meiotic chromosomes in various flowering plants, proposed a modification of Janssens's theory which removed this criticism. He suggested that the exchange between chromatids took place at the preceding pachytene stage, while the homologous chromosomes were closely associated. Exchange at this stage could not be directly observed, but the chiasmata seen subsequently would be the consequence of these exchanges, if one accepted, as Janssens had done, that throughout the length of each chromosome of the bivalent association, the pairing at the diplotene stage

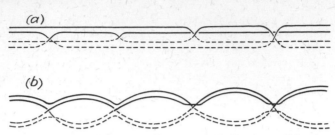

FIGURE 8.13   Diagrams to illustrate Belling's modification
of Janssens's hypothesis of exchange between
homologous chromosomes. Chromatid segments derived from one parent are shown with
a broken line and those derived from the
other parent by an unbroken line. (*a*) Pachytene stage. (*b*) Diplotene stage.

was always between daughter-chromatids derived from the same parent
chromosome (Fig. 8.13). Belling (1929) gave two primary reasons for
believing that the diplotene pairing was of this kind.

## § 8.13   Unequal chromosome pairs

Firstly, whenever homologous chromosomes differed in appearance at one
end, as had been observed by Carothers (1913), Wenrich (1916) and others
in various Orthoptera, and by Belling himself in *Aloe purpurascens*, the two
short chromatid-ends were always associated with one another at the
diplotene stage and similarly with the two long ones (Fig. 8.14(*a*)). This was
true whether or not the centromere was close to the end where the difference
occurred. Darlington (1932, pp. 274–277) took this argument a stage further.
He pointed out that since on the chiasmatype hypothesis sister-chromatids
are always associated at the diplotene stage, the loop or pair of arms containing the centromeres will have two paternal chromatids associated at one
centromere region and two maternal at the other. Since these centromere
regions move to opposite poles at anaphase I, paternal and maternal
centromeres will always show first-division segregation, just as Bridges &
Anderson (1925) had postulated on genetic evidence. If there is no chiasma
between the centromere and a region of the chromosome where the homologues differ, then this difference will segregate at the first division of meiosis,
as found by Carothers (1913) with *Brachystola magna*. On the other hand, a
single chiasma between centromere and inequality would lead to second-
division segregation for the region of difference.

An example of regular second-division segregation for an inequality was
found by Wenrich (1916) at spermatogenesis in the grasshopper *Phrynotettix
tschivavensis*. Chromosome-pair *B* was found in 11 individuals to show
heterozygosity for the presence and absence of a large terminal granule.
Wenrich observed a single medianly-placed chiasma in each *B* chromosome-
pair. Since the two chromatid-ends with the granule were always associated
with one another, and similarly with the two without the granule, and since

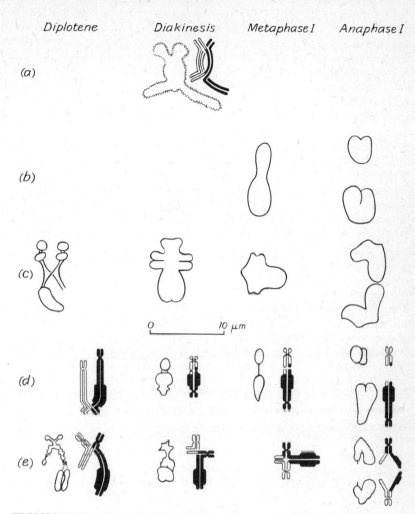

FIGURE 8.14   Unequal chromosome-pairs at stages of the first division of
meiosis.
(a) *Aloe purpurascens* (from Belling, 1928c).
(b) and (c) *Phrynotettix tschivavensis* chromosome-pair *C*
(from Wenrich, 1916).
(d) and (e) *Calliptamus palaestinensis* (from Nur, 1961).
(b) and (d) First-division segregation for the inequality.
(c) and (e) Second-division segregation for the inequality.
The diagrams to the right of the drawings in (a), (d) and
(e) show the interpretation of the observed structures
according to the chiasmatype hypothesis, with the black
and white segments indicating the parental origin.

Wenrich states that he found no indication of breaking or recombining of the parts of chromatids, he must have assumed that one paternal and one maternal centromere went to each pole at the first division. On the other hand, Darlington suggested there was first-division segregation of centromeres and that crossing-over had occurred at the position of the chiasma.

In two individuals of *Phrynotettix*, Wenrich found a similar inequality in chromosome-pair *C*, but here it was variable in behaviour (Fig. 8.14(*b*) and (*c*)). Out of 928 cells which he examined at metaphase I, 472 showed first-division segregation for the inequality and the remaining 456 (49·2%) showed second-division segregation. As with chromosome-pair *B*, the metaphase configuration was cross-shaped, implying a single medially-placed chiasma. Wenrich suggested that the centromeres were situated at the end of the chromosome showing the inequality when this segregated at the first-division of meiosis, and that they moved to the other end with second-division segregation. Darlington put forward the suggestion that the centromere was fixed in position, that it was medially placed, that the chiasma might occur in either arm, and that crossing-over preceded its appearance. If crossing-over occurred in the opposite arm to the inequality, first-division segregation of the latter would occur, and if in the same arm there would be second-division segregation. However, the centromere appeared to be subterminal and not median in position.

Nur (1961) has offered a solution to this dilemma. He discovered in another short-horned grasshopper, *Calliptamus palaestinensis*, that an inequality in a chromosome-pair, although often appearing terminal, was really interstitially placed (Fig. 8.14(*d*) and (*e*)). The centromere was sub-terminal in position and at meiosis in the male a single chiasma occurred regularly and appeared always to be in the long arm, just as in chromosome-pair *C* of *Phrynotettix* studied by Wenrich. Out of 410 cells in *Calliptamus* at the diplotene or diakinetic stage, Nur found that the chiasma appeared to be proximal to the inequality in 379 (92·4%) and distal to it in the remaining 31. Examination of 69 cells at anaphase I showed 63 (91·3%) with second-division segregation for the inequality and 6 with first-division segregation, in agreement with the observations at the earlier stage. It is presumed that the inequality in chromosome-pair *C* of *Phrynotettix*, studied by Wenrich (1916), was also interstitially placed although appearing terminal, and that chiasmata occurred proximal and distal to it equally often.

Several authors have made quantitative studies with unequal chromosome-pairs showing variability of behaviour like Wenrich's pair *C*. Their results are given in the upper part of Table 8.7. In each study the interval between centromere and region of inequality was short and never showed more than a single chiasma. With the exception of Haga's data for *Paris verticillata*, there was no significant difference (contingency $\chi^2$ value, for one degree of freedom, less than 4—see last column of Table 8.7) between the chiasma frequency observed at diakinesis or metaphase I, and the second-division segregation frequency observed at anaphase I or II. A remarkable feature of meiosis in *P. verticillata* is the regular occurrence of chiasmata close to the centromere in either or both arms of each chromosome.

It is therefore exceptionally difficult material in which to record the frequency of cells with a chiasma between the centromere and the inequality. In Haga's data, 667 cells had a chiasma in the unequal arm and 333 were thought to show chiasmata only in the other arm, but owing to the close proximity of the centromere to these chiasmata it is probable that some of them were in fact in the unequal arm. Hence the excess of second-division segregations (82·25%) over chiasmata (66·7%) is expected.

The data in the lower part of Table 8.7 refer to individuals heterozygous for the presence and absence of an exchange of part of an arm between two different (that is, non-homologous) chromosomes. Such individuals will have two chromosome-pairs each with an unequal arm. If the two chromosomes and their exchanged portions are similar in appearance, it may not be possible to recognize them individually, as was found by Kayano (1960*a*) with *Disporum sessile* and by Noda (1961) with *Scilla scilloides*. In any event, in such interchange heterozygotes the recognition of the unequal pairs at diakinesis or metaphase I requires skill, since the two pairs will commonly be held together by chiasmata to form a quadrivalent association. As with the simple inequalities discussed in the last paragraph, the intervals between centromere and inequality were short and except in chromosome I of *Acrida lata* (see table) never showed more than one chiasma. Without exception, no significant differences were found between the chiasma frequencies and the corresponding second-division segregation frequencies.

On the chiasmatype hypothesis, paternal and maternal centromeres segregate at the first division of meiosis, and the segregation of a chromosomal inequality at the second division can only be accounted for by postulating crossing-over between the centromere and the region of difference. Hence the finding that second-division segregation frequencies are statistically equal to chiasma frequencies is equivalent to finding that crossing-over and chiasma frequencies are equal, and so provides strong support for the chiasmatype theory. Furthermore, the invariable occurrence of pairing at late prophase and metaphase I between the two long segments, and similarly with the two short segments making up the inequality, even when a single chiasma has occurred between the inequality and the centromere, shows that these chiasmata are not merely points where the four threads have changed partners, but that breakage and crossing-over have occurred.

Inequalities in diplotene chromosome-pairs in oocytes of *Triturus cristatus* (Crested Newt) observed by Callan and Lloyd (1956) have provided direct evidence in support of the chiasmatype theory. The chromosomes at this stage are greatly elongated and have a remarkable 'lampbrush' structure (see § 17.6 and Pl. 17.1(*b*)), in which the two chromatids of which they are composed extend individually into lateral loops. The four chromatids in a chromosome-pair usually showed loops of similar appearance at corresponding positions, but in each meiotic cell in certain individuals a particular loop was absent from two of them. It was evident that these individuals were heterozygous for a difference at this point in the chromosome. The significant observation was that the two chromatids which possessed this particular loop were paired, and similarly with the two without the loop. A number of

TABLE 8.7   Frequencies of cells with and without a chiasma between the centromere and an inequality in a chromosome-pair, and frequencies of first- and second-division segregation for the inequality. In the right-hand column the contingency $\chi^2$ values are given from a comparison of the chiasma and segregation frequencies. $\chi^2$ values of less than about 4 indicate no significant difference between chiasma frequencies and second-division segregation frequencies.

| Type of Structural Heterozygote | Organism and Reference | Chromosome Pair | Diakinesis or Metaphase I | | | | Anaphase I or II | | | | |
|---|---|---|---|---|---|---|---|---|---|---|---|
| | | | Without Chiasma | With Chiasma | Total | Percentage with Chiasma | 1st Division Segregation | 2nd Division Segregation | Total | Percentage 2nd Division Segregation | $\chi^2$ $n=1$ |
| | *Mesocricetus auratus* Koller, 1938 | *X & Y* | 206 | 47 | 253 | 18·6 | 87 | 24 | 111 | 21·6 | 0·46 |
| | *Paris verticillata* Haga, 1944 | *D* | 333 | 667 | 1000 | 66·7 | 355 | 1645 | 2000 | 82·25 | 91·3 |
| Simple inequality | *Lilium formosanum* Brown & Zohary, 1955 | *A\** | 183 | 425 | 608 | 70·0 | 69 | 172 | 241 | 71·4 | 0·18 |
| | | | 114 | 120 | 234 | 51·3 | 67 | 81 | 148 | 54·7 | 0·43 |
| | | *I\** | 21 | 150 | 171 | 87·7 | 15 | 123 | 138 | 89·1 | 0·15 |
| | | | 72 | 273 | 345 | 79·1 | 43 | 149 | 192 | 77·6 | 0·17 |
| | *Allium fistulosum* Zen, 1961 | *S* | 151 | 618 | 769 | 80·4 | 61 | 237 | 298 | 79·5 | 0·087 |

| | | | | | | | | | | |
|---|---|---|---|---|---|---|---|---|---|---|
| *Lilium maximowiczii* Noda, 1960 | $J_1$ | 490 | 110 | 600 | 18·3 | 981 | 219 | 1200 | 18·25 | 0·0019 |
| | $J_2$ | 600 | 0 | 600 | 0 | 1200 | 0 | 1200 | 0 | 0 |
| *Disporum sessile* Kayano, 1960a | A ⎱ B | 1549 | 651 | {1100, 1100} | Mean 29·6 | 1987 | 853 | {1420, 1420} | Mean 30·0 | 0·12 |
| *Acrida lata* Kayano & Nakamura, 1960 | I | 0 | 50† | 50 | 97·0† | 2 | 28 | 30 | 93·3 | 0·60 |
| | II | 9 | 41 | 50 | 82·0 | 3 | 27 | 30 | 90·0 | 0·94 |
| *Allium fistulosum* Zen, 1961 | S | 59 | 208 | 267 | 77·9 | 17 | 50 | 67 | 74·6 | 0·33 |
| | F | 138 | 129 | 267 | 48·3 | 29 | 25 | 54 | 46·3 | 0·073 |
| *Scilla scilloides* Noda, 1961 | a ⎱ b | 1128 | 2820 | {1974, 1974} | Mean 71·4 | 278 | 678 | {478, 478} | Mean 70·9 | 0·097 |

Non-homologous interchange

*The two sets of data for chromosome A of *Lilium formosanum* refer to different individuals heterozygous for the same chromosomal deficiency, while those for chromosome I refer to the same individual in different years.

†In chromosome I of *Acrida lata* there were 47 cells with single and 3 with double chiasmata. Assuming there was no chromatid interference, only half the double chiasmata would lead to second-division segregation.

such differences between homologues were observed, and in each instance it was evident that sister-chromatids were paired at the diplotene stage, as expected on the chiasmatype theory.

## § 8.14   Polyploids

Belling's second main reason for believing that the diplotene pairing was between sister-chromatids was to account for configurations of three homologous chromosomes which he had observed at meiosis in triploid horticultural varieties of *Hyacinthus orientalis*. Belling (1929) and Darlington (1929*b*) discovered with triploid *Hyacinthus* and Newton and Darlington (1929) with triploid *Tulipa* that although for each chromosome all three homologues came together at the zygotene stage of meiosis, only two showed close pairing in any one region of the chromosome (cf. § 6.4). However, at intervals there were changes of partner. Thus a typical pachytene trivalent association is like that shown diagrammatically in Fig. 8.15(*a*) with two changes of pairing partner. Belling thought that the chiasmata seen at the diplotene stage in a trivalent chromosomal association could not have arisen by an exchange of pairing partner, but required crossing-over to explain their origin.

Here also the argument was taken further by Darlington (1930) who demonstrated with polyploid *Hyacinthus* that a chiasma was the consequence of previous crossing-over. This demonstration was based on the observation

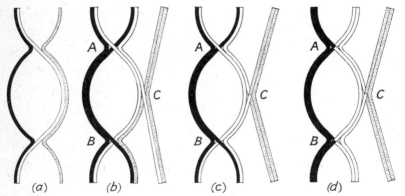

FIGURE 8.15   Diagrams of a trivalent chromosomal association to illustrate Darlington's demonstration that crossing-over must
precede chiasma-formation.
(*a*) Pachytene stage.
(*b–d*) Alternative explanations of diplotene configuration
with three chiasmata.
(*b*) With all chiasmata due to changes of pairing partner
without breakage (classical hypothesis).
(*c*) With chiasmata *A* and *B* classical, but *C* due to
breakage and crossing-over (chiasmatype hypothesis).
(*d*) With all chiasmata due to crossing-over (chiasmatype
hypothesis).

that a chromosome may show two chiasmata $A$ and $B$ (see Fig. 8.15(*b*)–(*d*)) with one homologue, and a third chiasma $C$ with another homologue at a point between $A$ and $B$. To account for such a configuration it is necessary to assume that at the earlier pachytene stage there were changes of pairing partner as in Fig. 8.15(*a*). On this assumption, chiasmata $A$ and $B$ must involve the same four chromatids, and therefore chiasma $C$ must involve breakage and crossing-over (see Fig. 8.15(*b*) and (*c*)). Critical evidence is not available for chiasmata $A$ and $B$ but there is no reason to suppose they differ from $C$. Such intermediate chiasmata with a third homologue are of frequent occurrence at meiosis in polyploid organisms. Since the pachytene association in all polyploids appears to be strictly confined to pairs over any given region of a chromosome, it follows that all such intermediate chiasmata with a third homologue must be the consequence of previous crossing-over.

## § 8.15  Chromosomal inversions

A third important source of critical information about chiasmata in relation to crossing-over followed from the discovery by McClintock (1931) that if a segment of a chromosome has become relatively inverted, individuals heterozygous for such an inversion show reversed pairing at the pachytene stage. Following *X*-irradiation of pollen, she obtained a plant of *Zea mays* heterozygous for a long inversion in chromosome 2. It occupied parts of both arms and included about two-thirds of the total length of the chromosome. Characteristic loops were formed at the pachytene stage in the pollen-mother-cells, with the two homologous chromosomes passing along the loop in opposite directions, such that homologous parts were in association (see Fig. 8.16(i) and (ii)). Later (McClintock, 1933) in untreated *Zea* she found a plant heterozygous for an inversion within the short arm of chromosome 8 and she showed that crossing-over within the inversion led to the production of a chromatid with two centromeres (dicentric) and another with none (acentric) (see Fig. 8.16(v)). This is true, provided, as in this instance, the inversion is paracentric, that is, the centromere does not lie within it. The dicentric chromatid formed a bridge joining the two chromosomes of the pair as they began to move apart at anaphase I and prevented their separation until it broke. The acentric chromatid appeared as a chromosomal fragment at the side of the spindle, where it lagged behind at anaphase I. In this plant heterozygous for the inversion in chromosome 8, she observed 25 cells with bridge and fragment at anaphase I out of 281 examined, whereas 123 cells from a control plant without an inversion had none.

These observations opened up the possibility of a direct comparison between chiasma frequencies within an inversion and cross-over frequencies in the same inversion, manifest as bridge and fragment at anaphase I. In making such a comparison it is necessary to take into account the consequences of double crossing-over. In her *Zea* plant heterozygous for the chromosome 8 inversion, McClintock (1933) found one instance of the chromosome-pair at anaphase I joined by two dicentric bridges and with two acentric fragments alongside. She explained this as due to a four-strand

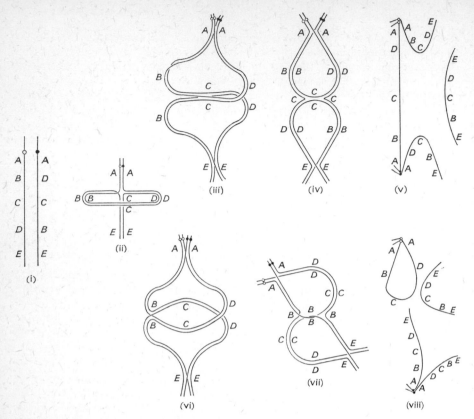

FIGURE 8.16   Diagrams to show stages of meiosis in a chromosome-pair heterozygous for an inversion in the middle of one arm. *A B C D E* represent different regions of the chromosome arm. The normal sequence
is alphabetical, but in the inversion the regions *B*, *C* and *D* are
reversed to give the sequence *A D C B E*. A hollow circle represents a
centromere derived from one parent and a solid circle from the other.

(i)     Leptotene stage.
(ii)    Pachytene stage with pairing in a loop such that homologous
        regions are associated.
(iii) and (iv) Diplotene stage with chiasmata at *A*, *C* and *E*. The
        chiasma at *C* is in the inversion and is of characteristic
        appearance, best seen in (iv). The cross-over at *C*
        involves the same two strands (or the other two) as
        that at *A*.
(v)     Anaphase I, with dicentric bridge and acentric fragment, such
        as would result from the configurations shown in (iii) and (iv).
(vi) and (vii) As (iii) and (iv), but the chiasma in the inversion is
        at *B* instead of *C*, giving an asymmetrical configura-
        tion, and the cross-overs at *A* and *B* involve three
        strands altogether instead of two or four.
(viii) Anaphase I, with dicentric loop and acentric fragment, such
        as would result from the configurations shown in (vi) and (vii).
The dicentric and acentric chromatids are shown interlocked at the
reversed chiasma in (vii), but not in (iv). Whether or not they will
be interlocked is thought to be a matter of chance.

126

double cross-over within the inversion. Smith (1935) found a wild plant of *Trillium erectum* which was heterozygous for a short inversion near the middle of the long arm of chromosome *D*. In about 9·5% of the pollen-mother-cells at metaphase I or anaphase I, a dicentric bridge and an acentric fragment about 6 $\mu$m long were present. In a further 2·5% of the mother-cells the fragment was present without the bridge, but then a dicentric loop could be seen at metaphase I, which gave a bridge at anaphase II. He explained the first-division loop by postulating one cross-over within the inversion and a second cross-over between the inversion and the centromere, such that one chromatid was involved in both cross-overs, that is, there was a three-strand double cross-over (see Fig. 8.16(vi)–(viii)). The proximal cross-over, that is, the one between inversion and centromere, then has the effect of converting the first-division bridge into a loop.

Smith's observations form the basis of a cytological method, referred to earlier in this chapter, of testing for chromatid interference. With no chromatid interference, a cross-over in a paracentric inversion and another proximal to it will give rise to equal frequencies of bridges at anaphase I (two- or four-strand relationships) and anaphase II (three-strand relationships between the two cross-overs). The marked excess of first-division bridges (9·5%) over loops (2·5%) in Smith's data for *Trillium erectum* may be due to the occurrence of cross-overs within the inversion without a proximal cross-over, rather than to an excess of two- and four-strand relationships over three-strand. On the other hand, Brown and Zohary (1955) following *X*-irradiation of pollen, obtained plants of *Lilium formosanum* heterozygous for a long inversion in one arm of chromosome *H*, where chiasmata in the inversion were regularly associated with proximal chiasmata in the same arm. Thus, 278 pollen-mother-cells from two individuals were examined at diakinesis and every one of the 252 cells containing chiasmata within the inversion also had a single proximal chiasma. The same number of anaphase I cells was counted and showed 131 cells with a dicentric bridge and 121 with a dicentric loop, in excellent agreement with the 1:1 ratio of two-strand plus four-strand: three-strand double cross-overs expected in the absence of chromatid interference (deviations 5, $\chi^2 = 0·40$, $n = 1$, $P$ is just over 0·5).

Crossing-over within a paracentric inversion in heterozygous form leads to other peculiarities in addition to the anaphase bridge or loop and fragment. Darlington (1936b), from studies of the chromosomes at spermatogenesis in the grasshoppers *Chorthippus parallelus* and *Stauroderus bicolor* showed that a cross-over near either end of the inversion gives an asymmetrical configuration at the diplotene and diakinetic stages (Fig. 8.16(vi) and (viii)), and Brown and Zohary (1955) drew attention to the characteristic morphology of the *reversed chiasmata* following crossing-over within an inversion. These chiasmata may or may not have the dicentric and acentric chromatids interlocked, depending on how the diplotene loops open out (see Fig. 8.16 (iv) and (vii)). In *Lilium formosanum* they could rarely distinguish a reversed chiasma with interlocked chromatids from a normal chiasma, but numerous examples were seen of reversed chiasmata where the chromatids were not interlocked, and also of asymmetrical configurations. The observation of

TABLE 8.8  Comparison of chiasma frequencies at diakinesis in an inversion in chromosome *H* of *Lilium formosanum* and fragment frequencies at anaphase *I*, from the data of Brown and Zohary (1955). The data in (*a*) and (*b*) relate to two different plants.

(*a*)

| | Diakinesis | | Expected Percentages of Cells with 0, 1 and 2 Fragments at Anaphase *I* | | |
|---|---|---|---|---|---|
| No. of Chiasmata | Observed Nos. of Cells (Total 145) | Percentage Cells | 0 | 1 | 2 |
| 0 | 17 | 11·7 | 11·7 | — | — |
| 1 | 67 | 46·2 | — | 46·2 | — |
| 2 | 61 | 42·1 | 10·5 | 21·0 | 10·5 |
| Anaphase I { Expected percentages | | | 22·2 | 67·2 | 10·5 |
| Expected numbers of cells (total 121) | | | 26·9 | 81·3 | 12·8 |
| Observed numbers of cells (total 121) | | | 21 | 80 | 20 |

$$\chi^2 = 5·5,\ n = 2,\ P = 0·1\text{--}0·05$$

(*b*)

| | Diakinesis | | Expected Percentages of Cells with 0, 1 and 2 Fragments at Anaphase I | | |
|---|---|---|---|---|---|
| No. of Chiasmata | Observed Nos. of Cells (Total 133) | Percentage Cells | 0 | 1 | 2 |
| 0 | 9 | 6·8 | 6·8 | — | — |
| 1 | 55 | 41·3 | — | 41·3 | — |
| 2 | 69 | 51·9 | 13·0 | 25·9 | 13·0 |
| Anaphase I { Expected percentages | | | 19·8 | 67·2 | 13·0 |
| Expected numbers of cells (total 197) | | | 38·8 | 132·6 | 25·6 |
| Observed numbers of cells (total 197) | | | 45 | 123 | 29 |

$$\chi^2 = 2·1,\ n = 2,\ P = 0·1\text{--}0·5$$

reversed chiasmata and of the chiasmata responsible for asymmetrical configurations provide direct support for the chiasmatype theory, since, like the intermediate chiasmata in trivalent associations discussed earlier, crossing-over must have preceded their formation.

The direct comparison between chiasma frequencies in a paracentric inversion, observed at diakinesis, and bridge (or loop) and fragment frequencies, observed at anaphase I, has been made by Brown and Zohary

(1955) with *Lilium formosanum*. Their data show an excellent agreement between chiasma and cross-over frequencies. In one plant of *L. formosanum* heterozygous for the chromosome *H* inversion 145 cells were examined at diakinesis. There were 17 without chiasmata in the inversion, 67 with one chiasma and 61 with two chiasmata. On the chiasmatype hypothesis and assuming there was no chromatid interference, two chiasmata within the inversion would be the consequence of two-strand, three-strand and four-strand double cross-overs in the ratio $1:2:1$. The two-strand double cross-overs would give no bridges and fragments, as though without crossing-over in the inversion, since the two cross-overs would in effect cancel out. The three-strand double cross-overs would give one bridge and fragment, and the four-strand double cross-overs would give two bridges and fragments. Hence, the expected frequencies of cells with 0, 1 and 2 fragments at anaphase I can be calculated (see Table 8.8($a$)). It was found that there was agreement between observed and expected numbers of fragments at anaphase I. A second plant also showed agreement: see Table 8.8($b$). Thus Brown and Zohary's data provide strong support for the chiasmatype theory.

## § 8.16   Additional cytological evidence for the chiasmatype theory

Critical cytological evidence for the occurrence of chiasmata as a consequence of previous crossing-over is available from several additional sources, but unlike the three main lines of evidence discussed above, these others have only rarely been observed, or have not been subjected to sufficient quantitative analysis. Darlington (1931$c$) reported figure-of-eight shaped chromosomal configurations at meiosis in *Oenothera* species, in which the chiasma at the centre of the '8' could only be interpreted as due to previous crossing-over between a pair of chromosomes which were heterozygous at both ends for interchanges of segments with non-homologous chromosomes. Mather (1933$b$) described an instance where two of the 12 chromosome-pairs at meiosis in a pollen-mother-cell of *Lilium regale* were interlocked such that a medially placed chiasma could only have arisen as a consequence of previous crossing-over, and in *L. henryi* and *L. japonicum* he several times observed (Mather 1935) a chiasma between a chromosome fragment and a normal chromosome, again implying previous crossing-over. Darlington (1935) observed coiling of chromosomes round one another on each side of a chiasma at the diplotene stage of meiosis in a species of *Fritillaria*, and deduced that the chiasma must have been the consequence of crossing-over, but this argument is valid only on the assumption that the coiling of the chromosomes round each other had arisen before they divided into chromatids.

Kayano (1960$b$) obtained an individual of *Disporum sessile* heterozygous for structural differences at both ends of the long arm of chromosome *A*: the presence or absence of an interchange with chromosome *B* at one end, and the presence or absence of a terminal knob (satellite) at the other. Study of this chromosome in 1100 pollen-mother-cells at metaphase I of meiosis showed 537 (48·8%) without chiasmata, 381 with 1 chiasma, 163 with 2 chiasmata, and 19 with 3 chiasmata. Assuming the chiasmatype hypothesis is true, and that there was no chromatid interference, the expected frequencies can be calculated of anaphase II cells in which the 4 chromosome $A$'s resembled those of the parent, or in which 2 of them showed recombination of the characters by which their ends differed and 2 did not, or in which all 4 showed recombination (see Table 8.9). The observed numbers of these

TABLE 8.9   Comparison of chiasma frequencies at metaphase I in the long arm of chromosome $A$ of *Disporum sessile* with the frequencies of different combinations at anaphase II of structural features at each end of the arm, from the data of Kayano (1960*b*).

| Metaphase I | | | Expected Percentages of Cells with Different Combinations at Anaphase II of the 4 Copies of Chromosome $A$ | | |
|---|---|---|---|---|---|
| No. of Chiasmata | Observed Nos. of Cells (Total 1100) | Percentage Cells | All 4 Parental | 2 Parental 2 Recombined | All 4 Recombined |
| 0 | 537 | 48·8 | 48·8 | — | — |
| 1 | 381 | 34·6 | — | 34·6 | — |
| 2 | 163 | 14·8 | 3·7 | 7·4 | 3·7 |
| 3 | 19 | 1·8 | 0·2 | 1·4 | 0·2 |
| Anaphase II { Expected percentages | | | 52·7 | 43·4 | 3·9 |
| Expected numbers of cells (total 80) | | | 42·2 | 34·7 | 3·1 |
| Observed numbers of cells (total 80) | | | 43 | 34 | 3 |

$$\chi^2 = 0·034, n = 2, P = 0·99\text{--}0·98$$

3 classes of anaphase II cells were in close agreement—indeed alarmingly close agreement—with the expected values (see table). The expected frequency of recombinant chromosome $A$'s is $\frac{1}{2}(100 - 48·8)\% = 25·6\%$, and the observed number in 1671 chromosome $A$'s examined at anaphase II was 419 or 25·1%, again in very close agreement ($\chi^2 = 0·23, n = 1, P = 0·9\text{--}0·5$).

## § 8.17   The hypothesis of crossing-over by breakage of classical chiasmata

The observations discussed above provide overwhelming evidence for a connection between chiasmata and crossing-over, but before drawing the conclusion that crossing-over is the cause of a chiasma, it is necessary to consider another hypothesis which relates chiasmata to crossing-over. Wenrich (1916), although not favouring breakage of chromosomes, thought that at the point where two chromatids cross each other in a chiasma, which he regarded merely as a point where chromatids change partners, a weakness of the strands might cause them to break and then recombine to give cross-over chromatids. This hypothesis of crossing-over through breakage of classical chiasmata has been revived on many occasions. It is open to the objection advanced by Muller (1916) against Janssens's theory, namely, that the precision needed in crossing-over is difficult to account for if the breakage and rejoining occur when the chromosomes are relatively short and thick, as at the diplotene stage. Indeed this difficulty is still more serious if the breakage occurs even later, as Wenrich's theory would suppose, since at diakinesis the chromosomes reach their maximum state of contraction. However, the theory appeared to find support from a reduction in the number of chiasmata between the diplotene stage and metaphase I in many organisms.

## § 8.18   Terminalization of chiasmata

This support was removed when Darlington (1931a), following a study of meiosis in *Primula sinensis*, showed that the reduction in numbers could be accounted for by postulating that the chiasmata moved to the ends of the

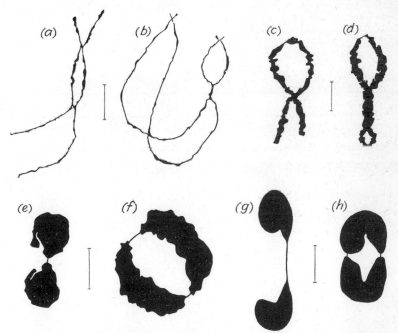

FIGURE 8.17   Drawings of chromosomes of *Primula sinensis* to show terminalization of chiasmata (from Darlington, 1931a). The vertical lines represent 1 μm.
(a) and (b) Early diplotene stage.
(c) and (d) Late diplotene stage.
(e) and (f) Diakinesis.
(g) and (h) Metaphase I.

chromosomes and replaced one another there, without breakage (see Fig. 8.17). This process of *terminalization,* as he called it, is found in some organisms, particularly those with small chromosomes, but not in others. In *P. sinensis* Darlington found from 2 to 5 interstitial chiasmata in each diplotene chromosome-pair, but at diakinesis and metaphase I all the chiasmata had terminalized and hence there were only two per bivalent association (or only one if no chiasmata had occurred in one arm). On the chiasmatype theory, terminalization implies that the position of the chiasmata will no longer coincide with the points of crossing-over (Fig. 8.18).

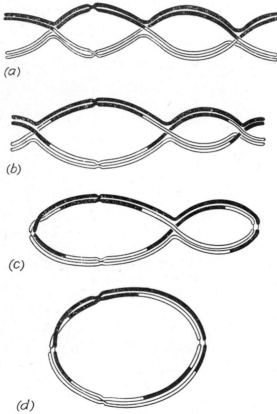

(a)

(b)

(c)

(d)

FIGURE 8.18   Diagram to show terminalization of chiasmata. The constrictions indicate the centromeres, and the shading the parental origin.

(a)–(d) Successive stages from early diplotene to diakinesis, showing how three interstitial chiasmata become two terminal chiasmata. The points where crossing-over has taken place remain unchanged.

The acceptance of the notion of chiasma movement led to the final discrediting of the idea that the chromosomes associate end to end instead of side by side in early prophase I of meiosis. This idea had been maintained, at least for some organisms, for nearly half a century. The end-to-end pairing had never been observed but was inferred from the end-to-end association seen at diakinesis and metaphase I in, for example, *Oenothera*.* Darlington's theory of terminalization of chiasmata explained these late prophase I terminal associations as of secondary origin from an earlier lateral association (Darlington 1929a, 1931c). The recognition of the terminal chiasma led Darlington to two further hypotheses: firstly (1929b) that it is chiasmata, either interstitial or terminal in position, that are responsible for maintaining the chromosomes in pairs between the end of the pachytene stage and the end of metaphase I; and secondly (1931b) that meiosis can be looked upon as a precocious mitosis, with the force which keeps the daughter-chromatids together in pairs during the prophase of mitosis also responsible for the zygotene and pachytene pairing of whole undivided chromosomes and for their diplotene separation again when the chromosomes divide into chromatids. This precocity hypothesis of meiosis now appears improbable (see § 11.5).

## § 8.19   Objections to the chiasma breakage hypothesis

Darlington (1931a) pointed out that crossing-over takes place both in organisms such as *Primula sinensis*, where the number of chiasmata diminishes between the diplotene stage and metaphase I, and in organisms such as *Zea mays* where it does not diminish. Moreover, the evidence presented earlier of agreement between chiasma and cross-over frequencies in *Zea mays* is inexplicable on the chiasma breakage hypothesis. Likewise, the critical evidence from various cytological sources, notably unequal chromosome-pairs, polyploids and inversions, that crossing-over precedes chiasma-formation, is clearly at variance with the chiasma breakage theory. However, Sax (1930) pointed out that the classical theory could be reconciled with such observations if one postulated a broken classical chiasma alongside each observed (classical) chiasma. This hypothesis is effectively the same as the chiasmatype theory as regards its consequences, since every observed chiasma is associated with crossing-over. However, whereas on the chiasmatype theory crossing-over is the cause of the chiasma, on Sax's suggestion crossing-over is the result of the disappearance of a classical chiasma near the observed one. There was no direct evidence for the latter suggestion, and with the accumulation of evidence that crossing-over precedes chiasma-formation, it became absurd to postulate a broken classical chiasma near every critical chiasma, and Sax (1936) accepted the chiasmatype theory.

---

* The multiple associations of chromosomes seen at meiosis in many species of *Oenothera* are due to heterozygosity for interchanges of chromosome arms between non-homologous chromosomes (cf. § 9.2).

## § 8.20   Male *Drosophila*

An objection to the chiasmatype theory which has frequently been raised is based on cytological and genetical observations with male *Drosophila*. Early in the genetic work with *D. melanogaster* it was discovered that there was no crossing-over in the male (see Morgan, Sturtevant, Muller and Bridges 1915). This was expected for sex-linked characters, since the male has only one *X*-chromosome, but was wholly unexpected for autosomal characters. Unfortunately, *Drosophila* is exceptionally unfavourable material for study of chiasmata. Darlington (1934*a*) found no chiasmata in the autosomes of male *D. pseudo-obscura*, the homologous chromosomes merely lying parallel to one another at diakinesis and metaphase I. In *Drosophila*, unlike most organisms, homologous chromosomes show pairing in the somatic cells as well as at meiosis, and the pairing of the autosomes at meiosis in the male was similar to this somatic association. On the other hand, Darlington found chiasmata between the *X*- and *Y*-chromosomes in the region near the centromeres where pairing takes place. In order to reconcile these observations with first-division segregation of the *X*- and *Y*-chromosomes, which is known to occur in *Drosophila*, Darlington had to postulate the regular occurrence of two chiasmata in close proximity and both in the same chromosome arm, and with a two- or four-strand relationship between their associated cross-overs. Such an elaborate explanation could hardly be said to support the chiasmatype theory. Cooper (1949) claimed to have seen chiasmata in the autosomes at meiosis in male *D. melanogaster*, and also in the somatic cells. Such chiasmata could not be associated with crossing-over, since none normally occurs in the male. Cooper concluded that these were classical chiasmata, without chromosome breakage and crossing-over. However, Slizynski (1964) has also examined the chromosomes of *D. melanogaster* males at meiosis and has found evidence that what appear to be chiasmata in the autosomes are merely places of superficial contact. Similar chiasma-like associations were also observed between non-homologous chromosomes. It thus appears that there are no chiasmata in male *Drosophila*, not even classical ones.

## § 8.21   The synaptinemal complex

Electron microscopy has confirmed that in male *Drosophila* the association between homologous chromosomes at meiosis is abnormal. Moses (1956) observed tripartite ribbons on electron micrographs of thin sections of spermatocytes of the crayfish *Procambarus clarkii* at prophase I of meiosis. He called them *synaptinemal complexes* (Moses, 1958) because, from study of them in *Plethodon cinereus* (a salamander), he found that they were associated with the pairing (synapsis) of the homologous chromosomes. Each lateral element was initially part of a single chromosome and was situated near the opposing faces of the individual chromosomes. The central element apparently arose when the two lateral elements were paired. The complex, or 'synapton' as it might more conveniently be called, is composed of protein (Moses and

Coleman, 1964) and seems to be a regular feature of the pachytene stage of meiosis. Its origin through the association of lateral elements formed in the individual chromosomes has been confirmed for many species. In the synapton the lateral elements lie in one plane and are 120–150 nm apart, depending on the species. Fine transverse threads cross the space between the lateral elements. Von Wettstein (1971), using ammoniacal silver ions to label the components of the synapton with silver grains, found that in the fungus *Neottiella rutilans* the central element arose in segments in the nucleolus. These segments took up their central position in the synapton as the lateral elements paired up. In this organism the lateral elements have transverse bands, alternately 5 nm and 10 nm thick, spaced at 19 nm intervals (Pl. 8.2). Woollam, Ford and Millen (1966) observed the number of attachments of synaptons to the nuclear membrane at pachytene in males of *Mus musculus*, *Microtus agrestis* and *Mesocricetus auratus* and from statistical analysis con-

PLATE 8.2   Electron micrograph of a longitudinal section through a pair of chromosomes at pachytene of meiosis in the fungus *Neottiella rutilans*, showing the synaptinemal complex lying between the two chromosomes. The lateral elements of the complex have regularly spaced transverse bands. For fuller discussion see von Wettstein (1971). C Chromosome. (After Westergaard and von Wettstein, *C. r. Trav. Lab. Carlsberg*, **37**, 239 (1970).)

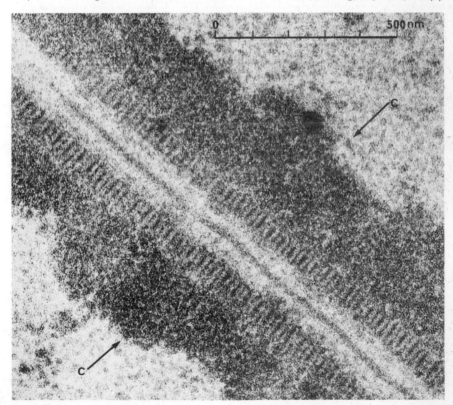

cluded that the synapton is attached to the membrane at both ends, but not elsewhere. Similar conclusions were reached by Wettstein and Sotelo (1967) from nuclear reconstructions from serial sections of spermatocytes of *Gryllus argentinus*. Gillies (1972) found from serial sections of isolated asci of *Neurospora crassa* that the synaptons of each of the seven pairs of chromosomes were attached to the nuclear envelope at both ends, with the exception of the nucleolar end of chromosome 2. The nucleolar organizer is near one end of this chromosome, and the nucleolus rather than the nuclear membrane seems to serve as anchoring site for this chromosome arm. Meyer (1961) found no synaptons at meiosis in male *D. melanogaster*, and they were also lacking (Meyer, 1964) in unpaired chromosomes of triploid females, and in females homozygous for a recessive mutant called $c(3)G$ in which chromosome pairing and crossing-over were absent (see also Smith and King, 1968). Moreover, in several other Diptera the presence of synaptons was associated with the occurrence of chiasmata, and lack of synaptons with their absence.

## § 8.22   Experimental test of the chiasmatype hypothesis

The data presented in this chapter have provided a substantial body of evidence that chiasmata are invariably the consequence of previous crossing-over between two of the four chromatids, and hence that Janssens's chiasmatype theory, in the modified form proposed by Belling, is true. The first direct experimental test of the chiasmatype hypothesis was carried out by Taylor (1965) by isotope labelling of chromosomes in *Romalea microptera*. His experiments, which provided the first evidence that crossing-over involves breakage and rejoining, are discussed in § 16.5, together with similar experiments by Peacock (1970) and Jones (1971). These studies have established the truth of the chiasmatype hypothesis.

# 9. *The classical theory of the gene*

## § 9.1 Multiple alleles

A discovery of great importance for the theory of the gene was made by Cuénot (1904). He found that the number of alternatives for a particular character which showed Mendelian inheritance was not limited to two. From the breeding of different strains of albino mice (*Mus musculus*), he obtained individuals with a black coat-colour, and others with a yellow coat. On crossing each of these with the normal grey he found that the black coat-colour was inherited as a Mendelian recessive and the yellow as a Mendelian dominant (with the complication that the homozygous yellow mice were apparently inviable). Furthermore, from crosses involving both the genes for black and for yellow fur, he found that these behaved as alleles towards one another, with yellow dominant to black. In other words, the genes for yellow, grey and black fur formed a series of alleles. There are six possible genotypes with respect to 3 alleles (the 3 homozygotes and the 3 heterozygotes), since a normal diploid individual necessarily cannot contain more than 2 alleles. Subsequently, the coat-colours grey with a light belly, and black with a light belly were found to belong to the same series of alleles. Normal mice are grey on the belly as well as on the back. Light belly-colour was found to be dominant to dark and grey back-colour dominant to black, so in order of descending dominance, the alleles are: Yellow; Grey with Light belly; Grey with grey belly, and black with Light belly (heterozygote Grey with Light belly); black.

Many other examples of multiple alleles were soon found. One of the first to be established in plants was through the work of Emerson (1911) with *Zea mays*. He had self-pollinated, and also inter-crossed, a number of strains of maize differing in the colour of the reproductive organs. He found that when he crossed pure-breeding strains having different colours in the axis (*cob*) of the fertilized female inflorescence, and in the ovary tissue (*pericarp*) surrounding the seed, only two classes appeared in $F_2$. Since the pericarp, like the cob, is parental tissue, all the grains on a maize ear will be alike as regards the pericarp colour, although there may be variation on the ear in the colour of the underlying tissues of the seed. The numbers of individuals in the two classes approximated to a 3:1 ratio, with the more frequent class having the same appearance as the $F_1$ generation. The deduction was clear: each pair of strains differed by one gene. Moreover, since different combinations were tested and all gave 3:1 ratios in $F_2$, it was evident that, in respect of the colour of these structures, the genes by which the strains differed were allelic. For example, he found that an $F_1$ plant with

137

red cob and pericarp gave an $F_2$ family consisting of 17 plants with red and 4 plants with white (colourless) cob and pericarp. Another red $F_1$ plant, from a different cross, gave in $F_2$ 45 plants with red and 12 with variegated (red and white) cob and pericarp. A third $F_1$ plant, with the latter characters, gave in $F_2$ 23 plants with variegated and 6 with colourless cob and pericarp. Evidently, red, variegated, and colourless cob and pericarp are determined by a series of 3 alleles, with red dominant to both the others, and with variegated dominant to colourless.

Over a dozen alleles are now known at this gene locus, giving different shades of red pigment (anthocyanin) and different distributions of it: for example, white cob and red pericarp; red cob and colourless pericarp; white cob and half-red pericarp (red at base, colourless at top of each grain); red cob and half-red pericarp. In every instance, the deeper shades of red were found to be dominant to the lighter ones, and the presence of colour to be dominant to its absence. Thus the heterozygote of the allele giving white cob and red pericarp with the allele giving red cob and colourless pericarp had red cob and pericarp. Moreover, dominance was found to be incomplete, a red-white heterozygote being paler than the red homozygote.

One of the best-known examples of multiple alleles was discovered by Morgan (1912) in *Drosophila melanogaster*. The white-eyed mutant of this normally red-eyed fly was found in 1910. The next year a male fly with yellowish-pink eyes described as eosin-coloured, occurred in an otherwise white-eyed family. On crossing eosin-eyed with normal red-eyed flies, the eosin character was found to show sex-linkage and to behave as a simple recessive to red. On crossing an eosin-eyed female with a white-eyed male the $F_1$ flies were all eosin-eyed, and in $F_2$ 1147 eosin-eyed and 344 white-eyed flies were obtained, in agreement with a 3:1 ratio (expected frequencies 1118·25 and 372·75, deviations 28·75, $\chi^2 = 2·95$, $n = 1$, $P = 0·1$–$0·05$). It was evident that the white and eosin eye-colours were inherited as a pair of alleles, with eosin dominant to white.* It was significant that in both $F_1$ and $F_2$ there were no wild-type (red-eyed) flies.

This result was in striking contrast to the results obtained when non-allelic mutants were crossed. For example, Morgan (1911*b*) had crossed a pink-eyed female with a male with bright red eyes which he described as vermilion-coloured. Both these eye-colours had been found to be inherited as recessive characters to the normal dull red eyes. All the $F_1$ generation flies from this cross had normal red eyes, and out of 1884 $F_2$ individuals, 1090 had normal red eyes, 397 had vermilion eyes, 313 had pink eyes and 84 had orange (= vermilion-pink) eyes. These figures approach a 9:3:3:1 ratio (1060:353:353:118) such as would be expected with two independently-inherited pairs of character-differences, but there is a significant shortage of flies with pink or vermilion-pink eyes, presumably due to a lowered viability of such individuals.

---

* There was a dosage effect for the intensity of the eosin colouring. Eosin-eyed males, and females heterozygous for eosin and white, both of which would evidently have only one eosin gene per nucleus, were found to have paler eyes than homozygous eosin females with two doses of the eosin gene per nucleus.

Further alleles at the 'white eye' locus were soon found and over a dozen are now known. They are all recessive to red and dominant to white, and include the colours described as 'cherry,' 'apricot' and 'ivory'.

The examples just described illustrate the two outstanding characteristics of a series of allelic mutants. First, all the alleles affect the same character and are non-complementary, that is to say, on crossing two of them, the $F_1$ generation is either intermediate in appearance between the two, or like one of them. On the other hand, non-allelic mutants, although they may affect the same character, can usually complement each other: each carries the dominant normal allele of the other's mutation (assuming the mutants are recessive), and on crossing them, the $F_1$ progeny are unlike either but have the normal (wild-type) phenotype of the organism.

Secondly, recombination studies indicate that allelic mutants appear to be due to genes located at the same positions in the linkage group. Thus, not only do they fail to show recombination with one another when crossed, but they each give the same recombination frequencies with other mutants. Thus, Morgan *et al.* (1915) found that the white/Red and eosin/Red eye-colour differences in *Drosophila melanogaster* both gave 1% of recombination with yellow/Grey body-colour, 33% with miniature/Normal wing-size, and 44% with Bar/normal eye-shape. On the other hand, non-allelic mutants show recombination with one another when crossed, giving in $F_2$ some wild-type and some double-mutant individuals, as well as the parental single-mutant types. Non-allelic mutants also differ from one another in their recombination frequencies with other mutants. The logical conclusion is that allelic mutants occupy the same gene locus in one of the chromosomes, whereas non-allelic mutants occupy different loci, either in the same or a different chromosome.

One of the best-known examples of multiple alleles concerns the four human blood-groups, O, A, B and AB. These correspond to the presence or absence of two antigens, A and B, in the erythrocytes (A—A only; B—B only; AB—both; O—neither). There is a reciprocal relationship between the presence of antigens in the red cells, and of antibodies in the serum; in other words, group O serum has antibodies against both A and B antigens, group A serum has antibodies against B, group B against A, and group AB against neither. Agglutination of cells occurs when a cell suspension and a serum are mixed, if they have an antigen and an antibody in common. Bernstein (1925) showed that the four blood-groups were inherited as Mendelian characters determined by three alleles A, B and O. Of the 6 genotypes (OO, AA, AO, BB, BO, AB), AA and AO are indistinguishable in phenotype and constitute blood group A, and similarly BB and BO are indistinguishable as group B. In Britain, the mean frequencies of the four blood-groups are approximately 46·7% for O, 41·7% for A, 8·6% for B and 3·0% for AB. Weinberg (1909) extended the Hardy-Weinberg theorem (see § 2.6) to include three alleles in a population. He showed that under the same conditions as for two alleles (that is, a large population, random mating, and no selection, migration or mutation), the frequencies of the 6 possible genotypes for three alleles are given by the terms of $(p + q + r)^2$, where $p$, $q$ and $r$ denote the frequencies of the three alleles. If $p$, $q$ and $r$ correspond to A, B and O, respectively, applying the British frequencies given above, $r^2 = 0\cdot467$, $p^2 + 2pr = 0\cdot417$, $q^2 + 2qr = 0\cdot086$, and $2pq = 0\cdot030$, from which good estimates for $p$, $q$ and $r$ are $p = 0\cdot257$, $q = 0\cdot060$ and $r = 0\cdot683$. Thus, $68\cdot3\%$ of British ABO chromosomes carry allele O, $25\cdot7\%$ carry allele A, and $6\cdot0\%$ allele B. Further alleles (subgroups of A) have since been recognized (see Race and Sanger, 1962).

## § 9.2   Mutation

The nature of the process of mutation whereby the various alleles at a gene locus are evidently derived from one another is clearly of fundamental importance to the theory of the gene. The term mutation was introduced into biological literature by De Vries (1901). He had made extensive studies of *Oenothera lamarckiana* (Evening Primrose) and found that new forms, differing appreciably in appearance from the original, apparently arose abruptly in small numbers in each generation. On the basis of these observations, he proposed a general theory of species formation by means of sudden discontinuous changes, to which he gave the name *mutation*. Study of the cytological and genetical behaviour of this and other species of *Oenothera* by many scientists during the succeeding 30 years established that these plants have an abnormal hereditary mechanism. It appears that successive exchanges of arms between non-homologous chromosomes have taken place during the course of their evolution, until all or nearly all the chromosomes have become involved. The plants are maintained heterozygous for these interchanges of chromosome segments by adaptations which prevent the survival, or in some instances even the formation, of the homozygotes. However, these adaptations occasionally break down, and this leads to the formation of small numbers of individuals homozygous, or at least partially so, for the chromosomal interchanges present in the parent. Owing to the presence of recessive genes which thereby become manifest, these homozygotes differ in appearance from the original plant, and moreover breed true for these characters. Thus it became evident that the 'mutations' in *Oenothera* described by De Vries were really the revealing of hereditary variation already present in the parent but normally hidden, rather than changes in the genes themselves. However, De Vries's idea of abrupt changes, even if unacceptable as a general hypothesis for the evolution of new species, was required to account for the origin of the character-differences which showed Mendelian inheritance. It was natural therefore that the term should have been transferred from the species to the gene.

The existence of multiple alleles implies that a gene could mutate in more than one way. This is illustrated by the behaviour of the white-eye alleles in *Drosophila melanogaster*, first studied by Morgan. As already mentioned, the original white-eyed fly arose from the normal red, and the original eosin-eyed fly from the white. But subsequently several instances were discovered of eosin eye-colour arising directly from red. Moreover, these steps were reversible, eosin occasionally mutating back to red, and also to white. An important feature of gene mutation was that the mutated gene was evidently replicated and passed to the progeny in the mutated form, since the altered phenotype was inherited. However, the rate at which mutations occurred was so low that it was exceedingly difficult to make detailed or quantitative studies of the process.

## § 9.3　Muller's grandsonless-lineage technique

Muller (1928*a*) described two ingenious techniques which he had devised for measuring the mutation rate in *Drosophila melanogaster*. Some 10 years previously he had come to the conclusion, on theoretical grounds, that lethal mutations might be sufficiently frequent to be studied quantitatively, since they were likely to occur considerably more often than those causing a visible change in the appearance of the organism. However, the detection of lethal mutations is more difficult than the detection of non-lethal ones. A mutation will in general not be manifest in a diploid organism such as a fly until the $F_3$ generation has been bred from the individual in which it occurred, since most mutations are recessive to their normal alleles. However, mutations occurring in the $X$-chromosomes of *Drosophila* are more favourably placed for detection, because they will become evident a generation sooner, causing, if lethal, the death of a fraction of the $F_2$ males. But even so, such mutations are not easily found, since the deaths will probably occur shortly after fertilization, and it will be necessary to try and distinguish a real shortage of males from a chance fluctuation from the normal 1 : 1 ratio of the sexes. With often fewer than 30 progeny per female, such a distinction may not be possible.

Muller's technique was to use a known recessive lethal gene in one of the $X$-chromosomes of a female in order to discover whether another recessive lethal gene was present in the other $X$-chromosome. The known lethal gene would necessarily be in the maternally-derived $X$-chromosome since the gene would have been lethal to the male parent with its single $X$-chromosome. A new lethal mutation, if present in the other $X$-chromosome, would presumably have arisen in the sperm which fertilized the egg from which the female developed. Such a mutation would cause the female to produce no sons, since, unless a cross-over occurred between the loci of the two lethal factors, the single $X$-chromosome in the male progeny would necessarily carry one or other of the two lethal genes. Crossing-over could be prevented, since strains of *Drosophila* were available in which it was largely suppressed in particular chromosomes. (This cross-over suppression was later shown to be due to the presence of inversions in heterozygous form occupying the greater part of the length of the chromosome. As shown in § 8.15, crossing-over within an inversion leads to the production of chromosomes with extensive duplications and deficiencies of chromosomal segments, and consequently to inviable gametes.) Thus, Muller was able to detect the occurrence of lethal mutations in the $X$-chromosomes of the reproductive cells of male *Drosophila* by the failure of these flies to produce any grandsons when their daughters carried a known lethal factor and a crossover suppressor in the other $X$-chromosome.

## § 9.4  Muller's accumulation technique

Muller also devised a method for counting lethal mutations which allowed them to accumulate over a number of generations until significant numbers were present. With the low mutation rates usually encountered, this method was likely to be preferable to the grandsonless-lineage method. A strain of flies was used in which there were two different recessive lethal (or sterility) genes, one in each chromosome of one of the pairs other than the $X$-chromosome. One of the pair also carried a cross-over suppressor: the other was the test-chromosome. In such a strain the known lethal (or sterility) genes would automatically be maintained in heterozygous condition, since the homozygotes would not survive (or would not leave any progeny). Any additional recessive lethal mutations which appeared would be similarly maintained and could be allowed to accumulate. Their existence in the test-chromosome could be revealed when desired by crossing one fly from each test lineage to another stock, and then interbreeding the progeny. Genes for several recessive characters were present in the test-chromosome, and consequently, in the absence of lethal mutations, each of these characters would be expected to be shown by one quarter of the $F_2$ generation following the outcross. However, if a lethal mutation was present, recessive characters linked to it would be absent.

Muller's first success with his lethal mutation counting techniques was the demonstration that a rise of temperature of 8°C produced a significant increase in the mutation rate. Using the accumulation method just described, 381 independent families kept at 19°C acquired a total of 12 lethal mutations in the test-chromosome in 16–17 generations occupying 11 months, and 359 lines kept at 27°C accumulated 31 lethals in the same chromosome in 18 generations during a period of 12 months. The mutation rates were therefore $\frac{12}{381 \times 16 \cdot 5} = 0 \cdot 19\%$ per chromosome per generation at 19°C, and $\frac{31}{359 \times 18}$ $= 0 \cdot 48\%$ at 27°C. The difference was significant.

This demonstration represented a considerable achievement, because numerous previous attempts by many experimenters to bring about a significant alteration in mutation rate by changing the external conditions in various ways had met with failure. Muller had found mutation rates varying up to 10 times greater in some experiments than others from unknown causes but presumably primarily due either to differences in physiological condition or genotype. Unless the conditions were carefully controlled, as in Muller's experiment, such variation would obscure the effects of particular environmental factors.

A 2·5-fold increase of rate with an 8°C rise in temperature such as Muller found, pointed tentatively to mutation as a multimolecular chemical change, since such reactions give increases of this order ($Q_{10} = 2$–3).

## § 9.5  Mutations induced by radiations

Muller's second success with his techniques for studying mutation quantitatively was an outstanding one. Using the grandsonless-lineage method, he showed (Muller 1927, 1928*b*) that X-rays enormously increased the rate of mutation. Males were irradiated and then crossed with females carrying the known lethal gene and the cross-over suppressor in one of their X-chromosomes. Any daughters from this cross who failed to produce any sons must have had another recessive lethal gene in their other X-chromosome, the one derived from their irradiated father (see Table 9.1). To reduce the labour of searching for these grandsonless lineages, the dominant gene for Bar eye-shape was incorporated in the chromosome carrying the known lethal gene and the cross-over suppressor. Only daughters with Bar

TABLE 9.1   Basis of Muller's grandsonless-lineage technique for detecting lethal mutations in the X-chromosome of *Drosophila melanogaster*.

| Female Parent | | Male Parent $XY$ Sperm | |
|---|---|---|---|
| | | $X^l$ | $Y$ |
| $X_{ClB}X$   Eggs | $X_{ClB}$ | $X_{ClB}X^l$ Bar female | $X_{ClB}Y$ Dies |
| | $X$ | $XX^l$ Normal female | $XY$ Normal male |

(*a*) Progeny of a female carrying in one of her X-chromosomes a cross-over suppressor ($C$), a recessive lethal mutation ($l$), and the dominant gene for Bar eye-shape ($B$), crossed with a male in which a lethal mutation has occurred in the X-chromosome of the sperm. $X$ = normal X-chromosome. $X_{ClB}$ = X-chromosome with cross-over suppressor ($C$), known lethal gene ($l$), and Bar mutant ($B$). $X^l$ = X-chromosome with new recessive lethal mutation.

| Female Parent | | Male Parent $XY$ Sperm | |
|---|---|---|---|
| | | $X$ | $Y$ |
| $X_{ClB}X^l$   Eggs | $X_{ClB}$ | $X_{ClB}X$ Bar female | $X_{ClB}Y$ Dies |
| | $X^l$ | $X^l X$ Normal female | $X^l Y$ Dies |

(*b*) Progeny of Bar female from (*a*) crossed with normal male. None of the sons survives.

eyes (and hence with the known lethal gene) were tested for their ability to produce sons. Non-lethal mutations in the $X$-chromosome can be detected at the same time as lethal ones, since the change of phenotype which they produce becomes evident in the grandsons of the irradiated males.

Muller found that male flies kept for 24 minutes at 16 cm from an X-ray tube (50 kV, 5 mA, 1 mm aluminium) gave 49 grandsonless lineages out of 676 tested, and others given the same irradiation for 48 minutes gave 89 such lineages out of 772 tested. The frequencies of lethal mutations in the $X$-chromosomes were therefore 7·2% and 11·5% respectively. Out of 198 control cultures, none were grandsonless, or, if a larger series of controls from other experiments is included, there were 5 lethal mutations in the $X$-chromosome of 6016 lineages tested, or 0·083%. The larger dose of X-rays thus gave a 140-fold increase in the mutation rate. As Muller had expected, the lethal mutations greatly outnumbered the non-lethal ones. Later work showed that, with both spontaneous and radiation-induced mutations, the proportion of lethal: non-lethal was about 10:1. Moreover, semi-lethal mutations giving much reduced viability (0·5—10% of normal) were rare: there were 4 with the 24-minute dose, and 12 with the 48-minute dose, or less than $\frac{1}{8}$ of the number of fully lethal mutations. Thus, the number of lethal mutations was reasonably clearly defined and likely to be a reliable guide to the total mutation-rate.

The visible mutations induced by X-rays included some at gene loci not previously known to have mutated, but others were allelic to, or apparently identical with, previously known spontaneous mutations such as white eye-colour. Some of the lethal mutations were allelic to visible ones, suggesting that the lethal ones did not differ in any fundamental way from non-lethal. There was great variety among the visible mutations and no suggestion whatever that X-rays were causing specific genes to mutate. All the indications were that X-ray induced mutations were essentially like spontaneous ones.

Muller's data suggested a direct proportionality between X-ray dose and number of mutations induced and this was later confirmed. Irrespective of the intensity at which the irradiation was administered, and also apparently of the wavelength, a dose of 1000 r gave about 2·5% of lethal mutations in the $X$-chromosome, 2000 r about 5%, and so on. There appeared to be no threshold dose, numerous small doses producing as many mutations as one large one of the same total size. The linear relationship between induced mutation rate and dosage pointed to a direct action of the radiation quanta or the released electrons upon the gene. However, different genes were found to mutate with significantly different frequencies, and this applied also to alleles. Thus, Timoféef-Ressovsky (1934), in reviewing experimental work on mutation, summarized his own extensive studies with the white-eye alleles in *Drosophila*, in which he had found that many of the various allelic changes possible, including reciprocal changes, occurred with significantly different frequencies under standard X-ray treatment. These findings again paralleled the observations on spontaneous mutation, where different genes, including different alleles, showed diverse rates of change.

Almost simultaneously with Muller's discovery of the mutagenic action of X-rays on *Drosophila*, Stadler (1928) discovered that X-rays, and also the gamma-rays from radium, caused mutation in *Hordeum vulgare* (Barley). Germinating grains were irradiated, and following self-pollination of the resulting plants, the grains from each ear were isolated. On sowing these grains, it was found that a number of those derived from one ear showed an abnormality, while other ears produced only normal progeny. It was evident that following the irradiation, one or more shoots (tillers) were heterozygous for a recessive mutation, while the remaining shoots on the same plant did not carry the mutation. Altogether 48 different mutations for seedling characters, such as white, virescent, and yellow leaf-colour, were found in the progeny of the irradiated grains, and none in the progeny of untreated plants.

Within a short time after Muller's initial discovery, mutation had been induced by radiation in a wide range of plants and animals, and it was clear that the phenomenon was a general one. It was found that all ionizing radiations, and also ultraviolet light, caused mutation. In every instance, a whole range of kinds of mutations were produced, and there was no suggestion of specific action. The findings with X-rays appeared to be typical in most respects of the results also found with the other mutagenic radiations.

An important discovery about the process of mutation was made by Stadler (1930) with *Hordeum*, and by Muller (1930b) with *Drosophila*. Stadler X-rayed germinating barley grains at 10°C, 20°C, 30°C, 40°C and 50°C and found no significant differences in the mutation rate. Muller X-rayed *Drosophila* males at 8°C and 34°C, and using his grandsonless-lineage technique, found 22 lethal or semi-lethal mutations in 67 lineages (33%) in the former, and 32 out of 120 (27%) in the latter. A repetition of this experiment using a smaller X-ray dose but larger samples gave 33 mutations out of 403 (8·2%) for those treated at 8°C, and 13 mutations out of 208 (6·2%) for the 34°C irradiation. It was evident that in both *Hordeum* and *Drosophila* over a large range of temperature, the mutation-rate was approximately constant.

The deduction from these experiments was that X-ray-induced mutation could not be an ordinary multimolecular chemical reaction, because such reactions have a large temperature coefficient. Rather it appeared that a unimolecular reaction was involved. The appreciable effect of temperature on the spontaneous mutation rate in *Drosophila*, which Muller had earlier found (see § 9.4) could be reconciled with the X-ray temperature data if it were supposed that whatever caused the spontaneous mutations acted indirectly through the medium of a multimolecular, and hence temperature-sensitive, chemical reaction. Calculations showed that natural radioactivity was insufficient to account for more than about one five-hundredth to one thousandth of the spontaneous mutations occurring in *Drosophila*. Moreover, if the frequency of spontaneous mutations in a given time was essentially the same in all organisms, as dependence on natural radioactivity would presumably demand, long-lived organisms such as *Homo* would of course accumulate much larger numbers of mutations per generation than would short-lived organisms such as *Drosophila*. For example, a lethal mutation in one X-chromosome out of every 1000 every 2–3 weeks (the *Drosophila* rate of 0·1% per generation) would imply that in 25 years about 1 in every 2 human X-chromosomes would acquire such a mutation. A consequence of this would be that the ratio of the sexes at birth would approximate to 2 girls : 1 boy instead of the equality (or slight preponderance of boys) that is found. It was evident that the spontaneous mutation rate was likely to be determined by the genotype and physiological condition of the organism, rather than by natural radioactivity, and that natural selection had probably acted to favour greater stability of the gene in long-lived organisms than in short. This conclusion implied that mutation rates could be influenced by chemical means.

## § 9.6   Mutations induced by chemicals

Despite the availability of Muller's highly efficient grandsonless-lineage technique for studying mutation quantitatively, it was 14 years after his discovery by this means of the mutagenic action of X-rays before clear evidence was obtained that mutations could be induced chemically. A variety of substances was tested by numerous experimenters, but the results were hardly statistically significant, until in 1941 Auerbach and Robson (1947) discovered that $\beta$-$\beta'$-dichloro-diethyl-sulphide (mustard-gas) was highly mutagenic.

A solution of mustard gas in cyclohexane was sprayed at 10-second intervals for a period up to 15 minutes into a stream of air passing over the flies. In the first experiment 1231 X-chromosomes were tested, that is to say, this number of daughters of treated males crossed to females carrying the known lethal gene was tested for their ability to produce sons. There were 90, or 7·3%, of these grandsonless lineages compared with 3 out of 1216, or 0·25%, in the untreated control. The treatment also gave rise to 11 semi-lethal mutations, whereas there was none in the control. A frequency of 7·3% of recessive sex-linked lethal mutations was comparable to that produced by 3000 r of X-rays, and was clearly highly significant.

Chemical mutagenesis was discovered independently by Rapoport (1946), who found that a solution of formaldehyde when mixed with the food of *Drosophila* in sublethal concentrations, gave 5·9% of sex-linked lethal mutations.

Following these initial discoveries with mustard-gas and formaldehyde, a wide range of chemicals was soon found to cause mutation, for example, urethane (ethyl carbamate), nitrogen mustard, manganous chloride and hydrogen peroxide. Numerous organisms were found to show mutations when treated with these substances, and their mutagenic action appeared to apply generally. The effect of these chemicals was similar to the effects of mutagenic radiations, that is to say, each produced a wide range of kinds of mutations and there was no evidence for specific action on particular genes.

## § 9.7   Structural changes in chromosomes

Before considering the significance of knowledge about mutation for the theory of the gene, another feature of mutagens must be considered. Muller (1927) discovered not only that X-rays caused mutation but also that they caused structural changes in linkage groups. This was revealed when linkage studies were made following irradiation. The recombination frequencies of genes were frequently less or more than had been found previously, suggesting that portions of hereditary material had been removed or added. Moreover, the sequence of genes was sometimes reversed, implying that a section had been turned round. In still other instances, genes showed linkage in inheritance where hitherto they had belonged to different linkage groups, and conversely, genes previously linked now showed independent inheritance. It was evident that part of one linkage group had become attached to another. Changes of all these kinds had been known previously, but they had appeared only very rarely. It seemed as if, as with mutation, X-rays were increasing the frequency of occurrence of phenomena which also occurred in their absence. It had been anticipated for some years that X-rays would be likely to cause chromosome breakage, but a remarkable feature of Muller's discovery was the frequency with which the hereditary material appeared to be joined up in new ways. Thus, Muller and Altenburg (1930) found no less than 117 instances of transfer of part of one linkage group to another in 883 lineages tested following a heavy dose of X-rays applied to *Drosophila* males.

Muller and Painter (1929) and Dobzhansky (1929) confirmed that structural changes in *Drosophila* linkage groups, following X-irradiation, were associated with corresponding changes in the chromosomes. Their work finally established that the order of the genes on the linkage map is the same as their actual physical order in the chromosome, although the relative distances often did not correspond. A large section of linkage map removed or transferred might correspond to only a small section of the chromosome, or the converse might be true. It was evident that the scale of the linkage maps varied greatly from region to region, when compared with the physical chromosome, crossing-over being more frequent in some regions than others.

Further confirmation that structural changes in *Drosophila* linkage groups involved rearrangement of the material of the chromosomes was obtained from study of the giant chromosomes in the nuclei of the salivary glands of the mature larvae. These chromosomes had been known in some other Diptera for over 50 years, before Painter (1934) discovered their value in genetics when he showed that structural changes in *Drosophila* linkage groups can be correlated with changes in the sequence of transverse discs or bands in these chromosomes. Corresponding parts of homologous chromosomes are closely paired in these salivary gland nuclei, and all the chromosomes are held together at their centromeres, the paired chromosome arms, of which there are five in *D. melanogaster* (the very short fourth chromosome apart), being cylindrical in shape, about 5 $\mu$m in diameter, and radiating for a distance of about 250 $\mu$m if straightened (Fig. 9.1 and Pl. 9.1). By contrast, in

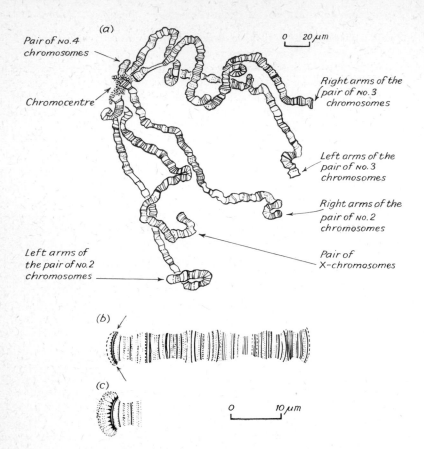

*(a)*

Pair of No.4
chromosomes

Chromocentre

0    20 μm

Right arms of the
pair of No.3
chromosomes

Left arms of the
pair of No.3
chromosomes

Right arms of the
pair of No.2
chromosomes

Left arms of
the pair of No.2
chromosomes

Pair of
X-chromosomes

*(b)*

*(c)*

0    10 μm

FIGURE 9.1   Drawings of the giant chromosomes found in the cell nuclei of the
salivary glands of Diptera.

(a) The complete set of salivary gland chromosomes of
*Drosophila melanogaster* (from Painter, 1934). The two
homologous chromosomes of each kind are fused
together throughout their length, and all the chromosomes are held together in the region of their
centromeres, forming the chromocentre.

(b) The no. 4 chromosome-pair of *Chironomus tentans*
(from Beermann, 1961). The arrows show the position of the gene concerned with the secretion of
granules in particular cells of the salivary glands.
In *C. tentans* this gene occurs in a mutant form such
that no granules are produced.

(c) The left-hand end of the no. 4 chromosome-pair
of *C. pallidivittatus* showing a puff formed at the position of the gene for granule-secretion.

PLATE 9.1  A squash preparation of salivary gland chromosomes from a mature
larva of *Drosophila melanagaster*, stained with acetic-orcein (cf. Fig. 9.1).
Magnification ca.  × 400. *C* = Chromocentre.

*(Courtesy of D. Briggs)*

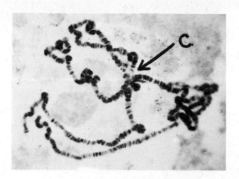

the reproductive cells at meiosis the chromosome arms measure about 2 *μ*m in
length. Over 1000 transverse bands occur in the salivary *X*-chromosome, and
there are similar numbers in the other four major chromosome arms. These
bands differ in thickness and spacing and have all been mapped. In individuals heterozygous for a structural change, the band sequences will differ
in the appropriate parts of the two halves of each chromosome pair involved.
From study of these differences, the precise nature of numerous structural
changes has been established.

When the salivary gland chromosomes of *Drosophila* individuals heterozygous for recessive lethal mutations were examined, an important discovery
was made. It was found that the lethal mutations were often associated with
minute deletions, a small number of adjacent transverse bands being
missing. Thus, Slizynska and Slizynski (1947) reported that in the *X*-
chromosome approximately one lethal mutation in five, by whatever
means it had arisen, was associated with a small chromosomal deficiency.
They had studied a total of 159 sex-linked lethal mutations and found 33,
or 20·7%, showed minute deletions. These mutations comprised 27 of
spontaneous origin (6 deficient), 24 X-ray induced (5 deficient), 4 following
neutron treatment (1 deficient), 21 from ultraviolet light irradiation (5
deficient) and the remaining 83 from mustard-gas treatment, of which 16
were deficient. The most frequent size of deficiency was the loss of 2
transverse bands.

These findings raised the question of what proportion of the remaining
lethal mutations were associated with deletions too small to be detected.
Were all lethal mutations really minute losses of chromosomal material?
Since lethal mutations, as indicated in § 9.5, do not appear to differ in any
significant way from viable mutations, the same questions could be asked of
mutations generally. These questions will be discussed in Chapters 13 and 14.

Since mutagenic radiations also caused structural changes, it was not
surprising to find that mutagenic chemicals behaved similarly. Auerbach

and Robson (1947) from study of genetic linkage in *Drosophila* following treatment of the flies with mustard-gas, and Oehlkers (1943) from study of *Oenothera* chromosomes at meiosis after keeping cut inflorescences in a solution of urethane (ethyl carbamate), were the first to demonstrate the induction of structural changes by chemical means. A wide range of chemicals has since been found to have this property of causing chromosome aberrations.

## § 9.8   The mechanism of induction of structural changes

Thoday and Read (1947) found that the effect of X-rays in causing structural changes in chromosomes was much influenced by the oxygen concentration at the time of irradiation. Root-tips of *Vicia faba* irradiated in the absence of oxygen gave only about one third of the aberration frequency that the same dose gave in oxygen. Kihlman (1961), in reviewing bio-chemical knowledge concerning the induction of structural changes, classified agents causing aberrations into three main categories.

First, some agents induce structural changes with a frequency which does not depend on the oxygen concentration, for example, alpha particles and ultraviolet light, where the effect is immediate, and alkylating agents such as nitrogen mustard and diepoxides, where the effect is delayed for some hours. These are highly reactive chemicals which presumably act more or less directly on the chromosomes, as the radiations are also presumed to do.

Secondly, there are agents which cause aberrations with a frequency which is much reduced in the absence of oxygen, but is unaffected by inhibitors of respiration. Examples of such agents are X-rays, with an immediate effect, and visible light in the presence of acridine orange, where the effect is delayed for some hours. Kihlman suggested that the oxygen reacts with free radicals produced by the radiations.

Thirdly, certain chemicals have been found to cause structural changes with a frequency which is much reduced, either by the removal of oxygen, or by the presence of respiratory inhibitors such as sodium azide. Examples are methylated oxypurines such as caffeine, where the effect is immediate, and maleic hydrazide where it is delayed for some hours. Evidently aerobic respiration is necessary for these agents to produce their effects.

It is evident that the mechanisms of induction of chromosome aberrations by radiations and chemicals are complex and diverse. Nevertheless, the end-results, in the shape of the range of different kinds of structural changes that are produced, are relatively uniform. Just as with mutation, mutagenic agents appear to produce the whole range of kinds of structural changes, although ultraviolet light has usually been found to produce relatively few interchanges between different chromosomes.

Second only to *Drosophila* as an organism for the study of structural changes has been *Zea mays*. Anderson (1936), in reviewing his own and others' work on such aberrations induced by irradiation in *Zea* chromosomes and linkage groups, drew attention to two peculiarities about these changes. First, all alterations involving two chromosomes or linkage groups were reciprocal, that is to say, if a piece of one chromosome was transferred to another, then a piece of the latter was also transferred to the former, and furthermore the rejoining was always at the points of breakage, never at the normal ends of the transferred segments. Secondly, all changes involving only one chromosome, such as inversions, duplications, deficiencies, and the production of ring-shaped chromosomes, involved interstitial segments. Deficiencies which appeared to be terminally placed, that is, loss of the end of a chromosome, were found on careful study to be in fact interstitial in position, the end segment being attached to the broken end of the main part of the chromosome.* It was evident that all these changes involved two points of breakage, followed by rejoining in new ways, and moreover, the normal ends of the chromosomes did not take part in this rejoining. These observations resembled those made with *Drosophila* referred to in § 9.7.

In order to account for these discoveries two hypotheses have been put forward. According to one, the initial event is chromosome breakage. On this hypothesis, it is necessary, in addition, to postulate that broken ends can rejoin, either in the original way or in new ways, but that normal chromosome ends cannot do so. According to the alternative hypothesis, the mechanism of induction of structural changes is by a process of non-homologous crossing-over or exchange, which is postulated as occurring either between different chromosomes, or between different parts of the same chromosome. This idea was first proposed by Belling (1927) with reference to an interchange of segments between non-homologous chromosomes, which he and Blakeslee had discovered in *Datura stramonium*. On this theory, the failure of normal chromosome ends to take part in rejoining is accounted for, without the necessity of postulating that these ends have special properties which prevent them from doing so.

Studies on the relation between the frequency of interchanges between different chromosomes and the dose of $X$- or gamma-rays showed that the frequency increased approximately as the square of the dose. A square law would be expected if two independent events were required to bring about the interchanges. These results were at first taken as evidence in support of the breakage-and-reunion theory. It was assumed that the two independent events which the square law demanded were the breaking of the two chromosomes by separate ionization tracks. Furthermore, it was assumed that, on the other hypothesis, the first step in non-homologous crossing-over would be the making contact between two chromosome segments, and that a single ionization-event in the vicinity of the paired segments would cause the process of exchange to occur. On this assumption, non-homologous crossing-over would give rise to a linear relationship between radiation dose and interchange frequency. Since the relationship was non-linear it was argued that the theory of non-homologous crossing-over was untenable. However, Revell (1955) pointed out that the square law does not necessarily invalidate the hypothesis of non-homologous crossing-over if one assumes that the two

---

* Some exceptions to this are now known: see for example § 9.11.

separate primary events are the triggering off, independently in each chromosome, of a series of changes leading to mutual crossing-over.

If the effect of radiations (and of mutagenic chemicals also, since their effects are similar) is to set in train a series of changes leading to non-homologous crossing-over, it is reasonable to suppose that this process will have features in common with normal crossing-over between homologous chromosomes at meiosis, and in particular that it will occur after the chromosomes have divided into chromatids, and that it will involve only one chromatid from each chromosome (or from each of the two participating segments of the same chromosome). Anderson (1936) made these assumptions for, having established that normal crossing-over occurs at the four-strand stage (see § 8.6), he favoured a similar situation for the hypothetical non-homologous process. The consequences of crossing-over between different parts of the same chromosome depend (a) on whether the bending back of the chromosome to bring non-homologous parts together is in the form of a

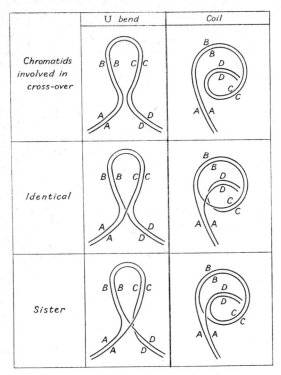

FIGURE 9.2   The four ways in which non-homologous crossing-over at the four-strand stage may occur between different parts of the same chromosome. The lines represent chromatids and the letters *A–D* denote different regions of the chromosome segment.

letter U, or of a coil, and (b) on whether the chromatids involved are identical
or sisters (see Fig. 9.2). Anderson pointed out that crossing-over between
different parts of the same chromatid would give rise either to an interstitial
inversion (Fig. 9.3(a)) or to an interstitial deficiency, the missing segment
forming a separate ring-shaped chromatid (Fig. 9.3(b)). The other chromatid
would be normal, since it would not have taken part in the exchange. If
the crossing-over was between different parts of sister chromatids, either a
pair of hairpin-shaped segments, each with a duplication and a deficiency,
would be formed (Fig. 9.3 (c)), or there would be a duplication of a segment
in one chromatid and a corresponding deficiency in the other one
(Fig. 9.3(d)).

On the alternative breakage-and-reunion theory, the configurations just
described are also possible, but a simple union of sister chromatids would be
expected much more frequently, following breakage of both chromatids at
approximately corresponding positions. This would give rise to a pair of
hairpin-shaped segments (Fig. 9.4(a)) similar to those derived from non-
homologous crossing-over (Fig. 9.3(c)) but differing from them in certain
respects. The differences are indicated by the lettering in the diagrams and
by the arrows which show the points of reunion.

In *Vicia faba* root-tip cells at metaphase, following X-irradiation in the
preceding resting-stage, Revell found that many of the pairs of hairpin-
shaped segments resulting from the irradiation showed constrictions at the

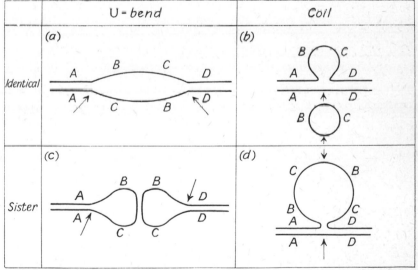

FIGURE 9.3   The configurations at metaphase of mitosis resulting from
non-homologous crossing-over between different parts of a
chromosome. The letters *A–D* correspond to those in Fig. 9.2.
It is assumed that the interval *BC* is relatively short, with the
result that the prophase pairing between the two *B* segments
has been lost, and similarly with the two *C* segments. Arrows
show the points where union has occurred.

positions of the arrows in Fig. 9.3(*c*), but not at the positions in Fig. 9.4(*a*). The constrictions were interpreted as showing the points of crossing-over, since similar constrictions were often seen at the position where rejoining had

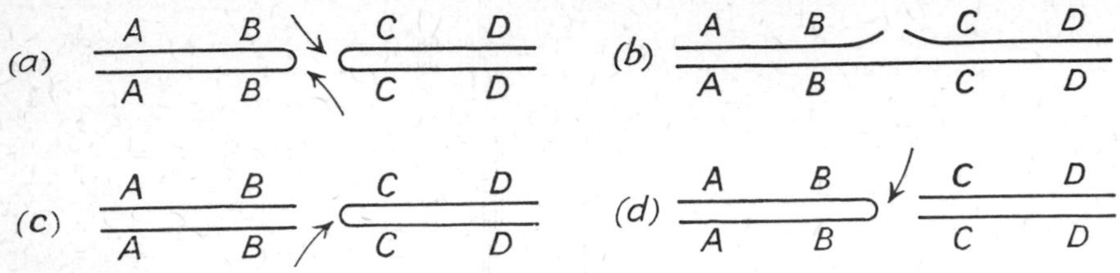

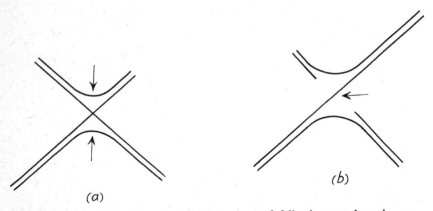

FIGURE 9.4   The types of chromosomal aberration expected on the breakage-and-reunion hypothesis (*a*) if all broken ends show union, and (*b–d*) if two broken ends fail to show union. Arrows indicate the points where union has occurred.

evidently occurred when different chromosomes had undergone interchange, that is, at the positions of the arrows in Fig. 9.5(*a*).

Furthermore, Revell found that the two hairpin segments of a pair were usually in close proximity to one another at metaphase. This would be expected if their construction was as in Fig. 9.3(*c*) where the two homologous segments labelled *B* would be expected to remain associated until metaphase, and similarly with those labelled *C*. If the construction of the hairpin segments was as in Fig. 9.4(*a*), they would be expected to have drifted apart by metaphase.

Anderson considered the genetic consequences of these structural changes, supposedly resulting from non-homologous crossing-over at the four-strand

FIGURE 9.5   The types of aberration expected following non-homologous crossing-over between two different chromosomes (*a*) following a complete exchange, and (*b*) if the free ends of one crossover chromatid failed to unite. Arrows show the points where union has occurred.

stage between different parts of the same chromosome, and pointed out that quantitative predictions capable of experimental testing would be complicated by viability phenomena, since chromosome segments lacking a centromere are soon lost from a cell, and such deficiencies often cause inviability, particularly to pollen-grains. Revell, on the other hand, has been concerned with the immediate cytological consequences, before differential viabilities take effect. Accurate quantitative predictions can then be made. Thus, Revell (1959) assumed that the two kinds of bending (U-shaped and coil) occurred with equal frequency, and that the exchanges involved either identical or sister strands equally often. The four configurations shown in Fig. 9.3 are then expected with equal frequency, and theoretically, by examining chromosomes fixed a few hours after irradiation, it should be possible to observe and count these aberrations. This should allow a direct test of the exchange hypothesis, since on the alternative breakage-and-reunion theory the pair of hairpin segments (Fig. 9.4(a), but resembling Fig. 9.3(c)) would be expected with greater frequency than the configurations shown in Fig. 9.3, for the reasons given above.

In practice, if the segment BC between the points of supposed non-homologous exchange is short, as appears usual, the configurations in Fig. 9.3(a), (b) and (d), often cannot be identified with certainty, and even the ring-chromosome (Fig. 9.3(b)) may be overlooked if it is small. Only the pair of hairpin segments (Fig. 9.3(c)) is readily recognized. Revell thus found that, at least in *Vicia faba* root-tips, a straightforward quantitative test of the hypothesis of non-homologous crossing-over could not be made. However, he suggested that in a minority of instances (7–10% in his experiments described below) the induced process of crossing-over was not completed, with the result that, although one cross-over chromatid was formed, the free ends of the reciprocal one failed to unite. Such a 'half cross-over' would be directly recognizable as a 'half-chiasma' when different chromosomes were involved (Fig. 9.5(b)), but the consequences when different regions of the same chromosome were concerned are diverse and are shown in Fig. 9.6. There are 8 alternatives depending on which way the chromosome bends (U or coil), which strands undergo exchange (identical or sister), and which fail to unite. The configurations shown in Fig. 9.6(a), (b), (d), (g) and (h) will appear as a break in one chromatid, those in Fig. 9.6(e) and (f) will appear as single hairpin-shaped segments, while that in Fig. 9.6(c) will appear more or less normal. Hence, on the assumption that the 8 alternatives occur with equal frequency, the occurrence of a break in one chromatid is expected to be 2·5 times as frequent as the occurrence of single hairpin segments. On the other hand, on the breakage-and-reunion theory, configurations with free ends would be the direct result of a failure to join after breakage. Simple chromatid breaks, as a result of a single ionization track and a subsequent failure to rejoin, as illustrated in Fig. 9.4(b), would be expected more frequently than any configurations involving two breakage events such as those illustrated in Fig. 9.4(c) and (d) and Fig. 9.6. Hence, single chromatid breaks would be expected much more than 2·5 times as often as single hairpin segments.

| CHROMATIDS INVOLVED | | U-bend | | Coil |
|---|---|---|---|---|

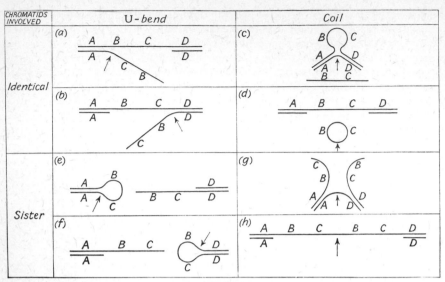

FIGURE 9.6   The configurations expected at metaphase of mitosis as a result of 'half crossing-over' between non-homologous segments of a chromosome. The letters *A–D* correspond to those in Figs. 9.2 and 9.3. Arrows show the points where union has occurred.

Revell (1959) examined 1513 cells at metaphase from *Vicia faba* root-tips treated with 50 r X-rays 5 hours before fixation, and found 32 examples of a break in one chromatid and 10 examples of single hairpin-shaped segments, in good agreement with a 2·5:1 ratio (expected frequencies 30 and 12, $\chi^2 = 0.47$, $n = 1$, $P = 0.5$). In a more extensive experiment, 5000 metaphases were examined following 65 r X-rays 5 hours before fixation, and the corresponding frequencies were 169 and 67 (expected values 168·6 and 67·4, $\chi^2 = 0.0038$, $n = 1$, $P = 0.95$). Revell found no less than 4155 non-staining gaps or 'false chromatid breaks' in the 5000 metaphases, many or all of which would formerly have been scored as breaks, before it was recognized that a thread so fine as to be beyond the resolution of the microscope may join chromatid segments.

These results, taken in conjunction with the qualitative evidence given above, provide strong support for the hypothesis of non-homologous crossing-over as an explanation of the origin of structural changes induced by radiations. Furthermore, Revell has found that hairpin-shaped segments and chromatid breaks increase in frequency in proportion to the X-ray dose raised to power 1·5. This conflicts with the predictions of the breakage-and-reunion hypothesis, where chromatid breaks would be expected to show a linear relation with dose, but is in agreement with the exchange hypothesis, assuming that non-homologous crossing-over between different regions of the same chromosome is sometimes due to a single ionization track and sometimes to two. Further support for the exchange hypothesis has been obtained by Heddle, Whissell and Bodycote (1969). They used ³H-thymidine labelling (see § 12.2) of chromosomes of *Potorous tridactylus* in tissue culture, and found that many, though not all, X-ray induced terminal deletions were really incomplete exchanges, as predicted by Revell. Evans and Scott (1969), also using ³H-thymidine labelling, studied chromosome aberrations in *Vicia faba* root-tips induced by nitrogen mustard. They concluded that although many of the aberrations resulted from exchanges, others were produced by faulty chromosome replication at the site of a single lesion.

The initial event in crossing-over is believed to be breakage of one of two longitudinally arranged sub-units of each participating chromatid, followed by association between these sub-units (see Chapter 16). There are several points of agreement between this postulated mechanism and the requirements of the Anderson-Revell chromatid-exchange hypothesis for the origin of structural changes (see § 17.5).

## § 9.9   The hypothesis of the chromosome as a string of beads

The classical concept of the gene, which grew up as soon as the chromosome theory was accepted, was to liken the chromosome to a string of beads, each bead representing a different gene. This idea incorporated all the salient features about genes, namely, that they behaved in inheritance as discrete particles which appeared not to influence one another, that they were linearly arranged in the chromosome, and that crossing-over between homologous chromosomes at meiosis apparently occurred between them.

The three primary ways of defining the gene, that is, in terms of its specific effect on a particular character or group of characters, in terms of its recombination with neighbouring genes as a result of crossing-over, and in terms of its change to a different allele as a result of mutation, were thought to represent different aspects of the same structure. Thus, the gene was considered to be a specific molecule or aggregate of molecules making up the bead, which in some way controlled the development of a particular character in the organism. Crossing-over was imagined as a process of mechanical exchange which could cause breakage between corresponding beads in homologous chromosomes followed by rejoining in new ways, but which necessarily left the beads themselves intact. Mutation was thought of as an abrupt change in the molecular structure of the gene of a kind which was stable and was copied when the molecule was duplicated, and thereby inherited, and which led to a change in the appropriate character of the organism.

Multiple alleles could also be incorporated into the bead hypothesis, since it was only to be expected that the molecular structure of a gene could be changed in many different ways. Structural changes would represent abnormal rearrangements of bead sequence, or the duplication or omission of some of them. Two other observations appeared to reinforce the bead concept. Firstly, the sequence of the genes in the chromosome seemed to be a random, fortuitous one, such that neighbouring genes were presumed to be concerned with totally unrelated characters in the organism. This heterogeneity supported the idea that neighbouring genes in no way influenced one another, and that their proximity was merely a consequence of the favouring by natural selection of a stringing together of the hereditary units as a more efficient means of distributing them with precision between daughter-cells at cell-division. Secondly, the structure of a chromosome as seen at the leptotene stage of meiosis had itself been likened to a string of beads. Belling (1928*b*) went so far as to postulate that each chrommere represented a specific gene, and others made a similar suggestion about the discs or bands in salivary gland chromosomes. Without going as far as this, many saw in the visible heterogeneity of the chromosome support for the bead hypothesis.

On the other hand, Raffel and Muller (1940) pointed out that although it was often assumed that the lines of demarcation between genes, as defined by crossing-over, breakage, mutation, function, and reproduction, would coincide with one another, there was no evidence for this, and the assumption had only a doubtful theoretical basis. Moreover, it was only an assumption that the lines of demarcation of the units defined by each of these criteria were necessarily invariable, non-overlapping or even well-defined. In other words, there was no evidence that the hereditary material was really sub-divided into intra- and inter-genic components, such as the bead idea implied. The gene was recognized at the functional level by its effect on the development of the organism, but it was not a necessary consequence of this that genes were differentiated from one another physically: the beads and the thread connecting them might both be made of the same material. Gold-schmidt (1938) had taken this idea to its logical conclusion when he suggested that it was the chromosome, which he thought of as a large chain molecule, which was the hereditary unit, that mutations were changes in this molecule at specific points, and that these changes disturbed the normal interplay of catalysed reactions. He considered that each point in the chain had a definite meaning in the chemical properties of the whole, as with all molecules, but that the unmutated (or wild-type) gene was merely a theoretical concept. That is to say, he regarded the normal (wild-type) condition as controlled by the whole chain as a unit, without the separate existence of individual genes.

Few geneticists were willing to go as far as Goldschmidt, with whom the string-of-beads idea had become a string without beads. Nevertheless, it was recognized that there was no evidence that the units of function, of recombination, and of mutation were co-extensive, such as the bead hypothesis assumed. Indeed, evidence against this hypothesis was accumulating.

## § 9.10  Evidence against the bead hypothesis: the position effect

The bead hypothesis suffered its first setback when Sturtevant (1925) made a most unexpected discovery. The dominant sex-linked gene called Bar, which causes a narrow eye in *Drosophila melanogaster* instead of the

normal round eye, was unusual in that it appeared to mutate with comparatively high frequency. Zeleny (1921) found that, out of 85,008 progeny
of homozygous Bar-eyed females crossed with normal males, 52 (or approximately 1 in every 1600) showed reversion to the normal round eye, and 3
(or approximately 1 in every 28,400) showed mutation to a more extreme
form dominant to Bar, which he called Ultra-bar. Ultra-bar itself showed
similar changes.

Sturtevant and Morgan (1923) found that these changes were confined
to the female, and moreover were not true mutations but were due to
crossing-over. They studied the progeny of homozygous Bar-eyed females
which were heterozygous for the recessive genes for *forked* bristles and for
*fused* wing-veins. These genes are placed 0·2 and 2·5 units, respectively,
away from Bar on either side. It was found that, whenever a reversion from
Bar to normal eye-shape occurred there was crossing-over between the loci
of the 'marker' genes *forked* and *fused*. Sturtevant (1925) found that this also
applied to the origin of Ultra-bar from Bar. Some of his data are given in
Table 9.2. He concluded that the normal round and the more extreme
Ultra-bar eye characters arose from Bar ($B$) through crossing-over occurring
obliquely so as to give two Bar alleles ($BB$) in one chromosome (Ultra-bar)
and none in the other (round). The round-eyed flies arising from homozygous Bar through crossing-over were entirely normal, both in appearance
and in their progeny. The lower frequency with which Ultra-bar arose
compared with round was attributed partly to oversight, since Ultra-bar is
not very dissimilar from Bar, and partly to a lower viability.

TABLE 9.2   Data of Sturtevant (1925) for the progeny of Bar-eyed ($B$) *Drosophila
melanogaster* with marker genes *forked* bristles ($f$) and *fused* wing-veins ($fu$) on either
side of the Bar locus. Plus signs indicate normal alleles.

$$\text{Parents:} \frac{+\,B\,+}{f\,B\,fu} \times f\,B\,fu. \qquad\qquad F_1\ \text{Progeny:—}$$

| Eye-shape | Numbers of Various Genotypes for Outside Markers | | | | |
|---|---|---|---|---|---|
| | Parental | | Recombinant | | Total |
| | $+\,+$ | $f\,fu$ | $+\,fu$ | $f\,+$ | |
| Homozygous Bar† | 10,631 | 7,909 | 187 | 264 | 18,991 |
| Heterozygous Bar† (females) | 0 | 0 | 0 | 2 | 2 |
| Normal round (males) | 1‡ | 0 | 2 | 1 | 4 |
| Ultra-bar | 0 | 0 | 0 | 2 | 2 |

  † Homozygous Bar and the heterozygote of Bar and normal can be distinguished
by the eye shape, which is more extreme in the homozygote.
  ‡ Thought to be a contaminant.

Sturtevant compared the appearance of the eyes in the various Bar genotypes, and made the surprising discovery that the heterozygote of Ultra-bar and round $\left(\dfrac{BB}{\cdot}\right)$ had narrower eyes (mean number of facets 45*) than the Bar homozygote $\left(\dfrac{B}{B}\right)$ (mean number of facets 68*), although each fly contained the same two Bar alleles. He concluded that two alleles were more effective in narrowing the eye when they lay in the same chromosome than when there was one in each chromosome. Muller, Prokofyeva and Kossikov (1936) and, independently, Bridges (1936) carried the analysis a stage further when they showed that the Bar phenotype was not due to a simple gene mutation but to a structural change. Study of the salivary gland chromosomes revealed that Bar-eyed flies had a chromosome segment ($S$) containing 6 transverse bands present in duplicate. The repeated section was adjacent to the original and in the same (not the reversed) sequence, giving the configuration $SS$. Ultra-bar flies had the same segment in triplicate ($SSS$), while the normal flies arising from the Bar-eyed ones through crossing-over had normal salivary gland chromosomes with the segment represented only once. Sturtevant and Morgan's discovery of oblique crossing-over was now explained on the supposition that the left-hand $S$ segment in one chromosome of homozygous Bar-eyed flies occasionally paired at meiosis with the right-hand $S$ segment in the homologous chromosome $\left(\dfrac{SS}{SS}\right)$. A cross-over in these paired segments would then give a gamete with the segment in triplicate ($SSS$) and another in which the segment was represented only once ($S$). This staggered pairing evidently happened in either direction (the right-hand segment on one homologue with the left-hand segment of the other, and *vice versa*), because the round-eyed flies derived from Bar could have either of the two complementary crossover genotypes for the marker genes, that is, either non-*forked fused* or *forked* non-*fused* in Sturtevant's data in Table 9.2. Sturtevant's earlier deduction that two Bar alleles in one chromosome were more effective than one in each, had to be replaced by the notion that the particular segment, $S$, if present in triplicate in one chromosome and represented once in the other $\left(\dfrac{SSS}{S}\right)$ resulted in a narrower eye than the same segment in duplicate in both chromosomes $\left(\dfrac{SS}{SS}\right)$.

It was evident that the appearance of an organism may be modified merely by altering the arrangement of the hereditary material in the chromosomes, without mutation and without loss or gain in the quantity of genetic material. On the bead hypothesis, the only way to accommodate Sturtevant's discovery of the effect of position on the phenotype was to postulate that the beads were not the isolated structures that had been supposed, but that neighbouring beads (genes) could either influence one another, or their

---

* Facet number was found to be smaller the higher the temperature during development over the range 15–30°C. These frequencies refer to flies grown at 25°C.

immediate products of activity could interact, thereby modifying the particular characters of the organism which they controlled. Although this initial example of an alteration of the phenotype through rearranging the hereditary material referred to an abnormal situation involving duplicated and triplicated segments, it was later established that the principle which it illustrates is of wide application and fundamental importance. It is discussed in § 13.3 and § 18.2.

### § 9.11   Evidence against the bead hypothesis: recombination between alleles

Work on a series of bristle mutants called *scute* of *Drosophila melanogaster* by a number of authors and extending over many years had suggested that the units of function, recombination and mutation might not correspond (cf. Raffel and Muller, 1940). More specific evidence for such lack of correspondence was obtained when Oliver (1940) discovered recombination in *D. melanogaster* between two allelic eye-character mutants called *lozenge*. The mutant alleles were both recessive to normal and sex-linked, and they gave rise in the homozygote to a reduction in the pigmentation and other effects on the eye. Oliver found that females heterozygous for the alleles, that is, with one in one chromosome and the other in the homologue, when crossed with males carrying either allele, gave small numbers of progeny with normal eyes, and, furthermore, these reversions were always associated with crossing-over in the vicinity of the alleles. This remarkable finding recalled Sturtevant and Morgan's discovery with Bar eye, but differed from it in an important respect: the normal-eyed progeny from the lozenge-eyed parents were associated with one particular crossover genotype, whereas with Bar, as indicated above, they were associated with either crossover combination of the marker genes.

Lewis (1941) made a similar discovery with two *star* eye-character alleles in the second chromosome. These alleles, one dominant (*S*) and the other recessive (*s*) to normal, caused a reduction in size and an increase in roughness of the eye-surface. Heterozygotes having *S* in one chromosome and *s* in the other had *star* eyes as expected with alleles. Females heterozygous for these two alleles and also for the recessive genes *arista-less* (*al*) and *held-out* wings (*ho*), which are closely linked on either side of the *star* locus, were crossed with star-eyed males and gave rise to 4 normal-eyed progeny among 31,106 star-eyed. The female parents had the genotype $\frac{al\ S\ ho}{+\ s\ +}$ (the + sign indicates the normal alleles), and all 4 normal-eyed progeny had evidently developed from eggs of constitution + + *ho*, or in other words they were cross-overs for the marker genes on either side, having the normal allele of *arista-less* together with the *held-out* mutation. In another experiment using a stock with increased recombination in this region, there were 12 normal-eyed flies from the same cross among 26,370 star-eyed, and again the normals were all non-*aristaless* and *held-out*.

It was evident that the sites of the two star alleles were not quite identical, and it appeared as if crossing-over occasionally took place between them. That the normal-eyed flies carried the mutation *held-out* but not *arista-less* implied that the dominant star allele was slightly nearer to the locus of *arista-less* than the recessive one, such that the genotype of the female parents could be rewritten as $\dfrac{al\ \ S\ +\ ho}{+\ +\ s\ +}$. In the production of the normal-eyed progeny a cross-over was presumed to have occurred at the position of the X: $\dfrac{al\ \ S}{+\ +}\times\dfrac{+\ ho}{s\ +}$ to give the $+\ +\ +\ ho$ genotype which was found. The complementary product of such crossing-over would have both alleles in the same chromosome ($al\,S\,s\ +$). Flies arising from eggs of this genotype were evidently not distinguished from the other star-eyed progeny. Since the normal-eyed recombinants were all of one genotype with respect to the marker genes, there was no reason to suppose that the star alleles were associated with a duplicated segment such as was known to occur with Bar. Lewis confirmed that there was no evidence from the salivary gland chromosomes of any structural changes associated with the star alleles.

The occurrence of recombination between the star alleles could be taken to imply that these were mutations of different genes. Lewis (1945) favoured this explanation and went so far as to re-christen 'star-recessive' as 'asteroid'. For reasons to be discussed in Chapter 13, it now appears that this decision was premature.

A discovery in many ways comparable to those of Oliver and of Lewis with *Drosophila* was made by McClintock (1944) when she discovered an anomaly in a series of alleles at the yellow-green-2 locus (*yg*-2) in *Zea mays*. Three mutants which produced yellow-green, pale yellow, and white seedlings, respectively, were found to be inherited as Mendelian recessives to the normal green plant-colour, and to behave as alleles, with descending dominance in the order quoted. The anomaly was that the heterozygote for the two intermediate members of the series, yellow-green and pale yellow, instead of having yellowish-green leaves as expected, was normal green.

Two alternative explanations of this observation are possible. If the mutants are regarded as alleles, then it is necessary to assume that different parts of the gene are abnormal in the yellow-green and pale yellow mutants, respectively, such that the normal parts can complement each other. Alternatively, it could be argued that these two mutants are not allelic, but are due to mutation of neighbouring genes concerned with a similar function. On this hypothesis, each would carry the dominant normal allele of the other's mutation, and hence the heterozygote would be normal.

The gene (or genes) concerned was known to lie close to one end of a particular chromosome (the short arm of no. 9). From an examination of this chromosome, McClintock found that whereas the green and the yellow-green plants had a normal chromosome, the pale yellow plants had a small terminal segment of the chromosome missing, and the white plants had a slightly larger terminal deficiency. These observations suggested that part of the gene (or one of the two on the two-gene interpretation) was missing in the pale yellow plants, and that the whole gene (or both of them on the two-gene theory) was missing in the white plants.

On either interpretation the bead hypothesis is undermined. On the one-gene theory, the alleles for yellow-green and pale yellow plant colour evidently differ from normal at points which do not correspond, so that in effect the gene has been split and, contrary to expectation, the parts found to show autonomy of action in the production of green leaf-colour in the heterozygote for these alleles. On the two-gene theory, neighbouring genes are presumed to interact in the production of the green colour in this heterozygote. A similar argument can be applied to the *Drosophila* data discussed earlier. There were indications that a comparable situation might apply with the human rhesus blood-group factors. About 85% of white people have been found to possess the rhesus antigen, such that their erythrocytes are agglutinated by the antibodies obtained by immunizing rabbits with the blood of the Rhesus monkey *Macaca mulatta*. The remaining 15% are rhesus negative and lack the antigen. Haemolytic disease of the newborn (erythroblastosis foetalis) has been shown to be due to rhesus blood group incompatibility between a rhesus negative mother and a rhesus positive child. The presence of the rhesus antigen was found by Landsteiner and Wiener (1941) to be inherited as a Mendelian dominant. A number of apparently allelic variants within both the rhesus positive (*Rh*) and negative (*rh*) blood groups have been recognized. R. A. Fisher (in Race, 1944) suggested that mutant differences at three closely-linked positions (*C/c*, *D/d* and *E/e*) were each responsible for the presence or absence of a specific antigen, of which *D* and *d* corresponded to rhesus positive and negative, respectively. Further mutant forms at what is possibly a single gene locus have since been recognized (see Race and Sanger, 1962).

It is evident that both the one-gene and the two-gene explanations of the results obtained by Oliver and by Lewis with *Drosophila* and by McClintock with *Zea* represented a breakdown of the classical theory of the gene, and probably likewise with a one gene (*Rh/rh*) or a three gene (*C/c*, *D/d*, *E/e*) explanation of the rhesus blood-groups. Were these instances to be regarded as exceptional or typical? This question will be discussed in Chapter 13.

# 10. *The theory of one gene: one enzyme*

## § 10.1   Introduction

The idea of a connection between genes and enzymes is a long-established one. Garrod (1909) showed that the rare human disease called alkaptonuria appeared to be associated with a failure of the breakdown of the benzene ring in the degradation of phenylalanine and tyrosine, with the result that 2,5-dihydroxyphenylacetic acid is excreted in the urine, instead of the normal fumaric and acetoacetic acids. Bateson and Saunders (1902) had shown, on the basis of data collected by Garrod, that alkaptonuria appeared to be inherited as if due to a single Mendelian recessive gene, and Garrod (1909) pointed out that the splitting of the benzene ring in normal metabolism was likely to be the work of a special enzyme which was presumed to be lacking in congenital alkaptonurics. A few years later this specific enzyme was isolated, and its absence from alkaptonurics confirmed. Bateson (1909, p. 266), in discussing the Mendelian unitary factors, came to the conclusion that 'the consequences of their presence is in so many instances comparable with the effects produced by ferments, that with some confidence we suspect that the operations of some units are in an essential way carried out by the formation of definite substances acting as ferments'.

Indications of how different genes may affect related steps in the development of particular characters in an organism were obtained early in the study of Mendelian heredity. Thus, Bateson, Saunders and Punnett (1905) crossed two pure-breeding white-flowered varieties of *Lathyrus odoratus* and were surprised to find that the progeny had purple flowers. When these $F_1$ plants were allowed to self-pollinate, and the $F_2$ generation grown, they found that out of 651 plants which flowered, 382 had coloured petals and 269 were white. These numbers fit a 9:7 ratio (expected numbers 366·2 and 284·8 respectively, $\chi^2 = 1·56$, $n = 1$, $P = 0·3$–$0·2$). Such a ratio would arise in $F_2$ if two independent but complementary dominant factors were required for the production of the flower pigment, the 9:7 ratio being a 9:3:3:1 ratio (cf. Chapter 3) in which the last 3 classes were indistinguishable. It was subsequently confirmed by further breeding tests that one parental white-flowered strain had the genotype *CCrr* and the other had the genotype *ccRR*, where *C* and *R* are the complementary dominant genes required for pigment formation.

One possible explanation of these results is that anthocyanin (the pigment in the coloured petals) was formed as the result of two successive biochemical steps, the end-product of the one forming the substrate of the other. If these steps were controlled by the two genes, respectively, and if the end-product of the first step was colourless, it would be necessary for both reactions to occur, that is, for both $C$ and $R$ to be present, for the pigment to be formed.

Bateson, Saunders and Punnett (1905) crossed a pure-breeding pink-flowered strain of *Salvia horminum* with a pure-breeding white-flowered variety. The $F_1$ plants had purple flowers. On self-pollination or inter-crossing these $F_1$ plants, 255 purple, 92 pink, and 114 white-flowered plants were obtained in $F_2$. These numbers are in agreement with a 9:3:4 ratio (expected numbers 259·3, 86·4, and 115·3 respectively; $\chi^2 = 0·45$, $n = 2$, $P = 0·8$), such as would be expected if the last two classes making up a 9:3:3:1 ratio were indistinguishable. Following confirmation of this explanation from further breeding, it was evident that flower colour in *S. horminum* was determined by two independently-inherited Mendelian factors, such that when both dominant genes were present a purple pigment was produced, while one alone ($A$) gave a pink pigment, and the other alone ($B$) the white colour which was also found in the absence of either dominant allele (*aabb*).

A possible explanation in biochemical terms of such a 9:3:4 ratio is that the two genes act sequentially in the synthesis of the purple pigment, gene $A$ leading to the production of a pink anthocyanin and gene $B$ modifying its molecular structure to give a purple anthocyanin. In the absence of $A$, no anthocyanin is formed and gene $B$ is then ineffective, the flower-colour remaining white. Bateson (1907*b*), using the metaphor 'higher and lower', proposed the term *epistatic* for the factor ($A$ in this instance) which stands above the other or *hypostatic* factor ($B$ in this instance) and determines whether it shall take effect.

Numerous examples of the occurrence in $F_2$ of modified 9:3:3:1 ratios such as 9:7 and 9:3:4 are known. They point to interaction in development between independently-inherited genes, but the precise nature of such interaction requires identification of the biochemical steps which the genes control. Extensive studies of the nature of such biochemical steps have been made, particularly with the pigments in the fur of mammals, the eyes of insects, and the petals of flowers. Thus, from work on the biochemistry and inheritance of the anthocyanin and anthoxanthin flower pigments, it has been established that individual genes appear to be concerned with specific chemical operations such as methoxylation and glycosidation at particular positions in the molecule (cf. Scott-Moncrieff 1936; Lawrence 1950). This is precisely the kind of result anticipated if there is a relationship between particular genes and specific enzymes.

## § 10.2   Nutritional mutants in fungi and bacteria

A major step forward in the study of biochemical genetics was taken by Beadle and Tatum (1941). They reversed the ordinary procedure, and instead of attempting to work out the chemical basis of known genetic characters, they set out to determine if and how genes control known biochemical reactions.

Conidia of *Neurospora crassa* and *N. sitophila* were X-rayed, and then used to fertilize a strain of the opposite mating-type. About 2000 of the resulting ascospores were germinated and cultured on a 'complete' medium containing agar, inorganic salts, malt extract, yeast extract and glucose. The ability of the cultures to grow on a 'minimal' medium was then tested. This medium comprised agar (optional), inorganic salts, biotin, and a disaccharide, fat or more complex carbon source. Three mutants were found which grew essentially normally on complete medium and scarcely at all on the minimal medium with sucrose as the carbon source. These strains were then tested systematically by adding particular vitamins, amino-acids, etc. in turn to the minimal medium to determine what substance or substances they were unable to synthesize. In this way it was found that the three mutants had lost the ability to synthesize pyridoxin, thiamin, and para-aminobenzoic acid, respectively. Study of the inheritance of these character-differences showed that each differed from normal by a single gene.

Taking advantage of the method which Beadle and Tatum had devised, Srb and Horowitz (1944) obtained by X- or ultraviolet irradiation 15 mutants of *N. crassa* which lacked the ability to synthesize arginine. When these mutants were crossed with wild-type, it was found that each differed from normal by a single gene. When they were crossed with one another it was found that a number of them gave only arginine-requiring progeny, or in other words appeared to be independent mutations of the same genes. On the other hand, others gave many arginine-independent progeny and were evidently not allelic. When allowance was made for the alleles, the 15 mutants were resolved into mutations of 7 different genes.

These authors also used a second method of testing for allelism based on heterokaryosis (see § 5.4). Beadle and Coonradt (1944) had shown that when two recessive mutants of *Neurospora* of the same mating-type but having different nutritional requirements were grown together on minimal medium, a cell fusion (but no nuclear fusion) occurred to form dikaryotic mycelium, and this heterokaryon had a growth rate similar to that of the normal fungus. In other words, each mutant could supply the other's deficiency. Many combinations were tested, for example, a strain requiring *p*-aminobenzoic acid and another requiring nicotinic acid. Confirmation of heterokaryon formation was obtained by isolating hyphal tips and breeding from them: in this instance, from one hyphal tip it was found that some of the progeny required *p*-aminobenzoic acid and others required nicotinic acid for growth. Beadle and Coonradt also demonstrated the value of such heterokaryons in testing for allelism. Two mutants requiring nicotinic acid for growth were shown to complement one another, or in other words, in the heterokaryon each kind of haploid nucleus could supply what the other lacked. It was concluded that they were due to mutations of different genes. (Later work has shown that this inference is not necessarily valid: see Chapter 13.) Using this method, Srb and Horowitz were able to show that certain of their arginine mutants were allelic and others not, and they found that this heterokaryon complementation test gave results in agreement with the progeny-testing method.

By testing the ability of the arginine mutants to respond to the related substances ornithine and citrulline, Srb and Horowitz found that those in 4 of the 7 classes (*A–D* in Fig. 10.1(*a*)) would respond to both these substances,

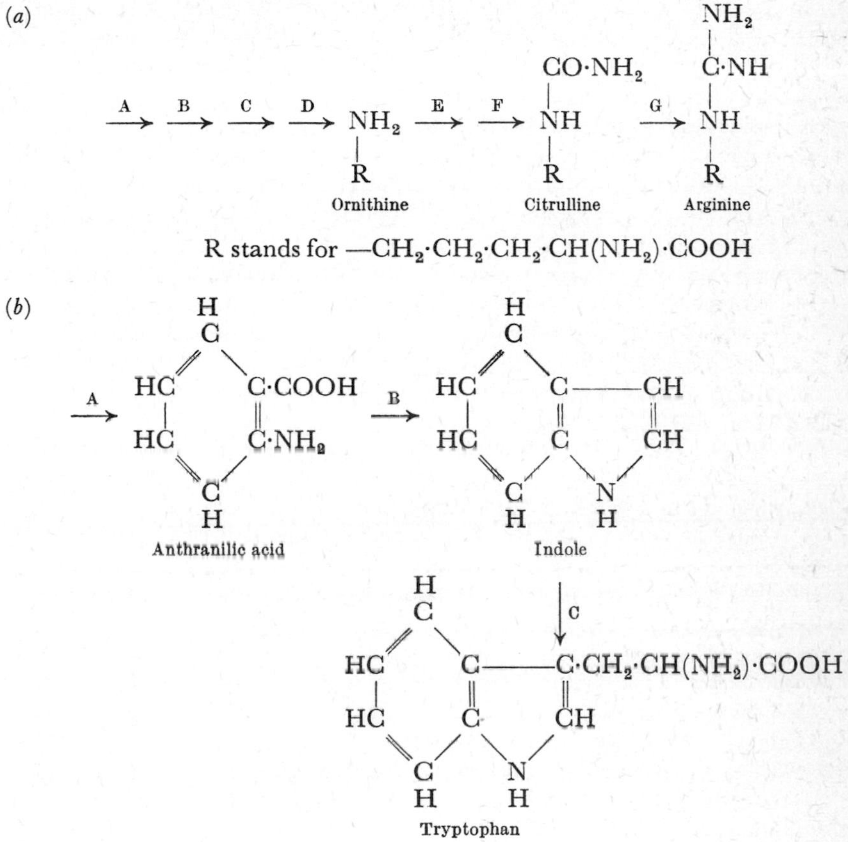

FIGURE 10.1   Steps in the synthesis of (*a*) arginine and (*b*) tryptophan in *Neurospora*.

those in a further 2 classes (*E, F* in the figure) would respond to citrulline but not to ornithine, and the 7th class (*G* in the figure) would respond to neither substance. They concluded that ornithine and citrulline were precursors, in the order quoted, in the biosynthesis of arginine, and that there was an ordered series of chemical steps, with each gene controlling a single step in a chain of reactions. The mutations were regarded as the loss of the ability to carry out the appropriate step, so that the sequence was blocked at that point. Since there were 7 different genes involved, it was assumed that there must be at least 7 steps in arginine synthesis. Much more is now known about these steps (cf. Fincham and Day, 1971).

Tatum, Bonner and Beadle (1944b) made comparable studies with tryptophan-requiring mutants of *Neurospora*. One (*A* in Fig. 10.1(*b*)) was found able to use either anthranilic acid or indole in place of tryptophan, while another (*B* in the figure), which was non-allelic with the first, grew well on indole but was unable to use anthranilic acid. It thus appeared that anthranilic acid was an intermediary in the synthesis of indole, and thence of tryptophan. Moreover, mutant *B* accumulated anthranilic acid in its tissues, as expected if the succeeding biosynthetic step was blocked. Mitchell and Lein (1948) discovered a third class of tryptophan mutant, *C*, which lacked the ability to couple indole and serine to produce tryptophan, and furthermore, cell-free extracts appeared to lack the appropriate enzyme, tryptophan synthetase, while the *A* and *B* mutants did not. Again, later work along the same lines has provided more information about the steps in tryptophan synthesis (see §13.7).

In view of the remarkable association found in *Neurospora* between mutation of specific genes and blocking of specific biosynthetic steps, it was natural for Beadle (1945) to assume that every biochemical reaction had a specific gene directing its course, and to suppose that the gene's primary and possibly sole function was in directing the configurations of protein molecules.

Critics of the one gene: one enzyme hypothesis pointed out that the method of analysis employed in the *Neurospora* work is a selective one favouring the detection of mutants with single deficiencies, because mutants with multiple defects would probably not survive. This criticism is not easily discounted. Nevertheless, it soon became evident that the apparent one-to-one relation applied very widely, and this tended to weaken though not to invalidate the argument that mutants with multiple effects were being overlooked. The fermentation of a number of sugars by *Saccharomyces cerevisiae* and other yeasts appeared to be controlled by specific genes, and in bacteria mutants were obtained comparable to those found in fungi. Gray and Tatum (1944), by X-ray treatment of cells, obtained a mutant of *Escherichia coli* requiring biotin for growth, and another requiring threonine. This led directly to the discovery by Lederberg and Tatum (1946) of recombination in *E. coli* (see § 16.3), which, in turn, enabled the genetic basis of the biochemical mutants to be established. It soon became evident that, as in fungi, each step in a biosynthetic pathway appeared to be controlled by a specific gene, and in a number of instances it was possible to demonstrate that a specific enzyme was apparently lacking.

Many instances have since been found of mutants associated with the lack of a specific enzyme: Fincham (1959b) listed 54 examples in organisms ranging from bacteria to man. A particularly clear example of a relation between a specific gene and a specific enzyme is due to Horowitz, Fling, Macleod and Sueoka (1961), who found 4 different forms of the enzyme tyrosinase in different wild-type strains of *Neurospora crassa*, and the differences were found to be due to allelic mutations.

## § 10.3   Apparent exceptions to the one gene : one enzyme hypothesis

Several apparent exceptions to the one gene-one enzyme hypothesis have been found, but on detailed study they have proved to be in keeping with the hypothesis.

Following irradiation of conidia of *Neurospora crassa* with ultraviolet light, Teas, Horowitz and Fling (1948) discovered a mutant which had a double

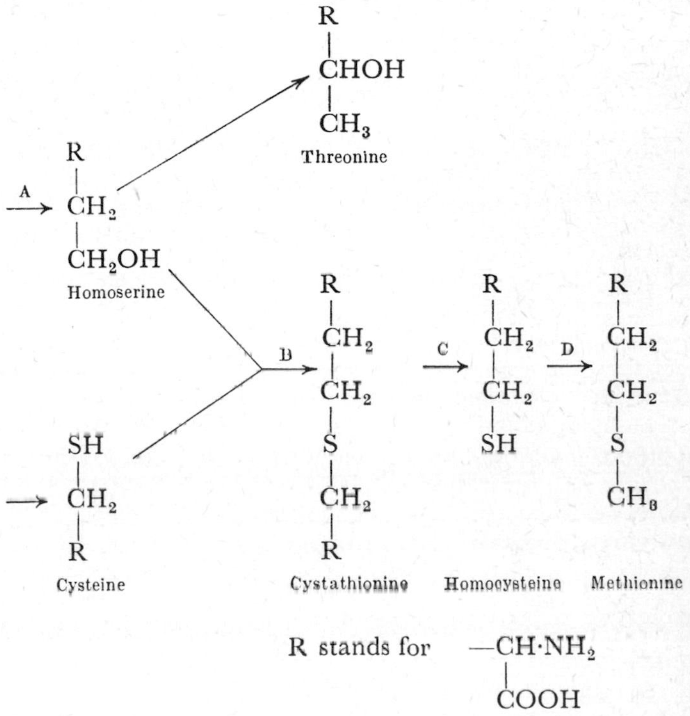

FIGURE 10.2   Steps in the synthesis of threonine and methionine in *Neurospora*.

requirement for growth: it was found to need both methionine and threonine. It was not due to two simultaneous mutations because a cross with wild-type showed that it differed from it by only one gene. Methionine-requiring mutants were already known controlling the steps *B*, *C*, and *D* in Fig. 10.2. Since the methionine requirement of the new mutant was satisfied also by homocysteine or by cystathionine, but not by cysteine, it was inferred that the double requirement was due to the blocking of a biosynthetic step in a common precursor of cystathionine and threonine. This was confirmed when it was discovered that homoserine alone was active in promoting growth. It was concluded that the mutant was blocked in step *A* in Fig. 10.2, and that threonine was derived from homoserine.

Another example of a double requirement was found to have a different explanation. A number of mutants of *N. crassa, Escherichia coli* and other organisms are known which require both isoleucine and valine for growth. It appears that the 4 final steps in the synthesis of these amino-acids are essentially the same (see Fig. 10.3), and that the same enzymes control these steps whether isoleucine or valine is being synthesised (see Wagner, Radhakrishnan and Snell, 1958).

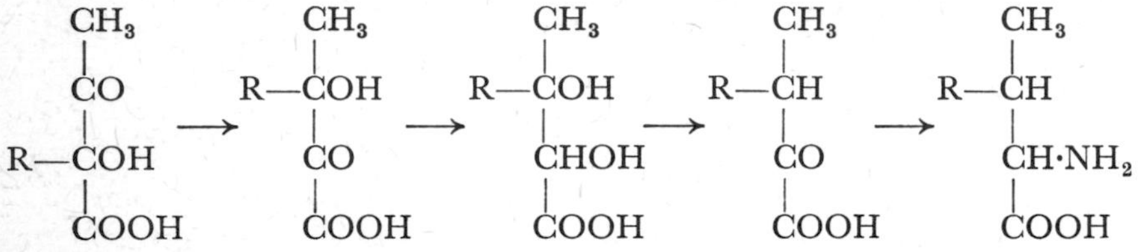

$$
\begin{array}{ccccccccc}
CH_3 & & CH_3 & & CH_3 & & CH_3 & & CH_3 \\
| & & | & & | & & | & & | \\
CO & & R{-}COH & & R{-}COH & & R{-}CH & & R{-}CH \\
| & \rightarrow & | & \rightarrow & | & \rightarrow & | & \rightarrow & | \\
R{-}COH & & CO & & CHOH & & CO & & CH{\cdot}NH_2 \\
| & & | & & | & & | & & | \\
COOH & & COOH & & COOH & & COOH & & COOH \\
\end{array}
$$

R stands for $-CH_3$　　　　in valine pathway

R stands for $-CH_2{\cdot}CH_3$ in isoleucine pathway

FIGURE 10.3　The four final steps in the synthesis of valine and isoleucine.

Early in the study of biochemically-deficient mutants in fungi it was discovered that apparent back-mutation to wild-type was sometimes due, not to a reversal of the original mutation, but to mutation of a different gene. The effect of this mutation was to suppress the effect of the original mutation. The suppressor mutation usually showed no linkage to the mutant it suppressed, the two separating from one another in many of the progeny. If the suppressor gene can take over the function of the gene it suppresses, this might imply that two genes can control one enzyme. However, more detailed study, such as has been made of suppressors of tryptophan synthetase deficient mutants (*tryp*-3) in *N. crassa* by Yanofsky and Bonner (1955), has revealed that the suppressor mutants are allele-specific. Thus, one (called $su_2$) was found to suppress the effect of allele no. 2 at the *tryp*-3 locus, two others ($su_3$ and $su_{24}$) suppressed both allele no. 3 and no. 24, while a fourth ($su_6$) suppressed alleles nos. 2 and 6. The suppressor mutations $su_3$ and $su_{24}$ appeared to be allelic with one another, but the others were due to mutation at unlinked loci. None was capable of suppressing all the *tryp*-3 alleles: $su_2$ and $su_6$ were tested against 25 *tryp*-3 alleles and it was found that only those mentioned above were suppressed.

It was evident that the suppressor genes were not capable of taking over the function of the *tryp*-3 gene. Instead, it seems likely that they act by restoring the functioning of this gene. Such restoration might come about in a number of ways, and there is evidence for diversity of action of different suppressor mutations (see § 14.15 and § 14.17). Irrespective of the precise mechanism of action, which is not known in most instances, it appears that suppressors are to be regarded as modifiers of the functioning of the mutant gene they suppress. Their existence therefore does not conflict with the idea that enzyme structure is essentially determined by one gene.

Another type of complication which has often been encountered in studies of nutritional mutants of fungi is competitive inhibition of growth of, for example, an arginine-requiring mutant by lysine, and *vice versa*. This has been observed by Doermann (1944) with *Neurospora crassa* and by Pontecorvo (1953) with *Aspergillus nidulans*. The explanation both for *Neurospora* (Roess and De Busk, 1968) and *Aspergillus* (Cybis and Weglenski, 1969), seems to be the inability of a permease for basic amino-acids, that is, an enzyme needed to catalyse the transport of these amino-acids across the cell membrane, to distinguish between lysine and arginine, with the result that there is competition for their uptake.

The examples given of multiple growth-requirements, suppressor mutations, and competitive inhibition, which on first discovery appeared to contradict the one gene:one enzyme hypothesis, have been found on detailed study not to do so. This has given considerable support to the hypothesis, which is based on the many examples now known of the lack of a specific enzyme (or its existence in a modified form) in mutant strains of various organisms. It has now been possible to specify the nature of the relationship between gene and enzyme in much more precise terms, and this is discussed in Chapters 13 and 14.

# 11. *The DNA theory of the chemical basis of heredity*

## § 11.1  Introduction

It has been known since the work of Miescher (1871, 1897) that the chief constituent of the cell nucleus is nucleoprotein, a combination of nucleic acid and basic proteins. It was later established that the nucleic acid which occurs in combination with protein in the chromosomes is deoxyribonucleic acid (DNA). DNA also occurs in the nuclei of bacteria, where it is associated with little or no protein.

Bacterial geneticists have used the term chromosome for the fine DNA thread found in the nuclei of bacteria and in bacterial viruses. However, these threads differ in a number of respects from the chromosomes of higher organisms. In addition to the chemical difference, the bacterial threads do not show the cycles of behaviour which chromosomes undergo at mitosis and meiosis. Other differences between bacteria and higher organisms in the organization of their DNA are referred to in later chapters. In order to avoid the confusion caused by this misapplication of the term chromosome to bacteria, the word *chromoneme*, meaning 'coloured thread', has been adopted in the present work for the DNA thread in bacteria and viruses. There is evidence that the primary division of living organisms is not into Plants and Animals, but into the Chromonemal and Chromosomal Kingdoms, since the differences between chromonemes and chromosomes appear to be associated with a number of other fundamental differences in cell structure and organization (see Appendix 1). In this primary subdivision, the Cyanophyta (Blue-green Algae) belong with the Bacteria. Recognition of these two kingdoms is important, because their differences are so great that conclusions about the mechanism of heredity, based on observations in the one, do not necessarily apply to the other.

For nearly 30 years after the acceptance of the chromosome theory of heredity, it was assumed more or less implicitly that the specificity of the gene resided in the protein part of nucleoprotein. The reason for this assumption was that proteins were known to occur in a very large number of highly specific forms, whereas nucleic acids were not thought to have a structure capable of much variety. DNA was found on hydrolysis to yield the purines, adenine and guanine; the pyrimidines, cytosine and thymine; phosphoric acid; and a sugar, deoxyribose. Knapp and Schreiber (1939) discovered, however, that the wavelength of ultraviolet light which gave the maximum frequency of mutations, following irradiation of the spermatozoids of *Sphaerocarpos donnellii*, was 265 nm corresponding to maximum absorption by DNA.

## § 11.2    Bacterial transformation

Further evidence that DNA was the carrier of hereditary characters was obtained by Avery, MacLeod and McCarty (1944) with *Diplococcus pneumoniae* (the pneumococcus). Griffith (1928) had discovered the phenomenon of genetic *transformation*. He had taken a virulent strain of the pneumococcus having a polysaccharide capsule surrounding the cell-wall giving the colonies a smooth (*S*) appearance. He heated this strain until all the cells were killed and then mixed with it a small quantity of living cells of an avirulent non-encapsulated strain giving rough (*R*) colonies, and injected the mixture into mice. He found that the mice frequently succumbed to the infection, that living virulent encapsulated (*S*) pneumococci could be isolated from them, and that this character combination of the bacteria was quite stable and was inherited. Control experiments using the living avirulent (*R*) culture alone, and others using the heated virulent (*S*) cells alone, showed no lethal effect of either on the mice, and no production of encapsulated (*S*) pneumococci.

Later, others demonstrated transformation *in vitro*, by growing the avirulent (*R*) cells in a fluid medium containing heat-killed virulent (*S*) cells. Subsequently, cell-free extracts of the *S* strain were obtained which could cause the specific genetic transformation from *R* to *S*. After many years' work, Avery and associates isolated in a highly purified form the substance responsible for this genetic change and showed that it was DNA.

Subsequently, many other characters were shown to be capable of transfer by DNA, both in this and a number of other species of bacteria. In 1952, the importance of DNA in heredity was also demonstrated with bacteriophage.

## § 11.3    Bacteriophage multiplication

Filter-passing agents pathogenic to specific bacterial strains were called bacteriophages before their viral nature, demonstrated by their infectivity and by their dependence for multiplication on their host, was established. Seven different phages, all active on a strain called *B* of *Escherichia coli*, have been given the type numbers 1 to 7 (*T*1, *T*2, etc.). Nos. 3 to 6 were isolated in America from sewage. The sources of the other three do not appear to have been recorded, but were almost certainly similar. These phages can be distinguished by their serological properties, by the range of strains of the host which they will attack, and by the shape and size of the clear area (plaque) which is produced in a bacterial colony by their destruction of the bacterial cells. The *T*-even phages (*T*2, *T*4, *T*6) are similar in many respects, and might be regarded as belonging to the same 'species', but each of the other four evidently belongs to a different species of virus.

The morphology of the *T*-even phages has been revealed in great detail
(see Fig. 11.1) from electron micrographs obtained by Brenner, Streisinger,
Horne, Champe, Barnett, Benzer and Rees (1959) using a new technique
involving phosphotungstate for negative staining. These viruses are alike in

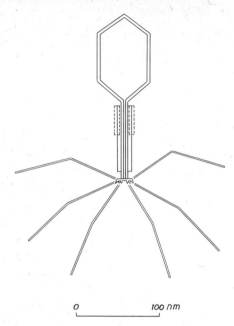

0                              *100 nm*

FIGURE 11.1   Diagrammatic representation of the structure of
the *T*-even phages of *Escherichia coli*, from the electron micrographs obtained
by Brenner *et al*. (1959).

appearance, and are tadpole-shaped with a head and tail. The head is a
bipyramidal hexagonal prism measuring about 100 × 65 nm. The contents
are enclosed by a membrane about 3·5 nm thick. The cylindrical tail, which
is about 100 nm long, consists of a core, surrounded by a contractile sheath.
The core, which is attached to the head, is a hollow cylinder of diameter
7 nm externally and 2·5 nm internally. The sheath, which is not attached
to the head, is about 80 nm long and 16·5 nm in diameter when extended.
The virus becomes attached to the host cell by the tip of the tail, and the
sheath then contracts towards the head to a length of about 35 nm and
diameter 25 nm externally and 12 nm internally, exposing the tail end of
the core. The sheath appears to be composed of helically arranged sub-units.
At the tip of the core, where contact with the bacterium is made, electron
micrographs have revealed a hexagonal base plate bearing six short spikes
and six long tail fibres measuring about 130 × 2 nm and with a kink in
the middle.

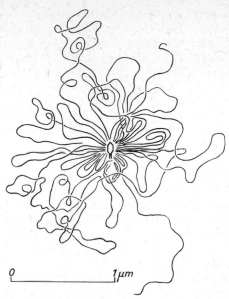

0 ————————— 1 μm

FIGURE 11.2  Appearance of *T*2 phage of *Escherichia coli* following osmotic shock and gentle diffusion of the DNA on a film of protein on water. The drawing is based on an electron micrograph obtained by Kleinschmidt *et al.* (1962).

Physical and chemical studies have established that the *T*-even phages are composed wholly of DNA and protein, and that the DNA, which accounts for 40% by weight of the total, occurs within the head. Brenner *et al.* (1959) have shown that the head membrane, the sheath, and the tail fibres are composed each of a different protein.

If *T*2 phage is suspended in a concentrated solution of sodium chloride and then the solution is rapidly diluted with water, the virus particles suffer an osmotic shock and are inactivated. Under the electron microscope tadpole-shaped ghosts are then seen—empty shells from which the contents have been removed. The shells will specifically adsorb to the bacteria although incapable of producing progeny virus. This adsorption appears to be determined by the tail fibres since detached fibres become attached to bacterial cells with the same specificity as the whole virus. The osmotic shock releases the DNA into the solution. This is shown by the vulnerability of the DNA to the enzyme deoxyribonuclease to which it was previously resistant, no doubt through protection by the protein membrane of the head. This release of DNA has also been beautifully demonstrated by Kleinschmidt *et al.* (1962) who floated the DNA on a film of protein on water and then examined the preparation under the electron microscope (see Fig. 11.2).

Hershey and Chase (1952*b*) isotopically labelled the protein and, in a separate experiment, the DNA of virus $T2$, using radioactive sulphur ($^{35}$S) for the protein and radioactive phosphorus ($^{32}$P) for the DNA. After infection of bacteria, unadsorbed virus was removed by centrifugation, and then the bacteria were agitated in a high-speed mixer in order to detach the virus particles from the bacterial cells-walls. It was found that at least 80% of the sulphur had remained at the cell surface and was removed by the agitation, while most of the phosphorus entered the cells. Despite this violent treatment the bacteria remained capable of yielding phage progeny, and it was found that little or none of the sulphur was incorporated in the progeny virus particles, while the phosphorus was transferred to the progeny to the extent of 30% or more. When labelled $T2$ was inactivated by osmotic shock, nearly all the sulphur was found in the ghosts. From these experiments, Hershey and Chase inferred that the bulk, if not all, of the protein of the virus takes no part in infection after the attachment of the virus to the bacterium. On the other hand, the DNA evidently entered the bacterium and played a part in the virus multiplication, since many of the same atoms appeared in the progeny. The structure of the virus bears out these conclusions. The elaborate tail appears to be a device for attaching the virus to the bacterial cell-wall, and injecting the virus DNA through it.

Hershey and Chase's experiments established that the genetic material of the virus is its DNA. This agrees with Avery, MacLeod and McCarty's discovery with *Diplococcus pneumoniae* in showing that hereditary characters may be carried by DNA alone, since little or no protein appears to be associated with the hereditary transmission. These discoveries gave support to the idea that in higher forms of life (chromosomal as distinct from chromonemal organisms) it might also be the DNA alone which was the chemical basis of heredity, with the protein part of the chromosome having some other function.

## § 11.4   The DNA content of nuclei

Boivin, Vendrely and Vendrely (1948) made chemical analyses of the DNA content of the nuclei of the thymus, liver, pancreas and kidney of *Bos taurus* (Domestic Cattle). The primary method used was hydrolysis, followed by estimation of the deoxyribose content by a specific colour reaction. The number of nuclei in each preparation was estimated, and the DNA content per nucleus calculated. This led to two remarkable discoveries. First, the nuclei of the various tissues were found all to have essentially the same DNA content of approximately 6·5 pg (1 picogram $= 10^{-12}$ g). Secondly, the DNA content of the sperm nuclei was found to be approximately half that of the nuclei of the diploid tissues, namely, 3·4 pg. It appeared that within a species the DNA content of non-dividing nuclei was independent of the tissue or the individual, and dependent solely on the chromosome content (haploid or diploid). Vendrely and Vendrely (1948), in reporting their results in more detail, pointed out that this was a strong argument in favour of the theory that the DNA is the carrier of the hereditary factors.

Mirsky and Ris (1949) obtained similar results for a number of other animal species, and moreover found significant differences in the DNA content of the nuclei of different species: in diploid tissues, the amount of DNA ranged from about 2 pg per nucleus in *Alosa* (Shad) to about 15 pg in *Rana* (Frog). They pointed out that, while supporting the DNA theory of heredity, these results did not necessarily mean that the gene consists of nothing but DNA.

Ris and Mirsky (1949) found that it was possible to obtain reliable information about the relative DNA content of different nuclei by measuring the intensity of pigmentation after staining by the Feulgen reaction. This staining method involves mild hydrolysis in warm dilute hydrochloric acid, followed by staining with decolorized basic fuchsin, and is specific for DNA, since it depends on the Schiff reaction with aldehydes derived from the deoxyribose. These authors showed the existence in the liver of *Rattus* (Rat) of three classes of nuclei with a ratio of intensity of Feulgen staining of 1 : 1·9 : 3·6. This was in agreement with expectation from the knowledge that diploid, tetraploid and octoploid nuclei occur in this tissue.

## § 11.5   The time of replication of DNA

Swift (1950) studied the DNA content of nuclei in developing tissues of *Mus* and *Ambystoma* using the Feulgen staining method. He found DNA values ranging between that of the normal diploid and twice this amount. He concluded that DNA synthesis occurs in interphase. A similar conclusion was reached for plant material by Howard and Pelc (1951*a*) who demonstrated autoradiographically that radioactive phosphorus ($^{32}$P) was incorporated into nuclei in the meristematic region of *Vicia faba* root-tips during interphase, but not during mitosis. Walker and Yates (1952) confirmed the interphase synthesis by measuring the absorption of ultraviolet light of wavelength 265 nm by nuclei in living cells in tissue cultures of particular organs of *Gallus*, *Mus*, *Oryctolagus*, *Rana* and *Triturus*. There is strong absorption by nucleic acids at this wavelength. The previous history of the individual cells had been recorded by phase-contrast photographs, and so the density of each nucleus on the ultraviolet negatives could be plotted against the time since the previous division of the cell. From this it was established that the doubling of the DNA took place in interphase and extended over a comparatively long period of time.

Swift (1950) showed that at meiosis, like mitosis, the replication of DNA took place in interphase before the nuclear division began. By measurement of the intensity of Feulgen staining, he showed that DNA synthesis was completed before prophase of the first division of meiosis in spermatocytes of *Mus*. This pre-meiotic synthesis was confirmed autoradiographically by Taylor (1953) using plant material. Following incorporation of radiophosphorus ($^{32}$P) into flower-buds of *Lilium longiflorum* and *Tradescantia paludosa*, he found that the period of DNA synthesis in the pollen-mother-cells of *Lilium* ended before the leptotene stage of meiosis, while in *Tradescantia* it extended to early leptotene, but was completed well before the zygotene stage when homologous chromosomes begin to associate. Confirmation of this time of DNA synthesis in relation to meiosis has been obtained from further autoradiographic studies made by Taylor (1959*b*) with *L. longiflorum*, and by Monesi (1962) with male *Mus musculus* (cf. Pl. 6.2(*a*)). These observations are important in connection with the mechanism of crossing-over (see Chapter 16).

It appears that the synthesis of the protein of the chromosome also takes place in the interphase before nuclear division. This has been shown by observing autoradiographically the uptake of radioactive sulphur ($^{35}$S) supplied in sulphate. By this means, Howard and Pelc (1951*b*) demonstrated that protein synthesis at mitosis in *Vicia faba* root-tips occurs at approximately the same time as DNA synthesis, and Taylor and Taylor (1953) showed the same thing for meiosis in flower-buds of *Lilium*. They found there was also some incorporation of $^{35}$S into chromosomes whenever the cell was metabolically active as indicated by uptake of $^{35}$S into cytoplasmic protein. Taylor (1959*b*) obtained similar results from autoradiographic study of the uptake of glycine labelled with carbon-14 ($^{14}$C) in the pollen-mother-cells of *L. longiflorum*.

If both DNA and protein synthesis are completed before nuclear division, whether mitotic or meiotic, Darlington's hypothesis (Darlington 1931*b*) that meiosis is to be regarded as a precocious mitosis appears untenable (see § 8.18).

## § 11.6   The base composition of DNA

Almost simultaneously with the discovery that the quantity of DNA per diploid nucleus was a specific character came the discovery that its composition showed a similar specificity. Chargaff, Vischer, Doniger, Green and Misani (1949) found that in *Bos taurus* the DNA of calf thymus had the same proportion of the four nitrogenous bases adenine, guanine, cytosine, and thymine, as beef spleen, while Vischer, Zamenhof and Chargaff (1949) found that the DNA of *Saccharomyces cerevisiae* (Yeast) and *Mycobacterium tuberculosis avium* had quite different base compositions from that of *Bos taurus* and from one another. The technique developed for these analyses involved hydrolysis, followed by chromatographic separation of the bases and their quantitative estimation by ultraviolet spectrophotometry.

Chargaff (1950), in reviewing these findings, drew attention to a remarkable feature, namely, that in every instance the number of molecules of the purine adenine liberated in the hydrolysis of the DNA was approximately equal to the number of molecules of the pyrimidine thymine, and likewise the molar quantities of the other purine, guanine, and the other pyrimidine, cytosine, were equal. On the other hand, the proportions of the two pairs varied widely depending on the species. On a molar basis, the adenine-thymine pair formed 30% of the total in *M. tuberculosis*, 58% in *Bos*, 62% in *Homo* and 64% in *S. cerevisiae*. These proportions were independent of the tissue or individual from which the DNA was obtained, and were evidently characteristic for each species (see Fig. 14.15).

Wyatt (1951) reported that in the DNA of *Bos taurus* about 6% of the cytosine was in fact 5-methylcytosine, and similarly in a number of other animal species the cytosine component included from 1% to 8% of 5-methylcytosine. In *Triticum* sp. (Wheat), he found that no less than 25% of the cytosine component was 5-methylcytosine. In every instance, however, the total molar quantity of cytosine and 5-methylcytosine was equal to the molar quantity of guanine. In the *T*-even viruses of *Escherichia coli*, Wyatt and Cohen (1952) found that the whole of the cytosine of their DNA was replaced by 5-hydroxymethylcytosine, which occurred in approximately equimolar quantities with guanine. On the other hand, the bacterium itself had a normal DNA composition with cytosine and no 5-hydroxymethylcytosine.

## § 11.7  The structure of DNA

The formulae for the chemical groups which make up DNA, that is, the four primary nitrogenous bases (adenine, guanine, thymine, and cytosine), the sugar (2-deoxy-D-ribose), and phosphoric acid, are given in Fig. 11.3. The glycosidic combination of base and sugar which occurs in DNA is called a *nucleoside*, and the phosphate ester of a nucleoside is known as a *nucleotide*. DNA consists of unbranched polymers of nucleotides. The backbone of these nucleotide chains is quite regular and consists of alternate sugar and phosphate groups joined in 3′,5′-phosphodiester linkages. Side groups consisting of one or other of the four bases are attached one to each sugar molecule. The covalent linkages by which the various groups are united to form the polynucleotide are shown in Fig. 11.4.

That DNA was composed of such nucleotide chains had been established and the filamentous shape of the molecule was known from electron microscopy, but no satisfactory detailed structure for the molecule as a whole had been proposed, until Watson and Crick (1953*a*) suggested a structure having several remarkable features. They postulated:

(*a*) that the number of polynucleotide chains in the molecule was two,

(*b*) that the chains followed right-handed helices, with 10 bases in each to one turn of the spiral,

(*c*) that the two chains were coiled plectonemically, that is, in an interlocked way, about the same axis,

(*d*) that the sequences of atoms in the two chains ran in opposite directions, or in other words, that one chain was inverted relative to the other,

(*e*) that the phosphates were on the outside of the double helix and the bases on the inside, with their planes set at right angles to the axis of the helix and spaced at intervals of 0·34 nm along it,

(*f*) that the two chains were held together by hydrogen bonding between the bases, which were joined together in pairs, a single base from one chain being hydrogen-bonded to a single base from the other chain, and

(*g*) that this bonding was of a highly specific kind, such that the purine adenine in either chain was associated with the pyrimidine thymine in the other, and likewise the purine guanine in either chain with the pyrimidine cytosine in the other.

Details of this proposed structure are shown in Fig. 11.5.

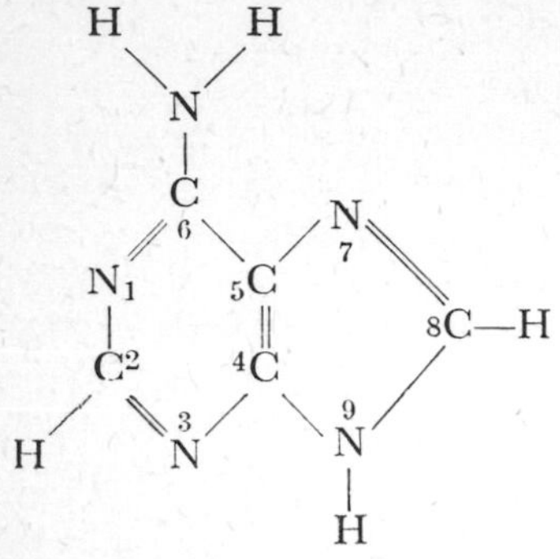

Adenine
(6–aminopurine)

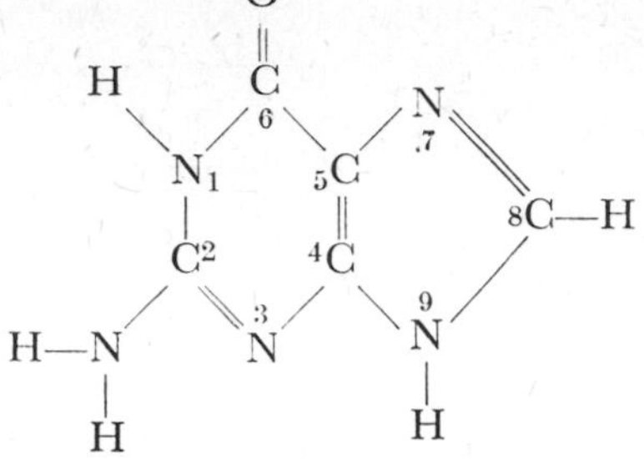

Guanine
(2–amino–6–hydroxypurine)

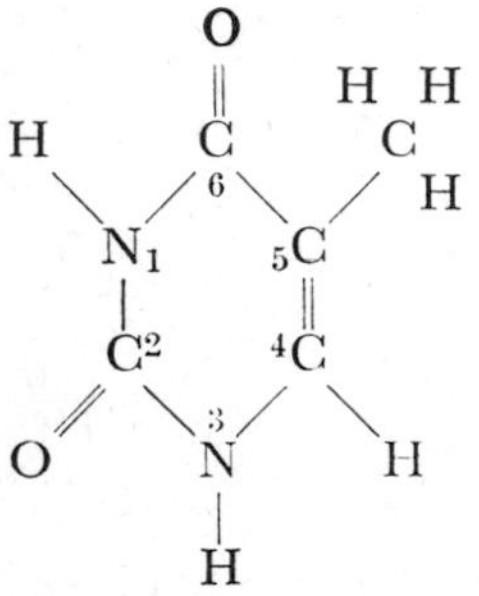

Thymine
(2,6–dihydroxy–5–methylpyrimidine)

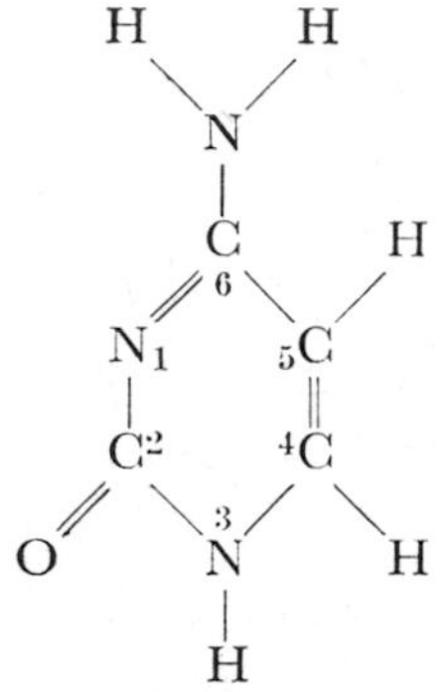

Cytosine
(2–hydroxy–6–aminopyrimidine)

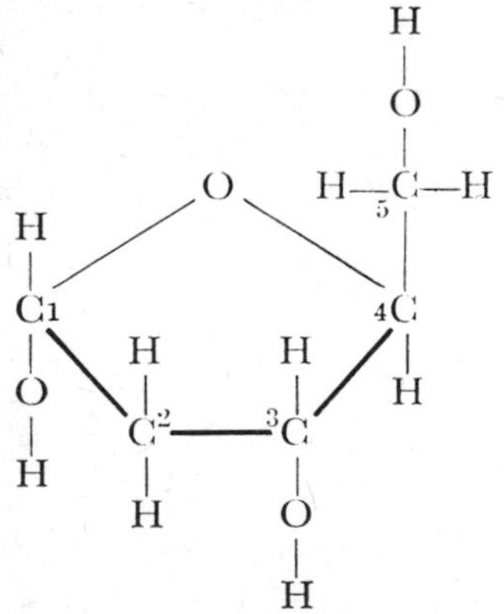

2–deoxy–D–ribose

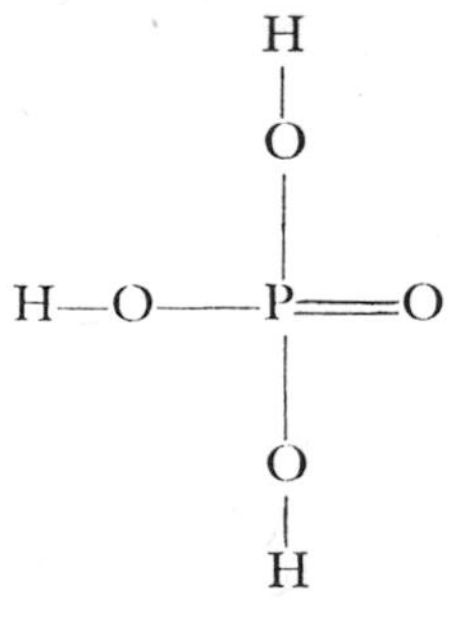

Phosphoric acid

FIGURE 11.3   The components of DNA.

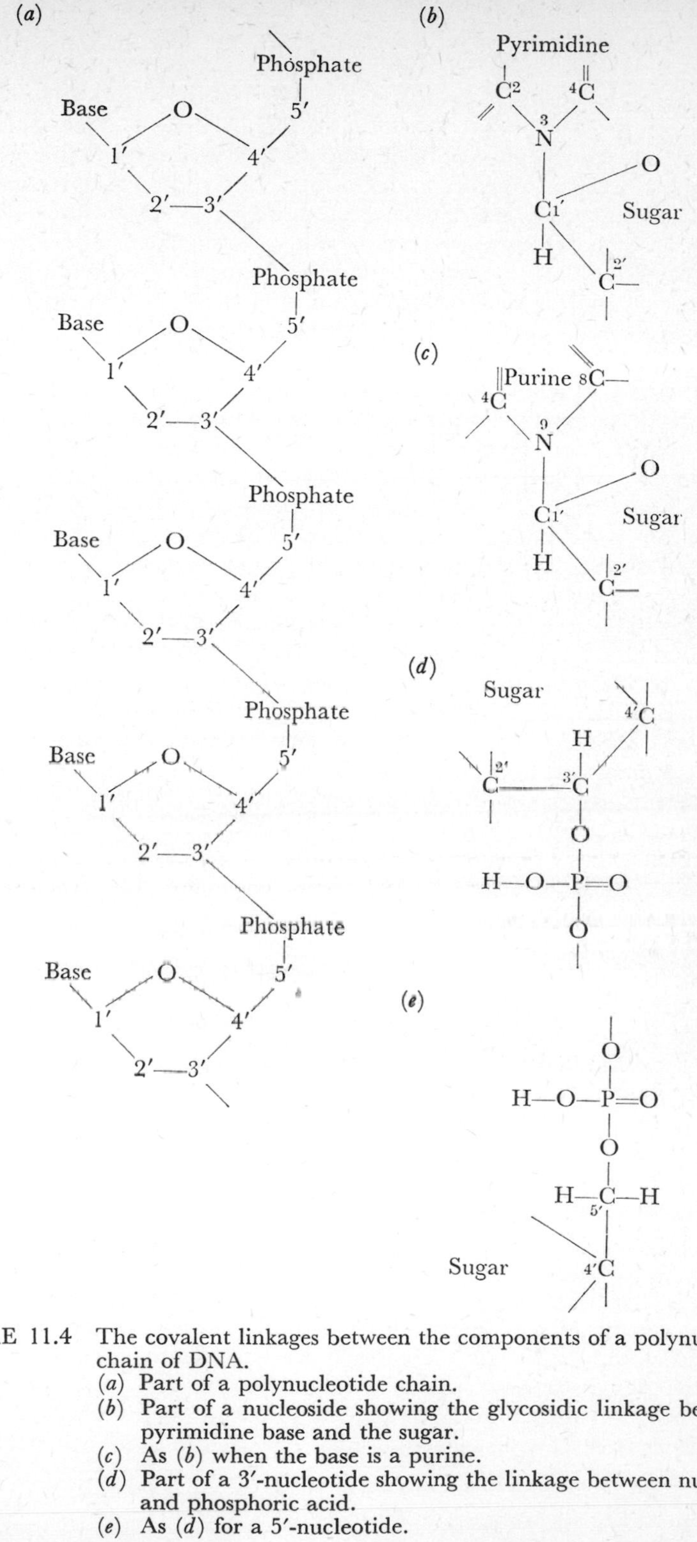

FIGURE 11.4    The covalent linkages between the components of a polynucleotide chain of DNA.
(a)  Part of a polynucleotide chain.
(b)  Part of a nucleoside showing the glycosidic linkage between a pyrimidine base and the sugar.
(c)  As (b) when the base is a purine.
(d)  Part of a 3′-nucleotide showing the linkage between nucleoside and phosphoric acid.
(e)  As (d) for a 5′-nucleotide.

The way in which these postulates for the structure were arrived at has been set out by Crick and Watson (1954). The primary evidence came from the X-ray crystallographic work of Wilkins and associates (Wilkins, Stokes and Wilson, 1953; Franklin and Gosling, 1953), the general nature of which had been made known to Watson and Crick prior to its publication. The specific base pairing was in agreement with the chemical evidence that certain bases regularly occur in equimolar quantities (see § 11.6), and the occurrence of hydrogen bonding within the molecule was supported by anomalous titration curves which DNA was known to give with acids and bases.

The method which Watson and Crick used, in attempting to find the structure of DNA, was to build models using the known interatomic distances and bond angles for the constituents of DNA, and see if each structure so produced was stereochemically feasible and if it would fit the X-ray data. Wilkins, Franklin *et al.* had found that DNA from the animals *Bos* and *Salmo* (Trout), the bacteria *Escherichia coli* and *Diplococcus pneumoniae*, and the virus *T2* of *E. coli*, all gave the same X-ray pattern. This led Watson and Crick to the idea that the sugar and phosphate groups have the same relative positions irrespective of the base attached to the sugar, since it was known that the DNA of different species had a different base composition (see § 11.6). The crystallographic data pointed to a helical structure of pitch 3·4 nm, and also to the supposition, first suggested by Astbury and Bell (1938), that the bases were set perpendicular to the long axis of the molecule and spaced at 0·34 nm, which is equal to their thickness. Density measurements from the X-ray patterns indicated that there were two polynucleotide chains.

Watson and Crick first attempted to construct models with the phosphates of the two chains near the helix axis, but they found that, because of the awkward shape of the sugar, there were relatively few configurations which the backbone can assume, and no satisfactory model could be devised along these lines. This led them to the belief that the bases form the central core and the sugar-phosphate chains the circumference of the molecule. The X-ray data suggested a molecular diameter of about 2 nm measured perpendicular to the length of the filament. This severely limits the types of model that can be constructed, and it appeared probable that each chain was nearly fully extended and made one revolution of the helix every 3·4 nm. The two chains would have to be intertwined because it is impossible to have paranemic coiling (a pair of helices which can be directly separated laterally) with two regular simple helices going round the same axis.

Watson and Crick considered it likely that the two chains would be held together by hydrogen bonds between the bases. These bonds are strongly directional in character, and can form only in the plane of the bases. However, since within one chain the bases appeared to be set partially on top of each other like a staggered pile of plates, hydrogen bonds can be expected to form only between bases belonging to different chains, thereby uniting the bases in pairs. If each sugar-phosphate chain is in the form of a regular

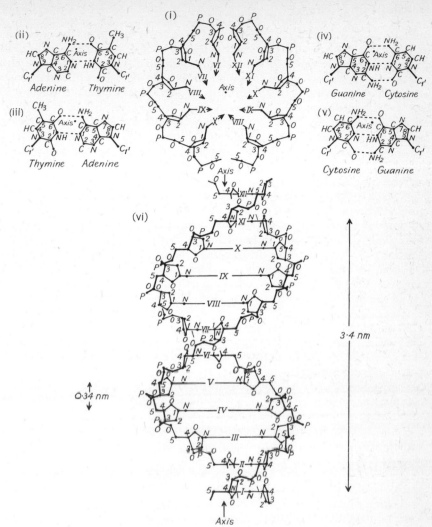

FIGURE 11.5    Diagrams to show the structure of DNA as proposed by Watson
and Crick (1953a). The diagrams incorporate the inferences from
later X-ray crystallographic work and are based on those of
Wilkins (1957).

(i) A cross-section of the molecule to show the two phosphate
sugar chains. N = Nitrogen. O = Oxygen. P = Phosphorus.
1, 2, 3, 4, 5 = Carbon atoms of deoxyribose. Hydrogen atoms
are not shown, nor the oxygen atoms attached to the phosphorus atoms, other than the oxygen atoms which form part
of the backbones of the chains. The Roman numerals indicate
the positions of successive base pairs numbered from below
upwards.

(ii) The adenine-thymine base pair.

(iii) The thymine-adenine base pair.

(iv) The guanine-cytosine base pair.

(v) The cytosine-guanine base pair.

(vi) The molecule seen from the side, lettered and numbered as in
(i). The phosphate sugar chains in the foreground are
thickened.

helix, as had been supposed, the link between sugar and base (the glycosidic bond) will always have the same orientation to the axis of the helix. The space available between the two chains will therefore be constant, and pairing between bases in opposite chains can only be done by joining bases of the right size. The space appeared to be such that the purine adenine in one chain could pair with the pyrimidine thymine in the other, and the purine guanine in one chain with the pyrimidine cytosine in the other, in agreement with the chemical evidence of their equality within each pair. For all other combinations there was either too little or too much space between the chains. The base pairing was thought to occur between positions no. 1 on the purine and the pyrimidine, and also between positions no. 6 on each, with the possibility of a third hydrogen bond at positions no. 2 in the guanine-cytosine pair (but not in the adenine-thymine pair). Hydrogen bonding could occur at a number of other places in isolated nucleotides, which can pair in many ways. The specific pairing is imposed by the regularity of the phosphate-sugar chains, and is therefore to be regarded, not as an intrinsic feature of these nucleotides, but as a property which is manifest by them when they occur in a polynucleotide chain.

The occurrence of 5-methylcytosine and 5-hydroxymethylcytosine in some sources of DNA (see § 11.6) is unaccounted for on the Watson and Crick model, but the 5 position is not involved in the hydrogen bonding with guanine, and the occurrence of equimolar quantities of guanine and of total cytosine, whichever derivatives are present, is in agreement with the specific pairing hypothesis.

The proposed structure imposes no restriction on the sequence of bases within a chain, but the specific base-pairing implies that the second chain will have a base sequence complementary to that of the first.

When models were constructed using the specific base pairs, further features of the structure became evident. It was found that the helices must be right-handed. Left-handed helices could be constructed only by violating the permissible separations of atoms. It was also found that the two glycosidic bonds of each base-pair must be symmetrically disposed about a line perpendicular to the axis of the helix. This means that each base-pair, if it could be detached and inverted, would fit between the chains just as well the other way round (see Fig. 11.5(ii)–(v)). It follows that all four bases may occur in both chains. Furthermore, one nucleotide chain must always be inverted relative to the other (but with the complementary base sequence). The sequence of atoms in the backbone of one chain will be

$$\ldots -C_3-C_4-C_5-O-P-O-C_3-C_4-C_5- \ldots$$

and in the same physical direction in the other it will be

$$\ldots -C_5-C_4-C_3-O-P-O-C_5-C_4-C_3- \ldots$$

(see Fig. 11.5(vi)).

The Watson and Crick model obtained immediate support from the work of Wilkins, Stokes and Wilson (1953) and of Franklin and Gosling (1953), who, independently of Watson and Crick, deduced that the structure was probably helical, that the phosphates were on the outside, and that there were two unequally spaced co-axial components. Nevertheless, some features remained hypothetical until the model had been tested with experimental X-ray evidence. This was obtained by Wilkins and associates, when Feughelman, Langridge, Seeds, Stokes, Wilson, Hooper, Wilkins, Barclay and Hamilton (1955) showed that the Watson and Crick structure was in all essentials correct, although some modification of details was necessary, notably the placing of the bases closer to the axis of the helix. It is this slightly modified version which is shown in Fig. 11.5.

## § 11.8   The biological implications of the structure of DNA

Watson and Crick (1953*b*, 1954) had drawn attention to the genetical implications of their proposed structure for DNA. That this structure was established as substantially correct did not necessarily imply that the biological hypotheses were also true. These hypotheses were:

1. that replication occurs by breakage of the hydrogen bonds, unwinding and separation of the two chains, and the formation of a new complementary chain alongside each, and
2. that the precise sequence of the bases in the nucleotide chain is the code which carries the genetical information.

The first of these hypotheses is discussed in Chapter 12, and the second in Chapters 13 and 14. As a corollary to the sequence hypothesis, Watson and Crick suggested that mutation may be due to a base occasionally occurring in one of its less likely tautomeric forms, so that at replication the wrong base is inserted at this position in the complementary chain. The molecular basis of mutation is discussed in § 14.2. Watson and Crick also suggested that pairing between homologous chromosomes at meiosis may depend on pairing between specific bases.

The brilliant idea underlying the Watson and Crick model of DNA was that there were two complementary nucleotide chains, and that replication of the gene involved each chain acting as a template or mould for the formation of its complement. Such a mechanism for copying the gene would not require gene specificity to be transferred to protein and then back to DNA again, such as previous hypotheses of self-duplication had assumed to be necessary. Watson and Crick's idea thus gave support to the theory that the chemical basis of heredity might be DNA alone. This hypothesis was based on Avery, MacLeod and McCarty's discovery that DNA alone, without protein, appeared to be the carrier of hereditary factors in *Diplococcus pneumoniae* (see § 11.2), and Hershey and Chase's comparable discovery for virus *T*2 of *Escherichia coli* (see § 11.3). Applied to higher forms of life, this would mean that the protein part of the nucleoprotein of the chromosome played no direct part in heredity. This inference could have been drawn by Knapp and Schreiber (1939) from their experiments on UV-irradiation of *Sphaerocarpos*, though such a conclusion would have been directly contrary to the general opinion current at that time.

# 12. *The theory of semi-conservative replication of DNA*

## § 12.1  Introduction

Delbrück and Stent (1957) gave the name semi-conservative to the mechanism of replication of DNA postulated by Watson and Crick (1953*b*), since each new DNA molecule would consist of one old nucleotide chain and one new one, complementary to the old. Watson and Crick (1953*b*, 1954) had visualized the replication process as involving breakage of the hydrogen bonds between the two chains, followed by uncoiling, and synthesis of a new complementary chain alongside each old one. They assumed that the uncoiling and separation of the two chains would occur progressively from one end of the molecule like the opening of a zip-fastener, and that the new synthesis would follow at once, and hence also occur progressively along the molecule. This mechanism would imply that the synthesis would extend from the same physical end in both chains. Owing to the inverted polarity of the two chains, this would mean that in a chemical sense the synthesis was occurring in opposite directions in the two chains. Fig. 12.1 is a diagrammatic representation of this replication process.

## § 12.2  Evidence for semi-conservative replication of chromosomes

One of the first indications that the replication of DNA might be occurring in the way that Watson and Crick had suggested was obtained by Taylor, Woods and Hughes (1957) with root-tips of *Vicia faba* (Broad Bean) labelled with tritiated thymidine. Thymidine is the nucleoside of thymine and deoxyribose and is incorporated exclusively into DNA. Tritium ($^3$H) has the great advantage over other radioactive isotopes that the $\beta$-particles which it emits are of such low energy that they will penetrate only for a distance of about 1 $\mu$m in photographic emulsion. This means that the position of the labelled atoms can be determined with corresponding precision.

*Vicia faba* was chosen because it has large chromosomes and because Howard and Pelc (1951*a*) had established the time of replication of the DNA in the meristematic cells of the root-tip in relation to mitosis. At room temperature mitoses succeed one another at about 24-hour intervals, but the divisions are not synchronized so that at any given time there are nuclei at all stages of the mitotic cycle. The DNA replication extends over an 8-hour period ending about 8 hours before metaphase. Accordingly, if *Vicia faba*

186

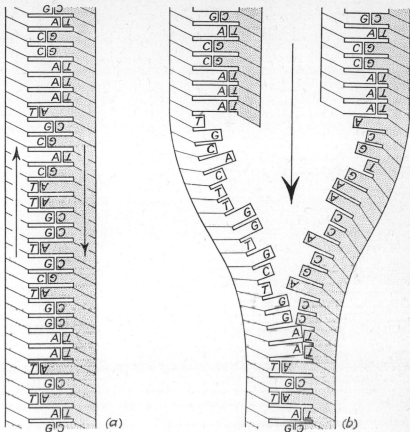

FIGURE 12.1   Diagrammatic representation of the method of replication
of DNA proposed by Watson and Crick.

(a) Part of a DNA molecule. The arrows indicate the
polarity of the two nucleotide-chains. Their spiral
coiling is not shown. $A$ = adenine, $C$ = cytosine,
$G$ = guanine, $T$ = thymine.

(b) The molecule in process of replication. The arrow
shows the direction in which chain separation and
synthesis occur.

seedlings are grown in tritiated thymidine solution for 8 hours, then thor-
oughly washed with water, and some 8 hours later the root-tips are removed,
fixed, stained by the Feulgen method, and squashed on microscope slides,
then any cells at metaphase should have incorporated tritiated thymidine at
the preceding DNA replication. Photographic film is then placed in contact
with the squashed cells, and the preparation is kept in the dark for about
two weeks before developing the film to find the position of the tritium.

The technique which Taylor, Woods and Hughes adopted differed
slightly from this. During the interval between washing to remove the

tritiated thymidine and fixing, the roots were placed in a non-radioactive mineral solution containing colchicine. There were two reasons for introducing colchicine. In the first place, colchicine causes the two chromatids of each chromosome to diverge, being held together only at the centromere. This facilitates observation of the presence or absence of the isotope in the individual chromatids. Secondly, colchicine inhibits the formation of the mitotic spindle, with the result that the chromosome number is doubled at each mitosis. Hence, the number of mitoses which have occurred since the roots entered the colchicine can be determined from the chromosome number. Taylor *et al.* kept some of the roots in the colchicine solution for 10 hours before fixing, and others for 34 hours.

Cells at metaphase after 10 hours in colchicine had the normal diploid chromosome number of 12, and the two chromatids of each chromosome were found to contain tritium, indicated by dark grains of silver above each chromatid where the photographic emulsion had been exposed. Each chromatid was uniformly labelled, and the two chromatids of a pair equally. Many of the cells at metaphase after 34 hours in colchicine contained 24 chromosomes, as was to be expected if a second mitotic cycle had intervened before the fixation. These 24-chromosome cells showed a remarkable feature: of the two chromatids making up each chromosome, one contained tritium and the other did not. A few of the metaphases after 34 hours in colchicine showed 48 chromosomes, indicating that two successive mitoses had occurred in the colchicine and implying that the 24-hour division cycle was by no means constant. In the 48-chromosome cells, half the chromosomes in each cell showed one labelled and one unlabelled chromatid, while the other half showed no label in either chromatid. The washing prior to placing the roots in the colchicine solution was evidently sufficient to remove the isotope solution, so that subsequent replications of the DNA took place in the absence of labelled thymidine.

These remarkable observations are shown diagrammatically in Fig. 12.2. It is evident that as far as its DNA is concerned, the chromosome is composed of two sub-units, one old and one new, which extend the whole length of the chromosome, and are such that the DNA of each sub-unit is conserved. This inference is shown in Fig. 12.3. The chromosome evidently replicates semi-conservatively, but it does not follow from this that the DNA molecule or molecules that form part of it necessarily also show this method of replication. The results are compatible with each of the following structures:

(*a*)  The chromosome contains a single giant DNA molecule which replicates semi-conservatively in the way Watson and Crick proposed.

(*b*)  The chromosome contains a number of DNA molecules each replicating semi-conservatively and held together in such a way that the new nucleotide chains form a unit which at the next replication separates from the unit formed by the old chains.

(*c*)  The chromosome sub-units contain one or a number of DNA molecules replicating conservatively, that is, maintaining the whole molecular structure through the replication process, and the new molecule or molecules form a unit which at the next replication separates from the unit formed by the old molecule or molecules.

Although their results are thus open to a variety of interpretations, a further observation led to more information about the nature of the chromosomal sub-units, and made it appear highly probable that the DNA molecules were replicating semi-conservatively. Taylor *et al.* noticed that in some of the 24-chromosome cells the tritium was in one chromatid for part of the length of the chromosome, and in the other chromatid for the remainder of the length. It appeared that an exchange of segments had occurred between the two chromatids (Fig. 12.2(iii)).

Taylor (1958) investigated these exchanges in more detail, using *Bellevalia romana* as his experimental material, since it has only 8 chromosomes in the diploid cells, and two of the 4 pairs of chromosomes are individually recognizable. Using essentially the same procedure as before but rather different timing (6–8 hours in the isotope, then 10 hours in a mineral solution free of isotope, and then 12–14 hours in colchicine solution before fixation), Taylor found that exchanges between chromatids were quite frequent, and moreover within one cell two homologous chromosomes often showed exchanges between their chromatids at corresponding positions. These twin exchanges, as he called them, evidently reflected a single exchange before the chromosomes had separated (Fig. 12.2(ii)).

All the exchanges observed appeared to involve both sub-units of each chromatid. The evidence for this was the absence of anaphase bridges and of unlabelled segments at the first mitosis. Exchange between individual sub-units, one from each chromatid, would be expected sometimes to generate these (see Taylor, 1959*a*).

The proposed mechanism for the origin of twin exchanges is shown in Fig. 12.3(ii). Immediately after the first replication, that is, the one in the presence of $^3$H-thymidine, an exchange takes place between the sister chromatids such that the rejoining of sub-units is always between a labelled and an unlabelled one. This pattern of rejoining is the only one that would be possible if the two sub-units making up each chromatid were structurally different and if rejoining could occur only between sub-units that were structurally alike. With this restriction on rejoining, single exchanges can arise only through an exchange following the second replication, as shown in Fig. 12.3(iii). Since by this time there are twice as many chromosomes as at the first replication, the expected ratio of single to twin exchanges is 2:1 if their frequency of occurrence per haploid chromosome set is the same at both replications.

If there is no restriction on the rejoining of sub-units, exchanges after the first replication will give on average a 1:2:1 ratio of invisible:single:twin exchanges. An invisible exchange would arise when the rejoining in both chromatids was between the labelled sub-unit from one chromatid and the labelled sub-unit from the sister chromatid, and likewise with the unlabelled sub-units. A single exchange would arise if the rejoining in one chromatid followed the twin-exchange rules and in the other chromatid the invisible-exchange rules. Thus, with random rejoining of sub-units, only $\frac{1}{4}$ of the first-division exchanges appear as twins, another $\frac{1}{4}$ not being detected and

Diploid nuclei    Tetraploid nuclei              Octoploid nuclei

(ii)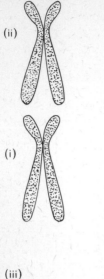

(i)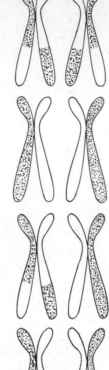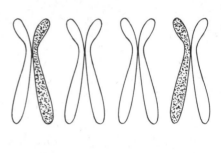

(iii)

(iv)

FIGURE 12.2    Diagram to illustrate the results obtained by Taylor and associates
by labelling the chromosomes of *Vicia faba* with ³H-thymidine. Each
drawing shows the appearance of a chromosome at metaphase of
mitosis. The two chromatids of each chromosome would normally
lie parallel but have separated, except at the centromere, as a result
of treatment with colchicine. The dots represent the silver grains
seen in the photographic emulsion overlying the chromosomes as a
result of exposure to the radiation from the ³H.

   (i) The normal findings, with both chromatids labelled after the
       replication in the presence of ³H-thymidine (diploid nuclei);
       with one chromatid labelled and one unlabelled after a further
       replication, this time in the absence of ³H-thymidine (tetra-
       ploid nuclei); and with equal numbers of unlabelled chromo-
       somes and chromosomes like those in the tetraploid nuclei,
       after a second replication in the absence of ³H-thymidine
       (octoploid nuclei).
  (ii) A twin exchange (tetraploid nuclei).
 (iii) A single exchange.
 (iv) Iso-labelling in a terminal segment in one chromosome (see
       § 17.4).

Exchanges and iso-labelling were also seen in the octoploid nuclei,
but these are not shown in the diagram.

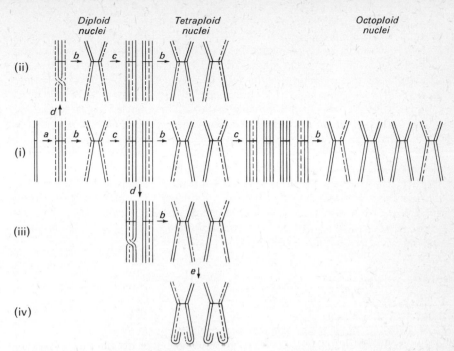

FIGURE 12.3   Diagram to illustrate the explanations proposed for the results
shown in Fig. 12.2.

(a) Replication in the presence of $^3$H-thymidine. The $^3$H is incorporated into the newly synthesized sub-units, which are shown
as broken lines on the outside of the old ones.

(b) Separation of the chromatids except at the centromere.

(c) Replication in the absence of $^3$H-thymidine. The newly synthesized sub-units are shown as unbroken lines on the outside
of each pair of old ones.

(d) Sister chromatid exchange, with rejoining of the new sub-unit
in one chromatid to the old one in the other, and *vice versa*.

(e) Sub-units folded back upon themselves.

The normal sequence of events (i) and the proposed mechanisms
for the origin of twin exchanges (ii), single exchanges (iii) and
iso-labelling (iv) are shown in separate rows of the diagram. In each
case the appearance after step *b* corresponds to that shown in
Fig. 12.2.

the remaining $\frac{1}{2}$ appearing as singles. Allowing for the doubled number of
chromosomes, and hence of exchanges, expected after the second replication,
all of which will again appear as singles, the expected ratio of single to twin
exchanges with random rejoining is $2\frac{1}{2}:\frac{1}{4}$ or 10:1.

Taylor (1959a) found 26 single exchanges and 14 twins in chromosome I
of *Bellevalia romana* when the roots were placed in colchicine from two hours
before the beginning of the $^3$H-thymidine treatment, in excellent agreement
with the 2:1 hypothesis. He inferred that the sub-units of the chromosome
are structurally different such that reunion can occur only between similar

ones. This inference, taken in conjunction with the semi-conservative replication which the chromosomes showed, was in such remarkable agreement
with Watson and Crick's proposals for the structure and mode of replication
of the DNA molecule, that it appeared highly probable that the DNA of the
chromosome was replicating in the way they suggested. The two nucleotide
chains of the DNA molecule (or molecules) in the chromosome would then
correspond to the two chromosomal sub-units. The inverted polarity of these
chains would preclude rejoining after an exchange except between chains
of the same polarity. It thus appears that the chromosome contains one
or a number of DNA molecules which replicate semi-conservatively such
that all the old nucleotide chains form one sub-unit and all the new chains
another. Herreros and Giannelli (1967) found 288 single and 128 twin exchanges in the chromosomes of human lymphocytes in culture, confirming
the 2:1 ratio. A different labelling pattern, shown in Fig. 12.2(iv), cast
doubt for some years on the validity of these conclusions about chromosome
structure and replication (see § 17.4).

## § 12.3   Evidence for semi-conservative replication of DNA

More decisive evidence about the mechanism of replication of DNA
molecules was obtained by Meselson and Stahl (1958) using DNA from
*Escherichia coli*. Their technique was to cultivate the bacteria for 14 successive
cell generations in a medium where the nitrogen was all the heavy isotope
$^{15}N$, and then abruptly to change to a normal $^{14}N$ medium. The growth of
the bacterial population was exponential and was recorded by cell counts.
Samples of about $4 \times 10^9$ bacteria were withdrawn from the culture just
before the transfer to $^{14}N$ and at intervals for several cell generations
afterwards. The generation time at the temperature of the culture, 36°C,
was approximately 50 minutes. The cell contents of each sample were
suspended in a concentrated solution of caesium chloride and centrifuged
at 44,770 revolutions per minute for 20 hours. After this time the opposing
processes of sedimentation and diffusion had set up a continuous increase of
density in the caesium chloride away from the axis of rotation, and the DNA
became concentrated at a point in this gradient which is dependent on its
own density. The position of the DNA was recorded by ultraviolet absorption
photographs. It had previously been established by centrifuging a mixture
of DNA from $^{14}N$ and $^{15}N$ bacteria that two separate bands then appeared
on the ultraviolet photographs due to the difference in molecular weight.

After one generation in the $^{14}N$ medium, all the DNA showed a band at a
position halfway between the $^{14}N$ and $^{15}N$ positions, while after 2 generations,
bands were present at this position, and at the $^{14}N$ position, and they were
of equal intensity (see Fig. 12.4). The interpretation of these results is shown
in the diagram. It is concluded that the DNA molecules consist of two sub-
units containing equal amounts of nitrogen and which survive intact for many
generations, and that each daughter molecule receives only one parental
sub-unit. This is in precise agreement with the expectations from the
Watson and Crick model for DNA replication, although the experiment does

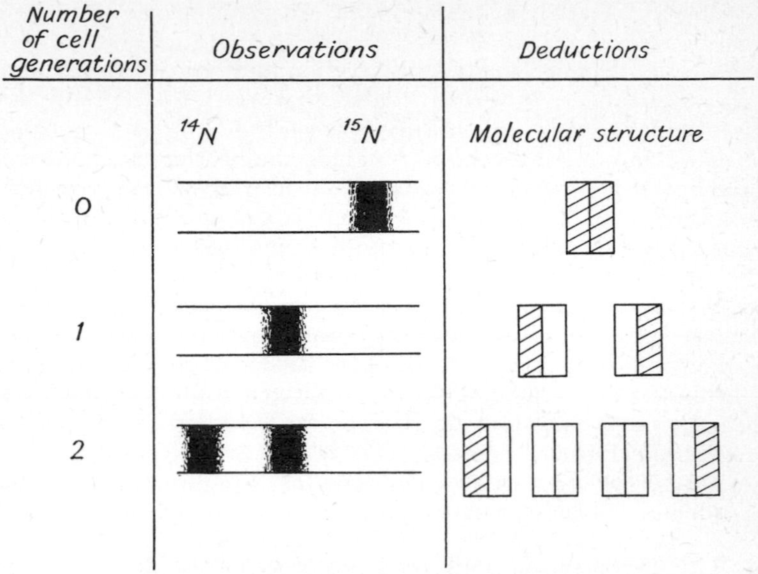

FIGURE 12.4    Meselson and Stahl's observations of the position of the DNA of *Escherichia coli* in a caesium chloride density gradient, after successive cell generations following transfer from a 15N to a 14N culture medium. The deductions concerning the mode of replication of DNA are shown on the right. The rectangles represent the two sub-units inferred to exist within each molecule, open rectangles containing 14N and shaded rectangles 15N.

not establish that the sub-units are single nucleotide chains. However, Meselson and Stahl took the DNA of intermediate density obtained after one generation in 14N medium, and heated it to 100°C for 30 minutes in the caesium chloride solution. Such heat treatment was thought to separate the nucleotide chains of DNA molecules. When the heated DNA was centrifuged, two bands were obtained instead of the previous one, and moreover they corresponded in position with the bands obtained when pure 14N and pure 15N DNA from *E. coli* were heated and then centrifuged. It was concluded that the DNA of intermediate density consisted of molecules with one heavy and one light polynucleotide chain, and that these were separated by the heat treatment.

Sueoka (1960) applied the density gradient centrifugation technique to a chromosomal organism: *Chlamydomonas reinhardii*. The generation time for mitotically dividing cells at 25°C was about 3 hours. Samples of the culture were removed immediately before transfer from 15N to 14N culture solution, and at 2-hourly intervals for the next 10 hours, and the DNA from each sample was centrifuged in concentrated caesium chloride solution. DNA from *Bos* was used as a density reference. It did not interfere with the observation of the *Chlamydomonas* DNA because of a difference in their densities. The results obtained by Sueoka with *Chlamydomonas* were the same as Meselson and Stahl had obtained with *Escherichia*, namely a single band midway between the $^{14}$N and $^{15}$N positions after one generation, and two bands of equal intensity, one at this position and one at the $^{14}$N position, after two generations. It was concluded that the DNA of *Chlamydomonas* replicates semi-conservatively as in *Escherichia*.

Djordjevic and Szybalski (1960) obtained similar results using a culture of human cells. Instead of cultivating them in a heavy nitrogen medium, they transferred the cells to a medium containing the nucleoside of 5-bromouracil and deoxyribose (5-bromodeoxyuridine). This base analogue is incorporated into DNA in place of thymidine (see § 12.4), and has a higher molecular weight, thereby allowing separation by density gradient centrifugation of normal DNA from DNA containing 5-bromouracil. The first generation cells after transfer to the base analogue gave an intermediate band in the centrifuge tube, and the second generation gave intermediate and heavy bands of equal intensity. Clearly the DNA molecules in the human cells were replicating semi-conservatively as in *Chlamydomonas* and *Escherichia*.

It is clear from these experiments that DNA replicates as Watson and Crick proposed insofar as each old chain has a new complementary chain.

### § 12.4    The synthesis of DNA by DNA polymerase I

Lehman, Bessman *et al.* (1958) purified an enzyme from *Escherichia coli* which would bring about the synthesis of DNA *in vitro*. The enzyme, now called DNA polymerase I, catalysed the incorporation of deoxyribonucleotides into DNA, using as substrate the four deoxynucleoside 5′-triphosphates, that is, of adenine, guanine, cytosine and thymine. Highly polymerized DNA was necessary as a primer and the presence of magnesium ions was also required. The triphosphates needed as substrate had been synthesized by enzymic phosphorylation by adenosine triphosphate (ATP) of the 5′-monophosphates. The existence in *E. coli* of a specific kinase for the phosphorylation of each of the four nucleotides had already been established.

Bessman, Lehman, Simms and Kornberg (1958) showed that in the polymerization process the omission of any one of the four nucleoside triphosphates reduced the reaction rate to about $\frac{1}{200}$, and omission of the DNA primer abolished it entirely (although it was later shown that some synthesis could occur after a lag period of 3 to 6 hours). They achieved a synthesis *in vitro* of up to 20 times the initial quantity of DNA, and they showed that an amount of inorganic pyrophosphate ions was released equal to the amount of nucleotide incorporated. The synthetic reaction is shown diagrammatically in Fig. 12.5. The necessity for all four nucleoside triphosphates was remarkable and, with the need for DNA as a primer, suggested that synthesis was by a template mechanism such as Watson and Crick had proposed.

Confirmation of the template action of the primer was obtained by Lehman, Zimmerman, Adler, Bessman, Simms and Kornberg (1958) who showed that the base composition of the synthesized DNA was the same as that of the primer. They experimented with DNA from *Aerobacter aerogenes*, *Bos taurus*, *Escherichia coli* and its *T*2 virus, and *Mycobacterium phlei*. The

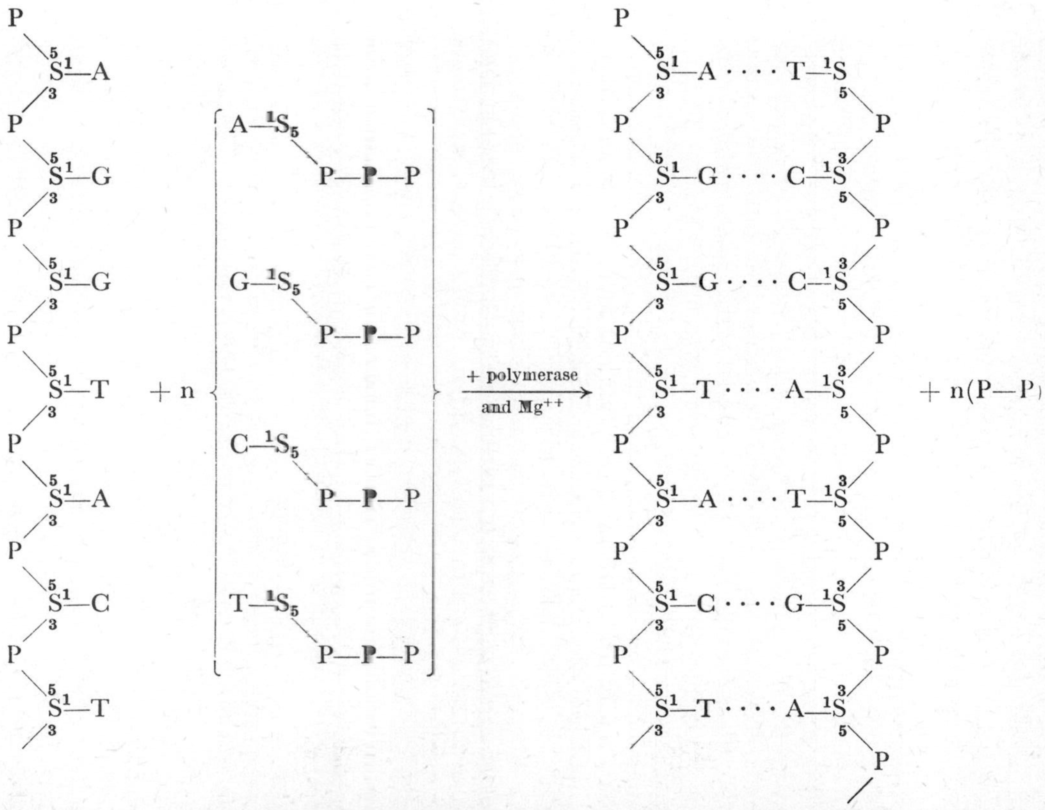

FIGURE 12.5    The synthesis of DNA from the four deoxynucleoside 5'-triphosphates, in the presence of DNA polymerase I, magnesium ions, and DNA of high molecular weight as a primer. Inorganic pyrophosphate ions are released in proportion to the number of nucleotides incorporated. A = adenine, C = cytosine, G = guanine, P = phosphate, S = sugar (deoxyribose), T = thymine.

adenine plus thymine contents of the DNA of these organisms, expressed as molar percentages of the total bases, cover a wide range, from 33% for *M. phlei* to 66% for *T*2, and in every instance the product of the polymerase reaction did not differ significantly in base composition from the primer. Moreover the components of each complementary pair of bases on the Watson and Crick model occurred with equal frequency, as in the primer. Variation in the proportions of the four nucleoside triphosphates in the reaction mixture did not affect their proportions in the product, which appeared to be determined solely by their proportions in the primer.

Dunn and Smith (1954) and Zamenhof and Griboff (1954) had discovered that 5-bromouracil and 5-iodouracil, although not occurring naturally, were incorporated into the DNA of *E. coli* and of its *T*2 virus in place of thymine (5-methyluracil) such that the total molar quantity of the uracil derivatives (including thymine) was equal to the molar quantity of adenine. This was in keeping with the base-pairing predictions of the Watson and Crick model, since these uracil compounds differ from thymine only in the substituents at position 5 of the pyrimidine ring, which is a part of the molecule not involved in hydrogen bonding with adenine. In view of the incorporation of these compounds into the DNA of living cells, Bessman, Lehman, Adler, Zimmerman, Simms and Kornberg (1958) tested the ability of DNA polymerase I obtained from *E. coli* to accept these unnatural bases (in the form of the deoxynucleoside triphosphates) as substrates for DNA synthesis. They found that uracil and 5-bromouracil were incorporated specifically in place of thymine; 5-methyl- and 5-bromo-cytosine in place of cytosine; and hypoxanthine (6-hydroxypurine) but not xanthine (2,6-dihydroxypurine) in place of guanine (2-amino-6-hydroxypurine). These results were in agreement with the specific base pairing in the Watson and Crick model, since the replacements are all by analogues with similar hydrogen-bonding capacities, although the failure of xanthine to be incorporated is unexplained. The specific replacement by nucleotides with similar base-pairing properties, revealed in this experiment, demonstrated that specific hydrogen bonding must be involved in DNA synthesis.

In order to study the very slight incorporation of nucleotides into DNA which was found to occur by the action of the DNA polymerase I when only one deoxynucleoside triphosphate was present instead of all 4, Adler, Lehman, Bessman, Simms and Kornberg (1958) devised an ingenious technique. The 5′-nucleotide of cytosine, labelled with $^{32}$P, was phosphorylated to obtain the nucleoside triphosphate, and then this was incorporated into DNA through the action of the polymerase. Sedimentation coefficients in the ultracentrifuge showed that the incorporated radioactive nucleotides had formed covalent linkages to molecules of the same average size as those of the *Bos* DNA primer. The DNA molecules containing the $^{32}$P were hydrolysed, first with deoxyribonuclease from *Micrococcus pyogenes*, and then with a phosphodiesterase from *Bos* spleen. These enzymes break down the molecule into 3′-nucleotides. It was found that all 4 of these 3′-nucleotides (that is, of adenine, guanine, cytosine, and thymine) showed radioactivity, indicating that the phosphate which had entered the DNA molecule attached by a 5′

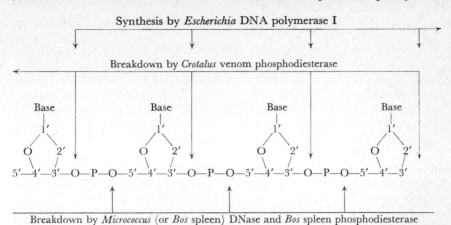

FIGURE 12.6　Diagram to show the action of various enzymes in DNA synthesis and breakdown. The *E. coli* DNA polymerase I adds 5′-nucleotides progressively to the 3′ end of the chain, and the *Crotalus* venom phosphodiesterase reverses this, detaching 5′-nucleotides progressively from the 3′ end. The *Micrococcus* deoxyribonuclease and *Bos* spleen phosphodiesterase cause breakdown into 3′- nucleotides.

linkage to the nucleoside of cytosine emerged attached by a 3′ linkage to any of the 4 nucleosides. This showed that the action of the polymerase was to attach 5′ nucleotides to the 3′ position at the nucleoside end (as distinct from the phosphate end) of a nucleotide chain, and to do this irrespective of the base in this terminal nucleotide (see Fig. 12.6). Similar results were obtained when the experiment was repeated using adenine or thymine in place of cytosine in the $^{32}$P-labelled 5′-nucleotide.

Confirmation that DNA polymerase I adds nucleotides at the nucleoside end of the chain was obtained by hydrolysis of the product with phosphodiesterase from the venom of the snake *Crotalus adamanteus*. This enzyme hydrolyses DNA nucleotide chains to 5′-nucleotides, and does so stepwise from the nucleoside end of the chain. It was found that nearly all the radioactivity was liberated when less than 3% of the nucleotides had been released, indicating that the venom phosphodiesterase reversed the action of the polymerase (see Fig. 12.6).

Richardson, Schildkraut and Kornberg (1964) showed that under the action of *E. coli* DNA polymerase I the two chains of DNA act one as template and one as primer, nucleotides being added at the 3′-hydroxyl terminus of the primer chain complementary to those in the other or template chain. It was found that DNA synthesis is inhibited by a 3′-phosphate terminus.

## § 12.5　Demonstration of the inverse polarity of the two nucleotide chains

By an ingenious experiment using essentially the same techniques as were used to establish the mechanism of the *in vitro* synthesis of DNA by *E. coli*

DNA polymerase I (see § 12.4), Josse, Kaiser and Kornberg (1961) demonstrated that the two nucleotide chains of the DNA molecule are relatively inverted in just the way Watson and Crick (1953a) had supposed. The principle underlying the experiment was to assemble the molecule using 5'-nucleotides and then dismantle it again, but into 3'-nucleotides. Radioactive phosphorus ($^{32}$P) introduced as part of, say, adenine 5'-nucleotides, would be recovered distributed among all four 3'-nucleotides with frequencies that could be determined and would depend on the source of the primer. The experiments differed from those used to establish the mechanism of synthesis (§ 12.4) in that all four nucleoside triphosphates were provided, with the result that extensive DNA synthesis could occur. In any particular experiment only one of the four bases was associated with the label in the phosphorus of the nucleotide, but each, in turn, was so labelled.

It is conventional to indicate 3'-phosphates with the phosphate symbol ($p$) to the right of the nucleoside symbol and 5'-phosphates with the phosphate symbol to the left. Thus $GpA$ stands for $G3p5A$, where $G$ and $A$ are the deoxynucleosides of guanine and adenine, respectively. One can imagine the experiment as the passing of the radio-phosphorus from the left hand of one nucleoside to the right hand of the adjacent nucleoside to the left. If the two chains of the molecule 'face the same way', passage of the radio-phosphorus will occur to the neighbour to the left in each chain. In other words, transfer from, for example, adenine to guanine will be expected to occur with the same frequency as transfer from thymine to cytosine ($GpA = CpT$). On the other hand, if the chains 'face opposite ways', passage to the left in each will mean passage in opposite directions physically. Hence, transfer from adenine to guanine would be expected to occur with the same frequency as transfer from cytosine to thymine ($GpA = TpC$). Thus, by determining the relative frequencies with which each base has each of the others to its left as its nearest neighbour, it should be possible to determine whether or not the two chains are relatively inverted. This is on the assumption that the nucleotide sequence is non-random.

Josse, Kaiser and Kornberg obtained such data using, in turn, as primer, DNA from 6 species of bacteria, 5 strains of bacterial viruses and *Bos* (calf thymus). Without exception, the results agreed with the hypothesis that the two chains were relatively inverted, and they contradicted any suggestion that the chains were of similar polarity. A selection of their data is given in Table 12.1, together with some of the data of the same kind obtained by Swartz, Trautner and Kornberg (1962) for a further range of organisms (2 species of plants, 10 species of animals and several more bacterial viruses). The 16 possible combinations of the 4 nucleosides are shown in the first column, and those expected to occur with equal frequency if the two chains are of opposite polarity are paired, and separated by horizontal lines. Those expected to occur with equal frequency if the chains were of the same polarity are joined by arrows. From the data the percentage frequencies of the two base-pairs have been obtained and are given at the top of the table for each species. From these values the expected frequencies of each of the 16 combinations have been calculated on the assumption that the nucleotide

TABLE 12.1   Some of the data of Josse, Kaiser and Kornberg (1961) and Swartz, Trautner and Kornberg (1962) for the frequencies of different base sequences in the DNA of various organisms. For explanation, see text.

| Base Sequence | Myco-bacterium phlei | Aerobacter aerogenes | Chlamydo-monas reinhardi | Homo sapiens | Haemo-philus influenzae | Para-centrotus lividus |
|---|---|---|---|---|---|---|
| $A + T$ | 32·6 | 44.3 | 46.2 | 59.5 | 61·8 | 64·2 |
| $G + C$ | 67·4 | 55·7 | 53·8 | 40·5 | 38·2 | 35·8 |
| A3p5A | 2·4 (2·7) | 5·9 (4·9) | 6·0 (5·3) | 9·7 (8·9) | 11·6 (9·5) | 11·0 (10·3) |
| T5p3T | 2·6 (2·7) | 6·1 (4·9) | 5·9 (5·3) | 9·7 (8·9) | 11·6 (9·5) | 10·2 (10·3) |
| A3p5T / T5p3A | 3·1 (2·7) | 5·3 (4·9) | 5·4 (5·3) | 8·1 (8·9) | 9·5 (9·5) | 10·4 (10·3) |
| T3p5A / A5p3T | 1·2 (2·7) | 3·6 (4·9) | 5·3 (5·3) | 6·7 (8·9) | 7·3 (9·5) | 9·0 (10·3) |
| A3p5G | 4·5 (5·5) | 5·6 (6·2) | 6·0 (6·2) | 7·0 (6·0) | 5·0 (5·9) | 5·3 (5·7) |
| T5p3C | 4·5 (5·5) | 5·7 (6·2) | 5·7 (6·2) | 7·1 (6·0) | 4·9 (5·9) | 5·9 (5·7) |
| T3p5C | 6·1 (5·5) | 5·7 (6·2) | 4·6 (6·2) | 5·7 (6·0) | 5·2 (5·9) | 5·9 (5·7) |
| A5p3G | 6·5 (5·5) | 5·8 (6·2) | 4·4 (6·2) | 6·1 (6·0) | 5·4 (5·9) | 5·7 (5·7) |
| A3p5C | 6·4 (5·5) | 5·2 (6·2) | 6·0 (6·2) | 5·4 (6·0) | 4·9 (5·9) | 5·8 (5·7) |
| T5p3G | 6·0 (5·5) | 5·2 (6·2) | 5·5 (6·2) | 4·9 (6·0) | 4·8 (5·9) | 5·3 (5·7) |
| T3p5G | 6·3 (5·5) | 6·9 (6·2) | 7·3 (6·2) | 7·4 (6·0) | 6·7 (5·9) | 6·7 (5·7) |
| A5p3C | 6·3 (5·5) | 6·7 (6·2) | 7·7 (6·2) | 7·4 (6·0) | 6·7 (5·9) | 6·7 (5·7) |
| G3p5G / C5p3C | 9·0 (11·4) | 6·7 (7·8) | 7·1 (7·2) | 5·0 (4·1) | 3·6 (3·7) | 3·3 (3·2) |
| C5p3C | 9·0 (11·4) | 6·5 (7·8) | 7·4 (7·2) | 4·7 (4·1) | 3·7 (3·7) | 3·8 (3·2) |
| G3p5C / C5p3G | 12·2 (11·4) | 10·3 (7·8) | 9·2 (7·2) | 4·3 (4·1) | 5·3 (3·7) | 3·1 (3·2) |
| C3p5G / G5p3C | 13·9 (11·4) | 8·8 (7·8) | 6·3 (7·2) | 1·0 (4·1) | 3·8 (3·7) | 2·0 (3·2) |

sequence is a random one. These frequencies, as percentages, are given in brackets after each observed percentage frequency.

In addition to the clear indication that the chains are of opposite polarity (equality within the pairs associated in the table, and frequent inequality between the pairs joined by arrows), the data show that each species has a unique pattern of base neighbours. Swartz, Trautner and Kornberg found evidence of fairly consistent differences between bacteria and chromosomal organisms in some of the sequence patterns, notably the low frequency of cytosine to the left of guanine ($CpG$) in chromosomal organisms (illustrated by *Chlamydomonas*, *Homo*, and *Paracentrotus* in the table) and the high frequency of guanine to the left of cytosine ($GpC$) in bacteria (illustrated by *Mycobacterium*, *Aerobacter*, and *Haemophilus* in the table). These peculiarities appear to be independent of the guanine-cytosine frequency, which varies widely in different species of both bacteria and chromosomal organisms (see Fig. 14.15). The shortage of $CpG$ in chromosomal organisms, particularly mammals, is discussed in § 14.13.

Josse, Kaiser and Kornberg found that, for any particular species, the same sequence patterns were obtained when DNA obtained directly from the organism was used as primer and when the DNA prepared enzymatically was itself used as primer. This indicated that the base sequence was preserved in the replication process. Swartz, Trautner and Kornberg found that a number of different tissues of *Bos* gave the same sequence pattern, and similar results were obtained with *Mus*. It was evident that, like DNA content (§ 11.4), and overall base composition (§ 11.6), the base sequence was species-specific and tissue-independent.

Confirmation that the two nucleotide chains of DNA are of relatively inverted polarity (antiparallel) has been obtained by Chargaff *et al.* (1965) by hydrolysis of the DNA of *Bos taurus*, and isolation of purine dinucleotides and pyrimidine dinucleotides by chromatography. It was found that the sequence $A3p5G$ was 1·3 times as frequent as the sequence $G3p5A$, and the inverted complements of these sequences had corresponding frequencies, $T5p3C$ being 1·3 times as frequent as $C5p3T$.

## § 12.6   The *in vivo* synthesis of DNA

The mechanism of DNA synthesis shown by DNA polymerase I extracted from *Escherichia coli* by Kornberg and associates, and discussed in § 12.4 and § 12.5, involves the addition of 5′-nucleotides to the nucleoside end of a nucleotide chain. This mechanism alone would not explain how both chains of a DNA molecule could replicate continuously from the same physical end, as Watson and Crick had postulated, since this would require the attachment of nucleotides at the phosphate end of one of the chains. Several authors have found evidence that at the level of resolution provided by the light microscope, or by widely spaced genes, replication *in vivo* can occur in the same physical direction alongside both nucleotide chains.

Cairns (1963) demonstrated this with *E. coli*. His technique was to grow the bacteria in the presence of ³H-thymidine for a known length of time and then find the position of the tritium by gentle extraction of the DNA followed by autoradiography. The film was exposed to the emissions from the tritium for about 2 months before being developed. It was already known that each bacterium in an exponentially growing culture syntheses DNA almost continuously. When bacteria were grown in the presence of tritiated thymidine for 3 minutes, it was found that two pieces of DNA each 60 to 80 $\mu$m long were labelled, and about twice this length was labelled with 6 minutes of treatment. At this rate of synthesis, up to about 800 $\mu$m of DNA could be covered in the cell generation time, which at 37° was about 30 minutes. If the DNA was extracted immediately after the labelling the two labelled pieces of DNA were close together, but if the bacteria were allowed to continue to grow for 15 minutes in unlabelled thymidine, the two labelled pieces of DNA were found to be further apart. It was inferred that two labelled chains were being formed simultaneously in the region of replication.

In a second series of experiments, Cairns grew the bacteria in the presence of the ³H-thymidine for a period of 1 hour. This would allow the occurrence of up to two complete replications of the DNA in the presence of the label. The autoradiographs so obtained revealed up to 900 $\mu$m of DNA labelled in an unbroken piece. This indicated that the whole of the bacterial DNA occurs in one piece, since assuming it contains two nucleotide chains and has the Watson and Crick structure, this length is in reasonable agreement with previous estimates of the DNA content of *E. coli* nuclei.

Furthermore, whenever a fork was visible, indicating that the DNA had been caught in the act of replicating, one limb of the fork showed twice the density of grains in the emulsion compared with the other limb and with the remainder of the molecule. This is what would be expected if the molecule replicates semi-conservatively, because the first replication in the presence of the tritium would label one nucleotide chain, and at the second replication in the presence of the tritium these labelled and unlabelled chains would separate and a new labelled chain would be laid down alongside each. Hence, at the end of the second replication one daughter molecule would be labelled in one of its chains and the other in both. Different molecules showed the second duplication at various stages of completion and it was evident that the replication was proceeding progressively from one end of the molecule to the other. Some further discoveries made by Cairns by auto-radiography of the DNA of *E. coli* are discussed in the next section.

Genetic evidence that, at the level of resolution provided by widely-spaced genes, replication of DNA *in vivo* can occur in the same physical direction alongside both nucleotide chains has been obtained by Yoshi-kawa and Sueoka (1963) with *Bacillus subtilis*. The frequency of genetic transformation of 10 different characters was studied, using the DNA from an exponentially growing population, and also of a culture in the stationary phase. On the assumption that the cells from the stationary phase have a complete chromoneme*, and not one that has partially replicated, comparison of the relative transforming activity of the DNA from the exponential and stationary growth phases should provide information about the mechanism of replication of the DNA. It was found that a genetic map could be constructed such that the relative transforming activity of a gene near one end was approximately double that of a gene near the other end, with a pro-gressive decline in between. This is what would be expected in an unsyn-chronized cell population if there was oriented replication from one end of the chromoneme to the other, with both chains undergoing replication and with little interval between the end of one replication and the beginning of the next. Genes near the replication origin would then be represented almost twice as often in the DNA of the exponentially growing population as genes near the replication terminus (but see p. 209).

All these results, obtained by various methods, are in agreement with the hypothesis that the DNA in these bacteria replicates progressively from one end to the other by synthesis of a new chain alongside each old one. These experiments have not established, however, whether replication *in vivo* occurs in both chemical directions, that is, by the addition of 5′-nucleotides at the free 3′-hydroxyl end of one chain and at the free 5′-phosphate end of the

---

* Chromoneme = bacterial chromosome—see § 11.1.

other. It is possible that DNA synthesis occurs only by the addition of nucleotides to the 3'-hydroxyl termini, as with the synthesis catalysed by DNA polymerase I. In at least one daughter-molecule the process would then need to be discontinuous, since the synthesis would be in the direction away from the growing-point. In this way short stretches could be synthesized in the 5' to 3' direction and subsequently be connected together by the formation of phosphodiester linkages. Okazaki *et al.* (1968), from study of the length of newly-synthesized chains, favoured discontinuous synthesis in both chains, the segments comprising 1000–2000 nucleotides.

Kelly, Atkinson, Huberman and Kornberg (1969) discovered that DNA polymerase I of *E. coli* could excise thymine dimers and other mismatched sequences from duplex DNA which had been irradiated with ultraviolet light (see § 16.6). DeLucia and Cairns (1969) made the remarkable discovery that a mutant of *E. coli* defective in DNA polymerase I could nevertheless multiply normally! The mutant (called *polA* 1) had less than 1% of the normal level of the enzyme and had increased sensitivity to ultraviolet light. These discoveries suggested that DNA polymerase I was involved in the repair of DNA rather than in its normal synthesis. Kelley and Whitfield (1971), using a temperature-sensitive DNA polymerase I mutant isolated by DeLucia and Cairns, showed that the mutant had an abnormal enzyme. This established that the *polA* mutations, which both mapped at the same locus, were indeed in the structural gene for DNA polymerase I.

Okazaki *et al.* (1970), working with the *polA* 1 mutant, concluded that the short DNA chains which they had previously observed were synthesized in association with the cell membrane, and Knippers (1970) and T. Kornberg and Gefter (1971) isolated DNA polymerase II from the membrane fraction. The enzyme requires a terminal 3' hydroxyl group and a complementary template chain, and it brings about synthesis in the 5' to 3' direction like DNA polymerase I. Gefter *et al.* (1971) obtained evidence for a third enzyme (DNA polymerase III) in *E. coli*.

Using [3]H-thymine instead of [3]H-thymidine, Werner (1971a) found evidence that thymine unlike thymidine is rapidly incorporated into large segments of DNA. He concluded that replication draws on thymine and the other bases, and so differs fundamentally from repair which uses thymidine and the other nucleosides as precursors, and he suggested (Werner, 1971b) that in replication the double helix is unwound well ahead of the enzyme and that the exposed chains select the deoxyribonucleotides in some activated form. Brewin (1972) argued that this activation might involve a carrier molecule of ribonucleic acid (see § 14.4) analogous to the mechanism known to operate in protein synthesis (see § 14.10). Evidence for the involvement of ribonucleic acid in the initiation of DNA replication in *E. coli* has been obtained by Lark (1972) and others: see p. 209.

Drake and Greening (1970) found that mutant forms of the DNA polymerase of virus *T*4 of *E. coli* suppressed certain kinds of mutation by particular chemicals. This enzyme catalyses the replication of *T*4 DNA. In addition to showing that the mutagenic specificities of chemicals (§ 14.2) partly depend on the characteristics of the replication enzyme, this discovery also showed that the enzyme is involved in the selection of the complementary nucleotides during replication.

## § 12.7   The mechanism of separation of the two nucleotide chains

From the evidence given in this chapter, there is reason to believe that DNA replicates semi-conservatively by the synthesis of a new chain alongside each old chain. As pointed out in § 12.1, in order to allow the parental nucleotide chains to separate prior to such replication, it would be necessary to break the hydrogen bonds between the complementary bases and to uncoil the chains. The breaking of the comparatively weak bonds would be a simple matter compared with the formidable problem posed by the uncoiling.

Levinthal and Crane (1956) suggested a possible solution. If the molecule rotated about the helix axis the chains could readily separate progressively from one end of the molecule, provided the direction of rotation was anti-clockwise as seen looking down on the end where separation was occurring, and provided the chains as they separated continued to rotate each about its own helix axis. As suggested by Watson and Crick, the molecule was visualized as shaped like a letter Y during the separation process with the crotch of the Y moving progressively downwards. Levinthal and Crane, using data for virus $T2$ of *Escherichia coli* multiplying at 37°C, made estimates of the viscous drag and of the time taken for separation to occur (100 seconds), and came to the conclusion that the energy required for the rotation of the molecule in order to separate the chains was only about one thousandth of that required for the formation of the new phosphate-sugar bonds in the polymerization of nucleotides to form the two new chains produced in the replication process. They assumed that the DNA of $T2$ consisted of molecules about 20 $\mu$m long (6,000 turns of the helix).

It is now known that the whole of the DNA of virus $T2$ occurs in a single molecule, which measures about 55 $\mu$m in length and contains about 150,000 nucleotide pairs, implying 15,000 turns of the double helix. This has been shown by measurement of the length of the thread when gently released from the head of the virus by extraction with phenol or after rupture by osmotic shock. Such measurement can be made either from autoradiographs (Cairns 1961) or from electron micrographs (Kleinschmidt *et al.* 1962; cf. Fig. 11.2). Since the molecule is known to replicate many times in the interval of about 30 minutes between infection of the host and release of the progeny virus, it is evident that an exceedingly efficient uncoiling mechanism must exist. Similar arguments apply to DNA replication in other organisms.

Kuhn (1957) considered that rotatory Brownian motion of the molecular axis was capable of achieving an essential uncoiling of a DNA molecule 3 $\mu$m long (900 turns of the helix) within one second and a complete separation of the chains of such a molecule within 50 to 80 seconds. Longuet-Higgins and Zimm (1960) have suggested that in addition there may be a decrease of free energy which would produce a small torque near the axis of the coil at the point of separation in the partly untwisted coil, and so drive the uncoiling.

Cairns (1963) from tritium autoradiographs of intact chromonemes of *Escherichia coli* (see § 12.6) made a quite unexpected discovery: each replicating molecule had the ends of the fork joined (see Fig. 12.7). Furthermore, from the autoradiographs the chromoneme appeared to be circular. Cairns (1964), using a slightly modified technique involving extraction of the tritium-labelled DNA with lysozyme, obtained autoradiographs showing the *E. coli* DNA as an intact circle of circumference approximately 1.1 mm. Usually the DNA was in process of duplication, having the form of two contiguous loops, as in Fig. 12.7. Ogawa, Tomizawa and Fuke (1968) isolated replicating DNA of virus $\lambda$ of *E. coli* and showed from electron micrographs that its structure when replicating was circular or, more strictly, $\theta$-shaped. Jaenisch, Mayer and Levine (1971) made a similar discovery with simian virus 40 (*SV*40) grown on cultured kidney cells of *Cercopithecus aethiops* (African Green Monkey). For rotation of such molecules about the axis of the double helix, there must be at least one break in one of the chains of the duplex, so that the unbroken phosphodiester bond in the other chain can act as a swivel. The simplest possibilities are a persistent break at the initiation point, or temporary breaks near the growing point. Champoux and Dulbecco (1972) obtained evidence for an enzyme in cells of *Mus musculus* which might serve as a swivel during DNA replication. The enzyme apparently breaks one chain of the DNA, forms a complex that allows the chains to rotate relative to the helix axis, and then reverses the reaction and seals the break.

Spatz and Crothers (1969) measured the rate of denaturation (separation of chains) of the DNA of viruses *T*2 and *T*7 of *E. coli* under various conditions, in order to find what determines the rate of unwinding of large DNA molecules. The optical density of the solution was used as a measure of uncoiling, which was brought about by raising the temperature. They concluded that even if there was only one single-chain break in a circular DNA molecule as large as that of *E. coli*, rotational diffusion could cause the DNA to turn sufficiently rapidly to account for the observed rate of DNA replication. This was on the assumption that whenever the Brownian motion rotated the double helix through the angle corresponding to one base pair, synthesis of a new base pair took place and prevented reverse rotation. They considered it unnecessary to postulate a special motor for unwinding, provided this ratchet mechanism was efficient.

### § 12.8   How replication is organized

Yoshikawa, O'Sullivan and Sueoka (1964) discovered that in rapidly growing *Bacillus subtilis* a second DNA replication may begin before the first has been completed. Cooper and Helmstetter (1968) put forward a general hypothesis to accommodate single or multiple replication points and the observation that there is more DNA per cell in fast-growing cells (see Fig. 12.7).

As already indicated, the 55 $\mu$m long chromoneme of virus *T*2 and the 1100 $\mu$m long chromoneme of its host, *E. coli*, each constitute one replicating unit. Jacob and Brenner (1963) proposed the term *replicon* for

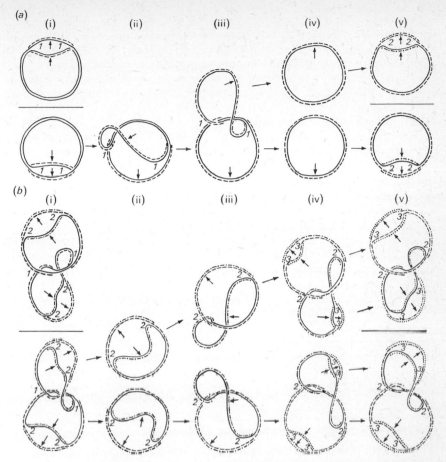

FIGURE 12.7  Diagrams based on those of Cairns (1964) of stages of replication of the DNA of *Escherichia coli* to illustrate the hypothesis of Cooper and Helmstetter (1968), according to which the DNA replication time is constant at 40 min and the interval between the end of replication and the occurrence of cell division is also constant at 20 min, but the cell doubling time and hence the interval between successive DNA replications can vary from 20 to 60 mins.

(*a*) Stages of replication at $12\frac{1}{2}$ min intervals ((i)–(v)) when the doubling time is 50 min.

(*b*) Stages of replication at $6\frac{1}{4}$ min intervals ((i)–(v)) when the doubling time is 25 min.

The numbers 1–3 show the growing-points of successive replications. An unbroken curved line indicates a nucleotide chain synthesized before the first numbered replication, a broken line a chain synthesized at replication no. 1, alternate dots and dashes at no. 2, and a dotted line at no. 3. An unbroken straight line indicates cell division. The cell divisions shown in diagrams (v) occur 20 min after the completion of replication no. 1. The arrows within the circles show the replication origin. There is now evidence that DNA replication proceeds in both directions from the origin (see p. 207) and the diagram has been modified to accommodate this.

such a unit of replication. It is established on good evidence that a fragment of a chromoneme, transferred to another cell (see § 16.3), is unable to replicate but may recombine with the intact chromoneme of the acceptor cell. Jacob and Brenner suggested that the capacity of chromonemes to replicate only as a whole and not when fragmented implies the presence and activity of specific determinants controlling replication (see § 15.9).

Chromosomes contain so much more DNA than the chromonemes of bacteria and viruses that it is not surprising that the replication mechanism appears to be differently organized. If the whole of the DNA in a chromosome consisted of one molecule it would have a length of many centimeters. It appears that a chromosome does not replicate progressively from one end to the other. This was first demonstrated by Taylor (1960) with cultured cells of *Cricetulus griseus*. It has been particularly clearly shown by Schmid (1963) with human cells in tissue culture. Each of the chromosomes has its own specific pattern of early and late replicating regions, as revealed by tritium labelling. Evidence for the constancy of these regions at the molecular level was obtained by Mueller and Kajiwara (1966). Using human cells in culture, synchronized in division by amethopterin treatment, they labelled with $^{14}$C-ribosylthymine the DNA which replicated early; at a subsequent division they again labelled the early-replicating DNA, but this time with 5-bromodeoxyuridine. The heavy DNA containing the 5-bromodeoxyuridine was separated from the remainder by density-gradient centrifugation and found also to contain the $^{14}$C, evidently in the other chain. This establishes that the early- and late-replicating segments are constant in position at the molecular level. This suggests that a chromosome may be made up of a series of replicons. Keyl (1965*a*, *b*) and Pelling (1966) have suggested that the molecular unit of replication in the chromosome is the chromomere. The evidence for this is presented in Chapter 18. As with the bacterial chromoneme, replication will require a single-chain break ahead of the replication fork to allow the molecule to rotate about its axis. In order to reconcile the existence of a number of replicons in a chromosome with the observations of Taylor and associates that old and new sub-units separate from one another throughout the length of the chromosome (see § 12.2), it is necessary to assume that, when replication has been completed in two adjacent replicons, old chain is linked to old chain and new to new across the replicon junction. The simplest way in which this linking could be maintained would be if the old chains were of one polarity and the new of the other.

Cairns (1966) treated human cells in culture with tritiated thymidine and then extracted the DNA using the same technique as for bacteria. From the autoradiographs he concluded that the replicating units were joined in series end-to-end. Huberman and Riggs (1968) confirmed and extended these observations using cells of *Cricetulus griseus* and *Homo sapiens* in culture. They made the unexpected discovery that the replicons appeared regularly to be in pairs back-to-back; that is, replication appeared to be initiated simultaneously at a common point of origin between the two replicons and then to proceed (at not more than 2·5 $\mu$m/min) in opposite directions along each. Before replication begins one chain of the DNA is presumably cut at

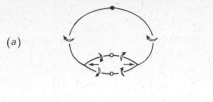

FIGURE 12.8   (*a*) Chromoneme of *Escherichia coli* with bidirectional replication
                      from a single origin.
               (*b*) Part of a chromosome showing bidirectional replication initiated
                      from two different points at different times.
               A white circle indicates a replication origin and a black circle a
               replication terminus. At these points chains can rotate independently
               of one another. Curved arrows show the direction of rotation and
               straight arrows the direction in which the replication fork is
               moving.

each replicon terminus as well as at the shared origin, in order to allow the
rotation needed to separate the chains (see Fig. 12.8). From autoradiographs
of replicating DNA of *Xenopus laevis* and *Triturus cristatus*, Callan (1972)
confirmed that replication is bidirectional from each point of origin. He made
the unexpected discovery, however, that the length of the replicating segments in *T. cristatus* is dependent on the tissue, the sequence embryonic cells–
somatic cells–spermatocytes being one of increasing length (see § 18.11).

Weintraub (1972) inferred bidirectional replication in *Gallus domesticus* by
an ingenious experiment. Synchronized cells were exposed to 5-bromodeoxy-
uridine for 5 minutes and then to [3]H-thymidine for 15 minutes. If replication
of the DNA is bidirectional, each segment containing bromine atoms will be
sandwiched between [3]H-labelled regions. Exposure of such DNA to UV will
cause it to break in the bromine segment and hence will halve the molecular
weight of the [3]H-labelled region. No such halving will occur with unidirec-
tional replication. Weintraub found the molecular weight was indeed halved
following UV exposure.

Schnös and Inman (1970) discovered from electron micrographs of
partially denatured molecules that the circular replicating molecules of virus
λ of *Escherichia coli* also usually had diverging growing-points from a fixed
point of origin. The partial denaturing was brought about by high pH or
high temperature and identified the positions of the growing-points, because
denaturing always begins in specific regions of the molecule believed to be
rich in adenine thymine base pairs. Masters and Broda (1971) made a
similar discovery with *E. coli* itself. They observed the frequency with which
virus *P*1 carried different host genes to progeny cells (a process called
generalized transduction—see § 16.3) and found that this frequency decreased in both directions from the replication origin. Bidirectional replication has also been claimed for virus *SV*40 of *Cercopithecus aethiops* (Jaenisch *et*

*al.*, 1971) and for polyoma virus of *Mus musculus* (Bourgaux and Bourgaux-Ramoisy, 1971) on the basis of electron micrographs of their DNA showing single-chain regions at both forks, which are therefore both interpreted as growing-points. With the polyoma electron micrographs the single-chain regions were inferred from the appearance of the DNA following treatment with a nuclease from *Neurospora crassa* which cuts single chains but not duplex regions.

Thus, it seems likely that DNA normally replicates in both directions from its starting point, whether the molecule is linear or circular. Single-chain breaks will always be required at the replication termini, or at least ahead of the forks, to allow rotation. Inman and Schnös (1971) with λ and Bourgaux and Bourgaux-Ramoisy (1971) with polyoma independently discovered that the single-chain regions at the growing-points were normally confined to one limb at one fork and to the other limb at the other fork, suggesting that at each fork replication is discontinuous in one limb but not in the other, and moreover with the kind of symmetry between the forks that is expected if replication takes place only in one chemical direction (see Fig. 12.9).

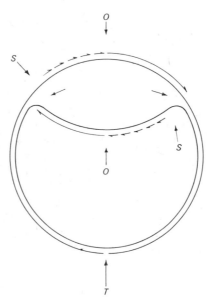

FIGURE 12.9    Diagram to show the structure of the replicating DNA of virus λ
                of *Escherichia coli* and of polyoma virus of *Mus musculus* inferred by
                Inman and Schnös (1971) and Bourgaux and Bourgaux-Ramoisy
                (1971), respectively, from the distribution of single chain regions on
                electron micrographs. O, replication origin; S, single-chain region;
                T, replication terminus. Half arrow-heads indicate the proposed
                direction of synthesis of the new chains, and unlabelled full arrows
                the direction of movement of the growing points. A single-chain
                break is shown at the replication terminus, but such breaks may
                occur at other positions ahead of the forks.

Nathans and Danna (1972) showed that the replication of *SV*40 DNA begins at a fixed point. The technique depends on the ability to separate replicating DNA molecules from non-replicating by sucrose gradient centrifugation, and on the ability to break the DNA at specific places into recognizable fragments. This breakage is made possible by a restriction endonuclease (see p. 368) from *Haemophilus influenzae* which cuts both chains of the molecule to give 11 pieces. These fragments were called A–K in order of decreasing length. By labelling the DNA with a pulse of $^3$H-thymidine before centrifugation and endonuclease digestion, the relative yield of each fragment could be measured. The non-replicating DNA showed increasing yield in the sequence C D A H E/F B G (omitting the smaller fragments I–K). This is evidently the order of synthesis from a fixed replication origin, since it is molecules that incorporated $^3$H-thymidine near the replication terminus that appear as non-replicating but labelled DNA. Conversely, the replicating molecules showed the opposite gradient, since they will consist predominantly of molecules in which synthesis has only just begun. Comparison of the temporal and physical fragment sequences will show whether replication proceeds in one or both directions.

From the labelling pattern in autoradiographs, Gyurasits and Wake (1973) concluded that DNA replication following germination of *Bacillus subtilis* spores is bidirectional. The contradiction between this conclusion and that of Yoshikawa and Sueoka (see p. 201) is unresolved.

Sugino, Hirose and Okazaki (1972) discovered, from experiments involving pulse labelling, density gradient centrifugation, and alkali and enzyme treatment, that newly-synthesized DNA fragments in *Escherichia coli* are covalently joined to RNA. This strengthens Okazaki's hypothesis (p. 202) and widens the evidence that RNA acts as a primer for DNA synthesis. The primer hypothesis was proposed by Kornberg and associates, initially for replication of the DNA of an *E. coli* phage. Subsequently, they found (Schekman *et al.*, 1972) that the initiation of DNA synthesis in virus $\phi$X174 of *E. coli* also requires RNA synthesis, and that the DNA is covalently joined to the RNA. Further studies (Westergaard *et al.*, 1973) indicated that the exonucleolytic activity of DNA polymerase I (see p. 202) may be responsible for the subsequent removal of the RNA primers during DNA replication.

# 13. *The theory of the genetic code*

## § 13.1  Introduction

Zamenhof, Brawerman, and Chargaff (1952) had found large differences in the base composition of the DNA of different species of bacteria, and this led them to suggest that there were specific differences in the sequences of nucleotides along the polynucleotide chains of DNA molecules. This idea received strong support from Watson and Crick (1953*b*), who regarded the precise sequence of the bases as the code which carried the genetical information. Their proposed replication mechanism for DNA was admirably adapted to maintain the same order of the bases in the daughter-molecules.

Dounce (1952) and, apparently independently, Gamow (1954) had the brilliant idea that the linear sequence of nucleotides in nucleic acids was responsible for determining the linear sequence of amino-acids in the polypeptide chains of protein molecules. That the linear construction of the genetic material might reflect the linear structure of specific polypeptides was a highly original idea of attractive simplicity. Moreover, it accommodated so well the one gene:one enzyme hypothesis, since all known enzymes are proteins. Gamow drew an analogy between polynucleotides as a long number written in a four-digital system (the four bases) and polypeptides as long words based on a 20-letter alphabet (the 20 different kinds of amino-acids), and he raised the question of how the four-digital numbers could be translated into such words.

Dounce and Gamow's idea has been variously called the sequence hypothesis, the co-linearity hypothesis, and the theory of the genetic code.

## § 13.2  The linear construction of the gene

One of the first questions raised by the co-linearity hypothesis is how the gene is constructed. It had been known ever since Sturtevant (1913) discovered that crossover frequencies fitted a linear map, that the genes appeared to be arranged in a line along the chromosome (§ 7.3). Did this linear construction extend to the intragenic organization?

Several instances of recombination between alleles had been discovered in *Drosophila melanogaster* (see § 9.11). These suggested that the classical bead hypothesis of the gene might not be valid. However, little significant information about the detailed structure of the gene was obtained until the advent of selective techniques for the recovery of recombinants. Such techniques were made possible by Beadle and Tatum's discovery of fungal mutants with specific nutritional requirements (§ 10.2). If two mutants are very closely linked and have the same biochemical requirement for growth, recombinants could be detected even if of very low frequency by crossing the mutants and germinating the progeny spores on minimal medium, since neither parental type of spore would be able to grow.

Roper (1950) applied this technique to 3 biotin-requiring mutants ($bi_1$, $bi_2$, $bi_3$) of *Aspergillus nidulans* which had been independently obtained by X-ray treatment, and found that crosses of one with another gave recombinants not requiring biotin (prototrophs) with frequencies of about 1 per 2,000 ($bi_1$ and $bi_2$) and 1 per 5,000 progeny ($bi_1$ and $bi_3$). It was deduced that the biotin-independent progeny were arising by recombination and not by back-mutation because of their genotype for linked character-differences. Furthermore, no prototrophs were obtained from crosses between strains with the same biotin alleles. The three mutants all appeared to be concerned with the same biochemical step, since they each required biotin or desthiobiotin for growth and did not respond to pimelic acid. Moreover, Roper (1953) found that heterokaryons obtained from pairs of them were still biotin-requiring. In other words, each mutant was unable to supply the other's deficiency. Similarly, rare diploid strains of *Aspergillus* which Roper obtained by a selective technique from the heterokaryons, were also biotin-requiring. These had one biotin mutation in one chromosome and the other in the homologous chromosome. By the definition of multiple alleles given in § 9.1, these mutants would qualify as allelic, since they are functionally similar, do not complement one another, and are due to mutations in essentially the same region of the linkage map. However, as with the *Drosophila* examples discussed in § 9.11, the discovery of recombination between such mutants raises the question of what is meant by the terms gene and allele. Pontecorvo (1952) tentatively concluded that the biotin-requiring strains were due to mutations at different sites distinguishable by recombination but within a single gene. On this supposition the gene as a unit in physiological action is a larger chromosome segment than either the unit of mutation or of recombination.

Pritchard (1955) investigated 7 closely-linked and apparently allelic adenine-requiring mutants of *Aspergillus nidulans* which had been obtained independently of one another. Twelve of the 21 possible combinations of the 7 alleles in pairs were tested for recombination by crossing the mutants and plating the progeny ascospores on medium free of adenine. In every instance adenine-independent progeny were obtained. Their frequencies depended on the alleles crossed and ranged from 1 in 1250 to 1 in over 150,000 progeny. Marker genes ($y$, yellow conidia, and $bi$, biotin-requiring), linked to the adenine locus on either side, enabled the behaviour of the neighbouring parts of the chromosome to be followed and confirmed that the prototrophs were arising by recombination and not mutation. Moreover, tests established that the mutants were stable and showed mutation back to adenine independence only with very low frequency. A majority of the adenine prototrophs had a particular crossover genotype for the marker genes, such as would be expected if crossing over had occurred between the alleles. For example, by crossing a strain having adenine-8 and yellow conidia with a strain having adenine-11 and biotin requirement, heterozygous asci of constitution $\dfrac{y\ 8\ +}{+\ 11\ bi}$ were obtained, where 8 and 11 represent the adenine alleles and plus (+) stands for the appropriate normal allele (green conidia, biotin independence) of the marker genes. A majority of the adenine-independent ascospores from this cross had the genotype *y bi* for the markers. It was deduced that the *ad*8 and *ad*11 sites of mutation were placed such that the genotype of the diploid nuclei in the young asci could be rewritten $\dfrac{y\ +\ 8\ +}{+\ 11\ +\ bi}$. Conversely, the most frequent class of adenine independent strains derived from asci of genotype $\dfrac{y\ 8\ +}{+\ 10\ bi}$ had the $+\ +$ genotype for the markers (green conidia, biotin independent), from which it was inferred that the formula for the diploid ascus nuclei was $\dfrac{y\ 8\ +\ +}{+\ +\ 10\ bi}$, with allele no. 10 to the right of allele no. 8. In this way a map giving the sequence of the 7 alleles was constructed. A map was also prepared from the recombination frequencies, both between the alleles and with the markers. The various mapping data were reasonably consistent with one another, and pointed to a linear construction for the adenine gene. Moreover, it was evident that the allele map was collinear with the gene map; for instance, the analysis above indicates a single linear structure with the sequence *y* 11 8 10 *bi*.

## § 13.3   The cis-trans position effect

Roper and Pritchard (1955) had obtained a diploid strain of *Aspergillus nidulans* heterozygous for two adenine alleles (nos. 8 and 16). It had been obtained from a heterokaryon of the two mutants and hence had one allele in one chromosome and the other in the homologous chromosome. Its genotype was $\dfrac{y\ +\ 8\ +}{+\ 16\ +\ bi}$, using the same notation as previously. This diploid, as expected with allelic mutants, was adenine-requiring. However, by attempting to grow conidia from it on a medium lacking adenine it was possible to select for adenine-independence among them. About 1 conidium in $10^7$ was found to show such independence, and from the behaviour of the linked character-differences (*y*, *bi*) it was evident that a process of mitotic crossing-over had taken place. Somatic crossing-over had been discovered by Stern (1936) as a rare occurrence in *Drosophila melanogaster*, and was found to occur between two strands at the four-strand stage (cf. § 8.6), but not to be associated with any reduction in chromosome number. From crosses involving the adenine-independent diploids, Roper and Pritchard found that some of them had the genotype $\dfrac{y\ +\ +\ bi}{+\ 16\ 8\ +}$, such as would be expected if reciprocal recombination had occurred between the alleles. This genotype was identified by the recovery of both *ad*8 and *ad*16 from the progeny. These alleles could be distinguished because *ad*16 allowed slow growth on minimal medium while *ad*8 failed to grow.

From this experiment it was evident that the $\frac{+\ +}{16\ 8}$ genotype (with the two normal alleles in one chromosome and the two mutations in the homologous chromosome) does not require adenine for growth, while the $\frac{+\ 8}{16\ +}$ genotype (with one mutation in each chromosome) does. This was an example of position effect, but it differed from Sturtevant's classical example of Bar eye in *Drosophila melanogaster* in that it did not involve a duplication of a segment of the chromosome (see § 9.10).

Haldane (1941), using a chemical analogy, introduced the terms *cis* and *trans* for the two geometrically isomeric types of double heterozygote previously described by Bateson's terms coupling and repulsion, respectively. Pontecorvo (1950) applied these terms to closely-linked mutations, where differences in the appearance of individuals with the cis and the trans configurations might be expected. Such differences had been discovered by Lewis (1945) with the *star* eye characters in *Drosophila melanogaster* described in § 9.11. He had found that flies with the double mutant in one chromosome and the normal alleles in the other $\left(\frac{S\ s}{+\ +}\right)$ had eyes similar to dominant Star $\left(\frac{S}{+}\right)$ and appreciably larger than those of the repulsion heterozygote $\left(\frac{S\ +}{+\ s}\right)$ with its very small and rough eyes. A classic example of the same phenomenon was due to the work of Green and Green (1949) with 3 recessive *lozenge* alleles of *D. melanogaster*. They had followed up Oliver's discovery of recombination between alleles at this locus (see § 9.11). Denoting the 3 alleles by 1, 2, and 3, Green and Green showed that flies of genotype $\frac{+\ +}{1\ 2}$, $\frac{+\ +}{1\ 3}$, or $\frac{+\ +}{2\ 3}$ had normal eyes whereas flies of genotype $\frac{+\ 2}{1\ +}$, $\frac{+\ 3}{1\ +}$, or $\frac{+\ 3}{2\ +}$ had the lozenge characters of glossy eyes with less pigmentation than normal. Several other examples of this cis-trans position effect were discovered in *Drosophila* shortly afterwards, and they included the classic example of multiple alleles, namely, the white eye locus (*w*). MacKendrick and Pontecorvo (1952) and Lewis (1952) found small numbers of recombinants with wild-type (red) eyes from crosses between certain pairs of alleles at this locus. All the red-eyed flies showed recombination for marker genes on either side of *w*. It thus began to appear as if recombination between alleles, together with the cis-trans position effect, might be normal features of the gene.

## § 13.4 The cistron

Genetic recombination in viruses was discovered by Delbrück and Bailey (1947) and by Hershey (1947) with the *T*-even phages of *Escherichia coli*. Hershey and Rotman (1948) studied recombination with a particular class of mutants which had arisen spontaneously in virus $T2$. In these mutants the clear areas (plaques) produced by the destruction of the bacteria were larger and with a sharper margin than those caused by normal (wild-type) virus. They were called $r$ mutants because breakdown (lysis) of the host cells was more *rapid* than normal. Hershey and Rotman found that when a mixture of two $r$ mutants of independent origin was added to a liquid culture of the bacterial host (strain $B$), such that there were about 5 times as many virus particles of each kind as there were bacterial cells, many of the bacteria were simultaneously infected with both strains of virus. When fresh bacteria were infected with the progeny virus particles released by the breakdown of the cell-walls of the doubly-infected bacteria, besides many mutant plaques, a certain number of small rough-edged (wild-type) plaques were produced. These had evidently arisen by recombination, because each $r$ mutant alone gave none. Moreover, their frequency of occurrence was characteristic for each pair of mutants. From the recombination frequencies, the mutants could be grouped into two classes, subsequently called I and II. There was much closer linkage within each class than between them (see Fig. 18.4). It was remarkable that every mutant tested by 'crossing' (mixed infection) with another gave rise to recombinants. Eleven $r$ mutants were studied in this way, and 47 out of the 55 possible crosses were made. Moreover, there was some indication that the recombination frequencies might fit a linear map.

It was subsequently found that the $rI$ and $rII$ classes of mutants (and also a third group) could be distinguished by their behaviour on other strains of *E. coli*. In particular, the $rII$ mutants, unlike $rI$ and $rIII$, were found not to grow on a strain called $K$ of *E. coli*. This inability was taken advantage of by Benzer (1955), since it allowed selection for wild-type recombinants between any two of the mutants, in much the same way as Roper (1950) had selected for recombinants between biotin-requiring mutants of *Aspergillus nidulans*. Benzer used virus $T4$ and crossed 8 $rII$ mutants of independent origin in 23 of the 28 possible pairs. Crossing was done by infecting strain $B$ of *E. coli* with an equal mixture of the two virus mutants, such that in total there were 6 times as many virus particles as bacterial cells. The progeny virus particles were tested for their ability to multiply on strain $K$ of the host. Although neither parent can propagate on $K$, a small proportion of the progeny from every cross produced plaques. The frequency of these was, nevertheless, much higher than could be attributed to back-mutation, and it was concluded that the plaques on $K$ were arising through recombination. This was confirmed by the finding that the recombination frequencies from the various crosses gave a reasonable fit with a linear map. The sites of the 8 mutations were found to be in the sequence

$$47 \quad 104 \quad 101 \quad 103 \quad 105 \quad 106 \quad 51 \quad 102$$

where the figures are the identification numbers of the mutants. These results are what might have been expected from Hershey and Rotman's earlier findings with $T2$ virus using the original non-selective technique.

Benzer tested whether the $T4$ mutants which he had mapped were allelic, by making mixed infections of them in pairs on strain $K$ of the host. Although none of the mutants alone would propagate on $K$, he found that certain combinations of them could supply each other's deficiency and so multiply and produce plaques. When strain $K$ is infected with an $r$II mutant, the virus is adsorbed to the host and the bacterial cells are killed, but no progeny virus is produced. With a mixed infection by two $r$II mutants, a situation comparable to the *Aspergillus* heterokaryons studied by Pontecorvo and associates evidently prevails, with two dissimilar hereditary contributions within one cell. When Benzer classified the behaviour of the $r$II mutants by this test for allelism, he made a remarkable discovery. All combinations in pairs of the 6 mutants to the left on the map, that is, nos. 47, 104, 101, 103, 105, and 106 failed to produce plaques, and were evidently alleles (see Fig. 13.1($f$)). He called this part of the map segment $A$. Similarly, the other two mutants, nos. 51 and 102, also failed to complement one another (Fig. 13.1($g$)): they formed segment $B$. But any mutant in segment $A$ when paired with either of the $B$ mutants produced plaques on strain $K$ of the host (Fig. 13.1($h$)). There thus appeared to be two segments within the $r$II region corresponding to independent functional units.

If $K$ bacteria were infected with a mixture of wild-type virus and an $r$II mutant (Fig. 13.1($a$) and ($b$)), the virus was found to multiply and both types appeared in the progeny. It was evident that the $r$II mutants were recessive to the normal condition, and that the wild-type virus could supply whatever was needed to allow the $r$II mutants to multiply. It was presumed that the *cis* configuration of a double mutant and wild type (Fig. 13.1($c$)–($e$)) would similarly allow multiplication to occur, irrespective of the position on the map of the mutant.

The situation which Benzer had found in virus $T4$ was evidently similar to that known for several genes in *Drosophila* and *Aspergillus*, with recombination between alleles and a cis-trans position effect. The $r$II mutants were particularly interesting, firstly, because of the proximity of two functional units, mutation of either giving the same phenotype, and secondly, because of the neat way in which the recombination and complementation data established this. Benzer suggested that the $A$ and $B$ segments might affect sequential steps in a chain of synthesis. Alternatively, the two segments might each control the production of a specific polypeptide chain, the two chains later being combined to form an enzyme. With either of these explanations, the end-product of the action of the two segments would be a single substance, the lack of which was presumed to cause the $r$II phenotype.

Benzer (1955, 1957) found that certain $r$II mutants gave no detectable wild-type recombinants with any of several other mutants, which nevertheless gave wild-type recombinants with each other. Such mutants can be represented as occupying a segment of the linkage map rather than a point. It was significant that back-mutation of such a mutant had never been observed, whereas some of the point mutations showed reversion to wild-type not infrequently. The most likely explanation for both the recombination behaviour and the stability of these mutants appeared to be that they had originated through the loss of a segment of the hereditary material covering

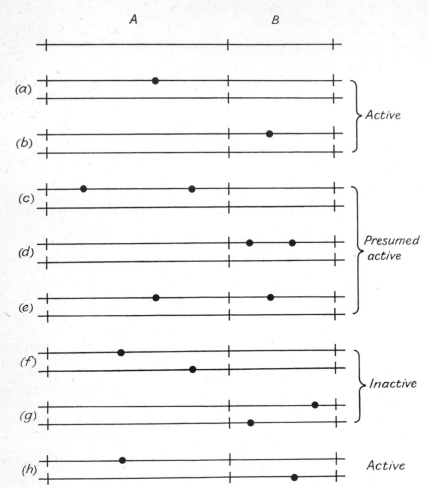

FIGURE 13.1   Diagram to show Benzer's results from mixed infection of
*Escherichia coli* strain *K* by two strains of virus *T*4. The
dots show the position of the *r*II mutations on the linkage
map. 'Active' means that extensive virus multiplication
took place.
(a), (b)  Single virus mutants mixed with wild-type virus.
(c)–(e)  Double mutants mixed with wild-type (that is,
      cis configuration).
(f)–(h)  Mixed infection by two single mutants (that is,
      trans configuration).
*A* and *B* are the two cistrons inferred to be present.

the length which they appeared to occupy on the linkage map. It will be recalled that mutation due to loss of a chromosome segment was already known in *Drosophila* (see § 9.7).

The use of these deletions greatly facilitated the mapping of new *r*II mutants. Moreover, by means of a series of overlapping deletions, a method of mapping point mutations was available independent of the mapping by recombination frequencies. Both methods were in agreement. Mapping was also facilitated by study of the behaviour of mixed infections on strain *K* of the host, because even if two mutants belonged to the same segment, a few plaques were in fact often produced. It appeared that a limited amount of multiplication in *K* bacteria was possible even with some *trans* configurations of alleles, and this allowed wild-type recombinants to form. In consequence, the number of plaques obtained was a measure of the distance apart of the mutant sites. A very rapid technique for mapping sites was therefore available, and Benzer (1961) reported that over 2400 spontaneous and induced *r*II mutations had been mapped, and that all the data fitted a one-dimensional map. The point mutations were at 200 different sites within the *A* segment, and at 108 sites within the *B* segment. By fitting a Poisson distribution to the numbers of sites which had mutated once and the numbers which had mutated twice, he estimated that, taking the whole *r*II region, there were at least another 120 sites at which mutation had not yet been observed.

Benzer (1957) had suggested that the functional unit of the hereditary material, defined by a comparison of the phenotype of the *cis* and *trans* configurations of a pair of mutations, be called a *cistron*. On his definition, a cistron would be a map segment corresponding to either the *A* or *B* segments of the *r*II region, within which there is no complementation between mutants. In other words, a cistron would be a segment of the hereditary material within which the cis configuration produced a normal (non-defective) phenotype, and the trans configuration produced a mutant phenotype. This definition amounts to equating the cistron with the gene defined as a unit of function. Alleles as previously understood would be mutations within the same cistron. Benzer proposed the terms *recon* for the unit of recombination, and *muton* for the unit of mutation. The recon was defined as the smallest element in the one-dimensional array that is interchangeable (but not divisible) by genetic recombination, and the muton as the smallest element that, when altered, can give rise to a mutant form of the organism. The introduction of these terms was a recognition that the bead hypothesis was inadequate, and that the term gene, hitherto used more or less indiscriminately for the unit of function, of recombination, and of mutation, was insufficiently precise.

Benzer drew attention to the remarkable parallelism between genetic maps, which appear to be one-dimensional at all levels of analysis, and the one-dimensional character of DNA. On the basis of (*a*) the total length of the linkage map for all the mutant characters in virus *T*4 that had been recognized and mapped, and (*b*) the total content of DNA, which was assumed to have the Watson and Crick structure, Benzer estimated that the *r*II cistrons probably contained several hundred or perhaps a few thousand nucleotide pairs, whereas the recon and the muton comprised a very small number of nucleotide pairs, and possibly only one.

Benzer's results with virus *T*4 implied that the functional units (cistrons) in the hereditary material could be defined unambiguously by the normal test for the allelism of closely-linked recessive mutants such as had been in use for many years. This test was to observe whether the trans (repulsion) heterozygote or heterokaryon had the mutant phenotype indicating alleles, or the wild-type phenotype indicating that the mutations were in different functional units (see § 9.1). However, experimental results with fungi suggested that the functional units of the hereditary material could not always be defined so easily.

Fincham and Pateman (1957) found that two apparently allelic mutants of *Neurospora crassa* of independent origin, both of which formed no detectable quantity of the enzyme glutamate dehydrogenase, would nevertheless produce the enzyme, although with lower than normal specific activity, when they were grown together as a heterokaryon. This remarkable discovery concerned amination-deficient (*am*) mutants which were unable to synthesize from ammonia the α-amino groups (see § 13.5) of a wide range of amino-acids, this being the reaction which is catalysed by glutamate dehydrogenase. The mutants were regarded as allelic because of the extremely low frequency (about 1 per $10^5$) with which wild-type progeny were obtained when they were crossed.

Similarly, Giles, Partridge and Nelson (1957) found that certain pairs out of 21 apparently allelic adenine-requiring mutants (*ad*-4 locus) of *N. crassa*, of independent origin, formed heterokaryons which were able to grow in the absence of adenine and which synthesized adenylosuccinase. This enzyme was not formed in detectable quantities by the individual mutants.

The outcome of these unexpected discoveries of complementation between mutants affecting the same enzyme, and the modifications to Benzer's concept of the cistron which they necessitated, are discussed in § 13.6.

## § 13.5   The effect of mutation on protein structure

From the evidence in the preceding sections of this chapter, and from similar results obtained subsequently for other genes, it appears that the first question raised by the genetic code idea, namely, how the gene is constructed, has been answered. As a unit of function it appears to be a linear structure, and to contain numerous smaller units capable of mutation and of recombination with one another. This is the construction that might be expected if the genetic material is DNA, with nucleotide sequence of paramount importance, and such that alteration of one or more nucleotides at any point incapacitates the whole gene. A second question raised by the Dounce-Gamow hypothesis of amino-acid sequence in polypeptides determined by nucleotide sequence is whether gene mutation can be shown to alter amino-acid sequence.

Neel (1949) established that the human disease called sickle cell anaemia is inherited as a Mendelian recessive. This disease is widespread in West Africa and in American negroes (see § 2.6), and is characterized by a severe anaemia which is usually fatal in childhood. In individuals with the disease, the erythrocytes change their form when the oxygen concentration is low, taking on the shape of a sickle instead of the normal biconcave disc. Neel found that the parents of such individuals were quite healthy, but always had a rather similar condition of the blood called sicklaemia or sickle cell trait. In sicklaemia the crescentic shape of the erythrocytes occurs only at lower oxygen concentrations such as are not encountered in the circulating blood, and there are no anaemic symptoms. From the regular occurrence of sicklaemia in the parents of individuals with sickle cell anaemia, it was inferred that sicklaemia was the heterozygous and sickle cell anaemia the homozygous condition for the sickling gene (cf. Fig. 13.4(*a*)).

Pauling, Itano, Singer and Wells (1949) found that the difference between sickling and normal erythrocytes was in their haemoglobins, which showed different electrophoretic mobilities. Moreover, sicklaemia haemoglobin was found to consist of about equal proportions of normal haemoglobin and sickle cell anaemia haemoglobin, as expected of the heterozygote. Ingram (1956) showed that there was a specific chemical difference between the two globins (the protein part of haemoglobin). He had devised a rapid technique for characterizing the chemical properties of a protein in considerable detail. Trypsin was used to split the peptide chains of haemoglobin and then the small peptides formed, which numbered about 30, were separated by a two-dimensional combination of paper electrophoresis and paper chromatography. Most of the peptides differ in net charge, and hence migrate at different speeds in an electric field on wet filter paper. The paper is then turned through 90° and a further separation brought about by their differing rates of migration in particular solvents. Finally, the position of the spots corresponding to the different peptides is revealed by staining with ninhydrin. Ingram called the resulting chromatogram the fingerprint of the protein.

Proteins are made up of one or more polypeptide chains. In haemoglobin there are 4, comprising 2 of one kind ($\alpha$-chains) and 2 of another ($\beta$-chains). Polypeptide chains consist of amino-acids linked together in a linear sequence by peptide bonds. In each of the 4 chains of the haemoglobin molecule there are about 150 amino-acids. All amino-acids have a carboxyl group ($-COOH$), an amino group ($-NH_2$), a hydrogen atom ($-H$), and a particular radical ($-R$), attached to one carbon atom called the $\alpha$-carbon (Fig. 13.2(i)). With 4 different groupings joined to the $\alpha$-carbon, there are 2 stereo-isomers of each acid (except glycine, where $R = -H$). Only the L-isomers (Fig. 13.2(i)) ordinarily occur in proteins. The radicals (R) are very diverse in structure and are listed in Fig. 13.2(iii) for the 20 amino-acids normally found in proteins. One of them, proline, is strictly-speaking an imino-acid: the R group is joined at its far end to the $\alpha$-carbon amino group. In the formation of the peptide bond, amino-acids link together by the attachment of the $\alpha$-carbon carboxyl group of one to the $\alpha$-carbon amino group of the next, with the elimination of a water molecule

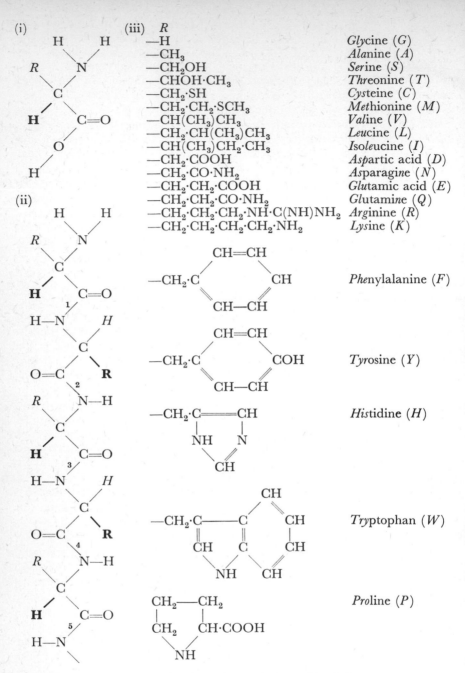

FIGURE 13.2  Structure of amino-acids and polypeptide chains.
(i) An amino-acid.
(ii) Part of a polypeptide chain. The peptide linkages are numbered from the amino group end of the chain.
(iii) The formulae of the 20 amino-acids normally found in proteins. In (i) and (ii), bold type and italics indicate projection in front of and behind the plane of the page, respectively. The letters in italics in the name of each amino-acid indicate the standard three-letter abbreviations for them, and the capital letter in brackets after each name indicates the standard single-letter abbreviation.

(see Fig. 13.2(ii)). A polypeptide chain will, therefore, have a free amino group at one end and a free carboxyl group at the other.

The primary structure of a protein is determined by the sequence of amino-acids in its polypeptide chain or chains. X-ray crystallography has shown that in proteins the polypeptide chain (or chains) is coiled in such a way that hydrogen bonding can occur between each peptide linkage (—CO·NH—) and the next-but-two in the chain. More specifically, there is hydrogen bonding between the oxygen atom of peptide linkage no. 1 and the hydrogen atom of peptide linkage no. 4, and similarly between O of 2 and H of 5, O of 3 and H of 6, and so on. The numbering of the bonds is from the free amino group end of the chain. This α-helix, as the coiled polypeptide is called, forms the secondary structure of a protein. Many proteins, including all those that are active in promoting chemical reactions, have the α-helix folded in an irregular way to give a characteristic shape to the molecule, which is called its tertiary structure. The association of a number of polypeptide chains in one protein molecule, as in haemoglobin, gives rise to a quaternary structure.

Ingram (1956) found that the fingerprint of haemoglobin *A* (the normal molecule) differed from that of haemoglobin *S* (from patients with sickle-cell anaemia) in respect of just one of the 30 or so peptides obtained by trypsin digestion. Trypsin breaks peptide links derived from the carboxyl groups of the amino-acids lysine and arginine. It was evident that the difference between the two haemoglobin molecules was confined to one small part of one of the polypeptide chains. Ingram (1957) and Hunt and Ingram (1959) analysed this peptide and showed that the amino-acid sequences were as follows (the abbreviations refer to the italicized letters in Fig. 13.2(iii)):

*A*    Val–His–Leu–Thr–Pro–Glu–Glu–Lys

*S*    Val–His–Leu–Thr–Pro–Val–Glu–Lys

The histidine is at the amino end and the lysine at the carboxyl end of the peptide. (All peptides formed by trypsin digestion will have either lysine or arginine at their carboxyl end.) It is evident that the two peptides differ by only one amino-acid, the glutamic acid residue at a specific point in *A* being replaced in *S* by valine. In each half-molecule of haemoglobin, comprising nearly 300 amino-acid residues, this single substitution constituted the only difference between haemoglobins *A* and *S* as regards their primary structure. For the first time, the effect of a single gene mutation was shown to be a change in one amino-acid in a polypeptide chain.

The tertiary structure of a polypeptide chain is determined exclusively by its primary structure. The tertiary structure is maintained by disulphide (—S—S—) covalent links between cysteine residues, and by hydrogen bonds or other interactions between amino-acid residues. The folding to give the tertiary structure occurs spontaneously during the synthesis of the polypeptide chain, which is known to occur progressively from the free amino end (see § 14.10). Although the change in the primary structure of haemoglobin due to the sickling mutation is confined to $\frac{1}{300}$th part of the molecule, its effects are profound. The replacement of two charged glutamic acid residues in normal haemoglobin (one in each half-molecule) by two uncharged valines alters the charge distribution on the surface of the molecule sufficiently to cause the abnormally low solubility of reduced haemoglobin $S$, which causes the sickling of the erythrocytes in the anaemia. This seemingly trivial change of one amino-acid in each half-molecule is thus wholly responsible for this lethal disease.

## § 13.6    The hybrid-protein hypothesis of allelic complementation

In attempting to explain the complementation between mutants of *Neurospora crassa* affecting the same enzyme, which had been discovered by Fincham and Pateman (1957) and Giles, Partridge and Nelson (1957) (see § 13.4), four observations were of importance. Firstly, and most remarkably, it was discovered that the patterns of complementation between alleles could be represented by a linear map. This was shown by Pateman and Fincham (1958) for *am*-1 mutants, Catcheside and Overton (1959) for *arg*-1 mutants, and Woodward, Partridge and Giles (1958) for *ad*-4 mutants of *N. crassa* (see Fig. 13.3). The sequence of the mutants on the complementation map is based on the supposition that the mutants have overlapping non-functional regions. These non-functional regions are indicated by the horizontal lines to the right of the tables of complementation data in Fig. 13.3. If two mutants are shown on the map by lines which overlap, this means their non-functional regions overlap and they fail to complement one another in a heterokaryon. The ability to represent allelic complementation by a linear map suggests that there is interaction between linear structures with defects in different regions. These linear structures need not be the genes themselves, but could be products of gene action.

Secondly, complementation between alleles seemed on the whole to be the exception rather than the rule. A majority of allelic mutants, including many which behaved genetically as if they were point mutations rather than deletions (cf. § 13.4), were found to be non-complementary. This suggested that the gene could not be subdivided into regions of distinct and non-overlapping function. Instead, it appeared that the linear complementation maps reflected the behaviour of gene products rather than the gene itself.

Thirdly, Fincham and Pateman (1957) found that the level of glutamate dehydrogenase activity through interallele complementation was comparatively low compared with that of wild-type fungus. Woodward, Partridge and Giles (1958) made a similar discovery with adenylosuccinase-deficient mutants. The enzyme activity rarely exceeded 25% of that of the wild-type.

Fourthly, Fincham (1959a) discovered that the enzyme resulting from interallele complementation was qualitatively abnormal. He found that a heterokaryon of alleles nos. 1 and 2 at the amination (*am*-1) locus produced a peculiar kind of glutamate dehydrogenase. It differed from wild-type enzyme in its low thermostability. Similar results were obtained with the enzyme produced by the heterokaryon of alleles nos. 1 and 3, and this enzyme had in addition a lower than normal affinity for glutamate.

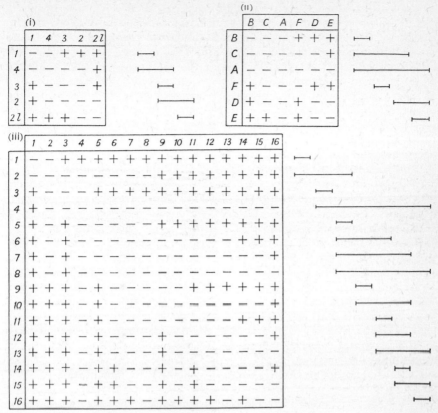

FIGURE 13.3 Results obtained from tests for complementation between allelic mutants of *Neurospora crassa*.
(i) Data of Pateman and Fincham (1958) for 5 alleles at the amination (*am*-1) locus.
(ii) Data of Catcheside and Overton (1959) for 6 alleles at the arginine-1 (*arg*-1) locus.
(iii) Data of Woodward, Partridge and Giles (1958) for 16 alleles at the adenine-4 (*ad*-4) locus.
The letters or numbers specify the mutants, a plus sign (+) indicates complementation, and a minus sign (−) indicates no complementation. The set of horizontal lines to the right of each table shows the one-dimensional complementation map derived from the data. Non-overlapping lines indicate complementation, and overlapping lines no complementation. In (ii), the sequence of *B* and *F* on the complementation map is arbitrary.

Fincham and Pateman (1957) suggested that allelic complementation might be due to the partial reconstitution of a component of the enzyme-forming system. It was assumed that two different defective nuclear products interacted in the cytoplasm of the heterokaryon. Giles, Partridge and Nelson (1957) also favoured nuclear product interaction as an explanation. Catcheside and Overton (1959) suggested that the complementation might be due to the enzyme consisting of an aggregate of two or more normally identical polypeptide chains. This would give an opportunity in a heterokaryon for the chains produced by each mutant to become associated, and through interaction of some kind, to restore the enzyme activity which the individual mutants lacked. They suggested that genes in which all the alleles failed to complement one another might control proteins consisting of only a single polypeptide chain. This possibility is exemplified by the histidine-6 (*his*-6) locus in *N. crassa*, where Catcheside (1960) found no complementation among 95 mutants tested in all pairwise combinations.

Strong support for this hybrid-protein hypothesis of allelic complementation was obtained when Fincham (1959*a*) found that the heterokaryon enzyme was qualitatively abnormal, and the hypothesis was confirmed by the demonstration of complementation *in vitro* between proteins isolated from mutants. Woodward (1959) obtained complementation by mixing extracts from pairs of mutants individually deficient for adenylosuccinase. Fincham and Coddington (1963) found that purified proteins from mutants nos. 1 and 3 at the *am* locus could interact to form a mixture of enzyme types very similar to that found in the heterokaryon of these mutants, and by labelling each mutant protein in turn with radioactive sulphur ($^{35}$S), Coddington and Fincham (1965) showed that the active interaction product contains sub-units from both mutant proteins, and that the ratio of the two is about 1:1 in at least a part of the hybrid material. The sedimentation characteristics of the hybrid enzyme, studied by sucrose density-gradient centrifugation, confirmed that the hybrid molecules contained the normal number of sub-units.

Clear evidence for the hybrid-protein model of complementation was also obtained by Schlesinger and Levinthal (1963) with mutants of *Escherichia coli* deficient for the enzyme alkaline phosphatase. This enzyme has been shown to consist of two normally identical sub-units, and so can be described as a dimer. Proteins serologically related to alkaline phosphatase were extracted from several mutants and purified. The sub-units of the protein were separated by reduction with thioglycollate in urea. The monomers derived in this way from different mutants known to complement one another *in vivo* were then mixed pairwise, after acidification, and it was found that partially active enzyme was obtained. It was shown by starch electrophoresis, by sedimentation in a sucrose gradient, and from the quantity of enzyme formed with different proportions of the parental monomers, that this enzyme was a hybrid dimer consisting of a monomer from each of the mutant proteins used in the reaction.

It thus appears that allelic complementation results from association between polypeptide chains derived from the two parent strains to produce active enzyme molecules. As Fincham (1959*a*) pointed out, this explanation of complementation means that Benzer's term cistron can be retained as the functional unit of the hereditary material, provided the definition of a cistron is modified such that the trans phenotype has to be non-defective at the enzyme level before two mutants are recognized as belonging to different cistrons. In other words, two mutants in the trans configuration in a heterokaryon or diploid will give normal enzymes if they are in different cistrons, and abnormal enzyme (or none) if in the same cistron.

Further aspects of allelic complementation are discussed in § 15.10.

## § 13.7   The one cistron:one polypeptide hypothesis

It is known that some protein molecules, such as the protein part of haemoglobin, contain dissimilar polypeptide chains. Are the amino-acid

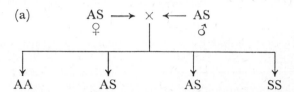

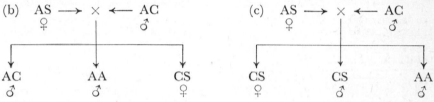

**FIGURE** 13.4

(a) The expectations on Neel's hypothesis of the inheritance of sickle cell anaemia.

(b) and (c). The two families in which Itano and Neel (1950) discovered haemoglobin C.

$A, C, S$. Three kinds of haemoglobin. $A$—Normal adult haemoglobin. $S$—Sickle cell haemoglobin. $C$—Another abnormal haemoglobin. The phenotypes of individuals with various haemoglobin combinations are as follows:

$AA$—Normal.

$AC$—Normal, but erythrocytes contain haemoglobin $C$ as well as $A$.

$AS$—Normal, but erythrocytes sickle at low oxygen concentrations (sickle cell trait).

$CS$—Mild form of sickle cell anaemia.

$SS$—Sickle cell anaemia.

sequences in these chains determined by the same cistron or by different cistrons?

Itano and Neel (1950), in studying the inheritance of sickle cell anaemia in man, encountered two families in which there were children with a mild form of the disease. It was found that in each instance only one of the parents had sicklaemia (sickle cell trait), whereas with the normal severe anaemia both parents show sicklaemia (see § 13.5 and Fig. 13.4(a)). Further investigation showed that the anaemic children had sickle cell haemoglobin (S) and also a new kind of haemoglobin (C) which differed electrophoretically from normal haemoglobin (A) and from haemoglobin S.

Moreover, the non-sickling parent in both families was found to have haemoglobin $C$ as well as $A$. The families and their haemoglobins are shown in Fig. 13.4($b$) and ($c$). It seemed likely that the $C$ and $S$ haemoglobins were inherited as a pair of alleles since each is recessive to normal and yet the individuals with both abnormalities ($CS$) are anaemic. Hunt and Ingram (1958) found that haemoglobin $C$, like haemoglobin $S$, had an amino-acid substitution in the $\beta$-chains of the molecule. Indeed, rather surprisingly, the substitutions were at identical positions, with the glutamic acid at this position in normal adult haemoglobin ($A$) replaced by valine in $S$ and by lysine in $C$.

Itano and Robinson (1960) found that an abnormality in the $\alpha$-chains appeared to be determined by a different genetic locus, which was probably not closely linked to the $\beta$-chain gene. The evidence for this came from study of the haemoglobin of relatives of an individual who was heterozygous for the $\alpha$-chain defect and for haemoglobin $S$. Direct evidence that the $\alpha$ and $\beta$ genes are on different chromosomes was obtained by Price, Conover and Hirschhorn (1972) by autoradiography of cytological preparations of human chromosomes to which rabbit haemoglobin messenger RNA labelled with $^3$H-uridine had been hybridized, using the method of Pardue and Gall (1969) (see § 18.9). Specific regions on chromosomes nos. 2 and either 4 or 5 (not distinguished) were labelled.

Itano and Robinson (1959) found that when a kind of haemoglobin called $I$, which is abnormal in the $\alpha$-chain of the molecule, is mixed with either haemoglobin $S$ or haemoglobin $C$ (both abnormal in the $\beta$-chain), two new kinds of haemoglobin could be produced as well as the original forms. At low pH the $\alpha$ sub-unit of the molecule separates from the $\beta$ sub-unit, and complete molecules are regenerated when the solution is neutralized. The new kinds of haemoglobin formed in this way consisted of normal haemoglobin $A$ and a doubly abnormal molecule combining the defects of both the original types. Baglioni and Ingram (1961) described a comparable situation *in vivo*— a person with four kinds of adult haemoglobin in her blood. These were the normal haemoglobin $A$; haemoglobin $G$, with a lysine residue in place of asparagine at a particular point in each $\alpha$-chain; haemoglobin $C$, with lysine in place of glutamic acid at a particular point in each $\beta$-chain; and haemoglobin $X$, which was found to have both the $G$ and $C$ substitutions. These four kinds of haemoglobin occurred with approximately equal frequency, suggesting that the mechanism for association of $\alpha$- and $\beta$-chains to form complete molecules did not discriminate between the two kinds of $\alpha$- or the two kinds of $\beta$-chain.

It is evident from these observations that the $\alpha$- and $\beta$-chains are determined by different genes, are synthesized separately, and are then assembled to give haemoglobin by the random association of $\alpha$- and $\beta$-sub-units. In individuals heterozygous for a defect in the $\alpha$-chain, the two $\alpha$-chains of each molecule always appear to be alike, and similarly with the $\beta$-chain. This is in contrast to the situation described in § 13.6, where association of mutant forms of normally identical sub-units appears to be responsible for the phenomenon of allelic complementation.

The one cistron:one polypeptide hypothesis has been found to apply widely. Indeed, so much so that the gene as a functional unit—in other words, the cistron—is now commonly defined as that part of the hereditary material that specifies the amino-acid sequence in one polypeptide. The earlier idea (Chapter 10) that one gene specifies one enzyme is now seen to be only an approximation to the truth. Not only are some proteins, such as that in haemoglobin, determined by more than one gene, but a single polypeptide may have more than one enzymic function. Indeed, the number of cistrons determining certain enzymic activities differs in different organisms. Tryptophan synthetase, the enzyme which catalyses the condensation of indole and serine to give tryptophan (see Fig. 10.1(*b*)), appears to be a double structure, one part (*A*) of the protein molecule having a site which combines with an indole compound (indoleglycerol phosphate), and the other part (*B*) a site which combines with a serine compound (serine + pyridoxal phosphate). In *Escherichia coli*, the *A* and *B* components are specified by different cistrons (Crawford and Yanofsky, 1958) (Fig. 15.1), but in *N. crassa* one cistron (*tryp*-3) determines both activities, which appear to be contained in different parts of a single polypeptide chain (Bonner, Suyama and DeMoss, 1960).

Another example of a bifunctional protein is provided by the enzymes responsible for steps 7 and 9 in the synthesis of histidine in *Salmonella typhimurium* (see Fig. 15.1(*b*)). Both of these steps appear to be catalysed by the same protein. Loper (1961) found that mutants lacking the specific dehydrase catalysing step no. 7 were due to mutation in any part of the gene that specifies the protein, but only those mutants due to mutation in one particular segment also lacked the phosphatase activity of step no. 9. Loper suggested that this protein might be active as the phosphatase under certain conditions, and in a different state of polymerization would be capable of the dehydrase activity.

Other examples of polypeptides with more than one function are discussed in § 15.10.

## § 13.8  Evidence for co-linearity of gene and polypeptide

Case and Giles (1960) studied the complementation in all pairwise combinations of 75 mutants of *Neurospora crassa* at the *pan*-2 locus. These mutants required pantothenic acid for growth. Of the 75 mutants, 23 showed ability to complement in one or more combinations, while the remaining 52 failed to complement in any combination. The complementation map of the mutants (which are numbered) is shown in Fig. 13.5(*a*). All the complementing mutants and 13 of the 52 non-complementing mutants were mapped genetically. This was done by crossing one with another, selecting for recombinants, and observing the frequency of recombination and the behaviour of outside marker genes. The sequence of the mutant sites on the linkage map so obtained is shown in Fig. 13.5(*b*). No attempt has been made to indicate the recombination frequencies, and the sites are shown uniformly spaced. On this genetic map, the complementing mutants are shown above the line, and the sample of non-complementing mutants below the line.

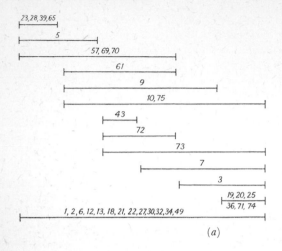

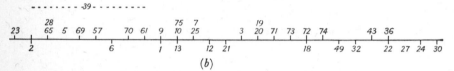

FIGURE 13.5   Data of Case and Giles (1960) for mutants at the pantothenic
acid 2 (*pan*-2) locus in *Neurospora crassa*. The numbers specify the
mutants.
(*a*) Complementation map.
(*b*) Recombination map. No attempt has been made to indicate
recombination frequencies and the mutants are shown uniformly spaced. Complementing mutants are numbered above
the line and non-complementing below. A further 39 non-
complementing mutants (comparable to the 13 on the bottom
line of (*a*)) were not mapped genetically.

Comparison of the complementation and genetic maps shows a distinct
tendency for the sequence of sites on the two maps to correspond. There are
some exceptions to this correspondence, but on the whole the relationship is
clear. This is what would be expected if complementation is due to non-
overlapping defects in otherwise identical polypeptide chains, and if the site of
modification of the polypeptide chain corresponds to the site of mutation in
the gene. Such a relationship between sites in DNA and in the corresponding
protein is what would be expected on the Dounce-Gamow hypothesis
that nucleotide sequence determines amino-acid sequence. The exceptions to
correspondence between genetic map and complementation map would be
accounted for if the gene determines the primary structure and the complementation reflects the tertiary structure of the protein, since the pattern of
folding to give the tertiary structure is likely to depend on interactions between
amino-acid residues in different parts of the polypeptide chain.
    Carlson (1961) obtained a circular complementation map for allelic

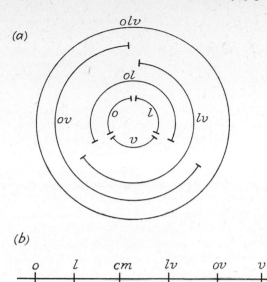

FIGURE 13.6    Complementation and recombination
maps for allelic mutants at the locus
called *dumpy* in *Drosophila melanogaster*
(from Carlson, 1961).
(*a*) Complementation map.
(*b*) Recombination map.
Mutations at this locus show one or
more of three recessive effects: an
obliquity (*o*) or truncation of the
wings; distrubed thoracic bristle patterns known as vortices (*v*); and a
lethal effect (*l*). *Comma* (*cm*) is a mutant
with weak *o* and *v* effects, and *humpy*
(*h*) is a mutant with exaggerated *o* and
*v* effects.

mutants at a locus called *dumpy* in *Drosophila melanogaster* (see Fig. 13.6). In
*Neurospora crassa*, circular complementation maps were obtained by Gross
(1962) with leucine-requiring mutants (*leu*-2) and by Kapuler and Bernstein (1963) with adenine-requiring mutants (*ad*-8 locus). With *ad*-8, if
the genetic map was drawn as a spiral it was found largely to correspond to
the complementation map. The significance of these circular complementation maps is uncertain. Kapuler and Bernstein suggested that the
tertiary configuration of the enzyme may be in the form of a spiral. Crick
and Orgel (1964) have concluded that the folding of the polypeptide chain
is the key to the relationship between the complementation and genetic
maps, but that there is no simple general way of deducing this tertiary
structure from them. In a hybrid dimer they believe that the potentially
misfolded region of one chain is corrected by the homologous region of the
other. Gillie (1966), from an analysis of existing complementation data, has

concluded that many loci with linear complementation maps are likely to be found to have non-linear maps when larger samples of mutants are tested. According to Crick and Orgel, each complementing mutant has a derangement of structure near the axis of symmetry of a polymeric protein. Gillie has extended this hypothesis and inferred that complementation maps are representations of the interfaces between the monomers making up a polymeric protein.

Study of the changes in the primary rather than in the tertiary structure of a polypeptide resulting from allelic mutations, in conjunction with genetic mapping of the mutants has provided conclusive evidence that gene and polypeptide are co-linear, or in other words, that the positions of the alterations in gene and polypeptide correspond. Sarabhai, Stretton, Brenner and Bolle (1964) studied a series of mutants of virus $T4$ of *Escherichia coli* affecting the protein coat of the head of the virus. These mutants were described as ambivalent ('amber') because they would grow on one strain of the host (strain $CR63$) but not on another (strain $B$) (see § 14.17 and § 15.3). In other words, host strain $CR63$ appeared to suppress the mutant character of the virus. Ten mutants induced by base analogues, and due to mutation at different sites in the cistron concerned with this protein, were mapped by means of their recombination frequencies with one another. Their sequence on the map is shown in Fig. 13.7($a$). The defective protein produced by these mutants when they infect strain $B$ of the host was purified and then digested with trypsin to give a series of peptides. Compared with normal virus, specific peptides were found to be missing in the mutants. When the mutants were placed in order according to the number of peptides present, it was found that the sequence was the same as that of the mutant sites on the genetic map. This establishes the truth of the co-linearity hypothesis, because the sequence in which the peptides occur in the polypeptide chain can be inferred from the progressive deficiencies of the mutants (see Fig. 13.7($b$)). Evidence was obtained that the free amino end of the chain is to the left in the diagram. Since it is known that synthesis of polypeptides begins at the amino end (see § 14.10), the inference is that in the mutants polypeptide synthesis is arrested before completion. Thus, the positions of the mutant sites on the genetic map appear to correspond with the positions where chain synthesis stops in the polypeptide, which is evidently co-linear with the gene. How it comes about that protein synthesis is allowed to go to completion when the virus mutants are grown on host strain $CR63$ instead of strain $B$, is discussed in § 14.17.

Yanofsky *et al.* (1964) and Yanofsky, Drapeau, Guest and Carlton (1967) obtained evidence of the truth of the co-linearity hypothesis by direct comparison of changes in gene and protein. A number of tryptophan-requiring mutants of *E. coli* were obtained by irradiation with ultraviolet light. Those which were unable to convert indoleglycerol phosphate to tryptophan in the presence of serine were selected for study. They were found to be defective in the $A$ protein of the enzyme tryptophan synthetase (see § 13.7). This protein is a single polypeptide chain containing 267 amino-acids (Guest *et al.*, 1967). From analysis of the $A$ protein produced (a)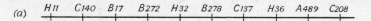

H11   C140   B17   B272   H32   B278   C137   H36   A489   C208

(b)

H11

C140

B17

B272          Cys

H32           Cys  His T7c

B278          Cys  His T7c  Tyr C12b

C137          Cys  His T7c  Tyr C12b  Try T6

H36           Cys  His T7c  Tyr C12b  Try T6  Pro T2a

A489          Cys  His T7c  Tyr C12b  Try T6  Pro T2a  Try T2

C208          Cys  His T7c  Tyr C12b  Try T6  Pro T2a  Try T2  Tyr C2

Normal        Cys  His T7c  Tyr C12b  Try T6  Pro T2a  Try T2  Tyr C2  His C6

FIGURE 13.7    Data of Sarabhai *et al.* (1964) for 10 mutants
               affecting the protein coat of the head of virus *T*4 of
               *Escherichia coli.*
               (*a*) Genetic map showing the sequence of the
                    mutant sites. No attempt has been made to
                    indicate recombination frequencies, and the
                    mutants are shown uniformly spaced.
               (*b*) The structure of the protein of the coat pro-
                    duced by each mutant, and by the normal virus.
                    The abbreviations refer to the peptides ob-
                    tained by digestion with trypsin.

by the mutants, it was found that each mutant differed from wild-type by
a single amino-acid substitution. Details of these substitutions are shown
in Fig. 13.8(i), where the letters refer to Fig. 13.2(iii) and identify the
amino-acids. The reference numbers of the mutants are given at the top of
the diagram. Nearly all the amino-acid substitutions found were within one
quarter of the polypeptide, which is represented by the sequence of 70
letters in the diagram.

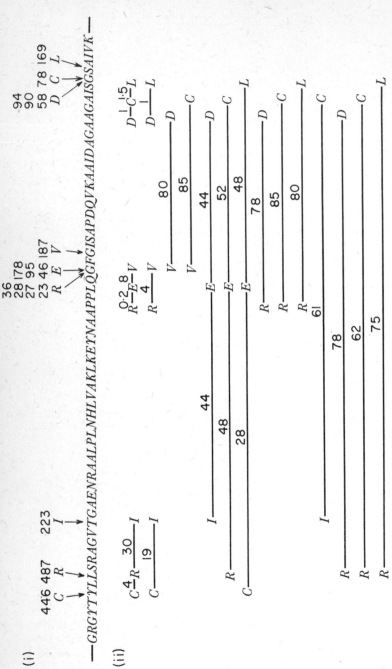

FIGURE 13.8   Data of Yanofsky *et al.* (1964, 1967) for tryptophan-requiring mutants of *Escherichia coli* defective in the A protein of the enzyme tryptophan synthetase.

(i) The diagram shows the sequence of amino-acids from no. 169 to no. 238 inclusive in the polypeptide chain of the protein, and the substitutions caused by the numbered mutants. The letters are the single-letter abbreviations for the amino-acids (see Fig. 13.2(iii)). There are 267 amino-acids in the entire polypeptide.

(ii) Genetic map of the mutants. The recombination frequencies are in arbitrary units. Where there is more than one mutant causing the same amino-acid substitution the recombination frequencies refer to mutants nos. 23, 46 and 58.

In Fig. 13.8(ii) is shown the genetic map of the mutants, which is based primarily on recombination frequencies, but supported by study of crosses with mutants having a segment of the gene deleted. The results show four features of particular significance. First, the genetic map of the mutant sites has the same sequence as the alterations to the polypeptide, thereby demonstrating the truth of the co-linearity hypothesis. It is evident that the position of a mutation in the gene determines the position of the corresponding amino-acid change. Secondly, certain of the mutants, for example, nos. 23 and 46, show recombination with one another and yet have an amino-acid substitution at the same position in the polypeptide. The significance of this discovery is discussed in § 14.13. Thirdly, the recombination frequencies (which are expressed in arbitrary units in the diagram), are roughly proportional to the number of amino-acids between the points of substitution in the polypeptide. This implies that the recombination frequency is related to the distance between the nucleotide substitutions in the DNA, and that this in turn is related to the distance between the amino-acid substitutions. Fourthly, certain mutants give consistently higher recombination frequencies than others, for example, nos. 23 and 187 compared with 46, and nos. 487 and 223 compared with 446. It is evident that recombination frequency is partly dependent on the nature of the mutations which are recombining. This remarkable conclusion is discussed further in Chapter 16.

Co-linearity of gene and polypeptide in a chromosomal organism has been demonstrated by Sherman *et al.* (1970) from a comparison of the genetic map of mutants of iso-1-cytochrome *c* in *Saccharomyces cerevisiae* with the position in the polypeptide of the amino-acid changes caused by the mutants.

All the essential pieces of evidence required to establish the Dounce-Gamow theory that nucleotide sequence specifies amino-acid sequence have now been obtained. When this idea was put forward there were no data available with which to test it, but it now appears that the gene is a linear structure with numerous sites of mutation within the functional unit; that mutations cause changes in amino-acid sequence in specific polypeptides; and moreover, that there is an exact point-by-point relationship between the position of the mutation in the gene and the position of the amino-acid substitution in the polypeptide. How nucleotide sequence in DNA is translated into amino-acid sequence in polypeptides is discussed in the next chapter.

# 14. *The theory of consecutive non-overlapping triplets of nucleotides as the amino-acid code*

## § 14.1   Introduction

When Dounce (1952) and Gamow (1954) independently put forward the basic idea that nucleotide sequence in nucleic acids might determine amino-acid sequence in polypeptides, they also suggested means by which this translation might be brought about. With four different nucleotides in DNA (having adenine, thymine, guanine, and cytosine respectively as bases), there are 16 possibilities for a sequence of 2 bases, 64 possibilities for a sequence of 3, 256 for a sequence of 4, and so on. Since there are 20 different amino-acids in proteins, it is evident that a minimum of 3 nucleotides is required to specify each. Both Dounce and Gamow therefore suggested that a sequence of 3 consecutive nucleotides was responsible for each amino-acid. Astbury and Bell (1938) had found from X-ray crystallography that the spacing between nucleotides in DNA was approximately the same as the spacing between amino-acid residues in a fully extended polypeptide chain. They considered that this correspondence was likely to be of biological significance, and this led both Dounce and Gamow to suggest that the nucleotide sequence was read at single nucleotide intervals. In other words, if a segment of the nucleic acid is represented by the letters $A\ B\ C\ A\ C\ D$, where $A$–$D$ specify the 4 individual nucleotides, the triplet $A\ B\ C$ would represent one amino-acid, $B\ C\ A$ the next, $C\ A\ C$ the third, and so on.

Gamow and Ycas (1955) pointed out that overlapping triplets such as these impose restrictions on the amino-acid sequences that are possible in polypeptides, and yet from an analysis of known sequences which they had made it appeared likely that all possibilities were permitted. They therefore rejected overlapping codes in favour of one in which the number of determining nucleotides exceeded by a factor of 3 the number of amino-acid residues in the synthesized protein, so that neighbouring residues did not share nucleotides. In other words, they favoured a non-overlapping triplet code.

From the range of dipeptides known to occur naturally, Brenner (1957) confirmed that a fully overlapping triplet code was unlikely to be correct. He analysed the published data on amino-acid sequences in polypeptides and found that of the 400 possible sequences in a dipeptide, 239 were already known. The data referred predominantly to various mammals (*Oryctolagus cuniculus, Balaenoptera borealis, Equus caballus, Bos taurus, Ovis aries, Sus scrofa, Homo sapiens*), but also included one bird (*Gallus domesticus*), one fish (*Salmo* sp.), one flowering plant (*Carica papaya*) and tobacco mosaic virus. If the code was the same in all these organisms, and was fully overlapping and based on trinucleotides, then there could not be more than 256 different dipeptides, since this is the number of ways of arranging a sequence of 4 nucleotides ($4^4$). Although near this limit, the known number did not exceed it. However, with full overlapping, only 4 different triplets can follow any given one. Thus, after a nucleotide trio represented as *A B C*, the possibilities are *B C A*, *B C B*, *B C C*, and *B C D*. In terms of amino-acid sequence, this would mean that the amino-acid represented by *A B C* could be followed only by four others. In order to allow for the possibility of more than 4 different amino-acids succeeding a particular one in the sequence, it would be necessary to postulate that a second triplet, for example *A B A*, represented the same amino-acid as *A B C*. If all the 20 possible dipeptides which begin with a particular amino-acid are found to occur naturally, a minimum of 5 different triplets would be required to code for this one amino-acid.

Brenner found that there was such diversity in the amino-acids which followed or preceded each particular one, that in total 70 different triplets would be required to code for all 20 amino-acids. This is more than the number possible with a triplet code ($4^3 = 64$). Hence, on the assumption that the organisms from which the amino-acid sequences were obtained had the same code, it was apparent that all fully overlapping codes based on trinucleotides could be ruled out. It could be inferred that if Astbury and Bell's discovery of a similar periodicity in DNA and protein was biologically important, its significance was unlikely to relate to the genetic code.

## § 14.2   The molecular basis of mutation

In order to understand the amino-acid code, it is essential to know in chemical terms the nature of the changes in DNA caused by mutation. Watson and Crick (1953*b*) had suggested that mutation might be due to a base in DNA occasionally occurring in one of its less likely tautomeric forms*, so that at replication the wrong base is inserted at this position in the complementary nucleotide chain. Following the discovery by Dunn and Smith (1954) and Zamenhof and Griboff (1954) that certain base analogues such as 5-bromouracil were incorporated into DNA in place of the specific bases which they resembled chemically (see § 12.4), Litman and Pardee (1956) discovered that 5-bromouracil had a powerful mutagenic effect on virus *T*2 of *Escherichia coli*. This led to a study by Benzer and Freese (1958) of the mutagenic effect of 5-bromouracil on the *r*II cistrons of the *T*4 virus. They made the outstanding discovery that 5-bromouracil did not enhance the spontaneously occurring mutations in a general way, but instead

---

* Tautomerism is the wandering of a mobile hydrogen atom from one multivalent atom to a neighbouring one within a molecule. Tautomeric forms of a molecule are usually in dynamic equilibrium with one another.

caused mutation at certain sites with characteristic frequencies. Their data relating to the end of the *B* cistron which adjoins the *A* cistron are given in Table 14.1. The pattern of mutations in the *A* cistron and the remainder of the *B* cistron was similar to that shown in the table.

Brenner, Benzer and Barnett (1958) made similar studies using the acridine proflavin as mutagen. Proflavin is known to interact with DNA, but unlike 5-bromouracil does not appear to become incorporated. A finding of very great interest was that proflavin causes mutation at a completely different series of sites from 5-bromouracil (see Table 14.1). The proflavin sites also differed largely from the spontaneously mutating ones. Furthermore, with proflavin there were single occurrences of mutation at many different sites, whereas with 5-bromouracil and also with spontaneous mutation there was more repetition of mutation at certain sites.

Freese and associates studied the *r*II mutations caused by other mutagens such as 2-aminopurine, nitrous acid, and hydroxylamine. The patterns of mutations caused by some of these mutagens on part of the *B* cistron are shown in Table 14.1, which is based on the collected data given by Benzer (1961). Many features of these patterns remain unexplained. Thus, it is not known why certain sites ('hot spots') show extremely high spontaneous mutation frequencies: a possible explanation is discussed in § 16.6. The mechanisms of action of the various mutagens are not understood in chemical terms with certainty. It is clear, however, that their actions are specific for certain sites. This is in direct contrast to the earlier studies on mutation, such as those described in Chapter 9, where the range of kinds of mutations found was independent of the mutagen used. It is evident from the results with *r*II mutants of virus *T4* that the specificity of mutagen action is at the level of mutant sites within the gene. The early studies on mutation were concerned with the gene as a whole, and at this level of organization the specific effects are lost, just as the totals of the various columns in Table 14.1 would give no indication of the specificity of action of the different mutagens.

Freese (1959) showed that the *r*II mutants produced by the base analogues 2-aminopurine and 5-bromodeoxyuridine (the deoxynucleoside of 5-bromouracil) could be induced to revert to wild-type by the action of these same substances, whereas proflavin-induced mutants and most spontaneous mutants could not. This confirmed the evidence from the differing site patterns that the base analogues on the one hand and proflavin on the other caused mutation in different ways.

Freese suggested that the base analogues caused mutation through mistakes in base pairing, either at the time of incorporation into DNA, or at a subsequent replication. The mechanism was thought to be similar to Watson and Crick's original suggestion, one base temporarily existing in a tautomeric form so that at replication the wrong purine or the wrong pyrimidine would be inserted (see Fig. 14.1(*b*) and (*c*)). Whatever the exact mechanism, the end-result would be that a purine would always be replaced by the other purine, and the corresponding pyrimidine by the other pyrimidine. Freese found that 5-bromouracil-induced mutations were reverted most readily by 2-aminopurine, and *vice versa*, suggesting that these base

TABLE 14.1 The table shows the numbers of mutations that have occurred spontaneously and with various mutagens at 50 different sites within part of the *rII B* cistron of *T*4 virus of *Escherichia coli*. (From Benzer, 1961.)

| Site | Spon-taneous | Nitrous Acid | Ethyl Methane Sul-phonate | 2-amino-purine | 2,6-diamino-purine | 5-bromo-uracil | 5-bromo-deoxy-cytidine | Pro-flavin |
|---|---|---|---|---|---|---|---|---|
| | | | | Mutagen | | | | |
| B1/1 | 1 | | | | | | | |
| 2 | 8 | | | | | | | |
| 3 | 3 | 1 | 2 | 1 | 1 | | | |
| 4 | 3 | | | | | | | |
| 5 | 4 | | | | | | | |
| 6 | 1 | | 2 | | 1 | | | |
| 7 | 2 | | | | | | | |
| 8 | 2 | | | | | | | 2 |
| 9 | 1 | | | | | | | |
| 10 | 32 | 14 | 7 | 9 | 1 | 28 | 13 | |
| 11 | 2 | | | | | | | |
| 12 | 1 | | | | | | | 1 |
| 13 | 1 | | | | | | | |
| 14 | | 1 | | | | | | |
| 15 | 1 | | | | | | | |
| 16 | 1 | | | | | | | |
| B2/1 | 1 | | | | | | | |
| 2 | 2 | | | | | | | |
| 3 | 6 | | | | | | | |
| 4 | 2 | | | | | | | |
| B3/1 | 5 | | | | | | | |
| 2 | 3 | | | | | | | |
| B4/1 | 1 | | | | | | | |
| 2 | 5 | 1 | | | | | | |
| 3 | | | | | | | | 1 |
| 4 | 1 | | | | | | | |
| 5 | 1 | | | | | | | |
| 6 | 1 | | 1 | | | | | |
| 7 | 1 | | | | | | | |
| 8 | | | | | | | | 2 |
| 9 | 1 | 8 | 5 | 1 | | 3 | 2 | |
| 10 | 51 | 4 | | | | | | |
| 11 | | | | | | | | 1 |
| 12 | | | | | | | | 1 |
| 13 | | | | 1 | | | | |
| 14 | 2 | | | | | | | |
| 15 | 1 | | 1 | 1 | | | | |
| 16 | | 1 | | | | | | |
| 17 | 17 | | | | | | | 1 |
| 18 | 3 | | | 4 | 1 | | | |
| 19 | 1 | | | | | 2 | | |
| 20 | 517 | 23 | 7 | | 3 | 5 | 5 | 1 |
| B5/1 | 1 | | | | | | | |
| B6/1 | | 1 | | | | | | |
| 2 | 1 | 3 | 4 | | | 3 | | |
| 3 | 1 | | | | | | | 1 |
| B7/1 | 1 | | | | | | | |
| 2 | 3 | | 1 | | | | | |
| 3 | 2 | 1 | | | | | | |
| 4 | 9 | | | | | | | |

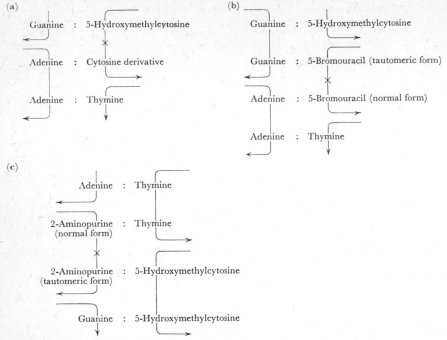

FIGURE 14.1   The hypothetical steps in the production of mutations in *T*4 virus of
*Escherichia coli* by (*a*) hydroxylamine, (*b*) 5-bromouracil and (*c*)
2-aminopurine. The steps outlined in (*b*) and (*c*) could also occur
in the reverse direction, though this appears to happen less often.
Arrows show the insertion of a base complementary to that in the
other chain, and the displacement of the latter at the next replication. *X* indicates where the mutagen has acted.

analogues acted predominantly in opposite directions. Later work (Freese
*et al.*, 1961) showed that mutants induced by hydroxylamine were similar
in properties to those induced by 5-bromouracil. Hydroxylamine is thought
to act by altering 5-hydroxymethylcytosine (which replaces cytosine in the
DNA of the *T*-even viruses—see § 11.6) so that it pairs with adenine instead
of guanine (Fig. 14.1(*a*)). At the next replication, thymine would be inserted
opposite the adenine, and the end-result would be that a guanine cytosine-
analogue base-pair was replaced by an adenine thymine base-pair. From its
behaviour in causing reverse mutation, it was concluded that 5-bromouracil
acts predominantly to give the same overall effect as hydroxylamine (Fig.
14.1(*b*)), while 2-aminopurine acts predominantly in the reverse direction,
causing the transition from an adenine thymine pair to a guanine cytosine-
analogue pair (Fig. 14.1(*c*)). On the other hand, nitrous acid was thought to
cause mutation through the oxidative deamination of either adenine or
cytosine (and its analogues) (see Fig. 14.2), with the result that transitions
from one base-pair to the other could occur more or less equally in both
directions. This hypothesis fitted the findings from the induction of mutations

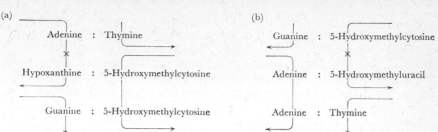

FIGURE 14.2    The hypothetical steps in the production of mutations in *T*4 virus of *Escherichia coli* by nitrous acid, through the oxidative deamination of (*a*) adenine to hypoxanthine, and (*b*) cytosine analogue to uracil analogue. Arrows show the insertion of a base complementary to that in the other chain, and the displacement of the latter at the next replication. *X* indicates where the mutagen has acted.

by nitrous acid and of their reversion by other mutagens, and *vice versa*. It is now known that the action of these mutagens is partly dependent on the replication enzyme (see § 12.6).

In contrast to these mutagens (nitrous acid, hydroxylamine, and the base analogues), which were thought to act by causing various patterns of *transition* (purine–purine and pyrimidine–pyrimidine substitution), Freese (1959) suggested that the second category of mutations, exemplified by proflavin-induced mutations, were due to the replacement of a purine by a pyrimidine, and *vice versa*. He called this hypothetical process *transversion*. However, Brenner, Barnett, Crick and Orgel (1961) postulated that proflavin causes mutation by the insertion or deletion of one or more base-pairs in the DNA, and not by transversion as Freese had suggested. Part of their evidence for this was that proflavin-induced mutants appeared completely to lack the normal activity of the gene, whereas mutants produced by other mutagens were often 'leaky', that is, showed some growth on strain *K* of the host (see § 13.4). With many virus characters, proflavin-induced mutants were not found, presumably because loss of the function of the gene controlling them was lethal. This difference in character of proflavin mutants, together with their unique site pattern and the inability of the base analogues to cause reversion, suggested that proflavin modified the gene in a fundamentally different way from the other mutagens. That this alteration to the gene was the addition or subtraction of base-pairs was confirmed by the experiments described in the next section.

For a more recent discussion of the molecular basis of mutation, see Drake (1970).

## § 14.3  Evidence that the code is read in triplets from a fixed origin

Crick, Barnett, Brenner and Watts-Tobin (1961) performed some ingenious experiments which appear not only to establish the nature of proflavin mutants, but also provide fundamental information about the genetic code. A single $r$II mutant of virus $T4$ of *Escherichia coli* induced by proflavin and situated at a site in segment $B1$ (cf. Table 14.1) formed the starting-point. As indicated in § 13.4, $r$II mutants are unable to grow on strain $K$ of the host, thus allowing selection for strains which have reverted to wild-type. A small number of revertants were found. When these were grown on strain $B$ of the host, nearly all were found to differ slightly in plaque morphology from wild-type, which suggested that the revertants did not arise by a precise reversal of the original proflavin-induced mutation, but by some other change which largely compensated for the original alteration to the gene. This was confirmed from the progeny of the revertants. It appeared that in nearly every instance the reversion was due to a second mutation at a different site from the original one. The suppressor mutations were in the same region of the gene, being restricted to segments $B1$ and $B2$ (cf. Table 14.1). This represents about $\frac{1}{10}$ of the whole $B$ cistron, only part of which is shown in the table. Each suppressor mutation, when separated through recombination from the proflavin mutant which it suppressed, was found to be an $r$II mutant itself.

To account for these observations, Crick *et al.* postulated that the initial proflavin-induced mutation was due to the addition of one nucleotide-pair at the mutant site, and that the suppressor mutations were due to the deletion of one nucleotide-pair. If the nucleotide sequence forming the gene was read from a fixed starting-point, such addition and subtraction of nucleotides would lead to misreading over the interval between the two sites but nowhere else. In consequence, abnormal amino-acids would be inserted in the segment of the polypeptide chain which corresponded with this part of the gene. This would explain why the suppressed mutant was not quite identical with the wild-type, and also why the site of each suppressor was near that of the original mutant. If the two sites were far apart, the polypeptide chain would be expected to be so abnormal that the phenotype would be mutant, and hence not found when selecting for revertants which would grow on strain $K$.

According to this hypothesis, the original proflavin mutation may be called a plus mutant (Fig. 14.3($c$)), and the suppressors of it may be called minus mutants (Fig. 14.3($b$)–($d$)). By repeating the original procedure, suppressors of the minus mutants were obtained (Fig. 14.3($h$) and ($i$)). They were found to be due to mutation at different sites near the site of the mutant they suppressed, just like the first group of suppressors. According to the hypothesis, this second family of suppressors should be plus mutants, and this was confirmed experimentally, because two plus mutants brought into the same piece of DNA through recombination had the mutant phenotype like each alone (Fig. 14.3($j$)), and similarly with two minus mutants (Fig. 14.3($k$)), but one plus and one minus mutant suppressed one another and gave the wild-type or nearly wild-type phenotype, provided they were not too far apart. By selecting for revertants of these plus mutants, a third family of suppressors was obtained and found to be minus mutants, as expected. Altogether about 80 $r$II mutants were obtained by this successive selection of suppressors, and their sites extended over about $\frac{1}{4}$ of the $B$ cistron.

An outstanding discovery was the finding that 3 plus mutants in the same

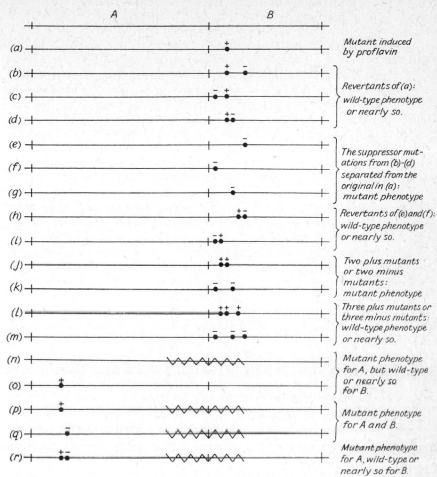

FIGURE 14.3   Results obtained by Crick *et al.* (1961) from study of suppressors of a proflavin-induced *r*II mutant of virus *T4* of *Escherichia coli*.

piece of DNA gave a near wild-type phenotype, and similarly with 3 minus mutants (Fig. 14.3(*l*) and (*m*)). On their hypothesis, this implied that the *coding ratio*, that is, the ratio of bases to amino-acids, was 3. The addition or deletion of 3 nucleotides within a short segment of the DNA had restored the function of the gene, indicating that the nucleotides were read in threes from one end. With the addition or subtraction of 1 or 2 nucleotides the gene is evidently misread to the end, resulting in the mutant phenotype, but with 3 added or removed it is misread only over the interval between the sites. Thus, their results suggested that a triplet of nucleotides was the coding unit or *codon* for an amino-acid.

Further evidence that the gene is read from a fixed starting-point was obtained from the use of a deletion known to cover parts of both *A* and *B* cistrons in the region where they join. The loss of this segment destroys the function of the *A* cistron, but not that of the *B*: it covers the part of the *B* cistron where the mutants used in the present study are situated. This is presumed to be a non-essential part of the *B* gene (Fig. 14.3(*n*)). A mutant in the *A* cistron which reverts freely with proflavin and is therefore presumed to be an addition (or deletion) of a nucleotide pair, destroys the function of the *A* gene without affecting the *B*, as might be expected (Fig. 14.3(*o*)). However, when this mutant and the deletion are combined in the same piece of DNA, it was found that the function of the *B* cistron was also destroyed (Fig. 14.3(*p*)). It was presumed that with the loss of the junction between *A* and *B* cistrons, both were being read as one, with the result that putting the reading out-of-phase in the *A* cistron affected the *B* as well. This was confirmed by the finding that a suppressor of the *A* cistron mutant restored the *B* cistron function (Fig. 14.3(*r*)). It thus appears that the gene (or nucleotide sequence derived from it) is read in trios of nucleotides starting from a fixed origin at one end.

It was assumed that the initial proflavin-induced mutant in this series of experiments had an addition of one nucleotide-pair at the mutant site in the *B* cistron. However, this mutant might equally well have been a deletion of one nucleotide-pair. This would not affect the argument except to inter-change plus and minus throughout. Crick and Brenner (1967) have evidence that the plus and minus signs, as originally used, may be correct. The initial mutant, and the series of suppressors obtained from it, might represent additions or deletions of 2 nucleotide-pairs instead of only 1. Although improbable, this remains a possibility, with its corollary that the coding ratio may be 6. However, independent evidence that the coding ratio is 3 has since been obtained (see § 14.11 and § 14.12).

Clear evidence for the validity of the arguments of Crick *et al.* (1961) has since been obtained (see § 14.13). Acridine mutagenesis has been discussed by Imada *et al.* (1970).

## § 14.4   RNA

In attempting to discover how the nucleotide sequence in DNA is trans-lated into amino-acid sequence in protein, it is essential to know whether the translation is direct, or whether other substances intervene in the process. The work of Brachet, Caspersson and many others (see Brachet, 1957) established that ribonucleic acid (RNA) participates in protein synthesis. One of the clearest demonstrations of this was when Brachet showed that treatment of cells with ribonuclease, which catalyses the breakdown of RNA, stops protein synthesis. Whereas DNA is largely confined to the nucleus, RNA occurs predominantly in the cytoplasm, although it is found in the nucleolus as well as in variable quantity in association with the chromosomes. Brachet demonstrated with the green alga *Acetabularia* (cf. § 5.3) that protein synthesis could continue in the cytoplasm in the absence of the nucleus. In many tissues a correlation has been shown between the amount of RNA present and the rate of protein synthesis. From observations such as these the idea developed that RNA provides a link between gene and protein. More specifically, it was suggested that information in the form of base sequence in DNA was transferred to RNA and thence to amino-acid sequence in proteins. The size of the RNA molecule (see below) is such that it could readily pass through the pores in the nuclear membrane.

The first clear evidence that RNA has the capacity to carry information

TABLE 14.2   The table shows the results obtained by Fraenkel-Conrat and Singer (1957) from study of two strains of tobacco mosaic virus, and of reciprocal chimaeras of their RNA and protein.

> $N$ = Normal *Nicotiana* strain of virus
> $P$ = *Plantago* strain of virus, sometimes called ribgrass virus.

The first and last columns of figures give the numbers of the various amino-acid residues in the $N$ and $P$ proteins, according to the data of Tsugita (1962). Both contain a total of 158 amino-acids. The other figures are Fraenkel-Conrat and Singer's values for the percentage weight of each amino-acid in the virus strains. The analyses were not corrected for destruction of acid-sensitive amino-acids during hydrolysis.

| Strain of Virus | RNA / Protein | $N$ / $N$ | $N$ / $P$ | $P$ / $N$ | $P$ / $P$ | |
|---|---|---|---|---|---|---|
| Serological character | | $N$ | $P$ | $N$ | $P$ | |
| Nature of disease produced by progeny | | $N$ | $N$ | $P$ | $P$ | |
| Serological character of progeny | | $N$ | $N$ | $P$ | $P$ | |
| | Glycine | 6 | 2·3 | 2·3 | 1·8 | 1·6 | 4 |
| | Alanine | 14 | 6·5 | 6·9 | 8·5 | 8·5 | 18 |
| | Serine | 16 | 9·0 | 8·8 | 8·1 | 8·1 | 13 |
| | Threonine | 16 | 8·9 | 8·9 | 7·5 | 7·2 | 14 |
| | Cysteine | 1 | Not determined (about 0·6%) | | | | 1 |
| | Methionine | 0 | 0·0 | 0·0 | 2·2 | 2·0 | 3 |
| Amino-acid composition of protein of progeny | Valine | 14 | 9·6 | 9·0 | 6·3 | 5·9 | 10 |
| | Leucine | 12 } | 14·2 | 14·3 | 12·2 | 12·2 | { 11 |
| | Isoleucine | 9 } | | | | | { 8 |
| | Aspartic acid | } 18 | 13·8 | 14·2 | 14·8 | 15·0 | 17 |
| | Asparagine | | | | | | |
| | Glutamic acid | } 16 | 12·4 | 12·1 | 17·3 | 16·4 | 22 |
| | Glutamine | | | | | | |
| | Arginine | 11 | 9·5 | 9·7 | 8·5 | 8·9 | 11 |
| | Lysine | 2 | 1·9 | 1·8 | 2·3 | 2·4 | 2 |
| | Phenylalanine | 8 | 7·2 | 7·1 | 5·4 | 5·3 | 6 |
| | Tyrosine | 4 | 4·1 | 4·3 | 6·2 | 6·3 | 7 |
| | Histidine | 0 | 0·0 | 0·0 | 0·7 | 0·7 | 1 |
| | Tryptophan | 3 | 2·8 | 2·6 | 2·2 | 2·2 | 2 |
| | Proline | 8 | 5·0 | 5·1 | 5·1 | 5·0 | 8 |

about protein structure came from work on tobacco mosaic virus. This virus contains no DNA. It is composed of RNA surrounded by a hollow cylinder of protein measuring 300 nm in length and 15 nm in external diameter. Gierer and Schramm (1956) separated the RNA and the protein and showed that the RNA by itself is infective, although to a much lesser degree (0·1%) compared with the intact virus. Fraenkel-Conrat (1956) mixed the RNA of one strain and the protein of another, and produced several very active chimaeras. The serological characters of these mixed virus preparations were those of the virus supplying the protein, but all the characters of the progeny were those of the virus supplying the RNA.

Fraenkel-Conrat and Singer (1957) investigated this remarkable discovery in more detail. One of the best examples concerned chimaeras derived from the normal virus and a strain originally isolated from *Plantago lanceolata*. On a Turkish variety of *Nicotiana tabacum*, the normal virus produces mottling of the leaves while the *Plantago* strain produces a distinctive ring pattern of diseased cells. Furthermore, the two strains differ appreciably in the composition of their protein, and in antigenic specificity. The protein of the

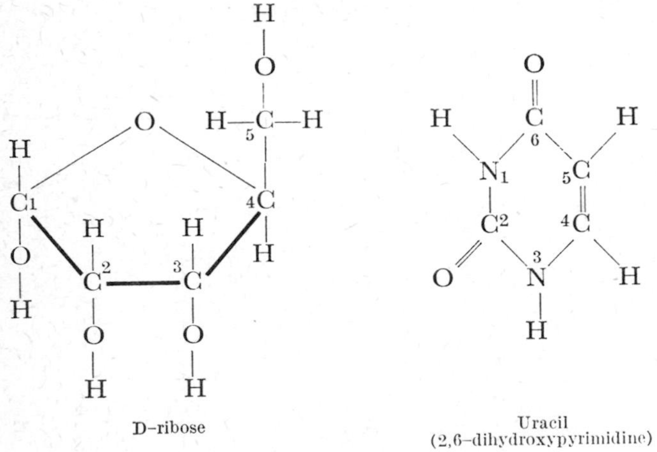

FIGURE 14.4   The components of RNA which differ from those of DNA (D-ribose in place of 2-deoxy-D-ribose, and uracil in place of thymine; cf. Fig. 11.3).

*Plantago* strain $(P)$ contains methionine and histidine which are absent from the normal *Nicotiana* strain $(N)$, and there are differences in frequency of nearly all the amino-acids (see Table 14.2). Indeed, the *Plantago* strain differs so much that it is regarded by some authors as a distinct species and is called ribgrass virus. Fraenkel-Conrat and Singer prepared reciprocal chimaeras, that is, RNA of the $N$ strain with protein of the $P$ strain, and *vice versa*. They found that the progeny of each had protein closely corresponding in amino-acid composition to that of the parent from which the RNA was derived (see Table 14.2). It appeared that the protein part of the virus made no contribution to the progeny, and that the specificity of the virus protein was determined solely by the RNA.

In view of this clear evidence that RNA can carry genetic information about protein structure, it is not surprising that the structure of RNA appears to be quite similar to that of DNA. The essential differences are that the sugar is D-ribose (Fig. 14.4(*a*)) instead of 2-deoxy-D-ribose, and one of the pyrimidines is uracil (Fig. 14.4(*b*)) in place of the thymine (5-methyluracil) of DNA. The other three bases (adenine, guanine and cytosine) are the same. The nucleotides are polymerized by 3′,5′-phosphodiester linkages as in DNA to form polynucleotide chains. These nucleotide chains are thought normally to occur singly, unlike DNA. However, there may be pairing of complementary bases between different parts of the same chain to give double-helical regions.

Grunberg-Manago and Ochoa (1955) isolated from *Azotobacter vinelandii* an enzyme, polynucleotide phosphorylase, which would bring about RNA synthesis *in vitro* from ribonucleoside 5′-diphosphates with release of orthophosphate. The reaction requires magnesium ions, but no primer is needed, and it is not necessary to have all 4 ribonucleoside 5′-diphosphates present. In view of the lack of requirement for primer, it is evident that this enzyme does not provide a mechanism for the maintenance of specific nucleotide sequences. An enzyme is also known which will break the 3′-linkages of RNA chains, and another which will break 5′-linkages, both comparable to the DNA enzymes (see § 12.4). Unlike DNA, RNA can also be broken down by hydrolysis with alkali, when it gives the nucleoside 3′-phosphates and also some 2′-phosphates.

Baltimore and Franklin (1962) working with Mengovirus in cultures of fibroblasts of *Mus musculus*, and Weissmann, Simon and Ochoa (1962) and August *et al.* (1964) studying viruses *MS*2 and *F*2, respectively, of *Escherichia coli*, demonstrated the existence of an RNA-dependent RNA polymerase which can bring about the replication of the RNA of these viruses. The enzyme was found to use the four ribonucleoside 5′-triphosphates as substrate and to require the presence of magnesium ions and an RNA template. The enzyme activity was induced following infection of the host cells.

Temin (1963) found that production of Rous sarcoma virus of *Gallus domesticus* from sarcoma cells was inhibited by actinomycin D. This is an RNA virus, but the action of actinomycin D was known to depend upon interaction with DNA (Goldberg, Rabinowitz and Reich, 1962). Temin inferred that the viral RNA is synthesized on a DNA template. Baltimore (1970) and Temin and Mizutani (1970) obtained strong support for this hypothesis with the discovery of an RNA-dependent DNA polymerase in the virus particles. Baltimore found such an enzyme also in Rauscher leukaemia virus of *Mus musculus*. The enzyme was detected through the incorporation of ³H-thymidine triphosphate into an acid-insoluble product that could be broken down by deoxyribonuclease. Spiegelman *et al.* (1970*a*) extended this discovery to the viruses of Moloney sarcoma and mammary tumour in *Mus*, leukaemia in *Felis catus* and myeloblastosis in *Gallus*, and they showed that all these RNA tumour viruses, as well as the original two, contained an enzyme that synthesized a DNA-RNA hybrid using the single chain viral RNA as template. Hybridization experiments (see § 14.6) confirmed that the DNA chain was complementary to the viral RNA. Subsequently they discovered (Spiegelman *et al.*, 1970*b*) that a DNA-dependent DNA polymerase also occurred in these virus particles. Mizutani, Boettiger and Temin (1970) independently found the DNA-dependent DNA polymerase in Rous sarcoma virus particles.

The discovery that some RNA viruses replicate their RNA by means of an RNA-dependent RNA polymerase, and that others bring about the same result through the action of a series of enzymes which first give rise to a DNA copy of the genome, has provided striking confirmation that the two nucleic acids have equal potentialities as carriers of hereditary information.

## § 14.5   Ribosomes

Tracer experiments with radioactive amino-acids have established that protein synthesis takes place in association with minute particles composed of RNA and protein. Roberts (1958) proposed the name *ribosome* for these ribonucleoprotein particles. They are found in the microsomal fraction when the components of cells are separated by differential centrifugation. Palade (1955) had shown from electron micrographs of thin sections of about 40

PLATE 14.1   Electron micrograph of polyribosomes (see § 14.11). Part of a sporogenous cell in a young anther of *Helleborus foetidus* ( × 63 000). Some of the ribosome aggregates are associated with the endoplasmic reticulum. At the top there is a cell-wall. (*Courtesy of P. Echlin*).

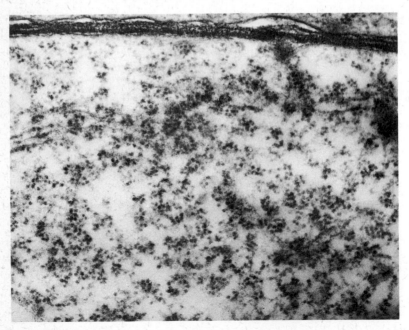

different mammalian tissues that these particles occur in large numbers in all the cells. Many were associated with the membrane system of the endoplasmic reticulum (cf. Pl. 14.1). In bacteria, however, where there may be several thousand ribosomes per cell, they are apparently not associated with membranes.

Ribosomes seem to be a fundamental constituent of all cells and to be very similar in all organisms. They are more or less spherical in shape and measure approximately 20 nm across. Ribosomes from bacteria are slightly smaller, however, than those of chromosomal organisms: their sedimentation coefficients are about 70 and 80 Svedberg units, respectively. Moreover, bacterial ribosomes contain about 50 % protein and 50 % RNA, while those of chromosomal organisms contain about 60 % protein. The ribosomes found in the chloroplasts and mitochondria of chromosomal organisms, however, are of bacterial type (see § 18.15).

At low concentrations of magnesium ions, ribosomes dissociate into two sub-units. In bacterial ribosomes these have sedimentation coefficients of about 50$S$ and 30$S$. The 50$S$ sub-unit yields RNA components of 23$S$ and 5$S$, and the 30$S$ sub-unit gives a 16$S$ RNA component. The 5$S$ component in *Escherichia coli* has been studied by Brownlee *et al.* (1967) who determined the entire sequence of the 120 nucleotides in its RNA, using techniques described in § 14.11. More than half the nucleotides are unpaired, but the two ends of the molecule have complementary sequences of bases and there are two other paired regions also. In man the 5$S$ component also consists of a single chain of 120 nucleotides, but there are many differences from the *E. coli* sequence (Forget and Weissman, 1969).

There is believed to be one copy of each of 48 different proteins in the ribosome of *Escherichia coli*, 21 in the 30$S$ sub-unit (Wittmann *et al.*, 1971) and 27 in the 50$S$ sub-unit (Mora *et al.*, 1971). It will obviously be a major task to find the specific functions of all these protein molecules. One of those in the 50$S$ sub-unit is an enzyme involved in polypeptide formation (see § 14.10).

The base composition of the RNA of ribosomes seems to be remarkably uniform: in bacteria there is about 26 % adenine, 20 % uracil, 32 % guanine and 22 % cytosine (molar proportions), irrespective of the DNA composition of the species from which it is obtained. This suggests that this RNA is not directly involved in the transfer of information from DNA to protein.

## § 14.6   The messenger RNA hypothesis

An exceptionally favourable situation for the study of the part played by RNA in protein synthesis is provided by the infection of *Escherichia coli* with the *T*-even viruses. Upon infection the synthesis of bacterial DNA stops abruptly, and the bacterial nucleus is rapidly destroyed, but synthesis of RNA and protein continue and are followed about 7 minutes after infection at 37°C by the production of virus DNA. Volkin and Astrachan (1957) determined the base composition of the newly-synthesized RNA by transferring infected bacteria to a medium containing labelled phosphorus ($^{32}$P), hydrolysing the RNA shortly afterwards, separating the 4 classes of nucleotides, and measuring their radioactivity. They found that the newly synthesized RNA was similar in base composition to the virus DNA (apart from having uracil in place of thymine, and cytosine in place of 5-hydroxymethylcytosine). This composition was different from that of the bacterial DNA, and from the overall composition of the RNA, 80% of which is in the ribosomes (see Table 14.3). These observations point to the newly synthesized RNA as a carrier of information from DNA to protein. It was notable also that this RNA showed a high turnover rate, with no net synthesis.

Hall and Spiegelman (1961) showed that the RNA synthesized after infection of *E. coli* by *T*2 virus could form hybrid molecules with the DNA of the virus. Owing to its difference in base composition, the newly-synthesized virus-specific RNA can be separated from the bacterial RNA by density-gradient centrifuging in sucrose solution. Hall and Spiegelman then

TABLE 14.3   Base composition of the DNA and RNA of *Escherichia coli* and its *T*-even viruses expressed in molar percentages.

| Base | *Escherichia coli* | | *T*-even Viruses | | |
| --- | --- | --- | --- | --- | --- |
| | | | | RNA specific to Virus | |
| | DNA | Total RNA | DNA | Volkin & Astrachan 1957 | Bautz & Hall 1962 |
| Adenine | 25 | 23 | 32 | 31 | 30 |
| Thymine (DNA) or Uracil (RNA) | 25 | 23 | 32 | 29 | 33 |
| Guanine | 25 | 31 | 18 | 22 | 21 |
| 5-Hydroxymethylcytosine (virus DNA) or Cytosine (RNA and bacterial DNA) | 25 | 23 | 18 | 18 | 16 |

mixed the virus-specific RNA with the virus DNA. The DNA had previously been heated to 95°C for 15 minutes, and then quickly placed in an ice-bath. Marmur and Lane (1960) had shown that such heating of DNA causes the complementary nucleotide chains to separate and that slow cooling allows them to re-associate again in complementary pairs, but quick cooling does not. The mixture of RNA and single-chain DNA was placed in a water-bath at 65°C and allowed to cool very slowly, taking 30 hours to reach 26°C. Hall and Spiegelman found that hybrid DNA-RNA molecules were thereby formed. Their evidence for this was based on radioactive isotope labelling using $^{32}$P in phosphate for the culture which provided the RNA, and $^{3}$H in thymidine for that which provided the DNA. In a control experiment, the labelled RNA and DNA were mixed, but not heated, and then centrifuged in caesium chloride solution for 5 days to give a density gradient. The contents of the centrifuge tube were then removed drop by drop in sequence and assayed for $^{32}$P and $^{3}$H. The $\beta$-particles emitted by these isotopes have different energies and so can be distinguished. It was found that the $^{32}$P and $^{3}$H were concentrated at different positions in the tube, as expected since RNA is denser than DNA. By contrast, after heating and slow-cooling the DNA-RNA mixture, the $^{32}$P showed a bimodal distribution, the new peak partly overlapping the $^{3}$H peak. It was inferred that the heating and slow cooling had allowed hybrid DNA-RNA molecules to form, and that these had a density close to that of DNA. The hybrid complex evidently contained considerably more DNA than RNA, presumably because the RNA was only a short segment.

Confirmation that hybrid molecule formation depends on complementary nucleotide sequences in the DNA and RNA was obtained by slow-cooling mixtures of the $T$2-specific RNA and previously-heated DNA of the host bacterium, *Escherichia coli*. No hybrid molecules were formed. Likewise there was no hybrid formation with the heated DNA of *Pseudomonas aeruginosa* nor with that of virus $T$5 of *E. coli*, which happens to have the same overall base composition as $T$2. Hall and Spiegelman's results thus provide a clear indication that the RNA formed after $T$2 infection is likely to have been synthesized in association with the virus DNA, since the RNA appears able to form double helices with single chains of $T$2 DNA, but not with other sources of DNA.

Weiss and Nakamoto (1961) and Hurwitz, Furth, Anders, Oritz and August (1962) described an RNA polymerase, both from bacteria and mammals, which has many features in common with DNA polymerase I described by Kornberg and associates (see Chapter 12). The RNA polymerase requires all 4 ribonucleoside 5'-triphosphates as substrate, and DNA as primer. Divalent metal ions are also needed, and pyrophosphate ions are liberated. This enzyme, first detected in the nuclei of the liver of *Rattus norvegicus*, has since been found in extracts of several bacteria and *Pisum sativum*. The reaction catalysed would appear to provide a means by which base sequence in DNA is transferred to RNA. Following RNA synthesis *in vitro* with this enzyme using 5'-triphosphates one of which was labelled with $^{32}$P, the RNA was broken down again by alkaline hydrolysis into 3'-phosphates (some 2'-phosphates were also formed), and the proportion of radioactive phosphorus associated with the 4 different nucleotides was determined. By labelling each nucleoside 5'-triphosphate in turn with $^{32}$P, the relative frequencies of the 16 possible sequences of a pair of nucleotides were estimated (cf. § 12.5). Using the DNA from various organisms in turn as primer, it was found that the base sequence in the synthesized RNA always resembled that of the DNA primer.

Chamberlin and Berg (1962) obtained similar results. That the DNA and RNA should have the same proportions of the 4 bases suggests that each of the complementary nucleotide-chains in the primer DNA was used for the synthesis of RNA chains. This was confirmed by Chamberlin and Berg by using as primer the DNA of virus $\phi X174$ of *E. coli*. This virus normally contains only one particular nucleotide chain of DNA (Sinsheimer 1959) with a molar base composition of 25% adenine, 33% thymine, 24% guanine, and 18% cytosine. The RNA synthesized when the $\phi X174$ single-chain DNA was used as primer had the complementary composition (34% adenine, 24% uracil, 19% guanine, and 23% cytosine). When DNA of this virus with normal complementary pairs of chains, obtained *in vitro* with the DNA polymerase, was used as primer for the RNA polymerase reaction, the resulting RNA had a base composition of 29% adenine, 29% uracil, 21% guanine, and 21% cytosine, which is identical with the DNA composition (except for uracil in place of thymine). This indicates that both DNA nucleotide chains were used for RNA synthesis. However, it cannot be inferred from this that RNA synthesis *in vivo* will necessarily occur in relation to both DNA chains. Indeed, evidence is given in § 14.7 which shows that RNA synthesis *in vivo* occurs in relation to one particular DNA chain only.

The idea that the short-lived RNA formed after virus infection constitutes a messenger between DNA and protein obtained strong support from some ingenious experiments by Brenner, Jacob and Meselson (1961). *Escherichia coli* was grown for several generations in a culture medium containing the heavy isotopes $^{15}N$ and $^{13}C$, so that all the nitrogen and carbon in the ribosomes would be of these heavy atoms. The bacteria were then infected with virus $T4$, and transferred immediately to a medium containing normal nitrogen ($^{14}N$) and carbon ($^{12}C$). In one experiment the bacteria were fed with phosphate containing radioactive phosphorus ($^{32}P$) from the second to the seventh minute after transfer, and in another experiment they were given sulphate containing radioactive sulphur ($^{35}S$) for the first 2 minutes of virus multiplication. The cells were subsequently disrupted and the ribosomes purified and centrifuged in caesium chloride solution to give a density gradient (cf. § 12.3). Control experiments showed that this treatment would separate any newly-synthesized ribosomes containing $^{14}N$ and $^{12}C$ from the old heavy ribosomes containing $^{15}N$ and $^{13}C$. It was found that no wholly new ribosomes were synthesized after the virus infection, and moreover, the newly-synthesized RNA indicated by $^{32}P$, and the newly-synthesized protein indicated by $^{35}S$, were found to be associated with the heavy band of ribosomes in the caesium chloride gradient. It was evident that at the time of cell disruption virus protein was being synthesized in old bacterial ribosomes under the influence of an unstable RNA fraction which was presumed to act as a messenger from the virus DNA. It appeared that the ribosomes were non-specialized structures which synthesize at a given time the protein dictated by the messenger they happen to contain. The function of the ribosomal RNA was not revealed by these experiments and remains unknown.

Nirenberg and Matthaei (1961) obtained evidence that the messenger RNA must be in the form of a single chain in order to function. In a cell-free system derived from *Escherichia coli*, a synthetic polyribonucleotide containing uracil but no other bases, and hence necessarily in the form of single chains, was found to lead to a specific type of protein synthesis (see § 14.12) whereas when the uracil chains were paired with adenine chains, no such synthesis occurred.

Evidence pointing to the existence of messenger RNA in chromosomal organisms was obtained by Ycas and Vincent (1960). They allowed *Saccharomyces cerevisiae* to grow in the presence of orthophosphate containing radioactive phosphorus ($^{32}$P), and found that the newly-synthesized RNA into which it was incorporated had a molar base composition of 32% adenine, 29% uracil, 19% guanine, and 20% cytosine. This is similar to the DNA composition: 31·5% adenine, 33% thymine, 18·5% guanine, and 17% cytosine, and quite different from the overall RNA composition (chiefly ribosomes) of 25% adenine, 28% uracil, 23% guanine, and 24% cytosine. The occurrence of messenger RNA in chromosomal organisms has been confirmed by numerous studies (see §§ 14.10–13). Gurdon *et al.* (1971) found that eggs and oocytes of *Xenopus laevis* provide a sensitive assay system for its identification. They injected haemoglobin messenger RNA derived from reticulocytes of *Oryctolagus cuniculus* (Rabbit) into the *Xenopus* cells and were able to demonstrate that haemoglobin synthesis took place, and continued for at least 24 hours.

## § 14.7 Evidence that one particular nucleotide chain of DNA is transcribed

Bautz and Hall (1962) isolated virus *T*4 messenger RNA from recently infected cells of *Escherichia coli* by absorbing the RNA at 55°C in a cellulose column containing virus *T*4 DNA in the single-chain condition following heating. The RNA was subsequently separated from the DNA by heating to 65°C, and then purified, and hydrolysed to find its base composition. The results obtained are shown in Table 14.3 and are in reasonable agreement with those of Volkin and Astrachan discussed in § 14.6 and obtained by a different method. In view of this agreement, Bautz and Hall attached real significance to a feature of their data of great interest, namely, that the total content of guanine and cytosine in the messenger RNA (37%) is approximately equal to the corresponding total for the virus DNA (36%), but the individual bases occur unequally in the RNA (21% guanine, 16% cytosine). These observations would be accounted for if the messenger RNA is always synthesized as the complement of one particular nucleotide-chain of the DNA, and if this chain contains 21% cytosine-analogue and 16% guanine. A similar argument can be applied to the data for the other pair of bases.

Evidence that only one specific DNA chain appears to be made use of when a gene function was also obtained by Champe and Benzer (1962) from study of the effect of 5-fluorouracil on the expression of certain *r*II mutants of *T*4 virus. This base analogue is thought to be incorporated into RNA in place of uracil, but then to show hydrogen bonding similar to cytosine. In consequence, an *r*II mutant which has uracil in place of cytosine at one point in its messenger RNA will lose its mutant character and show the normal phenotype if 5-fluorouracil has taken the place of the uracil.

The adenine in the DNA chain complementary to the uracil will be un-affected, so the mutant character will reappear in the progeny. Of 46 *r*II mutants thought from their reversion behaviour with mutagens (see § 14.2) to be due to a transition from a guanine cytosine-analogue base-pair to an adenine thymine pair, 29 were found to give the wild-type phenotype with 5-fluorouracil, while the remaining 17 did not respond. This result would be accounted for if the group of 29 mutants had one orientation of adenine and thymine with respect to the two DNA chains and the group of 17 mutants had the other orientation, and if the messenger RNA was synthesized as the complement of only one specific nucleotide chain of the DNA. The 29 mutants which responded to the uracil analogue were assumed to have adenine in this DNA chain at the mutant site, and hence normally would have uracil in the messenger RNA, while the other 17 were presumed to have thymine in the DNA chain and hence adenine in the messenger RNA.

Further evidence that only one of the two chains of the DNA is trans-cribed into messenger RNA has been obtained from several sources. In *Diplococcus pneumoniae* the two nucleotide chains of the DNA differ sufficiently in base composition to form separate bands when centrifuged in caesium chloride solution. Guild and Robison (1963) obtained preparations of the two fractions by this means. Using DNA from a novobiocin-resistant strain of the bacterium, they found that the heavy fraction (richer in the heavier purine, guanine, and the heavier pyrimidine, thymine) required about 45 minutes at 37°C to transform novobiocin-sensitive cells to resistance, whereas the lighter fraction modified the phenotype almost immediately. Since the cell generation time is about 40 minutes at 37°, it was inferred that the heavier chain needs to replicate before effective messenger RNA can be synthesized, whereas the lighter chain does not. It appeared likely that the messenger RNA from the gene for novobiocin-resistance was always comple-mentary to the lighter of the two DNA chains.

Fox and Meselson (1963) obtained virus lambda ($\lambda$) of *Escherichia coli* in which the thymine of the DNA was replaced by 5-bromouracil. Following mixed infection of the host with this and with normal $\lambda$, those progeny virus which had one chain of the DNA normal and the other containing 5-bromouracil were isolated by density-gradient centrifugation. Virus particles in which both chains of the DNA contain 5-bromouracil are rapidly in-activated by exposure to light. It was found that when the virus particles in which one chain contained 5-bromouracil and the other did not were exposed to light, half the population survived. This would be accounted for if one specific chain of the DNA was essential for the synthesis of messenger RNA which was needed for a function upon which the replication of the DNA was dependent. If this DNA chain happened to be the one containing 5-bromouracil, no messenger would be formed and the virus DNA would be unable to replicate. In the other half of the population of virus particles, the 5-bromouracil would be in the complementary DNA chain, damage to which would not prevent the formation of messenger RNA.

Studies with $\phi X$174 and certain other bacterial viruses in which the two chains of the DNA can be separated, either because of a difference in molecular weight or because all the mature virus contains only one specific chain, have shown that the messenger RNA formed after infection has a base composition similar to one chain of the virus DNA and hence complementary to the other. With virus $\phi X174$, progress has been made in understanding why the formation of messenger RNA *in vivo* appears to be complementary to one DNA chain, while *in vitro* it may be formed complementary to both (see § 14.6). This virus has its DNA in the form of a single molecule with the ends joined to form a closed circle. Hayashi, Hayashi and Spiegelman (1964) purified chromatographically the DNA of $\phi X174$ obtained during its replication in the host, *E. coli*. Although the mature virus contains only one specific nucleotide chain (see § 14.6), complementary chains are present during replication. It was confirmed both from electron micrographs and from sedimentation patterns that at least 90% of the virus DNA molecules retained their circular form. The RNA synthesized *in vitro* with the RNA polymerase using the circular double-chain $\phi X174$ DNA as primer was found to resemble one of the DNA chains in composition, whereas when the DNA was fragmented by means of a sonic oscillator, the RNA resembled both chains. This was established from the relative frequencies with which $^{32}P$, incorporated into the RNA in guanosine 5'-phosphate, was recovered in the 3'- (and 2'-) phosphates of each of the 4 nucleosides after alkaline hydrolysis, because the relative frequency of the 4 bases on the 3' side of guanine in one DNA chain happens to be quite different from that in the other chain. Confirmation of the results was obtained by heating and slow cooling the RNA and the single-chain DNA of the mature virus. The RNA synthesized with disrupted double-chain DNA as primer gave rise to hybrid molecules with the single-chain DNA, whereas the RNA formed with the unbroken double-chain DNA did not. The inference from these experiments is that messenger RNA produced from the unbroken two-chain DNA resembles the single DNA chain of the mature virus in composition. It is evident that fragmentation of the DNA causes breakdown of the mechanism which determines that the messenger RNA shall be formed as the complement of one specific DNA chain.

## § 14.8 The direction of transcription (synthesis of messenger RNA)

There are two possible directions for RNA synthesis: the nucleotides might be added (as nucleoside triphosphates with elimination of pyrophosphate) at the 5'-phosphate or at the 3'-hydroxyl end of the chain (Fig. 14.5($a$) and ($b$)). With the first alternative the triphosphate group at one end of the growing RNA molecule will be that of the last nucleoside to be added and so will be continually replaced as synthesis proceeds, while with the second alternative the initial nucleoside will retain its triphosphate group indefinitely. The direction of synthesis was established by making use of this and other differences between the alternatives.

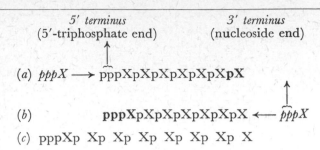

(c) pppXp  Xp  Xp  Xp  Xp  Xp  Xp  X

FIGURE 14.5   The diagram shows the two alternative ways in which RNA
synthesis might occur. X stands for any of the 4 ribonucleosides
and p before X means the 5′-phosphate and after X the 3′-phosphate
(see § 12.5). The initial nucleotide is shown in bold type and the
last in italics. (a) Growth in the 3′ to 5′ direction by addition of
the ribonucleoside triphosphate (pppX) to the 5′ terminus, with
release of pyrophosphate (pp) from the previous nucleoside triphosphate. (b) Growth in the 5′ to 3′ direction by addition of
pppX to the 3′ terminus, with release of pyrophosphate from that
triphosphate. (c) Components after alkaline hydrolysis, irrespective
of direction of synthesis.

RNA was synthesized *in vitro* by Bremer, Konrad, Gaines and Stent (1965)
using DNA-dependent RNA polymerase obtained from *Escherichia coli*, the
four nucleoside 5′-triphosphates as substrate, and the DNA of virus *T4*
as primer. The adenosine 5′-triphosphate (ATP) had been labelled with
tritium ($^3$H) in the adenine. During the RNA synthesis a 50-fold excess of
unlabelled ATP was added. The nucleotides at each end of the synthetic
RNA molecules were separated from one another and from the intervening
nucleotides by alkaline hydrolysis to give the 2′- and 3′-phosphates, followed
by paper electrophoresis. The separation of the nucleotides is possible because
of their diversity of composition: from the 5′ end of each molecule there will
be a nucleoside tetraphosphate (pppXp), from the 3′ end a nucleoside (X),
and from the intervening region nucleoside monophosphates (Xp), where
X stands for any of the four ribonucleosides and p before X means the 5′-
phosphate and after X the 3′-phosphate—see Fig. 14.5(c). Bremer *et al.*
found that the radioactivity of adenosine from the 3′ terminus was reduced
after dilution of the labelled ATP, but the radioactivity of adenosine tetraphosphate from the 5′ terminus stayed constant. They inferred that in the
growing molecule the 5′ terminus is constant, and additions are made to
the 3′ terminus or, in other words, RNA synthesis proceeds from the 5′ to
the 3′ end of the molecule (Fig. 14.5(b)).

Maitra and Hurwitz (1965) reached the same conclusion from a different
experiment. Guanosine 5′-triphosphate (GTP) labelled with phosphorus-32
in its gamma phosphate (that farthest from the ribose) was incorporated
into RNA *in vitro* for 5 min. using DNA from various sources in turn as
primer, and then a 30-fold excess of unlabelled GTP was added. Continuing
RNA synthesis did not diminish the amount of $^{32}$P already incorporated,
as expected if the initial nucleoside incorporated into RNA retained its

5'-triphosphate and growth was at the nucleoside or 3' end (Fig. 14.5(*b*)). Conversely, if the nucleoside end had been the initiation-point and the 5'-triphosphate the growing-point, subsequent synthesis should release the previously incorporated gamma-$^{32}$P, but this was not found.

Goldstein, Kirschbaum and Roman (1965) were able to show that messenger RNA synthesis *in vivo* also occurred in the 5' to 3' direction. They found that *E. coli* at 0°C synthesized messenger sufficiently slowly—1 nucleotide added every 13 seconds approximately—for differential labelling of the two ends of the molecule to be possible, using $^{14}$C-uridine. Radio-activity in the 3'-terminal nucleosides became constant very early after addition of the label, as shown by subsequent alkaline hydrolysis, whereas radioactivity in the internal nucleotides increased steadily, as expected if nucleotides are added to the 3' end. Progressively more and more of the internal nucleotides would be labelled as synthesis proceeded after addition of the label, while the nucleoside ends would all become labelled at once, assuming the nucleotide pool was small.

Evidence is presented in § 14.14 that the direction of translation of the nucleotide sequence in RNA into the amino-acid sequence in a polypeptide is the same as the direction of transcription from DNA into RNA.

## § 14.9 The initiation of transcription

Bremer, Konrad, Gaines and Stent (1965) found, using RNA polymerase *in vitro* with the DNA of virus *T*4 of *Escherichia coli* as primer, that adenosine triphosphate was preferentially used for chain initiation compared with the other three nucleosides. Similarly, Maitra and Hurwitz (1965) found that adenine was the most frequent 5'-terminal base of the RNA when the DNA of *E. coli* viruses *T*2 and *T*5 was the primer, but that guanine could also occur in this position and was more frequent than adenine when DNA of *E. coli* or two other bacteria or of *Bos* was used as primer. This discovery that the 5' end of the messenger frequently, and perhaps always, has a purine as base means that in the DNA chain which is transcribed a pyrimidine is preferentially chosen rather than a purine as the initiation-site for the synthesis of RNA. Maitra and Hurwitz suggested that a run of pyrimidines in DNA might be the signal for initiating transcription of that chain.

Support for this hypothesis was obtained by Kubinski, Opara-Kubinska and Szybalski (1966). They found that when the DNA of *Bacillus subtilis* was heated to separate the chains and then mixed with a synthetic RNA containing only guanine as base, one of the two chains of the DNA combined with the RNA and the other did not, allowing them to be separated in a caesium chloride gradient. It was inferred that the chain which formed a hybrid with RNA contains runs of cytosines. The significant observation was that this DNA chain, and not the other, also hybridized with messenger RNA from the bacterium, implying that the DNA chain with cytosine clusters was also the one which took part, predominantly at least, in trans-cription. Similar results were obtained with viruses *T*1 and *T*7 of *E. coli*. It is known that in virus $\phi X$174 (see § 14.7) and some other bacterial viruses the messenger RNA molecules are all transcribed from the same DNA chain.

In some other organisms, however, evidence has been obtained for cytosine-rich segments in both DNA chains. The hypothesis then predicts that there will be transcribing regions on both chains. This has been confirmed for virus λ of *E. coli* by Taylor, Hradecna and Szybalski (1967), who hybridized various λ-specific messenger RNAs with the chains of λ DNA which had been separated in a caesium chloride gradient, as they differ in density. Lambda DNA contains about 47,000 nucleotide-pairs, enough for some 40–50 genes, of which about half have been identified and mapped; moreover, Hogness, Doerfler, Egan and Black (1967) have established the orientation (that is, the direction of messenger synthesis) of several of these genes from genetic studies. Taylor *et al.* found, by the use of λ mutants deficient for particular regions of the chromoneme, that during the development of the virus a regular pattern of RNA synthesis occurred involving specific segments of each chain at specific times (Fig. 14.6). Their results

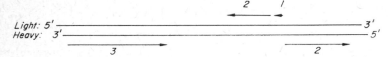

FIGURE 14.6   Diagrammatic representation of the light and heavy chains of the DNA of virus λ of *Escherichia coli* to show the transcription regions at different stages in the life-cycle of the virus:
(*1*) non-induced state, that is, the virus integrated into the host chromoneme; (*2*) early-acting genes; and (*3*) late-acting genes.
The arrows show the direction of messenger synthesis. The arrows labelled 2 and 3 each represent a number of different messenger RNA molecules.

were in agreement with those of Hogness *et al.* Moreover, the segments found to be transcribing regions were independently shown by their ability to hybridize with guanine-rich RNA to be segments containing cytosine clusters.

It is evident that the hypothesis that pyrimidine clusters are the sites for the start of transcription, although not conclusively established, has considerable evidence to support it. The site of initiation of transcription has been called the *promoter*, and has been identified genetically (see § 15.6).

Burgess, Travers, Dunn and Bautz (1969) discovered that a protein component which they called sigma (σ) could be separated from the RNA polymerase of *Escherichia coli* by chromatography on a phosphocellulose column. It was already known that RNA polymerase is a large molecule containing several different polypeptides. Burgess *et al.* showed that this core enzyme, without σ, could bring about RNA synthesis, and they suggested that a number of different σ factors might exist which determined the specificity of the initiation of transcription. Travers and Burgess (1969) confirmed that the σ factor is required for initiation. They measured the amount of $^{32}P$ incorporated into RNA from nucleoside triphosphates labelled with $^{32}$P in the $\gamma$ position, that is, the phosphorus farthest from the nucleoside. This was an assay of initiation because only the first nucleoside incorporated retains its triphosphate (see § 14.8). They found that in the presence of the $\sigma$ factor the increase in total RNA synthesis was directly proportional to the increase in the number of RNA chains initiated. Travers and Burgess also showed that when RNA synthesis has begun $\sigma$ is released and re-used by another polymerase molecule to initiate a new RNA chain.

Bautz, Bautz and Dunn (1969) found that the $\sigma$ factor of *E. coli* selectively promotes transcription of the genes of virus *T*4 that are expressed during the first minute after infection while in the absence of $\sigma$ the core enzyme caused initiation of RNA synthesis at random positions in both chains of the DNA of the virus. Bautz *et al.* concluded that $\sigma$ was a positive control element and that *T*4 was adapted to use this factor in the early stages of viral development. Travers (1969) showed that about 2 minutes after *T*4 infection the host $\sigma$ was replaced by a viral $\sigma$ which allowed the core enzyme to initiate RNA synthesis at a different set of *T*4 genes from those transcribed in the presence of the host $\sigma$. He subsequently found (Travers, 1971) that about 12 minutes after infection a second viral $\sigma$ factor appears which allows transcription of the viral genes which function late in development.

In addition to the substitution of different $\sigma$ factors into the RNA polymerase molecule at various stages of development of *T*4 virus, changes also occur in the core enzyme. This consists of two copies of an $\alpha$ polypeptide, one copy of a $\beta$ polypeptide and one copy of a $\beta'$ polypeptide. Early in *T*4 development a nucleotide is added covalently to the $\alpha$ chain and about 12 minutes after infection the $\beta'$ chain is modified (see Travers, 1971). These alterations to the core enzyme inhibit attachment of the host $\sigma$ factor and also allow the enzyme to bind more tightly to single-chain DNA. During the later stages of *T*4 development the viral DNA is replicating and it is evidently of adaptive value to allow transcription when the genes exist as single DNA chains.

The existence of diverse $\sigma$ factors controlling the specificity of RNA chain initiation implies that there must be corresponding diversity, which the $\sigma$ factors can recognize, in the nucleotide sequence at promoter sites.

## § 14.10   The adaptor theory

In considering how base sequence in DNA or RNA could be translated into amino-acid sequence in a polypeptide chain, Crick (1957) pointed out that it was difficult to see how this was possible by direct contact between the coding nucleotides and the specific amino-acid. He predicted that each amino-acid was likely first to be combined with its own specific *adaptor*, and that each of these adaptors would have a highly specific spatial pattern of hydrogen bonds capable of associating it with a particular sequence of bases in DNA or RNA. He further suggested that molecules based on sequences of nucleotides were likely candidates for the adaptors, and that the combination between each amino-acid and its own specific adaptor could be made by a special enzyme. Within a year, Hoagland, Zamecnik and Stephenson (1957) had discovered the existence of molecules of precisely the kind Crick had anticipated. They had found a low molecular weight RNA in the non-particulate fraction of an extract of the cytoplasm of liver cells of *Rattus norvegicus*. These RNA molecules, at first called soluble RNA to distinguish them from the RNA of the ribosomes, appeared to have the ability to bind amino-acids to themselves and to transfer them to peptide linkages. The names acceptor RNA and *transfer RNA* were subsequently proposed, because of this ability to accept amino-acids and transfer them to protein.

The primary steps in the process of protein synthesis, which were quickly established following the discovery of transfer RNA, appear to be as follows. Each amino-acid is first combined with adenosine triphosphate (ATP) and a specific enzyme to form a single complex consisting of the amino-acid, adenosine monophosphate (AMP) and the enzyme, with the elimination of pyrophosphate (Fig. 14.7(i)). Each of the 20 amino-acids has its own specific activating enzyme.

$$\text{(i)} \qquad E_1 + aa_1 + Appp \longrightarrow E_1(aa_1 \cdot pA) + pp$$

$$\text{(ii)} \qquad E_1(aa_1 \cdot pA) + RNA_1 \longrightarrow aa_1 \cdot RNA_1 + E_1 + Ap$$

$$\text{(iii)} \quad aa_1 \cdot RNA_1 + aa_2 \cdot RNA_2 \xrightarrow[\text{Gppp}]{\text{Tr}} aa_1 \cdot aa_2 \cdot RNA_2 + RNA_1$$

FIGURE 14.7   The three primary steps by which amino-acids are incorporated into proteins.   A = Adenosine (adenine ribonucleoside)
$aa_1$, $aa_2$ = Amino-acids nos. 1, 2
$E_1$ = Activating enzyme no. 1
$G$ = Guanosine (guanine ribonucleoside)
p = Phosphate
$RNA_1$, $RNA_2$ = Transfer RNA nos. 1, 2
Tr = Transfer enzymes

The second step is the interaction of the whole amino-acid–AMP–enzyme complex with a specific transfer RNA to form an amino-acid–RNA compound called an *amino-acyl transfer RNA*; AMP and the enzyme (amino-acyl transfer RNA ligase or synthetase) are released in the process (Fig. 14.7(ii)). There is at least one unique transfer RNA molecule for each amino-acid. It is customary to denote a transfer RNA by the name of its amino-acid as a noun (e.g. methionine transfer RNA) and to indicate the amino-acid–transfer RNA compound by the name of the amino-acid either as an adjective or abbreviated and hyphenated (e.g. methionyl transfer RNA or met-*t*RNA). Where more than one *t*RNA exists for the same amino-acid (see § 14.15) they are distinguished by a suffix.

The third and final step in protein synthesis takes place on the ribosomes. Takanami and Okamoto (1963) discovered, using a synthetic RNA containing uracil as base, that the messenger RNA becomes attached to the 30*S* sub-unit of the ribosome. Each transfer RNA–amino-acid complex evidently becomes associated through specific hydrogen bonding of certain of its nucleotides (*anticodon*) with the appropriate base sequence (codon) in the messenger RNA, thereby placing the amino-acids in the sequence dictated by the messenger. The formation of the peptide links is mediated by at least three (amino-acid non-specific) transfer enzymes. One of these transferases catalyses the binding of the amino-acyl transfer RNA to the ribosome; this reaction requires guanosine triphosphate as a co-factor (Fig. 14.7(iii)). Monro (1967) showed that the peptidyl transferase responsible for peptide bond formation is situated on the 50*S* sub-unit of the ribosome.

It is evident that the transfer RNA molecules correspond precisely with Crick's postulated adaptors.

The nature of the chemical bonds involved in the steps in protein synthesis is shown in more detail in Fig. 14.8. The link between AMP and the amino-acid is between the phosphate of AMP and the carboxyl group of the amino-acid (—P—O—C—). The transfer RNA molecules each consist of a single nucleotide chain containing about 80 nucleotides (see § 14.11). The end of this chain which becomes linked to the amino-acid has the same structure in all the transfer RNAs. The terminal base is adenine and the next two are both cytosine. The linkage between amino-acid and transfer RNA is between the ribose of this terminal nucleotide (adenylic acid) and the carboxyl group of the amino-acid (—C—O—C—). It has proved difficult to establish whether the link with the amino-acid is to the 2′ or the 3′ carbon atom of the ribose. Some experiments have suggested an equilibrium mixture of both isomers, others have pointed to 3′-amino-acyl-*t*RNA. The peptide linkage is formed by a covalent bond between the carboxyl carbon atom of the first amino-acid and the amino group of the second. The transfer RNA of the first amino-acid is detached from this carboxyl carbon in the process. The amino-acids are added successively in this way, each displacing the transfer RNA of the preceding one. The polypeptide chain therefore grows at its carboxyl end. In other words, the free amino end of the completed chain is where synthesis began and the free carboxyl end where it ended. This was demonstrated by Dintzis (1961) from study of the incorporation of tritium-labelled leucine into the haemoglobin of *Oryctolagus* reticulocytes.

Chapeville, Lipmann, von Ehrenstein, Weisblum, Ray and Benzer (1962) demonstrated the truth of the adaptor hypothesis. The amino-acid cysteine was converted into alanine by reduction with nickel hydride after the amino-acid had become attached to the cysteine-specific transfer RNA. It was then found that alanine was incorporated into polypeptides in place of cysteine. This was demonstrated with an *in vitro* protein-synthesizing system with which it has been shown that a synthetic RNA containing uracil and guanine will lead to the incorporation into polypeptides or certain amino-acids including cysteine, but not alanine (see § 14.12). This experiment shows that the messenger RNA code is recognized by the transfer RNA and not by the amino-acid attached to it, and thereby establishes the truth of the Crick-Hoagland adaptor hypothesis. Von Ehrenstein, Weisblum and Benzer (1963) using *Oryctolagus* reticulocytes, confirmed that in the same way alanine could be incorporated into a peptide of the α-chain of haemoglobin that normally contains cysteine but not alanine.

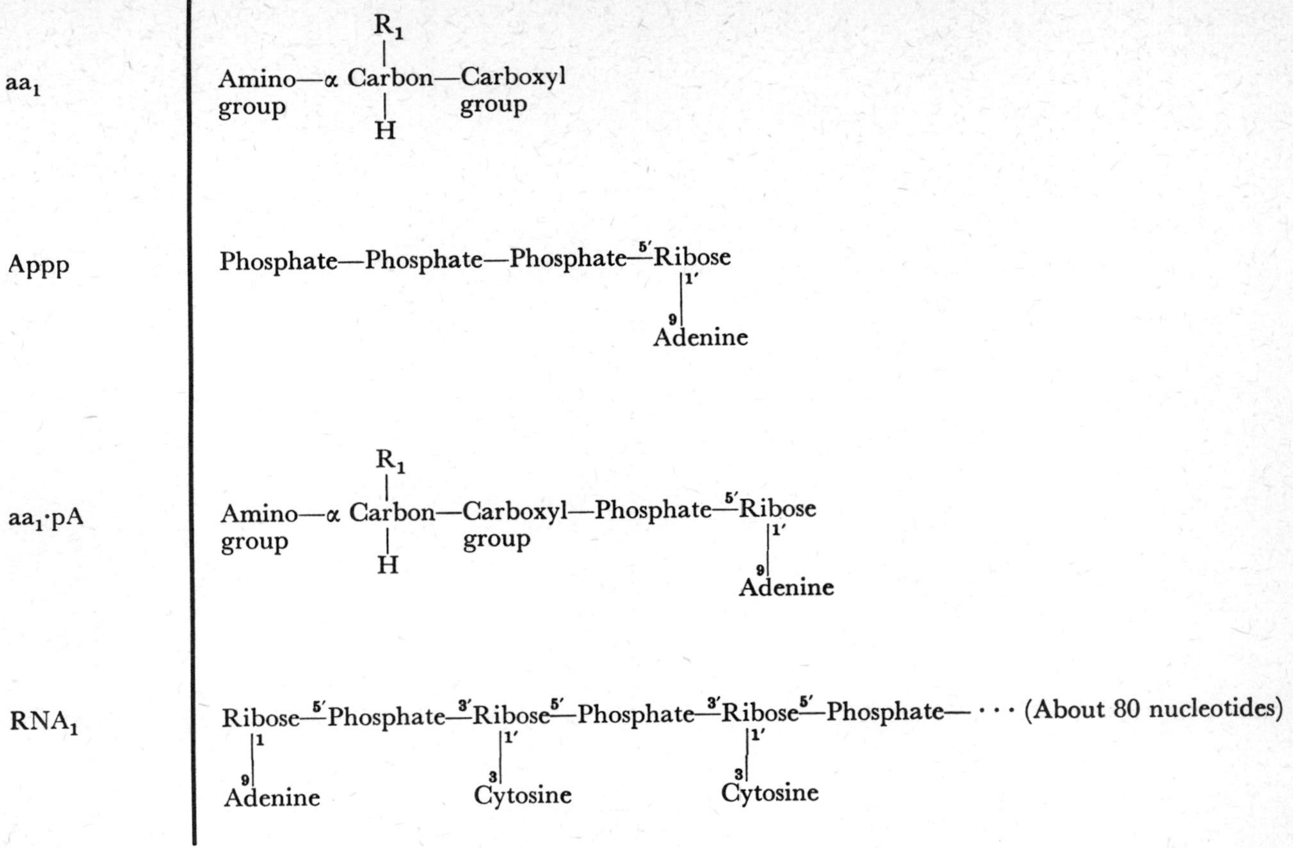

$aa_1$

$$R_1$$
$$|$$
Amino—$\alpha$ Carbon—Carboxyl
group        |        group
             H

$Appp$

Phosphate—Phosphate—Phosphate—5'Ribose
                                           1'|
                                        9
                                  Adenine

$aa_1 \cdot pA$

$$R_1$$
$$|$$
Amino—$\alpha$ Carbon—Carboxyl—Phosphate—5'Ribose
group        |        group                   1'|
             H                                  9
                                                Adenine

$RNA_1$

Ribose—5'Phosphate—3'Ribose—5'Phosphate—3'Ribose—5'Phosphate—····(About 80 nucleotides)
  1|                  1'|                  1'|
  9                  3                  3
  Adenine         Cytosine        Cytosine

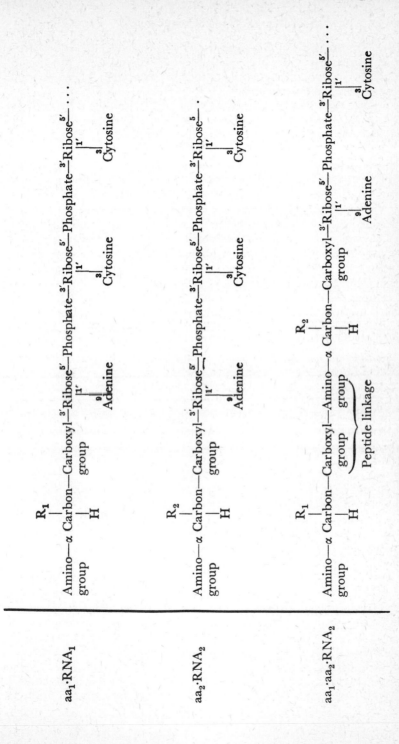

FIGURE 14.8   The structure of the intermediates in protein synthesis. The link between transfer RNA and amino-acid is shown as arising through the 3′ carbon atom of ribose but this is not yet conclusively established. The symbols in the left hand column have the same meaning as in Fig. 14.7.

## § 14.11   The structure of transfer RNA

With the establishment of the adaptor theory the analysis of the adaptors, or in other words, of transfer RNA, has become of paramount importance for understanding how the amino-acid code is translated. Their structure shows some remarkable features.

The complete nucleotide sequences of some of the transfer RNA (*t*RNA) molecules of *Escherichia coli*, *Saccharomyces cerevisiae*, *Torulopsis utilis*, *Rattus norvegicus* and *Triticum aestivum* have now been established. This was first achieved by Holley *et al.* (1965) for a *t*RNA for alanine in *S. cerevisiae* using partial digestion with two different ribonucleases, followed in each case by separation of the various oligonucleotides by chromatography and their identification. The segments formed by one ribonuclease overlapped those formed by the other because the two enzymes cut the molecule in different places: from the overlapping fragments the entire sequence could be determined. The molecule was found to consist of a chain of 77 nucleotides. Among possible configurations for it, Holley *et al.* suggested a clover-leaf shape (Fig. 14.9) because there were several places where a sequence of about 5 nucleotides was followed, after a region of unpaired nucleotides, by the complementary sequence in reverse order. Study of other *t*RNA molecules has given support for the clover-leaf model, but has revealed slight variations between one kind of *t*RNA and another in the lengths of the various parts (see Table 14.4). It seems, moreover, that although *t*RNAs normally have 4 arms (3 leaflets and a stalk to the clover-leaf) they may have a fifth (see Fig. 14.9 and Table 14.4).

Sanger, Brownlee and Barrell (1965) developed a two-dimensional technique for fractionating ribonuclease digests of $^{32}$P-labelled RNA. The oligonucleotides were separated by high-voltage ionophoresis, first on strips of cellulose acetate and then in a plane at right angles on diethylaminoethyl (DEAE) paper, the transfer being by a blotting technique. The positions of the oligonucleotides were then determined autoradiographically, the spots cut out, and the sequence of nucleotides in each fraction ascertained by further enzymic digestion. This technique has facilitated the rapid determination of nucleotide sequences in RNA.

One of the most remarkable features of transfer RNAs is the diversity of unusual nucleotides which they contain. Those found in the 7 *t*RNAs referred to in Fig. 14.9 and Table 14.4 are listed (as nucleosides) in Table 14.5. Over 30 different minor nucleotides, such as methylated purines of various kinds, or nucleosides with the ribose methylated are known in *t*RNAs. One of the most remarkable nucleotides is pseudouridylic acid which has uracil as its nitrogenous base but with the linkage to the ribose at position 5 in the uracil molecule instead of at position 3 (see Fig. 14.10). The nucleoside so formed is called 5-ribosyluracil or pseudouridine. Some *t*RNAs contain sulphur (see Table 14.5). The extraordinary diversity of nucleotides in *t*RNA simulates protein with its 20 different components rather than other nucleic acids. Nevertheless there is evidence that *t*RNA is formed like messenger RNA and ribosomal RNA by transcription of a

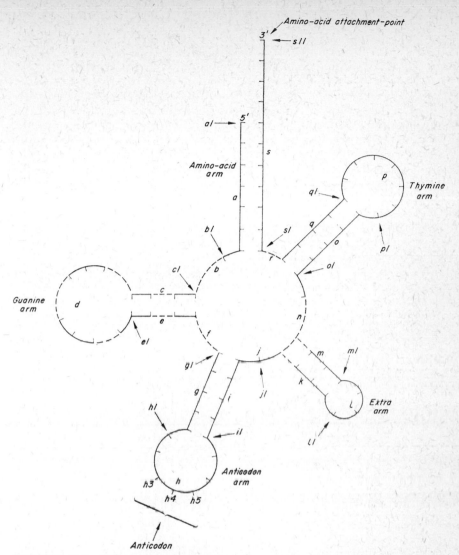

FIGURE 14.9   This is a diagrammatic representation of the structure of transfer RNA. The diagram is based on the nucleotide sequences of 7 *t*RNAs: for alanine (no. 1), serine (no. 1), tyrosine, valine (no. 1) and phenylalanine of *Saccharomyces cerevisiae*, and for tyrosine and formyl-methionine (see § 14.16) in *Escherichia coli*, from the results of Holley *et al.* (1965), Zachau, Dütting and Feldmann (1966), Madison, Everett and Kung (1966), Bayev *et al.* (1967), RajBhandary *et al.* (1967), Goodman *et al.* (1968) and Dube *et al.* (1968), respectively. The segments of the molecule are lettered in the 5′ to 3′ direction, that is, from the 5′-phosphate to the 3′-hydroxyl end of the chain, and the position of any nucleotide can be denoted by numbering them within each segment, also in the 5′ to 3′ direction: the first nucleotide of most of the segments is numbered in the diagram. Regions where the chain in one *t*RNA differs in length from that in another are indicated by broken lines. Regions of complementary pairing are shown by parallel lines.

nucleotide sequence from DNA (see § 18.7). The methylation or other modifications occur subsequently by means of specific enzymes. Johnson and Söll (1970) showed that pseudouridine could be formed *in vitro* at the poly-nucleotide level from uridine. Altman and Smith (1971) isolated the RNA of a potential tyrosine *t*RNA from *E. coli* before the modifications to the bases had taken place, and found uridine in place of pseudouridine and no methylation or other changes.

Transfer RNA molecules are synthesized as longer chains than the finished product. Altman and Smith found 41 additional nucleotides at the 5' end of the precursor molecule and 3 additional ones at the 3' end, but there was

TABLE 14.4  The table shows the numbers of nucleotides in the various segments of 7 different transfer RNAs from the results of the authors given in the caption to Fig. 14.9. The letters *a* to *s* refer to Fig. 14.9.

| Segment of Molecule | *Saccharomyces cerevisiae* Transfer RNA for | | | | | *Escherichia coli* Transfer RNA for | |
|---|---|---|---|---|---|---|---|
| | Ala$_1$ | Ser$_1$ | Tyr | Val$_1$ | Phe | Tyr | fMet |
| *Amino-acid arm* | | | | | | | |
| a 5' side | 7 | 7 | 7 | 7 | 7 | 7 | 7 |
| b Link to guanine arm | 2 | 2 | 2 | 2 | 2 | 1 | 2 |
| *Guanine arm* | | | | | | | |
| c Stem—5' side | 4 | 3 | 3 | 3 | 4 | 4 | 4 |
| d Loop | 10 | 10 | 12 | 11 | 8 | 11 | 9 |
| e Stem—3' side | 4 | 3 | 3 | 3 | 4 | 4 | 4 |
| f Link to anticodon arm | 1 | 1 | 1 | 1 | 1 | 0 | 1 |
| *Anticodon arm* | | | | | | | |
| g Stem—5' side | 5 | 5 | 5 | 5 | 5 | 5 | 5 |
| h Loop | 7 | 7 | 7 | 7 | 7 | 7 | 7 |
| i Stem—3' side | 5 | 5 | 5 | 5 | 5 | 5 | 5 |
| j Link* to k or n | 2 | 1 | 3 | 1 | 3 | 1 | 3 |
| *Extra arm* | | | | | | | |
| k Stem—5' side | 0 | 4 | 0 | 0 | 0 | 3 | 0 |
| l Loop | 0 | 3 | 0 | 0 | 0 | 3 | 0 |
| m Stem—3' side | 0 | 4 | 0 | 0 | 0 | 3 | 0 |
| n Link* to thymine arm | 2 | 2 | 2 | 3 | 2 | 3 | 2 |
| *Thymine arm* | | | | | | | |
| o Stem—5' side | 5 | 5 | 5 | 5 | 5 | 5 | 5 |
| p Loop | 7 | 7 | 7 | 7 | 7 | 7 | 7 |
| q Stem—3' side | 5 | 5 | 5 | 5 | 5 | 5 | 5 |
| r Link to amino-acid arm | 0 | 0 | 0 | 1 | 0 | 0 | 0 |
| *Amino-acid arm* | | | | | | | |
| s 3' side | 11 | 11 | 11 | 11 | 11 | 11 | 11 |
| TOTAL | 77 | 85 | 78 | 77 | 76 | 85 | 77 |

* When the extra arm (*k–m*) is absent *j* comprises purines and *n* pyrimidines.

evidence that there are normally more than 3 extra ones at this end. They detected a nucleolytic activity that seems to be responsible for trimming the ends from the precursor chain. From the nucleotide sequence of the extra piece at the 5′ end they inferred that nucleotides 41 and 42 (the cleavage point) were initially hydrogen-bonded to nucleotides 24 and 23 respectively, and they suggested that this protected the cleavage point from the endonuclease until the whole molecule had been transcribed, when the clover-leaf shape would be favoured and would lead to exposure of the cleavage point.

The minor nucleotides of transfer RNA occur chiefly in the unpaired regions, as can be seen from Table 14.5 where these regions are lettered *b*, *d*, *f*, *h*, *j*, *n* and *p* to correspond with Fig. 14.9 and the paired segments are lettered *c*, *g*, *i* and *o*. Minor nucleotides have not been found near the place of attachment of the amino-acid, that is, the 3′-hydroxyl end of the molecule and the part of the 5′-phosphate end with which it is paired to form the amino-acid arm (segments *a* and *s* in Fig. 14.9). There is a tendency for certain minor nucleotides to occur in the same region of different *t*RNAs. This is evident from Table 14.5 where for each nucleotide occurrences in the same region are grouped together on the same line of the table. It is evident also from the table that most of the minor nucleosides found in the *E. coli t*RNAs also occur at the same positions in at least one *t*RNA of *S. cerevisiae*, suggesting that these nucleotides may have persisted unchanged in position over the immense period of time since the chromonemal and chromosomal kingdoms diverged in evolution. It is clear, however, from the table that rather few of the minor nucleotides (indeed only two—and these neighbours) have been found at the same position in all 7 *t*RNAs. Nevertheless, when the normal nucleotides are also considered a fair number of constant features are evident (see Table 14.6). The -CCA sequence at the 3′-hydroxyl terminus was mentioned in § 14.10. Other constant features include the occurrence of (*a*) the sequence: a purine, one or more pyrimidines (often dihydrouracil), guanine,* guanine, in the first loop from the 5′-phosphate end of the chain, (*b*) the anticodon in the second loop, and particular associated bases which include uracil on the 5′ side and usually a minor purine on the 3′ side, and (*c*) a preponderance of pyrimidines in the loop nearest the 3′-hydroxyl end of the chain, including the sequence thymine–pseudouracil–cytosine. Further studies on *t*RNAs, including one or more from organisms as diverse as *Rattus norvegicus*, *Torulopsis utilis* and *Triticum aestivum* have maintained these constant features.

From the known function of the arms, or their base composition, they have been described in Fig. 14.9 and Table 14.4 as the amino-acid arm, the guanine arm, the anticodon arm and the thymine arm, with an extra arm in certain *t*RNAs. The guanine and thymine arms are sometimes called the dihydrouridine and the T$\Psi$C arms, respectively ($\Psi$ = pseudouridine).

Clark *et al.* (1968*b*) obtained one of the *t*RNAs of *E. coli* in microcrystalline form, and Blake *et al.* (1970) crystallized the mixed *t*RNAs of *S. cerevisiae*. The crystals in both studies were similar and enabled X-ray diffraction patterns to be observed. The crystallographic data promise to provide

* Nucleoside methylated in the ribose in several *t*RNAs.

TABLE 14.5   The table shows the minor nucleosides found in transfer RNAs for alanine (no. 1), serine (no. 1), tyrosine, valine (no. 1) and phenylalanine in *Saccharomyces cerevisiae* and for tyrosine and formylmethionine in *Escherichia coli* (for references see Fig. 14.9). The position of each nucleoside is indicated by the lettering and numbering system of Fig. 14.9, with identical or neighbouring positions for a nucleoside placed on the same line of the table.

| Nucleoside | Position in Transfer RNA | | | | | | |
|---|---|---|---|---|---|---|---|
| | *Saccharomyces cerevisiae* | | | | | *Escherichia coli* | |
| | $Ala_1$ | $Ser_1$ | Tyr | $Val_1$ | Phe | Tyr | fMet |
| **1.** *Purine nucleosides* | | | | | | | |
|   Unidentified | | | | | $h6$ | | |
|   (*a*) *Adenosine derivatives* | | | | | | | |
|     1-Methyladenosine | | | $p5$ | $p5$ | $p5$ | | |
|     $N^6$-Isopentenyladenosine | | $h6$ | $h6$ | | | | |
|     2-Thiomethyl-$N^6$-isopentenyl-adenosine | | | | | | $h6$ | |
|   (*b*) *Guanosine derivatives* | | | | | | | |
|     1-Methylguanosine | $b2$ | | | $b2$ | | | |
|     $N^2$-Methylguanosine | | | $c1$ | | $c1$ | | |
|     $N^2$-Dimethylguanosine | $f1$ | $f1$ | $f1$ | | $f1$ | | |
|     $N^7$-Methylguanosine | | | | | $j3$ | | $j3*$ |
|     2'-O-Methylguanosine | | $d5$ | $d6$ | | | $d5$ | |
|      | | | | | $h3$ | | |
|     Unidentified | | | | | | $h3$ | |
|     Inosine | $h3$ | $h3$ | | $h3$ | | | |
|     1-Methylinosine | $h6$ | | | | | | |
| **2.** *Pyrimidine nucleosides* | | | | | | | |
|   (*a*) *Cytidine derivatives* | | | | | | | |
|     $N^6$-Acetylcytidine | | $c3$ | | | | | |
|     5-Methylcytidine | | | | | $i2$ | | |
|      | | $n2$ | $n2$ | $n3$ | $o1$ | | |
|     2'-O-Methylcytidine | | | | | $h1$ | | $h1$ |
|   (*b*) *Uridine derivatives* | | | | | | | |
|     Ribothymidine | $p1$ | $p1$ | $p1$ | $p1$ | $p1$ | $p1$ | $p1$ |
|     5,6-Dihydrouridine | $d5$ | $d4$ | $d4,5$ | $d4$ | $d3,4$ | | |
|      | $d9$ | $d7,8$ | $d8,9,10$ | $d8,9$ | | | $d8$ |
|      | $n1*$ | | $n1$ | $n2$ | | | |
|     4-Thiouridine | | | | | | $b1$ | $b1$ |
|     2'-O-Methyluridine | | $j1$ | | | | | |
|     Pseudouridine | | | | $d1$ | | | |
|      | | | | $g1$ | | | |
|      | | $h1$ | | $h1$ | | | |
|      | | | $h4$ | | | | |
|      | $h7$ | $i1$ | $i1$ | | $i1$ | $i1$ | |
|      | $p2$ | $p2$ | $p2$ | $p2$ | $p2$ | $p2$ | $p2$ |

\* This nucleoside occurs in one form of the *t*RNA but not in another: see § 14.14.

(a)                                    (b)

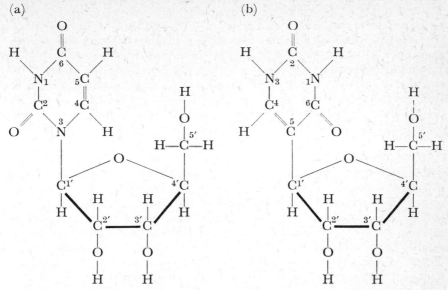

FIGURE 14.10   The structure of (*a*) uracil ribonucleoside (uridine), and (*b*) 5-ribosyluracil (pseudouridine).

TABLE 14.6   The table shows the nucleosides which occur constantly at certain positions in 7 transfer RNAs (for alanine (no. 1), serine (no. 1), tyrosine, valine (no. 1) and phenylalanine in *Saccharomyces cerevisiae* and for tyrosine and formylmethionine in *Escherichia coli*; for references see Fig. 14.9). The position of the nucleosides is indicated by the lettering and numbering system of Fig. 14.9. The alternatives in regions *d* and *n* given in brackets in the table occur in the same *t*RNAs.

| Position | Nucleoside | Position | Nucleoside |
|---|---|---|---|
| *b*1 | Uridine* | 3′ end of *n* | Cytidine*(or uridine) |
| *b*2† | With a purine as base* | *o*5 | Guanosine |
| 3′ end of *c* | Cytidine | *p*1 | Ribothymidine |
| | Adenosine | *p*2 | Pseudouridine |
| | Guanosine (or adenosine) | *p*3 | Cytidine |
| Sequence in *d* starting at *d*1, *d*2 or *d*3 | 1, 2 or 3 nucleosides all with a pyrimidine as base, often 5,6-dihydrouridine | *p*4 | With a purine as base |
| | | *p*5 | Adenosine* |
| | | *p*7 | With a pyrimidine as base |
| | Guanosine* | *q*1 | Cytidine |
| | Guanosine | *s*8 | With a purine as base |
| *e*1 | Guanosine | *s*9 | Cytidine |
| *h*1 | With a pyrimidine as base | *s*10 | Cytidine |
| *h*2 | Uridine | *s*11 | Adenosine |
| (*h*3–5 | Anticodon) | | |
| *h*6 | With a minor purine as base‡ | | |

\* A derivative in one or more of the *t*RNAs (see Table 14.5).

† Not present in *E. coli t*RNA for tyrosine (see Table 14.4).

‡ Except in *S. cerevisiae t*RNA no. 1 for valine and *E. coli t*RNA for formylmethionine (see § 14.16).

basic information about the three-dimensional structure of *t*RNA. Ichikawa and Sundaralingam (1972) obtained a new crystal form of phenylalanine *t*RNA from *S. cerevisiae*. The unit cell was the smallest yet found and implied that the *t*RNA molecule cannot be much more than 6.3 nm long and 2.3 × 3.3 nm wide. Levitt (1969) suggested that the clover-leaf was folded with the guanine arm stacked on the anticodon arm and, above and parallel to this, the amino-acid arm stacked on the thymine arm to form two regular double helices (Fig. 14.11). This arrangement was considered the most stable because both double-helical structures have one uninterrupted nucleotide chain—segments *e–g* and *q–s* respectively, in Figs. 14.9 and 14.11. The four purines which usually make up the sequence AG . . . GG (see Table 14.6) in region *d* (the guanine loop) were believed to form base pairs, respectively, with the following pyrimidines: (1) U at *b*1, (2) C at the 3′ end of *n*, (3) Ψ at *p*2 and (4) C at *p*3. These base pairs are numbered accordingly in Fig. 14.11. Support for the occurrence of base pair no. 2 was obtained by Cashmore (1971), who found that a mutant *t*RNA of *E. coli* with the sequence AA . . . GG in the guanine loop had the C at the 3′ end of *n* exposed instead of base-paired. This exposure was revealed by its reactivity to methoxyamine. The occurrence of base pair no. 2 was also supported by the discovery of three *t*RNAs with AA . . . GG as the normal sequence in the guanine loop and with U instead of C at the 3′ end of *n* (see Hirsch, 1971 and Cashmore, 1971). The three-dimensional structure of the anticodon loop is discussed in § 14.15.

Dudock *et al.* (1969) determined the entire nucleotide sequence of a *t*RNA for phenylalanine from *Triticum aestivum* and found that the guanine arm was the only major region of the molecule which was identical with the *t*RNA for phenylalanine in *Saccharomyces cerevisiae*. The specific enzyme in *S. cerevisiae* for attaching phenylalanine to this transfer RNA will also attach it to the *Triticum t*RNA, to the *t*RNA for phenylalanine in *E. coli*, and to *t*RNA no. 1 for valine in *E. coli* (Dudock *et al.*, 1971). These four *t*RNAs have identical stems to the guanine arm, with the base sequence GCUC in region *c* (Fig. 14.9) and the complementary sequence GAGC in region *e*. No other extensive part of the molecule was uniform in all four *t*RNAs, except the segments in the guanine and thymine loops which are constant in all *t*RNAs (see Table 14.6). Dudock *et al.* concluded that the stem of the guanine arm was the recognition site for the *Saccharomyces* phenylalanyl-*t*RNA synthetase. Yaniv and Barrell (1971) found that *t*RNAs nos. 1 and 2 for valine in *E. coli* differ by 22 nucleotides, yet both are activated by the same enzyme, though with different association constants. They suggested that the synthetase recognition site may be formed by the interaction of nucleotides in different parts of the chain in the folded clover-leaf which forms the tertiary structure of the molecule.

Gilbert (1963) suggested there was only one active site for the synthesis of the peptide bond on each ribosome, and that each *t*RNA molecule takes the place of the preceding one, adding its amino-acid to the chain and ejecting the previous *t*RNA as it does so. In Fig. 14.11, an amino-acyl-*t*RNA is supposed just to have added amino-acid no. 12 to a polypeptide chain,

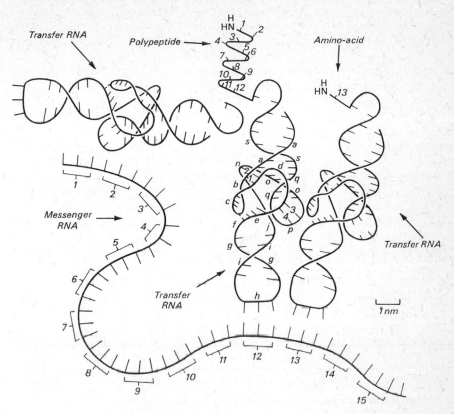

FIGURE 14.11   Diagram to illustrate the structure for transfer RNA
proposed by Levitt (1969), and to show how the triplet
sequence in messenger RNA determines which amino-
acyl transfer RNA molecules shall succeed one another
in taking up position on the ribosome (not shown),
and hence which amino-acids shall succeed one another
in the polypeptide. The transfer RNA which added
amino-acid no. 11 to the polypeptide has just been displaced by that which has brought no. 12, while that
bringing no. 13 has just been selected through the ability
of its anticodon to pair with triplet no. 13 in the messenger. In Levitt's model the clover-leaf of Fig. 14.9 is
folded with the guanine arm stacked upon the anticodon
arm and with the amino-acid arm stacked upon the
thymine arm. The broken lines indicate base pairing.
Those numbered 1–4 link the guanine loop with other
parts of the molecule and are described in the text. The
lettering of segments corresponds to Fig. 14.9.

thereby becoming a *peptidyl-tRNA* and displacing the *t*RNA which brought amino-acid no. 11. Gilbert suggested that this displacement of the previous tenant forces the messenger to move over the ribosome by one reading unit, thereby allowing the next amino-acyl-*t*RNA to take up position: in the diagram, this is the *t*RNA carrying amino-acid no. 13. Warner and Rich (1964), from study of haemoglobin synthesis in the reticulocytes of *Oryctolagus*, showed that each ribosome active in protein synthesis contained two *t*RNA molecules. They therefore modified Gilbert's proposals and suggested there were two adjacent active sites on the ribosome: one (called the *peptide site* by Bretscher and Marcker, 1966) for the peptidyl-*t*RNA carrying the growing polypeptide, and the other (the *amino-acid site*) for the amino-acyl-*t*RNA bringing the next amino-acid. Each time a peptide bond is formed an amino-acyl-*t*RNA becomes a peptidyl-*t*RNA and moves from the amino-acid site to the peptide site on the ribosome (see § 14.16).

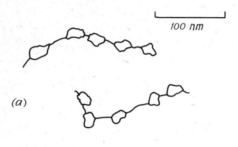

100 nm

(a)

(b)

150 amino-acids

← - - - - 450 nucleotides - - - - →

FIGURE 14.12   Ribosomes from reticulocytes of *Oryctolagus cuniculus* (Rabbit). The thread joining the ribosomes is thought to be messenger RNA for one of the polypeptide chains of haemoglobin.

(a) Appearance as seen in electron micrographs prepared by Slayter *et al.* (1963).

(b) Diagrammatic representation to show the progressive formation of the polypeptide chain on the ribosomes as they move along the messenger RNA (from left to right in the diagram).

Eventually, as polypeptide synthesis proceeded, the beginning of the messenger would be far enough away to attach a second ribosome. The aggregates of ribosomes which occur when protein synthesis is taking place (Pl. 14.1) would arise in this way, as suggested by Gierer (1963) and others, through a series of ribosomes being serviced by one messenger RNA molecule. Each ribosome would move along the messenger as its polypeptide chain was synthesized. Slayter, Warner, Rich and Hall (1963) studied the ribosome aggregates (polyribosomes) in the reticulocytes of *Oryctolagus* under the electron microscope. These cells synthesize haemoglobin predominantly, and so the messenger RNA is expected to be of a uniform length of not less than 150 nm, assuming 3 nucleotides to every 1 nm, a coding ratio of 3, and 150 amino-acids per polypeptide. The electron micrographs agreed with this expectation; there were usually 5 ribosomes held together by a thread 1 nm in width and 170–280 nm in length (see Fig. 14.12(*a*)).

Staehelin, Wettstein, Oura and Noll (1964) found the average length in nucleotides of the segment of messenger RNA between adjacent ribosomes in the ribosomal aggregates of liver cells of *Rattus*. By sedimentation analysis it was found that the molecular weight of the messenger RNA was proportional to the size of the polyribosome, and corresponded to 90 nucleotides per ribosome. On the assumption that the ribosome spacing is the same in *Oryctolagus* reticulocytes synthesizing haemoglobin, there must be approximately 450 nucleotides to the haemoglobin messenger RNA molecule since it usually extends to 5 ribosomes. As the haemoglobin polypeptides synthesized contain about 150 amino-acids, it follows that the coding ratio is 3 (see Fig. 14.12(*b*)).

If the mechanism of protein synthesis outlined in this and the preceding section is correct, the reading of the amino-acid code would be in consecutive non-overlapping triplets of nucleotides in messenger RNA. It is not necessary to assume that there is any recognition of the coding triplets in the transfer of nucleotide sequence from DNA to RNA.

## § 14.12   The deciphering of the code

A major step forward in the understanding of the amino-acid code was taken by Nirenberg and Matthaei (1961) when they discovered a direct method for deciphering it. They had found that a synthetic RNA in which all the bases were uracil would bring about the incorporation of phenyl-alanine into polypeptide chains. This was achieved with an *in vitro* protein-synthesizing system derived from disrupted cells of *Escherichia coli*. After centrifuging the cell contents, the ribosomes were mixed with supernatant solution containing amino-acyl transfer RNA and enzymes, and adenosine triphosphate (ATP) and an ATP-generating system were added. The synthetic RNA was obtained by means of the polynucleotide phosphorylase first isolated by Grunberg-Manago and Ochoa (1955) (see § 14.4). The amino-acid was labelled with [14]C and the subsequent presence of the label in polypeptide chains demonstrated.

Lengyel, Speyer and Ochoa (1961) synthesized an RNA using a mixture of the ribonucleoside diphosphates of uracil (U) and cytosine (C) in the proportion of 5:1. On the assumption that nucleotides are randomly incorporated into the chain in proportion to their abundance, out of every 216 triplets there will be 125 with the sequence UUU, 25 each with UUC, UCU, and CUU, 5 each with UCC, CUC, and CCU and 1 with CCC Lengyel *et al.* found that when added to the protein-synthesizing system, this RNA led to the incorporation of serine as well as phenylalanine into the polypeptide. There were approximately 4·4 phenylalanine residues to every 1 of serine, and so it was tentatively concluded that, if it is a triplet code, one of the 3 trinucleotides containing 2 uracils and 1 cytosine codes for serine. In the same way, a synthetic RNA containing uracil and adenine in the proportion of 5:1 was found to lead to the incorporation of phenylalanine and tyrosine in the proportion of 4:1, suggesting that the code for tyrosine contained 2 uracils and 1 adenine.

By an extension of this technique, provisional nucleotide triplets (without knowledge of the base sequence) were soon assigned to nearly all the amino-acids. A notable feature was that all these triplets contained uracil. This suggested that more than one triplet could code for the same amino-acid, since uracil was not known to occur in such excess in natural messenger RNA. The multiple coding* for individual amino-acids was confirmed when Gardner, Wahba, Basilio, Miller, Lengyel and Speyer (1962) found that most synthetic RNA molecules without uracil will also bring about amino-acid incorporation. Moreover, they showed that an RNA containing only adenine leads to a polypeptide composed of lysine residues. This was first demonstrated with an *in vitro* protein synthesizing system derived from liver cells of *Rattus*, and it was subsequently shown with the *E. coli* system. In conjunction with the original discovery that uracil RNA gives a phenyl-alanine polypeptide, this finding was of particular significance because adenine and uracil are complementary. The implication was that nucleotide sequence is transcribed from one particular chain of the DNA molecule, because the other chain would lead (through a complementary messenger RNA) to the formation of a different polypeptide. Thus an adenine triplet in DNA would give a uracil triplet in messenger RNA and hence phenyl-alanine in the peptide, while the complementary thymine triplet in the other DNA chain would give rise to an adenine triplet in messenger RNA and hence lysine in the peptide. As indicated in § 14.7 and § 14.9, there is experimental evidence that, in any particular region of the DNA, messenger RNA is indeed synthesized only as the complement of one specific chain of the DNA.

In confirmation of the alternative coding for individual amino-acids, Weisblum, Benzer and Holley (1962) demonstrated the existence in extracts of *E. coli* of two different transfer RNA molecules each specific for the same amino-acid (leucine). The two leucine transfer RNAs were separated by

---

* Multiple coding for individual amino-acids is often referred to as code *degeneracy*, but, as Tatum (1964) has pointed out, this is a misleading expression, since it has evolutionary implications which may not be valid.

countercurrent fractionation. This technique depends on the discovery that certain *t*RNAs have widely different partition coefficients in a two-phase solvent system composed of pH 6 phosphate buffer, formamide and iso-propanol. It was found that incorporation of leucine into polypeptides by one of these *t*RNAs was stimulated by a synthetic RNA containing uracil and cytosine, and incorporation by the other was stimulated by a uracil–guanine RNA.

Nirenberg and Leder (1964) described a method using *E. coli* extracts for measuring the interaction of transfer RNAs carrying their specific amino-acids with ribosomes carrying messenger RNA. The behaviour of synthetic polyribonucleotides of various lengths and compositions, acting as messenger, was studied. Two enzymatic methods had been devised for synthesizing these oligonucleotides. Trinucleotides with a 5'-terminal phosphate were found to be highly active in binding specific amino-acyl transfer RNAs to ribosomes, whereas similar dinucleotides, and also tri-nucleotides with 3'-terminal phosphate, were largely or entirely inactive. The trinucleotides with 5'-terminal phosphate containing only uracil (UUU), only adenine (AAA) and only cytosine (CCC) were found to direct the binding to ribosomes of *t*RNAs for phenylalanine, lysine and proline, respec-tively, in agreement with earlier work. Studies with this trinucleotide binding technique quickly established the identity of a majority of the coding triplets and, coupled with the work of Khorana *et al.* (1967) using poly-nucleotides with known repeating di- and trinucleotide sequences, solved the code (see Table 14.7 from Morgan, Wells and Khorana, 1966).

The technique developed by Khorana and associates was to prepare short deoxyribopolynucleotides of known sequence by chemical methods, to use

TABLE 14.7   The table shows the amino-acid code established from *in vitro* studies with *Escherichia coli*. The orientation of the trinucleotides follows the conventional sequence (cf. § 12.5), that is, with 5'-phosphates to the left and 3'-phosphates to the right of the symbols (A, C, G and U) for the ribonucleosides of adenine, cytosine, guanine and uracil.

| | | | | | | | |
|---|---|---|---|---|---|---|---|
| UUU<br>UUC | Phenylalanine | UCU<br>UCC<br>UCA<br>UCG | Serine | UAU<br>UAC | Tyrosine | UGU<br>UGC | Cysteine |
| UUA<br>UUG<br>CUU<br>CUC<br>CUA<br>CUG | Leucine | CCU<br>CCC<br>CCA<br>CCG | Proline | UAA<br>UAG | Stop† | UGA Stop†<br>UGG Tryptophan | |
| | | | | CAU<br>CAC | Histidine | CGU<br>CGC<br>CGA<br>CGG | Arginine |
| | | | | CAA<br>CAG | Glutamine | | |
| AUU<br>AUC<br>AUA | Isoleucine | ACU<br>ACC<br>ACA<br>ACG | Threonine | AAU<br>AAC | Asparagine | AGU<br>AGC | Serine |
| AUG Methionine* | | | | AAA<br>AAG | Lysine | AGA<br>AGG | Arginine |
| GUU<br>GUC<br>GUA<br>GUG | Valine | GCU<br>GCC<br>GCA<br>GCG | Alanine | GAU<br>GAC | Aspartic acid | GGU<br>GGC<br>GGA<br>GGG | Glycine |
| | | | | GAA<br>GAG | Glutamic acid | | |

* Also, polypeptide initiation (see § 14.16).
† Polypeptide termination (see § 14.17).

these synthetic polynucleotides as templates for DNA polymerase I action to obtain high molecular weight DNA with known repeating nucleotide sequence, and then to use DNA-dependent RNA polymerase to obtain a synthetic messenger RNA also of known repeating sequence. It was found that for DNA polymerase I to act it was necessary to have complementary sequences each of 8–12 nucleotides. The synthetic DNA formed had the duplex structure. It was prepared with not more than 3 different bases in the individual chains, so that by supplying only the appropriate ribonucleoside triphosphates the action of the RNA polymerase could be restricted to one particular DNA chain. Thus, a synthetic DNA with the base sequence TAC repeated many times in one chain and with GTA repeated in the other gave rise to RNA with the repeating sequence GUA when the triphosphates of these 3 nucleosides were supplied, and with the sequence UAC in the presence of the triphosphates of U, A and C. Having obtained RNA with such known sequences of nucleotides, Nirenberg and Matthaei's technique of cell-free polypeptide synthesis was used to assign codons to amino-acids. In this way Nishimura, Jones, Ohtsuka, Hayatsu, Jacob and Khorana (1965) found that a ribopolynucleotide containing the repeating sequence AAG directed the synthesis of 3 different polypeptides containing lysine, arginine and glutamic acid, respectively. Which of the 3 was formed in any particular instance evidently depended on whether the reading-frame was set to read the message as AAG, AGA or GAA. Nishimura, Jones and Khorana (1965) found that RNA containing the nucleosides U and C in alternating sequence directed the synthesis of a polypeptide containing serine and leucine in alternating sequence. This is again what is expected from a non-overlapping triplet code read consecutively, because the messenger will have UCU alternating with CUC.

Khorana and associates also applied Nirenberg and Leder's technique of trinucleotide-stimulated binding of amino-acyl transfer RNAs to ribosomes, using ribotrinucleotides prepared by chemical methods. By this combination of techniques they not only established the total structure of the amino-acid code but obtained direct proof that the coding ratio is three, that the code is non-overlapping and that contiguous triplets are read sequentially without omission of any bases.

These experiments established the code for *Escherichia coli in vitro*. How far the same code applies *in vivo* and to other organisms is discussed in the next section.

## § 14.13    Confirmation of the code

Study of the changes in polypeptides caused by mutations has provided information about the operation of the amino-acid code in living cells. Data are available for proteins in several diverse organisms, and all are in agreement with the code deduced from the cell-free studies described in § 14.12.

Tsugita and Fraenkel-Conrat (1962) and Wittmann (1962) studied the effects of nitrous acid in causing mutations affecting the coat protein of tobacco mosaic virus. Since nitrous acid is thought to cause mutation by altering cytosine to uracil, and adenine to hypoxanthine which is then expected to pair like guanine (see § 14.2), any changes in the protein of the virus coat caused by mutations induced with nitrous acid will be expected to be the result of one or other of these two base substitutions. These authors found that particular amino-acid substitutions such as phenylalanine in place of serine, and isoleucine in place of threonine, turned up repeatedly with nitrous acid treatment. Later results were reviewed by Wittman and Wittman-Liebold (1967). As pointed out by Crick (1967), there is good agreement between these substitutions and those predicted by changing cytosine to uracil or adenine to guanine in the code deduced from the results of protein synthesis *in vitro* using *Escherichia coli* components and synthetic RNA.

Yanofsky, Ito and Horn (1967) studied the changes due to mutations affecting the *A* protein of the enzyme tryptophan synthetase of *Escherichia coli* (see § 13.8 and Fig. 13.8). No less than 10 different amino-acid substitutions have been found at position 210 (Fig. 14.13(*a*)) and 4 at position 233 (Fig. 14.13(*b*)). Henning and Yanofsky (1962) and Guest and Yanofsky (1965) found that some of these strains yielded wild-type recombinants when

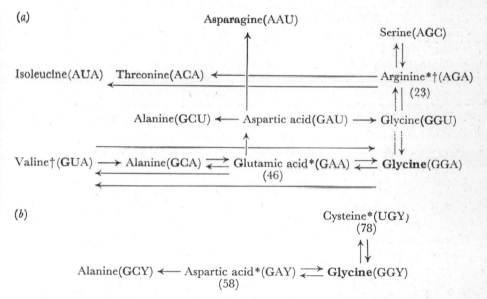

(*a*)

FIGURE 14.13   Diagram to show changes observed by Yanofsky and associates in the A protein of tryptophan synthetase of *Escherichia coli* as a result of mutation and recombination. The arrows show amino-acid changes observed through mutation (*a*) at position 210 and (*b*) at position 233 from the wild-type shown in heavy type. Numbered mutants correspond to Fig. 13.8; un-numbered mutants are full or partial revertants of these. Pairs of strains which have given wild-type recombinants on crossing are marked with asterisks or daggers. The probable RNA codons, on the basis of the *in vitro* code, are shown in brackets; Y = U or C.

crossed with one another, evidently through recombination within the nucleotides of DNA which provide the template for one codon in messenger RNA. The significant feature of the changes observed, either by mutation or recombination, is that they are all explicable in terms of single nucleotide changes, or single exchanges between nucleotides, on the basis of the *in vitro* code. The data thus support the code and indicate that mutations normally affect only one nucleotide.

Terzaghi, Okada, Streisinger, Emrich, Inouye and Tsugita (1966) found that the enzyme lysozyme produced by the normal (wild-type) form of virus *T*4 of *Escherichia coli* and the enzyme from a pseudowild strain carrying two proflavin-induced mutations differ by a sequence of 5 amino-acids at positions 36–40 in the molecule, which contains 164 amino-acid residues. The two sequences of amino-acids, and nucleotide sequences for the messenger compatible with both of them, are shown in the upper part of Fig. 14.14(*a*). One proflavin mutant (eJ42) had evidently lost a particular adenine nucleotide and the other mutant (eJ44) had gained a guanine nucleotide 15 nucleotides further along the molecule. The double mutant had a nearly normal phenotype (pseudowild-type) because the reading-frame was evidently set in register again after only 5 triplets had been misread. The results not only suggest that the code established for *E. coli in vitro* applies to *T*4 *in vivo*, but they confirm the hypothesis of Crick and associates (1961) that the translation of the genetic message is initiated at a given point where a reading-frame is set in register (see § 14.3). The results also confirm that proflavin acts as a mutagen by inserting or deleting nucleotides, that is, it causes frameshift mutations—a shift in the reading-frame causing a grossly different translation of the message beyond the site of the mutation. A further aspect of this remarkable experiment is discussed in § 14.14.

Okada, Terzaghi, Streisinger, Emrich, Inouye and Tsugita (1966) made similar studies with another proflavin mutant (eJ17) which, like eJ42, was found to give a pseudowild phenotype when brought into the same molecule as eJ44 by recombination. The amino-acid sequence of this double mutant is shown in the lower part of Fig. 14.14(*a*). It differs from the other double mutant by the addition of one amino-acid (serine). Mutant eJ17 evidently arose from the addition of two nucleotides (GU or UG). In the double mutant the addition of G by mutant eJ44 then restores the normal setting of the reading-frame, but adds one extra amino-acid to the molecule. Other examples of frameshift mutants in the lysozyme gene (see Fig. 18.4) of *T*4 are discussed by Streisinger, Okada *et al.* (1967) and Imada *et al.* (1970).

Brammar, Berger and Yanofsky (1967) studied frameshift mutants in the tryptophan synthetase *A* gene of *E. coli*. Two partial revertants, nos. 8 and 11, of mutant 9813 were found to have changes in the amino-acids at positions 174–8 and 173–4 in the molecule, respectively (see Figs. 13.8 and 14.14(*b*)). Nucleotide sequences compatible with the *in vitro* data on the amino-acid code could be assigned to these peptides, and these revealed that mutant 9813 was due to the loss of an adenine nucleotide in codon 174, while the partial revertants 8 and 11 had gained G in codon 178 and A in 173, respectively.

(a)

eJ42 eJ44   $\begin{cases} NH_2---Thr \mid Lys \mid Val \mid His \mid His \mid Leu \mid Met \mid Ala ---CO \cdot OH \\ 5'----ACZAAZGUCCAUCACUUAAUGGCX---3' \end{cases}$

                 −A ↑ (eJ42)               +G ↑ (eJ44)

Wild-type   $\begin{cases} NH_2---Thr \mid Lys \mid Ser \mid Pro \mid Ser \mid Leu \mid Asn \mid Ala ---CO \cdot OH \\ 5'----ACZAAZAGUCCAUCACUUAAUGCX---3' \end{cases}$

          +GU ↓ (eJ17)            +G ↓ (eJ44)

eJ17 eJ44   $\begin{cases} NH_2---Thr \mid Lys \mid Ser \mid Val \mid His \mid His \mid Leu \mid Met \mid Ala ---CO \cdot OH \\ 5'----ACZAAZAGUGUCCAUCACUUAAUGGCX---3' \end{cases}$

(b)

9813 PR8   $\begin{cases} NH_2---Tyr \mid Thr \mid Phe \mid Cys \mid Cys \mid His \mid Gly \mid Ala ---CO \cdot OH \\ 5'----UAYACCUUYUGCUGUCACGGXGCX---3' \end{cases}$

             −A ↑ (9813)             +G ↑ (PR8)

Wild-type   $\begin{cases} NH_2---Tyr \mid Thr \mid Tyr \mid Leu \mid Leu \mid Ser \mid Arg \mid Ala ---CO \cdot OH \\ 5'----UAYACCUAUYUGCUGUCACGXGCX---3' \end{cases}$

         +A ↓ $\binom{PR}{11}$    ↓ −A (9813)

9813 PR11   $\begin{cases} NH_2---Tyr \mid Asn \mid Leu \mid Leu \mid Leu \mid Ser \mid Arg \mid Ala ---CO \cdot OH \\ 5'----UAYAACCUUYUGCUGUCACGXGCX---3' \end{cases}$

FIGURE 14.14   Diagram to show the effects of frameshift mutants (a) on amino-acids 34–41 in the lysozyme molecule of virus *T4* of *Escherichia coli*, from the work of Terzaghi, Okada, Streisinger, Emrich, Inouye and Tsugita (1966) and Okada *et al.* (1966), and (b) on amino-acids 172–179 in the *A* protein of tryptophan synthetase of *E. coli*, from the work of Brammar, Berger and Yanofsky (1967). Nucleotide sequences compatible with the peptides are shown. X = A, C, G or U. Y = C or U. Z = A or G. Letters and numbers in brackets identify the mutants and their revertants. The base sequence *ZAAZA* near the left-hand end of the segment of the lysozyme gene in (a) is a mutational 'hot-spot' and is thought to consist of 5 adenines in a row (see § 16.7).

A remarkable feature of these studies on frameshift mutants, both in *T4* and its host, is that all the additions of nucleotides are duplications of nucleotides adjacent to the site of the addition (AA, GG and GUGU in place of A, G and GU, respectively). For further discussion of this feature of acridine mutagenesis, see Imada *et al.* (1970).

Crick (1967) examined the published data on the mutational changes known in human haemoglobin and found that the amino-acid substitutions could all be accounted for by single base changes, assuming that the code deciphered with the cell-free system from *E. coli* applied to man. Crick pointed out that all the mutations known at that time in haemoglobin start or finish with charged amino-acids (arginine, aspartic acid, glutamic acid, histidine or lysine) because of the method of detecting the mutants.

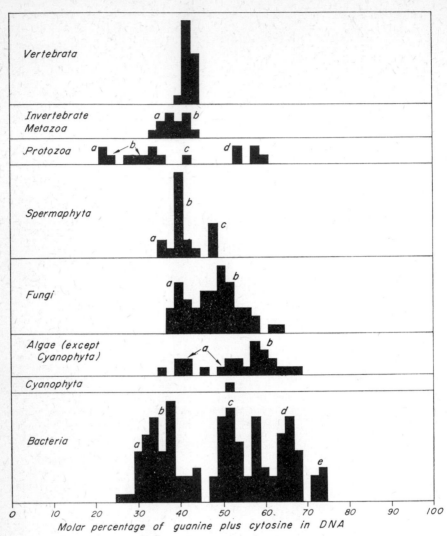

FIGURE 14.15    Histograms to show the molar percentage of guanine plus cytosine out of the total bases (adenine, cytosine, guanine and thymine) in the DNA of 275 organisms from the data tabulated by Sueoka (1964), supplemented with the results given by Jones and Thompson (1963), Tonomura, Malkin and Rabinowitz (1965), Storck (1966), Evans (1966) and Dutta, Richman, Woodward and Mandel (1967). The peaks in the histograms are caused by the following groups of organisms (see Appendix 1 for their classification):

**Bacteria:** (*a*) Bacillaceae, (*b*) Lactobacteriaceae and Parvobacteriaceae, (*c*) Enterobacteriaceae, (*d*) Achromobacteriaceae and Pseudomonadaceae, (*e*) Actinomycetales.
**Algae:** (*a*) Colourless Volvocales, Bacillariophyta *et al.*, (*b*) other Chlorophyta *et al.*

Adams, Jeppesen, Sanger and Barrell (1969) determined the sequence of 57 nucleotides in part of the RNA of virus $R17$ of *E. coli*. The fragment studied corresponded to amino-acids 81–99 of the protein of the viral coat, and provided direct confirmation of 17 of the 61 amino-acid codons. Min Jou, Haegeman, Ysebaert and Fiers (1972) determined the entire nucleotide sequence of the gene for the coat protein of virus $MS2$ of *E. coli*. The 387 nucleotides needed to code for the 129 amino-acid residues in the coat polypeptide provided confirmation of 49 codons. The absence of AUA as a codon for isoleucine and of UAU for tyrosine looked significant.

It is evident from these studies on tobacco mosaic virus, *E. coli* and its $T4$, $R17$ and $MS2$ viruses, and man, that at least part of the *E. coli in vitro* code applies *in vivo* and to organisms far removed from bacteria in evolution. There is a strong probability, therefore, that at least the greater part of the code will be found to be universal. The first evidence that the amino-acid code was likely to be the same in a wide range of organisms was obtained by von Ehrenstein and Lipmann (1961). They showed that amino-acyl transfer RNA derived from *E. coli*, when mixed with the ribosomes and attached haemoglobin messenger RNA from the reticulocytes of *Oryctolagus*, would bring about the synthesis of haemoglobin. In other words, the bacterial transfer RNA was interpreting the mammalian messenger correctly as judged by the fingerprint of the end-product.

That the amino-acid code should include several alternative coding units for most of the amino-acids provides an explanation for an otherwise puzzling phenomenon, namely, the remarkable diversity in the frequency of each base-pair in the DNA of different organisms, particularly among bacteria (cf. § 11.6). The range is from about $\frac{1}{4}$ to $\frac{3}{4}$ of the total. This variation is much greater than could be due to diversity in amino-acid composition of the constituent proteins. The molar percentage of guanine plus cytosine out of the total bases in the DNA of 275 organisms is shown graphically in Fig. 14.15. The uniformity in base composition of the more highly evolved organisms compared with protozoa, fungi, algae and bacteria is apparent. This may reflect the relative length of time that has elapsed since they evolved, or more probably the number of generations, since mutation-rate is related to generation-time (see § 9.5). With alternative codes for the same amino-acid, considerable change in nucleotide composition is possible

---

**Fungi:** (*a*) Mucorales and Endomycetales, (*b*) Eu-Ascomycetes (including Fungi Imperfecti) and Basidiomycetes.

**Spermaphyta:** (*a*) Liliaceae, (*b*) Dicotyledones and 1 conifer, (*c*) Gramineae.

**Protozoa:** (*a*) Acrasiales, (*b*) Ciliata (Holotricha), (*c*) Myxomycetales, (*d*) Flagellata (Zoomastigina).

**Invertebrate Metazoa:** (*a*) Echinodermata and 1 mollusc, (*b*) Crustacea, 1 annelid and 1 orthopteran.

**Vertebrata:** These consist of 4 fishes, 1 amphibian, 2 reptiles, 1 bird and 9 mammals (including man with 40% guanine + cytosine).

Data obtained by Edelman *et al.* (1967) and Stanier *et al.* (1971) (*Bact. Rev.*, **31**, 315 and **35**, 171) for about 70 isolates of Cyanophyta revealed GC contents ranging from 35–71 molar %.

without affecting the amino-acid composition of the proteins. It is to be expected, therefore, that if a group of organisms has been in existence for a sufficiently long period of time, the nucleotide composition of its DNA will diverge through the occurrence of successive mutations until it spans the permissible range. Figure 14.15 reveals how related organisms have similar DNA compositions; for example, the peaks in the bacterial histogram correspond to particular families (see caption to figure). Conversely, large differences in DNA composition, as between the ciliated and flagellated Protozoa, indicate wide evolutionary divergence.

Sueoka (1961b) has shown that part of the variation in DNA base composition in bacteria is to be attributed to alterations in the relative proportions of the various amino-acids in the proteins of the different species. He compared the DNA composition of 11 species with the amino-acid composition of their total protein. The molar proportion of guanine plus cytosine out of the total bases ranged from 35 % to 72 %. Sueoka found that asparagine and aspartic acid (which were not distinguished from one another), glutamine and glutamic acid (also not distinguished), isoleucine, lysine, phenylalanine and tyrosine were relatively more abundant in the species with a low proportion of guanine and cytosine than in those with a high content of this base-pair. Reference to Table 14.7 shows that these amino-acids (except for glutamine and glutamic acid) have a relatively low guanine–cytosine content for their coding nucleotides. Conversely, Sueoka found that alanine, arginine, glycine and proline were more abundant in species with a high proportion of this base-pair and from the table it is evident that these amino-acids are coded by nucleotides with a high guanine-cytosine content. The remaining amino-acids (apart from cysteine and tryptophan which were not estimated) showed no regular variation in frequency with DNA base content, and from the table it is evident that, except for leucine and methionine, they are coded by triplets which in the aggregate contain equal numbers of each base-pair. This comparison of amino-acid and nucleotide frequencies clearly shows that some of the diversity in DNA composition is related to diversity in protein composition. It also suggests that the code worked out for *E. coli in vitro* is similar if not identical for a range of bacteria *in vivo*.

Fitch (1967) surveyed 50 mutational changes in human haemoglobin and 229 in cytochrome *c* of various organisms. He assumed that the mutant forms of haemoglobin (cf. §§ 13.5 and 13.7) have been derived from the normal and that the diversity in the primary structure of the cytochrome *c* of 3 fungi, 2 insects, 1 fish, 2 reptiles, 4 birds and 8 mammals has evolved from a hypothetical ancestral molecule. For both proteins, applying the *E. coli* code, he found a marked excess of the change from guanine to adenine in the messenger compared with all the other 11 possible base substitutions including the complementary transition (cytosine to uracil). Another unexplained anomaly—the shortage of the *CpG* doublet in the DNA of chromosomal organisms—was mentioned in § 12.5. Subak-Sharpe *et al.* (1967) have inferred that in mammals (where *CpG* is particularly uncommon) selection may have acted to eliminate this doublet from within codons. This is possible because the amino-acids coded by triplets containing the

sequence *CG* (that is, serine, proline, threonine, alanine and arginine—
see Table 14.7) are also coded by triplets lacking this doublet. Fitch's
conclusions and those of Subak-Sharpe *et al.* imply that, for an unknown
reason, selection has favoured specific mutational changes in the particular
DNA chain which is transcribed into messenger RNA.

## § 14.14   The direction of translation (reading of message)

Thach, Cecere, Sundararajan and Doty (1965) established the direction
in which messenger RNA is translated into amino-acid sequence. A hexa-
nucleotide of composition AAAUUU (5′-phosphate terminus to the left,
3′-hydroxyl terminus to the right) was used in the cell-free system of protein
synthesis and found to generate a dipeptide of composition $NH_2$-lysine-
phenylalanine-COOH. Since polypeptide synthesis is known to begin at the
amino end (see § 14.10) and since AAA is known to code for lysine and UUU
for phenylalanine, it was inferred that the messenger is read in the 5′ to 3′
direction. Salas, Smith, Stanley, Wahba and Ochoa (1965) independently
obtained a similar result. Using the cell-free protein synthesizing system
they found that polynucleotides of structure (5′)AAA . . . AAC(3′), with
21 to 23 nucleotide residues, directed the synthesis of $NH_2$-lysine-lysine- . . . -
lysine-asparagine-COOH, from which they inferred that the message is read
from the 5′- to the 3′-end of the polynucleotide.

Yanofsky and associates established the co-linearity of gene and polypeptide using the *A* protein of tryptophan synthetase in *Escherichia coli* and
the corresponding gene (§ 13.8). They found that mutants *A23* and *A46*
had the amino-acids shown in Table 14.8 at position 210 in the polypeptide

TABLE 14.8   Data of Yanofsky, Ito and Horn (1967) for position 210 in the *A*
protein of tryptophan synthetase in *Escherichia coli*. The codons are inferred from the
more extensive changes given in Fig. 14.13.

| Strain | Amino-acid at Position 210 | Codon |
|---|---|---|
| *A23* | arginine | 5′ AGA 3′ |
| wild-type | glycine | 5′ GGA 3′ |
| *A46* | glutamic acid | 5′ GAA 3′ |

and gave wild-type recombinants when crossed (§ 14.13). These and other
changes in this codon due to mutation and recombination (see Fig. 14.13)
were compatible with the codons given in Table 14.8. Guest and Yanofsky
(1966) crossed *A23* with *A46* in conjunction with other linked mutants and
found from the behaviour of the linked mutants that the site of *A23* was the
nearer of the two to the end of the gene corresponding to the amino-terminal
end of the polypeptide. From its codon *A23* is evidently also the nearer of
the two to the 5′ end of the messenger. Thus the 5′ to 3′ orientation of the
messenger corresponds to the amino to carboxyl orientation of the polypeptide.

Terzaghi *et al.* (1966) in their study of frameshift mutants in virus *T4*
of *E. coli* (§ 14.13) found that codons compatible with the two sequences of

amino-acids (Fig. 14.14) could be obtained only if the triplets were orientated with the 5′ to 3′ polarity corresponding to the N-terminal to C-terminal polarity of the amino-acids. Since it had been shown that polypeptide synthesis proceeds from the N-terminal to the C-terminal end, the messenger must be translated from the 5′ to the 3′ end.

Thus, these results, obtained by various means, are in agreement with one another, and establish that translation takes place in the same direction as transcription (§ 14.8). The possibility thus arises that translation may begin before messenger synthesis has been completed, or even that the two processes are coupled in some way. This is discussed in § 15.3 and § 15.4.

## § 14.15   The wobble theory

Crick (1966) considered the possible base-pairing between codons on messenger RNA and anticodons on transfer RNA. He noted that in the first two positions of the codon (see Table 14.7) the four bases were clearly distinguished by the anticodon, and he inferred that the pairings in these two positions were the standard ones. Owing to the antiparallel direction of the chains, bases 1 and 2 in the codon will pair with bases 3 and 2 respectively in the anticodon (see Figs. 14.9 and 14.11).

The multiple coding for the same amino-acid is due chiefly to variation in the third position in the codon (Table 14.7), suggesting that at this position one transfer RNA molecule might recognize more than one nucleotide owing to alternative possibilities for base-pairing. Crick argued that, if the first two bases in the codon paired in the standard way, the pairing in the third position might be close to the standard ones. He assumed that base no. 3 in the codon is always paired with the anticodon and furthermore that at this position, as Table 14.7 shows, the pyrimidines (U and C) can sometimes be distinguished from the purines (A and G), for example, in coding for histidine and glutamine. From this he deduced that, among the various possible abnormal base-pairs, pairing between one pyrimidine and another did not occur. He considered that the likely rules for pairing were as follows. In addition to the standard pairs (which include cytosine in the codon pairing with hypoxanthine in the anticodon) slight wobble allows (*a*) uracil in the codon to pair with guanine or hypoxanthine in the anticodon, (*b*) adenine in the codon to pair with hypoxanthine in the anticodon, and (*c*) guanine in the codon to pair with uracil in the anticodon (see Fig. 14.16). There is support for the pairing of guanine with uracil from the data on nucleotide sequence in transfer RNAs and 5$S$ ribosomal RNA, where single G·U pairs have been found in otherwise normally base-paired regions.

The wobble hypothesis predicts that an amino-acyl transfer RNA which recognizes a triplet ending in C will also recognize at least one other triplet. A similar prediction applies to a triplet ending in A. The data in Table 14.7 agree with these predictions, for there are no amino-acids corresponding to such triplets alone. The hypothesis also predicts that when an amino-acid (such as valine) is coded by all 4 bases in the third position, there must be at least 2 transfer RNAs for this amino-acid. These would have either G and U respectively at the first position in the anticodon (corresponding to

Third place in codon              First place in anticodon

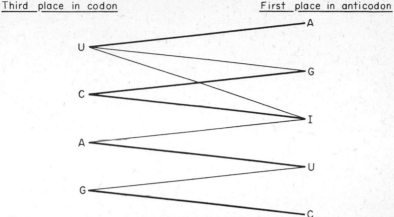

FIGURE 14.16    Diagram to show the bases which can form hydrogen bonds
                between the third place in the codon in messenger RNA and
                the first place in the anticodon in transfer RNA according to
                the wobble hypothesis of Crick (1966). The letters A, C, G, I
                and U stand for the nucleosides adenosine, cytidine, guano-
                sine, inosine and uridine, respectively, corresponding to the
                bases adenine, cytosine, guanine, hypoxanthine and uracil.
                Heavy lines show the standard base pairs and the thinner
                lines the extra pairs permitted.

U/C and A/G in the codon), or I (inosine—the nucleoside of hypoxanthine)
and C respectively (corresponding to U/C/A and G in the codon).

The anticodons of a number of transfer RNAs are given in Table 14.9,
together with the codons predicted by the wobble hypothesis. These are all in
agreement with the code established for *Escherichia coli in vitro*. Moreover Söll,
Cherayil and Bock (1967) tested the binding of amino-acyl transfer RNAs to
ribosomes in the presence of trinucleotides known to be codons for the
respective amino-acids, and found that particular *t*RNAs can often recognize more than one codon differing in the third base. Furthermore, the
multiple recognition patterns were those predicted by the wobble hypothesis:
for example, the major serine *t*RNA in *Saccharomyces cerevisiae* showed
binding with the trinucleotides UCU, UCC and UCA, but not with the
other serine codons (UCG, AGU and AGC). This is what was expected
on wobble theory from the known anticodon (see Table 14.9).

Convincing evidence for the multiple recognition by one kind of transfer
RNA of codons differing in the third base was obtained by Söll and
RajBhandary (1967). Their technique was to synthesize a polypeptide in
the cell-free system using ribopolynucleotides of known repeating sequence
as messengers and with purified amino-acyl *t*RNA of one particular kind.
They found, for example, that phenylalanyl-*t*RNA from *S. cerevisiae* translates
the codons UUU and UUC with equal efficiency, as predicted from its
anticodon (see Table 14.9).

Söll, Cherayil and Bock (1967) found that, although the amino-acid
code is apparently the same in *S. cerevisiae* and *E. coli*, the patterns of codon

TABLE 14.9   The table shows the anticodons of a number of transfer RNAs, and the codons with which they are predicted to pair on the wobble hypothesis, assuming that the modified nucleotides pair like those from which they are derived. For references to the sources of the information in the table, see the text and Staehelin (1971). The letters after the name of the amino-acid accepted by the *t*RNA indicate the species: *E, Escherichia coli*; *R, Rattus norvegicus*; *S, Saccharomyces cerevisiae*; *To, Torulopsis utilis*; *Tr, Triticum aestivum*; and the suffix after the letter identifies the *t*RNA when two or more with different anticodons are known or expected for the same amino-acid: f, formyl-methionine, implying the *t*RNA for polypeptide initiation (see § 14.16); m, methionine at internal positions in polypeptides; $\Psi$, Pseudouridine.

| Amino-acid accepted by *t*RNA | Anticodon | | | Codons predicted |
|---|---|---|---|---|
| Alanine ($S_1$) | I | G | C | GCU, GCC, GCA |
| Aspartic acid ($S$) | G | U | C | GAU, GAC |
| Glutamic acid ($E_2$, $S_3$) | U* | U | C | GAA, GAG† |
| Glutamine $\begin{cases} E_1 \\ E_2 \end{cases}$ | U* | U | G | CAA, CAG† |
|  | C | U | G | CAG |
| Isoleucine (*To*) | I | A | U | AUU, AUC, AUA |
| Leucine ($E_1$) | C | A | G | CUG |
| Methionine $\begin{cases} E_f, S_f \\ E_m \end{cases}$ | C | A | U‡ | AUG, G‡UG |
|  | C§ | A | U | AUG |
| Phenylalanine $\begin{cases} E \\ S, Tr \end{cases}$ | G | A | A⎫ | UUU, UUC |
|  | G‖ | A | A⎭ | |
| Serine ($R_1$, $S_1$) | I | G | A | UCU, UCC, UCA |
| Tryptophan (*E*) | C | C | A | UGG |
| Tyrosine $\begin{cases} S, To \\ E \\ E** \\ E†† \end{cases}$ | G | $\Psi$ | A⎫ | UAU, UAC |
|  | G¶ | U | A⎭ | |
|  | C | U | A | UAG |
|  | U | U | A | UAA, UAG |
| Valine $\begin{cases} E_1 \\ E_2 \\ S_1, To_1 \end{cases}$ | U‡‡A | C | | GUA, GUG |
|  | G | A | C | GUU, GUC |
|  | I | A | C | GUU, GUC, GUA |

* Probably a derivative of 2-thiouridine.
† The *t*RNA fails to pair with this codon.
‡ Wobble at third position in anticodon (see § 14.16).
§ Probably N⁴-acylated cytidine.
‖ 2'-O-Methylguanosine.
¶ Unidentified guanosine derivative.
** Amber suppressor (see § 14.17).
†† Amber and ochre suppressor (see § 14.17).
‡‡ Uridine-5-oxyacetic acid.

recognition by transfer RNAs show some differences. Thus, in *E. coli* one valine *t*RNA recognizes GUU and GUC and another GUA and GUG, while in *S. cerevisiae* the major valine *t*RNA recognizes GUU, GUC and GUA (Table 14.9). A similar difference is shown by the serine and alanine *t*RNAs. Such results were expected because the *t*RNAs of *S. cerevisiae* were known to be richer in inosine than those of *E. coli*. (Inosine is the nucleoside needed in the first place of the anticodon for the recognition of U, C and A in the third place of the codon—see Fig. 14.16.) These differences, however, between the anticodons of *S. cerevisiae* and *E. coli* are not found for every amino-acid that is coded by all 4 bases in the third position. Thus, in both organisms arginine shows the 3 and 1 pattern (one *t*RNA recognizing U, C and A in the third position and the other recognizing G) and glycine the 2 and 2 pattern (U and C *versus* A and G). With glycine in *S. cerevisiae*, however, a third *t*RNA is known which recognizes GGG alone.

Multiplicity of transfer RNAs for the same amino-acid is known of yet another kind: Holley *et al.* (1965) found that at position *n*1 in alanine *t*RNA no. 1 of *S. cerevisiae* there was variability, this base being either uracil or dihydrouracil. Zachau *et al.* (1966) found two forms, I and II, of serine *t*RNA no. 1 in *S. cerevisiae*, which, unlike the previous example, would require to be coded by different genes: they had the same anticodon but differed in the bases at *l*2, *p*4 and *p*6. Likewise, Goodman *et al.* (1968) found there were two forms of *t*RNA for tyrosine in *E. coli* which differed in the bases at positions *l*2 and *l*3, and Dube *et al.* (1968) found variability at *j*3 in formyl-methionine *t*RNA. Söll and associates also found two *t*RNAs in *E. coli* for each of the codon pairs UUU and UUC (phenylalanine), AUU and AUC (isoleucine), and AAA and AAG (lysine). They suggested that the multiplicity of equivalent *t*RNAs could be of great biological significance as a safeguard against the often lethal effect of a mutation in a *t*RNA gene. This multiplicity also explains the origin of suppressor mutations (see below).

Contrary to the wobble hypothesis, amino-acyl transfer RNAs specific to codons ending in A are known for glutamic acid in *Saccharomyces cerevisiae* (Yoshida *et al.*, 1970) and *Escherichia coli* (Ohashi *et al.*, 1970) and for glutamine in *E. coli* (Folk and Yaniv, 1972). In each of these *t*RNAs, the first nucleoside in the anticodon is believed to be a derivative of 2-thiouridine. The substitution of a sulphur atom for an oxygen atom on carbon atom no. 2 of this nucleoside prevents pairing with guanine, but the likely significance of the 2-thiouridine derivative, as pointed out by Ohashi *et al.*, is to prevent U at the first position in the anticodon allowing the *t*RNA to pair with a codon ending in U or C. With glutamic acid, glutamine and certain other amino-acids (see Table 14.7) such an occurrence would be lethal as it would allow mis-translation. Another exception to the wobble theory is discussed in § 14.17.

Brody and Yanofsky (1963) found that a suppressor gene can alter the primary structure of a protein. Mutant *A*36 leads to the insertion of arginine instead of glycine at position 210 (Fig. 13.8) in the *A* protein of tryptophan synthetase in *E. coli*. This makes the enzyme non-functional and hence imposes a requirement for tryptophan on the bacterium. A mutation of a suppressor gene, which was not linked to the *A* gene, caused mutant *A*36 to become wild-type in phenotype, and it was found that glycine was inserted again in place of the arginine. Brody and Yanofsky suggested that the suppressor gene probably interfered with the translation of the *A* gene messenger RNA into polypeptide by altering the specificity of the glycine or arginine activating enzymes or transfer RNAs.

A special class of suppressor mutations in which the mutant suppressed fails to synthesize a complete polypeptide is discussed in § 14.17. It has now been established that suppression of such mutants can arise through the occurrence of mutation in a gene for a transfer RNA at the position of the anticodon, a duplicate unmutated gene also being present.

The occurrence of more than one transfer RNA for the same amino-acid, in addition to its safeguard against lethal mutations, may have been favoured for another reason. Twardzik, Grell and Jacobson (1971) discovered that a suppressor of *vermilion* eye colour mutants in *Drosophila melanogaster* acted by altering a tyrosine transfer RNA. Chromatography on a reverse phase column of extracts of the flies revealed that insects homozygous for the mutation at the suppressor locus (called *su(s)* as it is sex-linked) lacked one of the kinds of tyrosine *t*RNA found in the wild-type. The *su(s)* gene is believed to specify an enzyme which modifies this tyrosine *t*RNA no. 2 during its maturation. The vermilion eye colour is caused by lowered activity of tryptophan pyrrolase. This enzyme catalyses the conversion of tryptophan to kynurenine, one of the steps in the synthesis of a brown pigment normally present in the eyes of *Drosophila*. Jacobson (1971) discovered that tryptophan pyrrolase activity could be restored to extracts of vermilion-eyed flies by treatment with a ribonuclease, but that subsequent addition of tyrosine *t*RNA no. 2 inhibited the enzyme again. The wild-type enzyme was unaffected by ribonuclease. He concluded that this transfer RNA associates with tryptophan pyrrolase, and that in vermilion mutants the enzyme is abnormal and its activity is inhibited by the *t*RNA. Removal of the *t*RNA, either by ribonuclease or through its failure to mature as a result of homozygosity for a *su(s)* mutation, releases the mutant tryptophan pyrrolase from inhibition. Jacobson suggested that the significance of multiple transfer RNAs for the same amino-acid may be in the control of development which they exercise through the inhibition of specific enzyme activities.

On the basis of the known nucleotide sequences, and evidence from X-ray diffraction on the conformation of base-paired regions, Fuller and Hodgson (1967) derived a model for the anticodon arm. Earlier X-ray diffraction studies (Arnott *et al.* 1966) of double helical RNA from reovirus and other sources had suggested that the base-pairs may be tilted 17° instead of being set perpendicular to the helix axis as in DNA. The tilting would allow 11 nucleotide-pairs per turn of the helix in 2·8 nm, compared with 10 pairs in 3·4 nm in DNA. Fuller and Hodgson assumed that the base-paired region of the anticodon arm (segments *g* and *i* in Fig. 14.9) has this tilted double helical structure. In their model, 5 of the 7 nucleotides of the anticodon loop (*h*3-*h*7 in Fig. 14.9) are stacked in this tilted conformation so that they lie on the same helix as the nucleotides in the adjoining double helical region (segment *i*). This model represents a unique solution to the problem of maximizing base-stacking in the anticodon loop, and moreover it accounts for the wobble in pairing at position no. 1 in the anticodon, because this is the last nucleotide (*h*3) of the 5 in the stack and the helix begins to become distorted at this point for the return loop (nucleotides *h*2 and *h*1). The alternative pairing at position 3 in the codon would thus arise through distortion of the anticodon and not of the codon.

It is evident that Crick's wobble theory is not only supported by experimental evidence from the use of specific amino-acyl transfer RNAs in the systems used to establish the amino-acid code, but it is also in keeping with a likely three-dimensional structure for the anticodon loop. The theory has been extended to account for variability at the first position in the codon in the special case of the transfer RNA responsible for initiating polypeptides (see § 14.16).

## § 14.16   Polypeptide initiation

Marcker and Sanger (1964) discovered that within cells of *Escherichia coli*, and also in cell-free systems derived from *E. coli*, formylation of the α amino group of methionine $(NH_2 \cdot CH(CH_2 \cdot CH_2 \cdot SCH_3)CO \cdot OH)$ can take place, and that this occurs after the methionine has become attached to its transfer RNA, thus giving rise to N-formyl-methionyl transfer RNA $(OHC \cdot NH \cdot CH(CH_2 \cdot CH_2 \cdot SCH_3)CO \cdot O \cdot tRNA)$. No other amino-acid showed formylation. Marcker (1965) established that the formylation was catalysed by a specific enzyme, a transformylase, and he suggested, among various possibilities, that formyl-methionyl-*t*RNA could function as a polypeptide initiator.

This hypothesis was confirmed by Adams and Capecchi (1966), who labelled formyl-methionyl-*t*RNA with ⁸H in the formyl group, and studied the incorporation of the label into proteins synthesized with the *E. coli* cell-free system, using the RNA of virus *R*17 of *E. coli* as messenger. They found that the labelled formyl groups were incorporated into at least two of the 3 proteins coded by *R*17 RNA. Since it seemed likely that the formylation prevented the amino group of the methionine from taking part in peptide bond formation, the formylated methionine could be incorporated into a polypeptide only as the N-terminal amino-acid. Adams and Capecchi concluded that formyl-methionyl-*t*RNA was the initiator of protein synthesis. They were puzzled, however, because the N-terminal amino-acid of the coat protein of the intact virus particle is alanine. The coat protein made *in vitro* was digested with pronase and this resolved the paradox: they found that although all the labelled formyl groups were in N-formyl-methionine, the amino-acid next to this was alanine. They concluded that *in vivo* the initial formyl-methionine of the coat protein is removed enzymatically, yielding protein with N-terminal alanine. Webster, Engelhardt and Zinder (1966) obtained similar results using the RNA of virus *F*2 as messenger.

Clark and Marcker (1966) showed there were two methionine transfer RNAs in *E. coli*. They separated them by counter-current distribution and found that none of the attached methionine of the faster moving one (met-*t*RNA$_1$) could be formylated, whereas all that of the slower (met-*t*RNA$_2$) could be. Using the cell-free system, they found that the tRNA$_2$, whether carrying methionine or formyl-methionine, could initiate polypeptides, whereas met-*t*RNA$_1$ inserted methionine only in internal positions in a polypeptide. It was evident that the specificity of the *t*RNA$_2$ as a polypeptide initiator resided in the structure of this transfer RNA and not in the formyl group. Clark and Marcker found that the internal met-*t*RNA (that is, no. 1) would bind only to trinucleotides of sequence AUG, while the initiating met-*t*RNA (no. 2) bound to GUG as well as AUG. Clark *et al.* (1968*a*) found that a fragment from *g*1 to *j*2 (Fig. 14.9) of the initiating met-*t*RNA specifically bound to ribosomes in the presence of AUG or GUG. This was the first direct evidence for the position of the anticodon in any *t*RNA.

Bretscher and Marcker (1966) obtained evidence that the initiating met-*t*RNA binds to the peptide site (see § 14.10 and Fig. 14.17) of the ribosome, while the internal met-*t*RNA binds in the normal way to the amino-acid site. The antibiotic puromycin is believed to inhibit protein synthesis by substituting for an amino-acyl-*t*RNA and reacting with peptidyl-*t*RNA bound to the peptide site of the ribosome. Bretscher and Marcker found that the internal met-*t*RNA, when bound to ribosomes with the triplet AUG, was comparatively insensitive to puromycin, whereas the initiating met-*t*RNA when so bound was found to be sensitive to the antibiotic; this sensitivity was manifest whether the methionine was formylated or not.

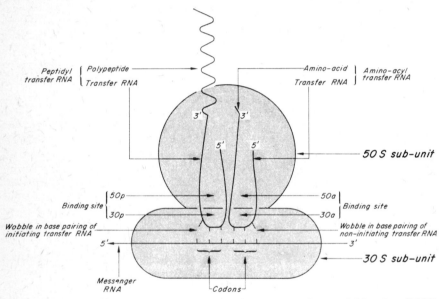

FIGURE 14.17   The diagram is based on that of Bretscher and Marcker (1966) and shows a ribosome during protein synthesis. The binding-sites for amino-acyl and peptidyl transfer RNAs on the ribosomal sub-units are indicated by *a* (amino-acid) and *p* (peptide).

Stanley, Salas, Wahba and Ochoa (1966), and two other groups independently discovered two or more factors associated with the ribosomes, and hitherto discarded when purifying them, which experiment showed were involved in the initiation of protein synthesis. Nomura and Lowry (1967) discovered that in the presence of the RNA from virus *F*2 of *E. coli* as messenger the initiating met-*t*RNA will bind to the 30*S* ribosomal sub-unit but not to the intact 70*S* ribosome. No other amino-acyl-*t*RNA showed this effect. They proposed that the first step in protein synthesis was the recognition of the initiation codon, or of specific sequences adjacent to it, by the 30$S$ particles and their associated initiation factors, followed by the formation of a complex consisting of a 30$S$ particle, messenger RNA and formylmethionyl-transfer RNA (fMet-$t$RNA$_f$). They obtained evidence that the other sub-unit —the 50$S$ particle—then joins the initiation complex and allows amino-acyl-$t$RNAs other than the initiating met-$t$RNA to bind with the ribosome. Direct support for this hypothesis was obtained by Ghosh and Khorana (1967) using synthetic messengers containing AUG or GUG: they found that the 30$S$ sub-unit binds the initiating met-$t$RNA in the presence of one of these codons, and then the joining of the 50$S$ sub-unit to the 30$S$ forms a second site which now binds non-initiating amino-acyl-$t$RNAs (Fig. 14.17).

Mangiarotti and Schlessinger (1966), from study of ribosome metabolism, concluded that the 30$S$ and 50$S$ sub-units join together only in association with protein synthesis and that these particles separate and return to a pool of sub-units when the polypeptide synthesis has been completed. This model of ribosome behaviour is in keeping with Nomura and Lowry's hypothesis of how protein synthesis is initiated. Support for Mangiarotti and Schlessinger's idea was obtained by Kaempfer, Meselson and Raskas (1968). Bacteria labelled with heavy isotopes ($^{13}$C, $^2$H, $^{15}$N) were transferred to a medium containing the normal isotopes. During growth in this light medium the heavy ribosomes disappeared and were replaced by hybrid ribosomes, as shown by density-gradient centrifuging. It was found that the sub-units themselves were stable, remaining intact and evidently forming part of one ribosome after another.

Bretscher (1968$b$) suggested that the amino-acid ($a$) and peptide ($p$) binding sites extend to both sub-units of the ribosome, and that the movement of a transfer RNA from $a$ to $p$ occurs in two steps. An amino-acyl-$t$RNA in the amino-acid site (30$a$ and 50$a$ in Fig. 14.17) becomes a peptidyl-$t$RNA through the formation of the peptide bond; its 30$S$-binding region then moves to the peptide site on the 30$S$ sub-unit (30$p$), while its 50$S$-binding region remains in 50$a$; displacement of the 50$S$ sub-unit (to the left in Fig. 14.17) relative to the 30$S$ would make this possible. The anticodon region and associated messenger would move at the same time as the 30$S$-binding region of the $t$RNA, so the next codon would now be set to register with the anticodon of an appropriate new amino-acyl-$t$RNA. This is the stage to which the initiating met-$t$RNA would lead directly by binding to site 30$p$ and, with the joining of the 50$S$ sub-unit, to site 50$a$. By a reverse movement of the 50$S$ particle relative to the 30$S$, the 50$S$-binding region of the peptidyl-tRNA (or initiating met-$t$RNA) would then move from 50$a$ to 50$p$. Site 50$a$, in conjunction with 30$a$, would then be free to accept a new amino-acyl-$t$RNA, in preparation for the formation of the next peptide bond. According to Dintzis (1961) the synthesis of a haemoglobin polypeptide of about 150 amino-acids takes $1\frac{1}{2}$ minutes at 37°, so the entire cycle of events between the formation of one peptide bond and the next would occupy just over half a second.

In view of the binding of the initiating met-$t$RNA with the triplet GUG as well as AUG, Bretscher and Marcker (1966) suggested that when an amino-acyl *t*RNA is in the peptide site of the ribosome the first base in the codon may not be specifically recognized, in contrast to amino-acyl *t*RNA in the amino-acid site, where it is the third base which shows the wobble in pairing (see § 14.15 and Fig. 14.17). Support for this inverted wobble was obtained by Dube, Marcker, Clark and Cory (1968). They established the complete nucleotide sequence of the initiating *t*RNA of *E. coli*, using the technique of Sanger, Brownlee and Barrell (see § 14.11), and found that the anticodon was CAU. Assuming that for this *t*RNA the wobble theory applies to the third base of the anticodon, the occurrence of uracil at this position means the codons predicted are AUG and GUG (see Fig. 14.16 and Table 14.9), in agreement with observation.

The lengths of the various segments of the initiating *t*RNA were found by Dube *et al.* to be as shown in Table 14.4 (column headed fMet.), and the identity and positions of minor nucleosides to be as shown in Table 14.5. The nucleotide composition and sequence did not reveal, however, which part is responsible for the binding to the peptide site, nor where the specificities for met-*t*RNA synthetase and transformylase lie. A peculiarity of the molecule is that the nucleotide at the 5′ terminus (*a*1 in Fig. 14.9) is not paired, but this is not necessarily of special significance: the *t*RNA no. 1 for alanine in *S. cerevisiae* shows no pairing at position *a*6. Dube *et al.* hoped that study of the nucleotide sequence of the other met-*t*RNA of *E. coli* would reveal what sites are responsible for their differences in behaviour, but this hope has not been realized: Cory *et al.* (1968) found that they differ in 39 of their 77 nucleotides, with differences in all parts of the molecule.

Sundararajan and Thach (1966) discovered that when the triplet AUG was incorporated into longer polynucleotides it suppressed the reading of codons which partially overlapped its sequence, and promoted the reading of the adjacent codon on the 3′ side. For example, they found that the sequence AUGUUU . . . stimulated the binding of *t*RNAs carrying methionine (and therefore registering with the codon AUG) and phenyl-alanine (codon UUU), but not those carrying valine (codon GUU). They also found that AUG had this ability to fix the reading-frame even when it was not at the 5′ end of the chain. Thus, AAAUG was found strongly to stimulate the binding of *t*RNAs carrying methionine but not those carrying lysine (codon AAA), whereas AAACG coded very well for lysine. It thus appears that the triplet AUG determines the phase in which the message is read.

The small RNA viruses such *R*17 and *Qβ* of *E. coli* are particularly suitable for studying polypeptide initiation because they have only a single chain of RNA which functions directly as messenger. Moreover, they have only 3 genes: in sequence along the RNA from the 5′ end, these code for the matura-tion protein, the coat protein and the *β* sub-unit of the viral replicase, respectively. The maturation protein is a structural component of the viral particle. Steitz (1969) devised a technique for isolating the polypeptide initiation sequences. Ribosomes were bound to $^{32}$P-labelled viral RNA in the presence of f-met-*t*RNA$_f$, thus ensuring that the ribosomes were bound only at the initiation regions. Treatment with a ribonuclease then removed those parts of the molecule not protected by ribosomes, and the methods outlined in § 14.11 were applied to the initiation regions in order to determine their nucleotide sequence. Her results for virus $R17$ are given in Table 14.10, together with the corresponding data for virus $Q\beta$ obtained by Hindley, Staples and associates.

Surprisingly, the six initiation sequences do not possess a common run of bases, nor do they have a common secondary structure such as a sequence of nucleotides followed by complementary ones in reverse order to generate a base-paired loop. It may be significant that the six AUG triplets which function as start signals are preceded by a pyrimidine, while the three that do not (see Table 14.10) are preceded by a purine. But this cannot be the whole explanation of how certain AUGs are recognized as start signals and others not, because AUG occurs frequently within genes, both in and out of phase, and will often follow a pyrimidine. Some potential initiation sites are believed to be in folded regions of the chain with base pairing (see § 15.5). Mild disruption of the RNA of virus $F2$ of *E. coli* by formaldehyde was found by Lodish (1970) to allow at least 3 polypeptides to be initiated, none of which corresponded to known $F2$ proteins. The formaldehyde evidently exposed AUG triplets that are normally hidden by the secondary structure of the RNA.

As the three genes of virus $R17$ do not produce their products in equimolar amounts, it seems likely that the ribosomes, with their associated initiation factors, can discriminate between the three signals. Dube and Rudland (1970) showed from binding experiments that after infection with virus $T4$ the ribosomes of *E. coli* recognize the initiation signals of $T4$ messenger RNA more efficiently than those of host messenger or of the RNA of viruses $R17$ or $F2$. When the initiation factors were separated from the ribosomes of infected cells and combined with the ribosomes of uninfected cells it was found that ribosomes with $T4$ specificity were generated. Evidently the messenger selection is brought about by the initiation factors. Ribosomes with initiation factors from uninfected cells bind more efficiently to the start sequence of the coat-protein gene of virus $R17$ than to the starting points of the other two genes. On the other hand, in the presence of the initiation factor from $T4$-infected cells, Steitz, Dube and Rudland (1970) found that the ribosomes bind chiefly to the initiation sequence of the gene for the maturation protein. Modification or replacement of host initiation factors evidently takes place following $T4$ infection, and the diversity of initiation sequences in viruses $R17$ and $Q\beta$ seems to be related to the specificity of particular initiation factors.

Caskey, Redfield and Weissbach (1967) separated two methionine-accepting transfer RNAs from cells of *Cavia porcella* (Guinea Pig) and discovered that, although the cells contained no transformylase activity, one of the transfer RNAs (met-$t$RNA$_f$) could be formylated by *E. coli* transformylase but the other (met-$t$RNA$_m$) could not. A similar discovery was made with *Saccharomyces cerevisiae* by Takeishi, Ukita and Nishimura (1968). Smith and Marcker (1970) discovered that met-$t$RNA$_f$ from *S. cerevisiae* and from *Mus musculus* incorporated methionine exclusively into amino-terminal positions

**TABLE 14.10** Table showing the nucleotide sequence near the beginning and end of genes in viruses *R17* and *Qβ* of *Escherichia coli*, from the work of Steitz (1969), Hindley and Staples (1969), Staples, Hindley, Billeter and Weissmann (1971) and Staples and Hindley (1971) on initiation sequences, and of Nichols (1970) and Weiner and Weber (1971) on termination sequences. The letters A, C, G and U stand for the nucleotides of adenine, cytosine, guanine and uracil, respectively, in 5′ to 3′ sequence in the single chain of RNA of the phages, and n stands for nucleotides. Single dots indicate the triplet reading frame within genes and its extrapolation outside them. Lines above and below triplets indicate potential start and stop signals, respectively, for polypeptide synthesis.

| Phage | End of gene, and protein for which the gene codes | Before, between and after the genes | Beginning of gene, and protein for which the gene codes |
|---|---|---|---|
| *R17* | ——— | 5′ end......(112 n)...... CC.UAG.GAG.GUU.UGA.CCU.AUG.CGA.GCU.UUU.AGU... | Maturation |
| *Qβ* | ——— | 5′ end...(40 n)...UCA.CUG.AGU.AUA.AGA.GGA.CAU.AUG.CCU.AAA.UUA.CCG... | Maturation |
| *R17* | Maturation | ...CC.UCA.ACC.GGG.GUU.UGA.AGC.AUG.GCU.UCU.AAC.UUU... | Coat |
| *Qβ* | Maturation | ...CGU.AAA.GCG.UUG.AAA.CUU.UGG.GUC.AAU.UUG.AUC.AUG.GCA.AAA.UUA.GAG... | Coat |
| *R17* | Coat | ...AAC.UCC.GGU.AUC.UAC.UAA.UAG.AUG.AUG.CCG.GCC.AUU.CAA.ACA.UGA.GGA.UUA.CCC.AUG.UCG.AAG.ACA.ACA... | Replicase |
| *Qβ* | Coat | ...UGA...(c 600 n).............AG.UAA.CUA.AGG.AUG.AAA.UGC.AUG.UCU.AAG.ACA.G... | Replicase |
| *R17* | Replicase | ... 3′ end | ——— |
| *Qβ* | Replicase | ... 3′ end | ——— |

of polypeptides, while the met-$t$RNA$_m$ incorporated methionine into internal positions only. Confirmation that met-$t$RNA$_f$ is the initiating transfer RNA was obtained by Brown and Smith (1970) who showed with a cell-free system involving components from *Mus* that there was a unique interaction between met-$t$RNA$_f$, the AUG codon in messenger RNA and 80$S$ ribosomes. Met-$t$RNA$_m$ showed no such interaction. Furthermore, in the presence of all the $t$RNAs, synthetic messengers were not translated unless AUG was present near the beginning. Jackson and Hunter (1970) found that the $\alpha$ and $\beta$ polypeptides of haemoglobin in *Oryctolagus cuniculus* are initiated with methionine as N-terminal amino-acid, but this methionine is removed during the early stages of polypeptide synthesis. A similar discovery was made by Wigle and Dixon (1970) for protamine synthesis in *Salmo gairdnerii* (Rainbow Trout).

Sherman *et al.* (1970) obtained mutants in the gene ($cy_1$) which codes for iso-1-cytochrome *c* in *Saccharomyces cerevisiae*. Certain of these mutants gave intragenic revertants with abnormal numbers of amino-acid residues at the amino-terminus of the polypeptide, as shown in Table 14.11. Each of the

TABLE 14.11   Data of Sherman *et al.* (1970) for the amino-acid sequence at the amino-terminus of iso-1-cytochrome *c* in *Saccharomyces cerevisiae*. The table shows the normal sequence and that of revertants of certain $cy_1$ mutants. These revertants all had polypeptides of abnormal length.

| Source | | Amino-acid sequence at amino-end |
|--------|---|------------------------------------|
| Normal | | NH$_2$-Thr-Glu-Phe-Lys-Ala-Gly- |
| Revertants | *a* | NH$_2$-Met- Ile-Thr-Glu-Phe-Lys-Ala-Gly- |
| | *b* | NH$_2$-Met-Leu-Thr-Glu-Phe-Lys-Ala-Gly- |
| | *c* | NH$_2$-Met-Arg-Thr-Glu-Phe-Lys-Ala-Gly- |
| | *d* | NH$_2$-Val-Thr-Glu-Phe-Lys-Ala-Gly- |
| | *e* | NH$_2$-Ala-Gly- |

mutants which reverted in this way mapped at the same position and gave rise to only one of the alternatives *a–d*. Any of the mutants, however, could give rise to revertant *e* or to revertants with the normal amino-acid sequence. It was inferred that the mutants were defective in the initiator codon, that this codon was AUG, and that it immediately preceded the codon for the amino-terminal threonine residue of the normal polypeptide, because revertants *a–d* had an amino-acid at this position which could result from a single base change in AUG, namely, UUG or CUG (leucine), GUG (valine), AGG (arginine) and AUU, AUC or AUA (isoleucine). Revertants *a–d* evidently arose through the introduction of a new initiation codon immediately before the normal one, and revertant *e* through mutation of the lysine codon AAG to AUG. The amino-terminal methionine residue is evidently excised from the normal polypeptide and from revertants *d* and *e*.

These elegant experiments on the $cy_1$ gene provide striking confirmation of the conclusions about polypeptide initiation in chromosomal organisms reached from study of transfer RNAs. It is apparent that the mechanism of initiation does not differ in any essential way from that in chromonemal organisms. Indeed, the only difference detected is the lack of formylation of the initiating methionine.

## § 14.17  Polypeptide termination

Benzer and Champe (1962) devised an ingenious technique for recognizing mutants which blocked the synthesis of a polypeptide. Mutant *r*1589 in the *r*II region of virus *T*4 of *E. coli* is a deletion that includes the junction of the *A* and *B* genes, and part of each (see § 14.3). The *A* function is thereby lost, but the *B* function remains: this was shown by the fact that *r*1589 will complement with any *r*II *B* mutant that has an intact *A* gene, thereby allowing growth on strain *K* of the host. (Both *A* and *B* activities are needed for such growth.) Evidently the lost end of the *B* gene is non-essential. Crick *et al.* (1961) showed, however, that the *B* function can be turned off by transferring (through recombination) certain deletions and proflavin-induced mutants into the *A* fragment (see Fig. 14.3(*p*) and (*q*)). They suggested that in *r*1589 the *A* and *B* fragments are transcribed into a single messenger RNA, which is then translated into a single polypeptide. A frameshift mutation in the *A* gene would then affect the *B* gene as well, and likewise with a mutant that interrupted polypeptide formation.

Benzer and Champe applied this test for polypeptide-terminating mutations in the *A* gene to mutants described as ambivalent because they were inactive in one strain of *E. coli* but active in another. The host strain which suppressed the viral mutants had arisen by mutation. The viral mutants reverted on treatment with 2-aminopurine, and were therefore thought to have arisen by base substitution and not by addition or deletion of nucleotides (§ 14.2). Five such mutants, when combined with *r*1589 as a double mutant, showed no *B* gene activity. It was concluded that they blocked polypeptide synthesis. All were suppressed by the same mutant strain of the host. It was inferred that when the host contained a suppressor, the coding unit from the virus would be translated as an amino-acid, thereby permitting protein synthesis to continue. This misreading implied that the suppressor mutation affected the translation mechanism, most probably modifying either a transfer RNA or its activating enzyme. A likely possibility was that the suppressor mutation affected the fit of a transfer RNA to a coding unit. As indicated below, this anticodon hypothesis has now been confirmed for two suppressors.

Garen and Siddiqi (1962) likewise discovered mutants which apparently led to premature stopping of polypeptide synthesis, and the effect of which was suppressed by mutation of another gene. The suppressible mutants were in the gene for alkaline phosphatase in *E. coli* and the polypeptide termination was deduced from the lack of any enzyme protein in these mutants. Benzer and Champe found that the suppressor of the phosphatase-deficient mutants also suppressed one group of the ambivalent *r*II mutants which they had studied.

Confirmation that in ambivalent ('amber') mutants of *E. coli* polypeptide synthesis ends prematurely was obtained from study of the effect of such mutants on the protein of the head of virus *T*4 (see § 13.8).

Weigert and Garen (1965) studied the reversion of an amber mutant of the alkaline phosphatase gene of *E. coli*. The amino-acid at the position in the polypeptide corresponding to its premature end in the mutant was identified in 21 revertants, and from the range of amino-acids found and the known codon assignments for them, it was inferred that there was an RNA triplet indicating the end of the polypeptide, and that this triplet was UAG.

Brenner, Stretton and Kaplan (1965) also found that the RNA nucleotide triplet coding for polypeptide termination in amber mutants was UAG. They used mutants of virus *T*4 in which the synthesis of the protein of its head was stopped prematurely by such mutants and they identified the amino-acids which occurred at these points when the virus was grown on a mutant bacterial strain (an amber suppressor) which allowed the protein synthesis to go to completion. From the codons assigned to these amino-acids they deduced the codon for polypeptide-end, since only single nucleotide change would be expected to distinguish it from each of these amino-acid codons. This was the same argument as Weigert and Garen had used.

Brenner, Stretton and Kaplan found that in another group of mutants of virus *T*4 called 'ochre' mutants, the mutant character of which was suppressed by different strains of *E. coli* from those which suppressed amber mutants, the RNA codon for the end of the polypeptide was UAA. The suppressors of ochre mutants were found to be too weak to allow the isolation of ochre mutants of the protein of the head of the virus. The method that had been used to identify the amber triplet could not therefore be applied to the ochre mutants, nor could the hypothesis that ochre mutants result in polypeptide termination be tested. That UAA was the ochre triplet was shown, however, by study of the production and reversion of *r*II ochre and amber mutants using chemical mutagens, in conjunction with the observation that ochre mutants can be converted into amber mutants by mutation. Brenner and Beckwith (1965) found that, although the *E. coli* strains in which amber mutants were suppressed failed to restore *r*II activity to ochre mutants, the converse was not true: strains suppressing ochre mutants also suppressed amber mutants. They inferred that the ochre suppressors recognize both the ochre (UAA) and amber (UAG) triplets, whereas amber suppressors recognize UAG only.

Brenner and Stretton (1965) obtained evidence that amber and ochre mutants cause polypeptide termination not by producing a shortened messenger RNA in transcription from DNA but in the translation from RNA to polypeptide. They studied the effects of frameshift mutants in the *r*II region of virus *T*4 in conjunction with amber or ochre mutants. Strains were obtained in which an amber or ochre mutant was situated between a plus and a minus frameshift mutant. The amber and ochre mutants had no effect when so placed: they did not cause polypeptide termination, nor did they affect the reading of the remainder of the gene. It was inferred that amber and ochre mutants were base substitutions sensitive to the phase of reading the message, and that they were recognized in protein synthesis and not in messenger RNA formation.

Using the cell-free protein synthesizing system derived from *E. coli*, Morgan, Wells and Khorana (1966) obtained negative evidence that the triplet UGA, like UAA and UAG, does not code for any amino-acid. By the techniques described in § 14.12 they obtained a synthetic RNA with the repeating sequence GAUGAUGAU .... Despite the phasing effect of AUG (see § 14.16) this messenger induced the production of a polypeptide containing aspartic acid, corresponding to the frame set to read GAU, as well as a methionine polypeptide corresponding to AUG, but there was no polypeptide corresponding to UGA. By analogy with amber and ochre mutants, mutants in which the triplet UGA has been substituted for a triplet coding for an amino-acid are called *opal* mutants.

Brenner, Barnett, Katz and Crick (1967) confirmed from genetical studies that the triplet UGA did not stand for any amino-acid. They found that a particular mutant induced by 2-aminopurine and situated near the left-hand end of the *B* gene of the *r*II region of virus *T*4 (cf. Table 14.1) was not suppressed by any amber or ochre suppressor and was therefore not UAG or UAA. The mutant was converted to an ochre mutant by hydroxylamine, and the ochre triplet so produced did not require any replication for expression. This suggested that a change from G to A had occurred in the messenger (cf. § 14.2), and, as the mutant was not an amber, its sequence was evidently UGA. Brenner *et al.* argued that it was unlikely that this polypeptide-terminating sequence functioned in transcription from DNA to RNA rather than in translation from RNA to protein because, like other terminating codons, the effects of UGA depended on its being read in phase: when placed in a shifted frame (between a plus and a minus frameshift mutant) its effect vanished.

Sambrook, Fan and Brenner (1967), following treatment of *E. coli* with a chemical mutagen, isolated a host mutant which specifically suppressed opal (UGA) mutants in the *r*II *B* gene of virus *T*4. The suppressor had no effect on ochre or amber mutants at the same sites. This enabled the authors to use the opal suppressor to select for mutation from ochre (UAA) to UGA in the β-galactosidase gene of *E. coli*. It was known that UAA and UAG mutants of this gene show a polar effect which seems to be a consequence of polypeptide termination: see § 15.3. UGA mutants of the galactosidase gene were found also to be polar. Amber and ochre suppressors failed to restore the galactosidase activity destroyed by the UGA mutants. Sambrook *et al.* concluded that UGA, like UAA and UAG, results in polypeptide termination. In view of the lack of strong suppressors for UAA (in contrast to the other two triplets), they suggested that UAA was the triplet commonly used by *E. coli* to signal the end of a polypeptide.

The ability of amber suppressors to suppress mutations in more than one gene implies that they probably act at one of the steps in the translation of RNA into protein. Capecchi and Gussin (1965) obtained support for Benzer and Champe's hypothesis that amber suppression was due to a mutation affecting an anticodon. They used a cell-free system with RNA from a suppressible mutant of the RNA virus *R*17 as messenger, and showed that no functional coat protein was synthesized unless serine *t*RNA from a suppressor strain of *E. coli* was present. The hypothesis was confirmed by Goodman, Abelson, Landy, Brenner and Smith (1968), who showed that the mutation giving rise to a particular amber suppressor in *E. coli* changes the anticodon of a tyrosine *t*RNA from GUA to CUA (G is a guanosine derivative). As indicated in Table 14.9, this change allows the transfer RNA to register with the triplet UAG in the messenger instead of the tyrosine codons UAU and UAC, and so to insert tyrosine in the polypeptide at the position corresponding to the amber triplet. There was no other change in the nucleotide sequence of the *t*RNA. As indicated in § 14.15, two forms of tyrosine *t*RNA were found to occur in *E. coli*, both having the same anticodon but differing in the nucleotides at positions *l*2 and *l*3 in Fig. 14.9. In order to isolate sufficient material of the amber suppressor *t*RNA, use was made of an *E. coli* virus called $\phi$80, a defective strain of which carries the amber suppressor gene. By infecting the host with this virus, cells containing large numbers of copies of this gene and its product were obtained. This technique also enabled this *t*RNA to be labelled isotopically for nucleotide-sequence determination.

Altman, Brenner and Smith (1971) converted the amber suppressor with the CUA anticodon in a tyrosine *t*RNA into an amber and ochre suppressor by treatment with hydroxylamine as mutagen, and showed that it now had UUA as the anticodon. The suppression of both amber and ochre mutants, which it showed, is expected on the wobble hypothesis for this anticodon (see Table 14.9). Other amber suppressors are thought to result from changes in serine and glutamine *t*RNAs, corresponding to reading the codon UAG as if it were UCG or CAG. This conclusion is based on the work of Weigert, Lanka and Garen (1965) and Kaplan, Stretton and Brenner (1965), who discovered that amber codons in the phosphatase gene of *E. coli*, and the gene for the protein of the head of virus *T*4, respectively, can be translated as serine, glutamine or tyrosine, according to which of three suppressor genes is responsible for suppression.

Duplicate genes for a particular *t*RNA seem to be necessary for the suppression of mutations causing premature polypeptide termination: the unmutated gene would allow translation of the normal codon for that *t*RNA. Goodman *et al.* found there were at least 3 genes for tyrosine *t*RNA, all normally giving rise to *t*RNAs with the same anticodon, although, as already mentioned, one had a slightly different nucleotide sequence elsewhere in the molecule.

The work of Goodman *et al.* not only established a mechanism for amber suppression, but provided direct experimental evidence for the position of the anticodon in *t*RNA, and showed, furthermore, that at least the first nucleotide of the anticodon is not necessary for recognition of the *t*RNA by its specific amino-acyl synthetase.

Hirsch (1971) found that the opal suppressor studied by Sambrook *et al.* differed from normal in a tryptophan *t*RNA, but surprisingly the only difference was A instead of G at $e3$ (Fig. 14.9) and not in the anticodon which remained as CCA (Table 14.9). According to wobble theory this anticodon should not recognize UGA. Hirsch obtained evidence that the normal tryptophan $t$RNA can read UGA, and that the suppressor strain can do this more efficiently.

Nichols (1970) determined the nucleotide sequence at the end of the gene for the coat protein in virus $R$17 of *E. coli* and made the surprising discovery that there were two consecutive stop signals (see Table 14.10). Lu and Rich (1971) obtained evidence that such double stop signals might be common in *E. coli*, because they found that tryosine-inserting amber and ochre suppressors frequently added a tryosine residue to the carboxy-terminus of polypeptides. This would not have been detected had there not been a second stop signal immediately following the one which was suppressed by tryosine insertion. They suggested that the double stop had selective value because mutation in a stop signal, causing it to fail to function, would almost always be detrimental. In contrast to the double stop at the end of the coat gene of virus $R$17, however, Weiner and Weber (1971) inferred that at the corresponding position in virus $Q\beta$ there was only one stop signal. They obtained evidence that a natural read-through of this stop signal occurred regularly and gave rise to a coat protein variant substantially larger than normal. This variant forms 8–12% of the gene product, but only 4% of the protein of the virus particle, the normal coat protein evidently being favoured in viral assembly. The stop signal was believed to be UGA, because in the presence of an opal suppressor the yield of the read-through protein rose by a factor of 3.5. The discovery by Hirsch (1971) that the tryptophan $t$RNA can read UGA would account for this read-through. Unlike virus $R$17 some 600 nucleotides are believed to intervene in virus $Q\beta$ between the coat and replicase genes (see Table 14.10) to account for the extensive over-running observed.

Hawthorne and Mortimer (1963) discovered some suppressor mutations in the yeast *Saccharomyces cerevisiae* which suppressed simultaneously certain mutants of several different genes. Magni and Puglisi (1967), from study of reverse-mutation, concluded that the genetic basis of this *super-suppression* in *S. cerevisiae* was probably modification of a transfer RNA so that it could translate a polypeptide-terminating codon and this has now been confirmed by Gilmore *et al.* (1968) from the amino-acid replacements resulting from super-suppression of a stop mutant* in the structural gene for cytochrome *c*.

The mechanism of releasing a polypeptide from the transfer RNA which brought the final amino-acid has been studied by Capecchi (1967), who isolated a protein component required for the process. He used an amber mutant in the gene for the coat protein of virus $R$17. In a cell-free system, RNA from this mutant directs the synthesis of a peptide containing only 6 amino-acids, and by starvation of a chosen one of them, synthesis of the coat protein fragment can be stopped at any prescribed point. The substrate for the release factor was found to be the ribosome–messenger RNA–

---

* Polypeptide-terminating mutants are often misleadingly called 'nonsense' mutants.

peptidyl-*t*RNA complex, implying that release occurred while the peptidyl-*t*RNA was still attached to the ribosome. A hypothetical polypeptide-terminating *t*RNA was not found. Capecchi raised the possibility that the stop codons (UAA, UAG and UGA) might be read, not by a transfer RNA, but by the release factor, or even by the ribosome itself. The further possibility was considered that these codons function because they cannot be read.

Bretscher (1968*a*) using an amber mutant which gave a hexapeptide fragment of the coat protein of RNA virus *F*2—a system similar to Capecchi's —showed that polypeptide termination is an active process and does not occur simply because a codon cannot be read. He purified the 6 transfer RNAs needed to synthesize the fragment, and used a cell-free protein-synthesizing system containing these 6 *t*RNAs and no others. Under these conditions the RNA of the amber mutant allowed the free hexapeptide to form, whereas with the RNA of the wild-type virus the hexapeptide remained attached to its transfer RNA. The codon in the wild-type gene at the position of the amber mutant was CAG for glutamine. This could not be read in the *in vitro* system owing to the absence of the glutamine *t*RNA, but this failure to read the CAG codon did not lead to release of the peptide from its transfer RNA. Evidently the UAG triplet of the amber codon is an active signal. Moreover, the method of purifying the 6 *t*RNAs would have eliminated a hypothetical polypeptide-terminating *t*RNA, and yet release of the hexapeptide occurred with the amber mutant. Bretscher concluded that there was no transfer RNA for polypeptide termination and that the stop codons were read in some other way.

Scolnick, Tompkins, Caskey and Nirenberg (1968) separated Capecchi's release factor into two components: R1 corresponding to codons UAA and UAG and R2 corresponding to codons UAA and UGA. Klein and Capecchi (1971) purified the release factors and showed that after recognizing the appropriate stop codon each hydrolyses the peptidyl-*t*RNA ester bond, thereby releasing the polypeptide chain. Each release factor, of which there were estimated to be about 600 molecules per cell, was found to consist of a single polypeptide of molecular weight about 45,000. It is remarkable that the 3 stop codons should be read by protein molecules when the other 61 are read by transfer RNAs.

## § 14.18   Conclusion

The theory of consecutive non-overlapping triplets of nucleotides as the amino-acid code is now established and all the 64 triplets have been allocated, 61 to amino-acids and 3 as stop signals in polypeptide synthesis. There are strong indications that the code deciphered for *Escherichia coli* is likely to be the same in all organisms. Much is known of the remarkable mechanism by which nucleotide sequence in DNA is first transcribed into messenger RNA and then translated into amino-acid sequence by means of transfer RNAs, acting in association with ribosomes. It seems likely that several hundred different protein molecules are needed in order to synthesize even one such protein! These proteins, as well as being involved in large numbers in ribosome construction and activity, including the formation of the peptide bond itself, also catalyze transcription, the methylation and other changes in particular nucleotides in each transfer RNA, the synthesis of amino-acids, the attachment of each amino-acid to its transfer RNA, and the formylation of methionine after it has been joined to the polypeptide-initiating transfer RNA. How this interlocked mechanism evolved is a fascinating riddle.

# 15. *The theory of the operon*

## § 15.1  The bacterial operon

Demerec and associates (1955) obtained large numbers of nutritionally-deficient mutants of *Salmonella typhimurium*. Initially, ultraviolet light was used as mutagen, but subsequently it was found that the spontaneous mutation-rate was adequate, since mutants were readily selected from large populations of cells. The selection technique involved incubating the bacteria in minimal medium containing penicillin, which is lethal to growing cells but harmless to auxotrophic mutants, which were not metabolizing because they were unable to grow on minimal medium.

When two mutants were grown in mixed culture, recombinants were often obtained. It was established that these were occurring as a result of transduction, that is, transfer of small fragments of the hereditary material from one cell to another by a virus (*P*22) (cf. § 16.3). By means of this recombination, detailed mapping of mutant sites was possible. It was found that there were numerous linearly-arranged sites of mutation within each functional unit (cistron). This was comparable to results with *Aspergillus nidulans* and the *T*-even viruses of *Escherichia coli* discussed in Chapter 13. However, an unexpected discovery made by Demerec and Demerec (1956) with tryptophan-requiring mutants, and by Hartman (1956) with histidine-requiring mutants, was that cistrons concerned with successive steps in the synthesis of these amino-acids were closely linked, and moreover were often arranged on the linkage map in the same order as the steps which they control in the biosynthetic pathway. This was in complete contrast to the results with chromosomal organisms, where, for example in *Neurospora*, the genes concerned with the various steps in the synthesis of specific amino-acids appeared to be scattered more or less at random through the linkage-groups.

Series of closely-linked genes arranged in the same order as the steps which they control in a biochemical sequence have since been found to be of frequent occurrence in bacteria, including the actinomycete *Streptomyces coelicolor* (Hopwood and Sermonti, 1962). Such linkage, however, is not found for all the steps in every pathway, and appears to be lacking in *Pseudomonas aeruginosa* (Fargie and Holloway, 1965).

Ames and Garry (1959) discovered that the amount of each of the enzymes concerned in histidine synthesis in *Salmonella* could vary over a wide range depending on the growth conditions, but the ratio of the activity of one enzyme to that of another was constant. They suggested that the cluster of genes acted as a unit of regulation. The rate of production of the enzymes decreased (repression) as the amount of the final end-product, histidine, increased, and the integrated action of histidine on the group of enzymes has been called *co-ordinate repression*.

Jacob, Monod and associates (see Jacob and Monod 1961, 1962) found that mutants of *Escherichia coli* unable to make use of lactose were due to mutations in one or other of a series of 3 closely-linked genes, *z*, *y* and *a*, each concerned with the production of a different enzyme (β-galactosidase, galactoside permease, and thiogalactoside transacetylase, respectively—see Fig. 15.1(*a*)). These enzymes are normally produced only in response to an inducer (lactose or other β-galactoside). In addition to mutants deficient in the ability to synthesize one of these enzymes, other mutants were found which formed all the lactose enzymes constitutively, that is, without the need of inducer. These mutants were found to be of two kinds.

One class of constitutive mutants was due to mutation at sites placed at one end of the series of lactose genes. Jacob *et al.* (1960) called this region an *operator* and the neighbouring group of genes which it affected in an integrated way they called an *operon*. Jacob and Monod (1961) found that the rate of

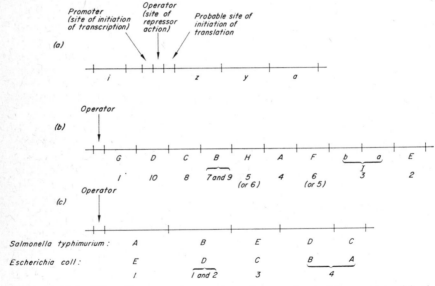

FIGURE 15.1    The diagrams show maps of some bacterial operons: (*a*) the lactose operon in *Escherichia coli* and its regulatory gene, *i*; (*b*) the histidine operon in *Salmonella typhimurium*; and (*c*) the tryptophan operons in *S. typhimurium* and *E. coli*. Genes are indicated by letters, and the sequence of action in the biosynthetic pathway of the enzymes for which they code is indicated by numerals. Diagram (*c*) is approximately to scale according to estimates by Imamoto and Yanofsky (1967) of the size of the *E. coli* tryptophan genes, based on the sub-unit molecular weights of their products; gene *A* is known to have 801 nucleotide-pairs, since it codes for a polypeptide containing 267 amino-acids (see § 13.8).

production of the three enzymes was much increased by lactose, but that the ratio of the activities of the different enzymes was constant. This *co-ordinate induction* was attributed to control by the operator. The co-ordinate repression, for example, of the histidine enzymes in *Salmonella* by histidine, was similarly explained. By study of heterozygotes for both an operator and an operon mutant in the *E. coli* lactose system, Jacob and Monod found that the operator directly influenced the genes of the neighbouring operon. Mutation in one particular cistron of this operon causes a failure of formation of the enzyme $\beta$-galactosidase, and as already mentioned, mutation in the operator causes the lactose enzymes to be formed constitutively (operator constitutive, $o^c$) instead of only in response to inducer. In cells heterozygous for these mutants (partial heterozygotes occur following conjugation—see § 16.3), the operator mutant was found to affect the lactose operon in its own chromoneme*, but not that in the homologue. Denoting the $\beta$-galactosidase-deficient mutant by $z$, they found that the cis configuration $\left(\dfrac{z \; o^c}{+ \; +}\right)$ was normal, that is, $\beta$-galactosidase was formed only in response to inducer, but in the trans configuration $\left(\dfrac{+ \; o^c}{z \; +}\right)$ $\beta$-galactosidase was formed constitutively.

The second class of constitutive mutants was found by Pardee, Jacob and Monod (1959) to be due to mutation at a gene locus called $i$ (inducer) which was also closely linked to the lactose operon. Double heterozygotes for $i$ and $z$ mutants were found to be normal, that is, they formed $\beta$-galactosidase only in response to inducer, whether the mutants were in the cis $\left(\dfrac{z \; i}{+ \; +}\right)$ or the trans $\left(\dfrac{+ \; i}{z \; +}\right)$ configuration.

Another kind of mutant of the $i$ gene, called $i^s$, or super-repressed, was found not to form the *lac* enzymes even in the presence of a $\beta$-galactoside.

## § 15.2   Jacob and Monod's theory

Jacob and Monod (1961) proposed that the $i$ gene specified the structure of a repressor molecule which could bind to the operator of the lactose operon and thereby prevent the formation of the products of the $z$, $y$ and $a$ genes. It was suggested that the repression occurred by blocking the transcription of messenger RNA from the genes of the operon. It was further suggested that a $\beta$-galactoside (the inducer) binds to the repressor molecule so as to remove it from the operator, thereby allowing transcription and translation of the genes of the operon to occur.

According to this hypothesis the operator-constitutive ($o^c$) mutants have a defect in the operator such that the repressor can no longer bind to it. Conversely, the repressor-negative ($i^-$) mutants have a defect in the repressor which prevents it from binding to the operator. With either class of mutant, large amounts of the *lac* enzymes will be synthesized even in the absence of inducer, in agreement with observation. That the $o^c$ mutants affect only

---

* Chromoneme = bacterial (or viral) chromosome—see § 11.1.

the immediately adjacent genes, while the $i^-$ mutants act in the trans configuration as well, is in agreement with this hypothesis, because the $i$ gene product (the repressor) could diffuse to the lactose operator on another chromoneme as readily as to that on its own, but diffusion does not apply to operator mutations. In $i^s$ mutants, which never form the *lac* enzymes, the repressor was presumed to have lost its affinity for the inducer.

An essential feature of this hypothesis for the control of enzyme activity is that messenger RNA should be extremely unstable, such that enzyme synthesis stops within 2 or 3 minutes of stopping the synthesis of the messenger. The instability would presumably be on account of rapid breakdown by ribonucleases (see § 15.4). For the lactose operon there is evidence from the work of Hayashi *et al.* (1963), Attardi *et al.* (1964) and others that the amount of messenger RNA present can vary, depending on the presence of inducer or other conditions. The messenger RNA was isolated by the technique described in § 14.6 of hybridization with homologons DNA.

Gilbert and Müller-Hill (1966) isolated the repressor of the lactose operon by an ingenious technique. An $i$ gene mutant was obtained which had a greater than normal affinity for isopropylthiogalactoside (IPTG), a substance which is not used by the bacterium but will act as an inducer. An extract of the mutant was allowed to reach dialysis equilibrium with radioactive IPTG. More radioactivity was found in association with the extract within the dialysis sac than outside. This showed that a component of the extract could bind the inducer, and by using the binding as an assay, it was possible to purify the repressor. It was found to be a protein with a molecular weight of 150,000–200,000. In a control experiment, an extract of an $i^s$ mutant did not concentrate IPTG. From the yield of repressor after the first steps of purification, Gilbert and Müller-Hill estimated there were about 10 repressor molecules per copy of the $i$ gene in the cell, assuming there are two sites for the inducer on each repressor molecule.

The repressor molecule which controls the activity of virus $\lambda$ of *E. coli* has also been isolated and, like the *lac* repressor, found to be a protein. This virus can exist in two states: integrated in the host chromoneme as the dormant provirus, or free in the cell (see § 16.3) where it can multiply rapidly and kill the cell. Jacob and Monod (1961) proposed that the inactivity when in the proviral state of all but one of the 40 or 50 genes in the virus is due to a repressor molecule determined by the one active gene called $c_I$. The messenger from this gene is indicated by the arrow labelled 1 in Fig. 14.6. The repressor molecules were believed to act at operators situated at the tails of the arrows labelled 2 in the figure, and to do so not only in the provirus but also in any other $\lambda$ virus that enters the cell, with the result that the cell is immune to destruction by the virus. Ptashne and Hopkins (1968) confirmed the existence of these two operators, which are situated on either side of $c_I$ and just to the right of gene $N$ in Fig. 16.4. Ptashne (1967$a$) eliminated most of the protein synthesis of host cells carrying $\lambda$ provirus by ultraviolet irradiation to inactivate their DNA, and he then added a high concentration of $\lambda$, giving about 30 virus particles per cell, and the culture was supplied with radioactive ($^3$H) leucine. The irradiation followed by the heavy

$\lambda$ infection should mean that a high proportion of the label incorporated into protein would be in the repressor molecules. It was necessary to use defective virus, unable to kill the host, because initially there are not enough repressor molecules to cope with 30 $\lambda$ particles at once. Another similar culture was set up using in turn various $\lambda c_I$ amber mutants which prevent synthesis of the $c_I$ product. This culture was supplied with $^{14}$C-leucine. After an hour's growth, fractionation of the proteins from the two cultures on a diethylaminoethyl (DEAE) ion exchange column revealed similarity, except for a major peak of $^3$H not shown by $^{14}$C: this protein peak was therefore presumed to be the product of the $c_I$ gene. Subsidiary experiments confirmed this. Temperature-sensitive $c_I$ mutants which produce modified repressor *in vivo* gave a different chromatographic pattern for this protein peak on the DEAE column, thus identifying this protein as the $c_I$ gene product and hence as the $\lambda$ repressor. The repressor was found to be an acidic protein with an estimated molecular weight of 30,000.

Ptashne (1967$b$) mixed $\lambda$ DNA with labelled $\lambda$ repressor and from analysis on a sucrose gradient he found repressor bound to the DNA. With DNA of a $\lambda$-434 hybrid virus, which contains most of the $\lambda$ genes but has virus 434's sites for accepting the repressor and so is sensitive to 434's repressor but not $\lambda$'s, he found no such binding to $\lambda$ repressor. He also showed that the $\lambda$ repressor cannot bind to denatured $\lambda$ DNA, that is, when its two chains have been separated. This work has thus established that the repressor binds specifically and with high affinity to $\lambda$ DNA, in precise agreement with one of the primary features of the Jacob–Monod model, namely, that the repressor binds directly to the DNA of the operator. Pirrotta and Ptashne (1969) isolated the repressor of virus 434 and found it to be a more basic protein than the $\lambda$ repressor. It was found to bind to DNA containing the 434 operators.

Gilbert and Müller-Hill (1967) found that the *lac* repressor also binds directly to the operator DNA. The repressor was labelled with radioactive sulphur ($^{35}$S) by growing the bacteria in labelled medium, and then it was mixed with DNA containing the *lac* operator, and the mixture centrifuged in a glycerol gradient. The DNA had been obtained in quantity by using a defective strain of $\phi$80-$\lambda$ hybrid virus which contained the *lac* operon part of its host's DNA. The DNA, being larger, sedimented faster than protein, but carried some of the labelled protein bound to it, and this was evidently the *lac* repressor. Some further experiments confirmed this: first, addition of IPTG specifically released the protein from the DNA; second, DNA from $\phi$80-$\lambda$ carrying an operator-constitutive ($o^c$) mutant of the *lac* operon showed only a weak affinity for the protein; and third, virus DNA not carrying the *lac* operon showed no binding of protein.

Gilbert and Müller-Hill's remarkable experiments established not only that the product of the $i$ gene is a protein, but moreover that it specifically binds to the DNA of the *lac* operator, in direct agreement with the Jacob–Monod theory. No binding was obtained when the DNA chains had been separated by denaturing. Gilbert and Müller-Hill concluded that the repressor protein binds to a specific sequence of duplex DNA. They argued that this sequence must be at least 11 or 12 base-pairs long to give the necessary specificity, allowing about $3 \times 10^6$ base-pairs in the *E. coli* chromoneme ($4^{11} = 4.2 \times 10^6$ approx). Estimates of the binding constants led to the conclusion that the inducer must actively trigger the release of the repressor from the operator.

Gilbert and Müller-Hill's experiments with the *lac* repressor and Ptashne and Pirrotta's with the $\lambda$ and 434 repressors have greatly strengthened Jacob and Monod's theory. Many less simple hypotheses had been proposed, such as that the repressor blocks translation rather than transcription, by binding to the messenger RNA or to transfer RNAs, or by destroying the messenger. Further support for Jacob and Monod's theory is provided by the isolation by Parks *et al.* (1971) of the repressor of the galactose operon, and the demonstration that it is a protein, that it binds specifically to the DNA of the galactose operon, and that the binding is prevented by galactose and other inducers of the operon.

Jacob and Monod recognized two kinds of genes: *structural genes*, such as $z$, $y$ and $a$ in the lactose operon, which determine the amino-acid sequence of specific proteins, and *regulatory genes*, such as $i$ and $c_I$ which control the functioning of the structural genes. Gilbert, Müller-Hill and Ptashne have confirmed this distinction, and have shown that the $i$ and $c_I$ regulatory genes are the structural genes for the repressor proteins. Garen and Otsuji (1964) pointed out that a regulatory gene might not produce a repressor directly, but might be involved in a reaction required for repressor formation. They identified a regulatory gene of this kind affecting the synthesis of alkaline phosphatase (see § 15.7).

Smith and Sadler (1971) analysed over 800 operator constitutive ($o^c$) mutations of the lactose operon of *E. coli*. They discovered that almost all the mutants were base substitutions, and all of them were only partially constitutive for the lactose enzymes. Moreover, the mutants fell into 7 discrete classes as regards the constitutive rate of $\beta$-galactosidase synthesis. The classes were numbered I to VII in order of decreasing rate. Each class could be divided into two subclasses, $a$ and $b$, differing in their maximum levels of expression of the *lac* enzymes ($a$ higher than $b$). Sadler and Smith (1971) mapped 26 of the $o^c$ mutations. They found there was no recombination between mutants of the same subclass, implying that each subclass corresponded to a particular site in the operator. Furthermore, there was a remarkable symmetry in the subclass map. Omitting the mutants of class I which were not mapped, the sequence of the subclass sites was as follows: $z$ gene, VII$b$, VI$a$, V$a$, II$a$, III$b$, IV$b$, IV$a$, III$a$, II$b$, V$b$, VI$b$, VII$a$, $i$ gene. In other words, the operator is a double structure, one half being the mirror image of the other, with mutations at corresponding positions in the two halves giving the same constitutive rate. There are three possible base substitutions at a mutant site, and these would be expected to affect the binding of the repressor differently. It is surprising, therefore, that all the $o^c$ mutants at a site behaved alike. Sadler and Smith thought it likely that two of the three possible substitutions at each site prevented the *lac* operon from functioning and so were not detected as mutants. This inference was supported by their discovery (Smith and Sadler, 1971) that an acridine-type mutagen failed to induce $o^c$ mutations, presumably because additions or deletions of nucleotides in the operator also prevented the operon from functioning.

The *lac* repressor protein is believed to be a tetramer (Riggs and Bourgeois, 1968). Sadler and Smith discussed the possibility that two sub-units of the repressor bind one to each segment of the operator. Gierer (1966) had pointed out that, by analogy with transfer RNA, an operator might contain reversed duplications, for example:

$$5'—G-T-T-C-A-C-T-C-T-G-A-A-C—3'$$
$$\cdot \quad \cdot \quad \cdot \quad \cdot \quad \cdot \quad \cdot \quad \cdot \quad \cdot \quad \cdot \quad \cdot \quad \cdot \quad \cdot \quad \cdot$$
$$3'—C-A-A-G-T-G-A-G-A-C-T-T-G—5'$$

This could form a secondary loop in each chain:

Gierer suggested that the repressor interacted with such loops. The mirror-image structure of the operator found by Sadler and Smith is in keeping with Gierer's model.

Sobell *et al.* (1971) suggested that the binding of actinomycin to DNA, which is known to inhibit RNA synthesis, may reveal general features of repressor-operator interaction. Actinomycin consists of a phenoxazone ring (two benzene rings joined by nitrogen and oxygen atoms), with identical cyclic polypeptides attached to each benzene ring. These polypeptides contain 5 amino-acid residues. From X-ray crystallography of the complex of actinomycin D and deoxyguanosine, Sobell *et al.* showed that the phenoxazone ring system intercalates between adjacent $G \cdot C$ and $C \cdot G$ base pairs where the sequence in each chain is $GpC$, not $CpG$, and that the two cyclic 5-peptide chains lie in the narrow groove of the DNA with hydrogen bonding to the guanine residues. It is of interest that actinomycin, and the nucleotide sequence to which it binds, have two-fold symmetry.

With pathways which are co-ordinately repressed by the end-product, as in the histidine and tryptophan operons, Jacob and Monod proposed that the product of a regulatory gene cannot by itself bind to the operator but must first combine with the end-product of the pathway. In other words, the end-product functions as a co-repressor.

Myers and Sadler (1971) discovered a mutation of the *i* gene which had the remarkable property of reversing the normal behaviour of inducers. They called the mutant $i^{rc}$ for repressor constitutive, because in its presence galactosides acted as co-repressors, shutting off the synthesis of the lactose enzymes. The $i^{rc}$ mutant evidently has some defect in the synthesis of repressor, and IPTG is needed to complete one of the steps in its formation. This mutant is of interest in showing that induction and repression may be closely related in evolution. Another example, the converse of $i^{rc}$, is provided by the regulation of the enzymes which catalyze the synthesis of arginine in *E. coli*. The structural genes for these enzymes do not form an operon but are mostly at scattered positions in the chromoneme. A regulator gene, *argR*, produces a repressor which is believed to be activated by the end-product of the pathway, that is, arginine, and to act at operator sites alongside each structural gene. Jacoby and Gorini (1969) found that a mutant, *argR*[B], of the regulator gene occurs in strain B of *E. coli*, and has an abnormal repressor molecule the formation or functioning of which is reduced at high arginine concentrations. In other words, in strain B arginine acts as an inducer, decreasing the amount of active repressor, whereas normally arginine acts as a co-repressor.

## § 15.3   Polar mutants

Franklin and Luria (1961) and Jacob and Monod (1962) found that some mutations in the $\beta$-galactosidase gene ($z$) of *Escherichia coli* in addition to preventing the synthesis of this enzyme, gave rise to reduced rates of synthesis of the other two lactose enzymes. Some permease ($y$) mutants had a reduced level of transacetylase, but $\beta$-galactosidase was unaffected. Thus, the mutants were polarized in their effects, only enzymes made by genes on the distal side of the mutant site with respect to the operator being reduced in amount. Beckwith (1964*b*) showed that a class of mutants which Jacob and Monod had called operator zero ($o^0$) mutations, because they prevent the formation of all the enzymes of the operon, appear to be polar mutants with the mutant site situated in the $\beta$-galactosidase gene near the operator, and with exceptionally strong effects in lowering the levels of activity of the $y$ and $a$ genes.

The histidine operon in *Salmonella typhimurium* has been intensively studied by Hartman, Ames and associates (see Hartman, Hartman, Stahl and Ames, 1971). Over 1500 histidine-requiring mutants have been isolated. The operon has been found to consist of at least 10 genes specifying the enzymes of the pathway (Fig. 15.1(*b*)). There is at least one instance of a pair of genes determining one enzyme, and one gene specifies two enzymes, a dehydrase (no. 7) and a phosphatase (no. 9). It appears that different parts of the same protein catalyze different reactions (see § 13.7). Numerous polar mutants have been found, and the direction of their effect indicates that the operator is at the left-hand end of the operon in Fig. 15.1(*b*).

The trytophan operons in *S. typhimurium* and *E. coli* are illustrated in Fig. 15.1(*c*). They are alike as regards the sequence of genes coding for the enzymes of the pathway. Polar mutants place the operator at the left-hand end in the diagram in both species. Moreover, the *A* gene in *E. coli*, which codes for one of the two polypeptides of tryptophan synthetase, has been orientated with respect to the direction of messenger synthesis (see § 14.14) and this confirms that the operator is to the left.

Jacob and Monod (1961) favoured the idea that a single messenger RNA molecule was formed corresponding to an entire operon, because this would provide a neat explanation of the co-ordinate induction or repression of the enzymes of the operon. Attardi, Naono, Rouvière, Jacob and Gros (1964) isolated the messenger RNA corresponding to the lactose operon of *E. coli* by means of the technique developed by Hall and Spiegelman (1961) whereby messenger RNA forms hybrid molecules with homologous DNA when a heated mixture is slowly cooled (see § 14.6). Attardi and associates found the sedimentation coefficient of the messenger RNA was such as to suggest that it corresponded to the entire operon. Likewise, Martin (1964), using differential labelling with $^{14}$C and $^{3}$H to distinguish the RNA of a mutant deficient for the *S. typhimurium* histidine operon from that of a non-deficient strain, found by chromatography on a methyl albumin column, followed by centrifuging in a sucrose gradient, that the histidine operon messenger RNA had a sedimentation coefficient appropriate for the entire operon. Imamoto, Morikawa and Sato (1965) isolated the tryptophan messenger RNA of *E. coli* and found by density gradient centrifugation that it had a sedimentation coefficient of 33$S$, which again is appropriate for the whole operon. Mutants with deletions of parts of the operon had smaller messengers. By sedimentation in a sucrose gradient, Kiho and Rich (1965) measured the relative size of the ribosome aggregates (polyribosomes) showing $\beta$-galactosidase activity in *E. coli*. They found that strains with deletions in the *y* and *a* genes had smaller polyribosomes. They concluded that the *lac* messenger was multigenic in translation as well as transcription, that is to say, it did not break up into gene-length segments for polypeptide formation.

Beckwith (1964*a*) discovered that some of the polar mutants of the lactose operon of *E. coli* were suppressed by amber suppressors (§ 14.17) and were evidently polypeptide-terminating mutants. Extensive studies have since been made of the effects of such stop mutants in three operons: in the lactose operon by Newton, Beckwith, Zipser and Brenner (1965), in the tryptophan operon of *E. coli* by Yanofsky and Ito (1966) and in the histidine operon of *S. typhimurium* by Martin, Silbert, Smith and Whitfield (1966). All the results were similar and may be summarized as follows:

(*a*) The effects of amber (UAG) mutants were indistinguishable from those of ochre (UAA) mutants, and both invariably reduced the relative rates of synthesis of all proteins specified by genes situated on the distal side of the stop mutant with respect to the operator. As mentioned in § 14.17, Sambrook, Fan and Brenner (1967) found that opal (UGA) mutants were also polar.

(*b*) For any particular stop mutant, all the genes on the non-operator side of the one containing the mutation had their activities lowered to the same extent.

(*c*) Suppressors of stop mutants suppressed the polar effect to the same degree and with the same specificity as they suppressed polypeptide termination.

(*d*) Within each gene there was a gradient of polarity, such that stop mutants situated near the operator end of a gene were more strongly polar (that is, reduced to a greater extent the rates of synthesis of the enzymes coded by the operator-distal genes) than those near the non-operator end, the gradient beginning afresh in each gene.

Yanofsky and Ito (1966) discovered that two stop mutations present simultaneously in different genes of the operon showed polar effects independently, the percentage reduction in the relative rates of activity of distal genes caused by the second mutation being superimposed on the reduction resulting from the first mutation. On the other hand, two such mutations in the same gene behaved as if only one were present.

Zipser and Newton (1967) showed in an ingenious way that the gradient of polarity within a gene is dependent on the distance of the stop mutant from the distal end of the gene and not on the distance from the operator end. They found that the polarity caused by stop mutants in the $z$ gene of the lactose operon was unaffected by deletions of segments of $z$ on the operator side of the stop mutant, but was greatly reduced by deletions situated within $z$ on the distal side. Such deletions reduced the distance to the non-operator end of the gene.

Ito and Crawford (1965) discovered there was no polar effect with mutations in the tryptophan operon of *E. coli* that gave rise to cross-reacting material, that is, protein that could be recognized immunologically as similar to the normal product of the gene. Conversely, polar mutants produced no cross-reacting material. Likewise, Martin *et al.* (1966) found no polarity with mutations in the *C* gene of the histidine operon of *S. typhimurium* that gave cross-reacting material. Such mutations are thought to be base substitutions causing an amino-acid substitution at one point in the polypeptide. Unlike stop mutants, these 'mis-sense' mutants do not respond to amber or ochre suppressors and are often leaky, that is, allow some growth to occur.

Martin *et al.* (1966) found that another class of mutants, in addition to stop mutants, were polar. These were frameshift mutants (§ 14.3 and § 14.13). They were recognized by their failure to revert with base-substituting mutagens, their lack of suppression through mutation of another gene (such as gives rise to amber and ochre suppressors), and by their lack of cross-reacting material. Martin *et al.* suggested that the polarity observed with frameshift mutants was due to a stop codon created by the change of reading-frame. Martin (1967) obtained support for this idea from study of revertants of frameshift mutants. Such revertants will be expected to have an addition of 1 or 2 nucleotides in proximity to a corresponding deletion (cf. Fig. 14.14) and not to be polar: in order to be polar the plus and minus frameshifts would need to straddle the postulated stop codon generated by the first frameshift, but this would not produce a revertant because the second frameshift would not be read. The revertants were indeed non-polar.

It thus seems likely that polar effects in operons may all be due to stop codons arising within a gene, either directly through an appropriate base change, or indirectly through a frameshift following upon the addition or deletion of a number of nucleotides other than 3 or a multiple of 3.

The discovery that stop mutants and frameshift mutants show a reduced level of activity of operator-distal genes but not a complete absence of these gene products (except when the stop mutant is at the operator end of the first gene) implies that there must be polypeptide-initiating regions at the beginning of each gene of an operon. The existence of re-initiating regions within a messenger has been demonstrated: see § 14.16.

## § 15.4   The messenger breakdown hypothesis

Attardi *et al.* (1964) found that two strongly polar mutants situated at the operator end of the *z* gene of the lactose operon of *E. coli* apparently contained no *lac*-specific messenger RNA after induction. This finding seemed to conflict with Beckwith's (1964*a*) discovery that these mutants could be suppressed by stop-mutant suppressors and therefore were blocked in translation rather than in transcription. Beckwith suggested that the messenger RNA transcribed from the lactose region was particularly unstable in these mutants and so escaped detection. The gradient of polarity within *z* found by Newton *et al.* (1965) would then be accounted for if messenger breakdown depended on the distance between the site of the mutation and the start of the next gene, perhaps because ribosomes normally protected the messenger from ribonuclease attack.

Imamoto and Yanofsky (1967) found from DNA–RNA hybridization that polar mutants of the tryptophan operon of *E. coli* were deficient in the messenger corresponding to the genes on the operator-distal side of the mutated gene, and moreover, the proportion of such short messengers was greater with the more strongly polar mutants. From pulse labelling experiments, Morikawa and Imamoto (1969) concluded that the messenger RNA for the tryptophan operon in *E. coli* is broken down in the 5' to 3' direction. This is the same direction as transcription and translation, and means that as an RNA polymerase molecule moves along the operon synthesizing the messenger, a cluster of ribosomes and associated amino-acyl transfer RNAs could move along the newly-formed messenger keeping pace with its synthesis and translating its nucleotide sequence into polypeptide, and these ribosomes could be followed by ribonucleases progressively destroying the messenger again. Morse and Yanofsky (1969) studied the kinetics of tryptophan messenger RNA synthesis and breakdown in wild-type *E. coli* and in *E. coli* with stop mutants in the genes of the tryptophan operon. They concluded that the polar effect of stop mutants results from rapid destruction of the messenger just beyond the mutant site, where it is prematurely exposed by the release of the ribosomes, with the result that in a fraction of the messenger chains the breakdown then immediately follows synthesis and prevents ribosomes becoming attached at the start of subsequent genes.

This hypothesis of messenger breakdown triggered by stop mutants was confirmed from study of a polarity suppressor called *suA*. Scaife and Beckwith (1967) found that, unlike amber, ochre and opal suppressors, *suA* restored permease and transacetylase activity when there was a polar mutant in *z*, without restoring β-galactosidase activity. Moreover, this effect was found whether the polar mutant was a base substitution or a frameshift. Scaife and Beckwith concluded that *suA* does not repair the polar mutant defect, but directly modifies the polarity mechanism. Morse and Primakoff (1970) found that the polarity suppression by *suA* applied to the tryptophan as well as the lactose operon, and was associated with a failure to break down the messenger distal to the polar mutant. Kuwano, Schlessinger and Morse (1971) obtained evidence that the *suA* strain of *E. coli* lacked an endonuclease which they believed to be responsible for initiating breakdown of messenger RNA in normal *E. coli*. The destruction of the messenger is thought to be brought about by ribonuclease V, an exonuclease acting in the 5′ to 3′ direction (Kuwano, Schlessinger and Apirion, 1970).

Miller, Hamkalo and Thomas (1970) obtained beautiful electron micrographs of genes of *Escherichia coli* in action. They used a bacterial mutant with fragile cell walls. The cells were burst osmotically by rapid dilution of the culture medium. The electron micrographs showed numerous messenger RNA molecules attached to the bacterial chromoneme and forming a gradient of increasing length along the gene. There were up to 40 ribosomes associated with each messenger and they extended right down to the junction of the messenger with the DNA. This directly confirmed that translation starts as soon as the messenger begins to be synthesized and long before transcription is complete.

## § 15.5   The secondary structure hypothesis

Lodish and Zinder (1966) discovered polarity of gene expression in an RNA virus. They found that an amber mutant in triplet no. 6 of the coat gene of virus *F*2 of *Escherichia coli* led to a failure to synthesize the viral RNA polymerase, while an amber mutant near the middle of the coat gene did not have this effect. Gussin (1966) made a similar discovery with virus *R*17, and suggested that the secondary structure of the RNA at the start of the polymerase gene might make it unavailable for ribosome binding until synthesis of the coat protein had proceeded sufficiently far for the ribosomes to induce the RNA to unfold and expose the start signal. With an amber mutant early in the coat gene such exposure would not happen since the ribosomes would be detached from the RNA at the amber mutant.

The polarity mechanism in these RNA viruses was shown to be different from that in their host when Morse and Primakoff (1970) found that *suA*, which suppresses the polarity of the lactose and tryptophan operons (see § 15.4) had no effect on virus *F*2 polarity.

As already mentioned (§ 14.13), Min Jou *et al.* (1972) determined the entire nucleotide sequence of the coat gene of RNA virus *MS*2, a close ally of *F*2 and *R*17. Their work has given strong support to the secondary structure hypothesis to explain polarity of gene expression in these viruses. Their model for the secondary structure of the coat gene of *MS*2 is shown in Fig. 15.2. It was obtained by choosing the base-pairing scheme with the greatest thermodynamic stability, starting from the products of partial nuclease digest. There are numerous places where the chain is believed to be folded back upon itself to give base-paired segments, and the overall pattern is stellate or flower-like. According to this model, the initiating AUG triplet of the polymerase gene is paired with parts of triplets 27 and 28 of the coat

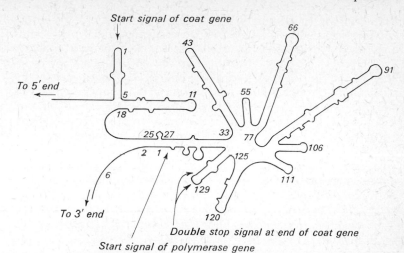

FIGURE 15.2   Model proposed by Min Jou, Haegeman, Ysebaert and Fiers (1972) for the secondary structure of the coat gene of RNA virus *MS2* of *Escherichia coli*. Parallel lines indicate base-paired regions and the numbers show the positions of some of the 129 triplets of nucleotides which make up this gene. A few triplets at the start of the polymerase (replicase) gene are also numbered (cf. Table 14.10).

gene, and will therefore not be recognized until translation of the coat gene has reached this region. This is in good agreement with the findings that an amber mutant at triplet no. 6 in the coat gene is polar but at triplet no. 50 or later is non-polar.

## § 15.6   The promoter

Jacob, Ullman and Monod (1964) discovered that when a deletion in the lactose operon of *E. coli* covers the operator and the neighbouring end of the *z* gene, activity of the remaining genes (*y* and *a*) is possible only if the deletion extends beyond *i* (Fig. 15.1(*a*)) into another operon such as the purine operon or the tryptophan operon. If *y* and *a*, in the absence of the operator and hence not repressed, could function by connecting them to any region of DNA, one would expect to find deletions starting in *z* which ended either in the operator or in whatever length of DNA may exist between the operator and *i*, but no such deletions were found. Jacob *et al.* concluded that there was a site in the neighbourhood of the operator which was necessary to allow the *lac* genes to be expressed. Removal of this site inactivates the operon, unless the corresponding site of another operon is substituted by means of a long deletion.

Jacob *et al.* also found that $o^c$ mutations, which destroy the function of the operator so that the lactose genes function without the need of inducer, did not reduce the maximal rate of operon expression. This implied that the operator, as the site of repressor action, was not the site which determines this rate. They proposed the term *promoter* for the site for the initiation of operon expression.

Ippen, Miller, Scaife and Beckwith (1968) isolated mutants which co-ordinately reduced the rate of expression of the *lac* genes. Like $o^c$ mutants (§ 15.1), these mutants were *cis-dominant*, reducing the rate of expression of the operon on the same chromoneme but not on another one. It was concluded that the mutants had an alteration in a site essential for the initiation of either transcription or translation of the operon. The mutants were mapped by the use of a series of deletions of various extent and, contrary to previous ideas, found to lie between the regulator gene (*i*) and the operator (Fig. 15.1(*a*)). This position implies that they were promoter mutants, that is, defective in transcription rather than in translation, because it seems unlikely that the operator is translated into protein. Mutations of the operator ($o^c$) are not suppressed by amber or ochre suppressors, which are known to act at the level of translation by reading the stop codon as an amino-acid (§ 14.17). If the operator were translated, some amber or ochre $o^c$ mutants would be expected, but none has been found. Ippen *et al.* suggested that the promoter might act as a binding site for RNA polymerase, the promoter mutants reducing the site's affinity for the enzyme. The repressor, in binding to the operator, could then block the movement of the RNA polymerase into the structural genes of the operon.

Support for the idea that the promoter is a binding site for RNA poly-merase was obtained by Arditti, Scaife and Beckwith (1968), who found that single transitions in the promoter (substituting one purine for the other, and similarly with its pyrimidine complement) reduced the activity of the *lac* operon 15-fold. Such marked effects of a base change would be expected with protein-DNA interactions.

By comparing the binding of λ repressor to the two λ operators in the presence and in the absence of RNA polymerase, using wild-type λ DNA and also λ DNA containing a mutation in the left-hand promoter ($p_L$), Chad-wick *et al.* (1971) found that the binding of the λ repressor to the λ operators was hindered when RNA polymerase was bound to the adjacent promoter. They therefore suggested that the repressor blocks transcription by prevent-ing the attachment of the polymerase at the neighbouring site. The lactose promoter and operator, however, do not overlap, for Chen *et al.* (1971) found that *lac* repressor does not inhibit the binding of RNA polymerase. A *lac* promoter mutant, on the other hand, caused competitive binding.

Blattner and Dahlberg (1972) allowed messenger RNA synthesis in virus λ of *E. coli* to proceed *in vitro* for not more than 1 minute, so that only the initial nucleotide sequences of the messengers were formed. By hybridizing this RNA with λ DNA, the sites of transcription were mapped. They were four in number and included the two operons (leftward and rightward) controlled by λ repressor. Each of the four kinds of RNA molecules was found to have a unique nucleotide sequence. The position of the startpoint ($s_L$) of RNA synthesis for the leftward operon was established by hybridization experi-ments using mutants of λ with various segments of DNA deleted. The un-expected discovery was made that $s_L$ is to the left of the sites of operator ($o_L$) and promoter ($p_L$) mutants of the leftward operon, and hence that the pro-moter (as the binding site of RNA polymerase) and the operator are not transcribed. This conclusion was confirmed by the observation that the RNA starting sequence in the leftward operon was unaffected by $o_L$ and $p_L$ mutants. Likewise, the initial sequence of the rightward operon messenger was unaffected by an $o_R$ mutant. Study of electron micrographs of heteroduplex molecules (§ 16.5) prepared *in vitro* by annealing nucleotide chains of deletion mutants, in conjunction with measurements of the recombination frequencies of these mutants with the $o_L$ and $p_L$ mutants, enabled Blattner *et al.* (1972) to estimate that the interval between $s_L$ and $p_L$ was about 200 nucleotides, or 8 times the diameter of the RNA polymerase molecule (9 nm). They suggested that the enzyme moves along the DNA from $p_L$ to $s_L$ without transcription taking place, or alternatively, that the two sites are brought together physically.

Bauerle and Margolin (1967) discovered that in the tryptophan operon of *S. typhimurium* there is a secondary initiator of gene expression between genes $B$ and $E$ (Fig. 15.1(c)). Their main findings were as follows. The first two genes of the operon ($A$ and $B$) form one co-ordinate unit of expression and the other three ($E$, $D$ and $C$) another, which is not, however, wholly independent of $A$ and $B$. Polar mutations in $A$ depress the activity of $B$ more severely than that of $E$, $D$ and $C$. Deletions extending into $A$ from the operator end have no $B$ activity, but retain a reduced level of expression of $E$, $D$ and $C$ without control, however, by tryptophan. All the genes are inactive when deletions extend from the operator end beyond the $B$-$E$ boundary. Deletions of the $B$-$E$ junction render the remaining parts of the operon co-ordinate. Bauerle and Margolin concluded that there were two sites for initiation of gene expression, one at the operator end and the other between genes $B$ and $E$. These sites were believed to be promoters, that is, to initiate transcription, the first ($p1$) giving a single messenger for all five genes, and the second ($p2$) an $E$–$D$–$C$ messenger*. Callahan, Blume and Balbinder (1970) showed from genetic studies with deletion mutants that the main promoter ($p1$) is situated on the opposite side of the operator from the first structural gene ($A$). This is the same sequence as in the lactose operon.

Many inducible enzymes in bacteria, including the lactose, galactose and arabinose enzymes of *E. coli*, show catabolite repression, that is, enzyme induction is inhibited by glucose or related substances. Pastan and Perlman (1968) found a mutant of the *lac* promoter in which the lactose enzymes were not repressed by glucose. Silverstone *et al.* (1969) made a similar discovery— a partial deletion of the promoter which made the lactose operon insensitive to catabolite repression. The existence of these mutants showed that the inhibition by glucose occurred at transcription.

It is known that cyclic 3′,5′-adenosine monophosphate (cyclic AMP) overcomes the repression of inducible enzymes by glucose. Cyclic AMP is a ribonucleotide of adenine in which the phosphate group is joined to the ribose at both the 3′ and the 5′ positions (see Figs. 11.4(d),(e) and 14.4). Zubay, Schwartz and Beckwith (1970) and Emmer, de Crombrugghe, Pastan and Perlman (1970) purified a protein to which cyclic AMP binds. This receptor

---

* A similar internal promoter exists in the tryptophan operon of *E. coli*. Jackson and Yanofsky (1972) mapped it, using short deletions, and made the unexpected discovery that it was situated within gene $D$ (see Fig. 15.1(c)) near its junction with gene $C$.

protein, and the cyclic AMP which has such a high affinity for it, were believed to be part of a positive control system for activating catabolite-sensitive genes. This hypothesis was confirmed for the lactose operon by de Crombrugghe *et al.* (1971) using a cell-free system. They concluded that the *lac* promoter was a double site, one part for the binding of RNA polymerase and the other for the action of cyclic AMP and its receptor protein. Glucose, or more strictly a glucose breakdown product (catabolite), lowers the cyclic AMP level and so prevents the activation of catabolite-sensitive genes.

## § 15.7   Other bacterial systems for the control of gene activity

The negative control system by repression, now known to occur in *E. coli* with the lactose operon and virus λ, is not the only mechanism of control of gene action in bacteria.

Englesberg, Irr, Power and Lee (1965) studied the genetic basis of the use of L-arabinose by *E. coli* and found that the system differs in important respects from the lactose control mechanism. Genes for a kinase (*B*), an isomerase (*A*) and an epimerase (*D*) are closely linked in the order given and believed to form an operon. The enzymes function in the breakdown of arabinose in the sequence *A*, *B*, *D*. A regulatory gene (*C*), also closely-linked in the order *CBAD*, is believed to produce a repressor which attaches to the *ara* operator, from which it is removed by arabinose. Here the resemblance to the lactose system ends, because the repressor–arabinose complex is thought to have a positive function as an *activator*, promoting synthesis of the arabinose enzymes, which cannot occur merely in the absence of repressor. Furthermore, this activator promotes co-ordinate synthesis not only of the products of the three structural genes of the *ara* operon but also of a permease determined by a gene (*E*) situated in a different part of the linkage map. Englesberg *et al.* suggested that positive (activator) control is required for the co-ordinate control of genes in different operons. They assumed that gene *E* is part of another operon, and that the activator is needed to remove it from control by this other system.

From genetic studies using deletion mutants, Englesberg, Squires and Meronk (1969) inferred that the *C* gene product can exist in two functional states, repressor and activator, which attach to separate controlling sites, namely, an operator (*o*) and an initiator (*i*), respectively. These sites are in the sequence *C o i B A D*. The repressor state is epistatic to the activator state, but an inducer such as arabinose alters the equilibrium between the two in favour of the activator form. Greenblatt and Schleif (1971) demonstrated that the *C* gene product shows the same functional duality *in vitro*.

The control of alkaline phosphatase synthesis in *E. coli* shows a different system again. Garen and Otsuji (1964) found there were three regulatory genes. One of these (*R*1) is closely-linked to the phosphatase structural gene and, from earlier studies, is thought to control the formation of an inducer required for the synthesis of the phosphatase. The other two (*R*2*a* and *R*2*b*) are closely-linked to one another in a different part of the chromoneme and are thought to control the formation of an enzyme that can convert the inducer into a repressor when a high concentration of orthophosphate is present. The product of gene *R2a* was shown to be a protein. Its formation was found to be repressed under the same conditions as the phosphatase, and so Garen and Otsuji considered that both were regulated by the same repressor. They suggested that this might explain how regulatory genes were themselves regulated. For the regulatory gene for the lactose operon, however, promoter mutants are known which modify the number of repressor molecules in the cell. It is believed that the relatively low activity of this gene is controlled by its promoter.

Harris and Sabath (1964), from study of the synthesis of the enzyme penicillinase in *Bacillus cereus* in response to inducer (cephalosporin *C*), found that the antibiotic actinomycin *D* inhibited RNA synthesis within 1 or 2 minutes of its addition to the culture, but that synthesis of the enzyme continued for the cell generation-time (about 20 minutes). They concluded that the messenger RNA for penicillinase was stable and that in consequence the regulation of the synthesis of the enzyme could not occur by controlling messenger formation.

These results suggest that there are several different ways in which enzyme synthesis may be controlled and that, although the primary control is at transcription, a secondary mechanism also exists at the level of the translation of messenger RNA into polypeptide. Two molecular mechanisms for the control of translation have already been discussed: initiation factors (§ 14.16) and the secondary structure of the RNA (§ 15.5). A third mechanism is discussed in § 18.6.

## § 15.8   Viral operons

Evidence pointing to the existence of operons, defined by the occurrence of multigenic messenger, has been obtained for bacterial viruses. In virus *T*4 of *Escherichia coli* several pairs of closely-linked genes are known in which amber mutants in one lower the activity of the other, and these genes are concerned with the same function: Stahl, Murray, Nakata and Craseman (1966) found that amber mutants in gene 51 depress the activity of gene 27 (see Fig. 18.4), both of which are concerned with the formation of the base-plate of the tail of the virus (King, 1968). Likewise, Stahl *et al.* found that amber mutants in gene 34 depressed the activity of gene 35. According to Edgar and Lielausis (1968) these genes affect the last two steps, 35's product before 34, in the formation of the tail-fibres. Terzaghi (1971) demonstrated the polar effect of amber mutants in gene 34 on the activity of gene 35 by *in vitro* complementation, that is, mixing the viral components from cells infected separately with *T*4 containing amber mutants in genes 34 and 35. Amber mutants which mapped near the opposite end of gene 34 from gene 35 depressed the activity of gene 35 much more than ambers of 34 situated closer to 35. Edgar and Lielausis found that amber mutants in gene 13 lower the activity of gene 14, both of which concern the final steps in the formation of the head of the virus.

In the viruses *λ* and *T*4, genes concerned with related functions are frequently grouped together (see Figs. 14.6 and 18.5). The comparative rarity, however, with which amber mutants in one gene affect another suggests that these groups of genes usually do not have multigenic messengers, and so should not be regarded as operons. Stahl and Murray (1966), from complementation studies between mutants in viruses *T*2 and *T*4, argued that in the evolutionary divergence of these two closely-related viruses selection had favoured the maintenance of particular combinations of mutations in different genes, and so had favoured close linkage of these genes.

## § 15.9  The replicon

As a derivative of the idea of interactions between the products of regulatory genes and operators in the control of gene action, Jacob and Brenner (1963) made comparable postulates for the control of DNA replication. They suggested that replication is controlled in such a way that it occurs independently in each unit or *replicon*. Bacterial and viral chromonemes appear to be single replicons, but a chromosome contains a number of them (see § 12.8). Jacob, Brenner and Cuzin (1964), in elaborating the replicon concept, pointed out that the enzymes which make DNA and RNA copies from a DNA primer function in an uncontrolled way *in vitro*, but that *in vivo* both replication and transcription appear to be controlled with such precision and specificity that they cannot be regulated by the flow of their precursors. Jacob *et al.* postulated that DNA replication required, first, a structural gene controlling the synthesis of a specific *initiator*, which might be a specific DNA polymerase or a priming enzyme able to separate the chains of the double helix, and secondly, an operator of replication, or *replicator*, which would be a recognition site upon which the initiator would act, allowing the replication of the DNA attached to the replicator. This model for control of replication is based on the repressor model for the control of gene action, but positive regulation of replication is postulated instead of the negative regulation, by removal of a repressor, proposed for the lactose enzymes. The initiator and replicator were considered to be specific for each replicon, so that, for instance, the initiator for a bacterial virus would not recognize the replicator for its host. This hypothesis would account for the observation that chromonemal fragments, introduced into a bacterial cell by conjugation, transformation, or transduction, fail to replicate unless incorporated into the chromoneme of the acceptor cell (see § 16.3), since the fragments would be unlikely to include the replicator.

## § 15.10  The possible occurrence of operons in chromosomal organisms

The operon concept was developed by Jacob, Monod and associates to explain the integrated control of the action of genes in a biosynthetic pathway in bacteria. In chromosomal organisms, the genes concerned with successive steps in reaction chains appear, in general, to be scattered through the linkage groups. This is well-established for *Neurospora crassa* and *Aspergillus nidulans*, and is indicated, for instance, by the frequency of 9:3:3:1 ratios (or modifications such as 9:7) in the $F_2$ generation in studies on the inheritance of anthocyanin pigmentation in flowering plants. Recently, however, some examples of operon-like organization have been discovered in fungi.

De Serres (1964) pointed out that frameshift and stop mutants will be expected to fail to complement with allelic mutants situated distal to them in translation. If the polypeptide coded by the frameshift or stop mutant has a sufficient length of normal amino-acid sequence, however, it may not necessarily fail to complement with mutants situated proximal to it. For example, if 10 allelic mutants are numbered 1–10 in the order of their positions within the gene, the numbering being from the end of the gene where translation starts, then a stop mutant situated between mutants 8 and 9 will fail to complement with mutants 9 and 10, but may complement with nos. 1–8. De Serres found examples of such polarized complementation among mutants of the *adenine-3B* gene of *Neurospora crassa*. Polar mutants were particularly frequent with spontaneous and X-ray-induced mutants, which no doubt often have additions or deletions of nucleotides causing frameshift. Radford (1969a) found polarized complementation at the *pyrimidine-3* genetic region of *N. crassa*, using mutants obtained with the acridine mustard ICR170, which is believed to cause frameshift mutations. The *pyr-3* region codes for two enzymes: carbamyl phosphate synthetase and aspartate transcarbamylase. Both functions appear to be controlled by a single protein, so the *pyr-3* region is to be regarded as only one gene. Radford (1969b) found that, unlike *pyr-3* mutants induced by ultraviolet light or nitrosoguanidine, none of the ICR170 mutants lacked the synthetase but possessed the transcarbamylase activity. He inferred that any *pyr-3* frameshift mutation which destroys the synthetase activity simultaneously destroys the transcarbamylase activity, and therefore the synthetase must be controlled by the part of the gene that is translated first, and the transcarbamylase by the later part. In agreement with this conclusion, all mutants, however obtained, which lacked the transcarbamylase activity but not the synthetase activity, mapped near the end of the gene translated last as defined by the polarity of complementation. Evidently the protein specified by *pyr-3* has its two functions at least partially restricted to different parts of the primary structure.

The histidine biosynthetic pathway in *Neurospora crassa* involves the same steps as in *Salmonella typhimurium*, but unlike *Salmonella* where the genes coding for these enzymes form an operon (Fig. 15.1(*b*)), most of the steps in *Neurospora* are catalyzed by the products of genes at scattered positions in the linkage groups. Ahmed, Case and Giles (1964) found, however, that the *histidine-3* region in the right arm of linkage group I of *N. crassa* specified enzymes concerned with steps 2, 3 and 10. Many *hist-3* mutants were defective in only one of these enzymes, complemented those defective in either of the other two, and mapped in discrete segments of the *hist-3* region in the sequence 3–2–10. Fink (1966) found a similar situation in *Saccharomyces cerevisiae*, where the corresponding region is called *hist-4*. In both fungi, functional integration of the 3–2–10 region was indicated by the occurrence of non-complementing mutants lacking all three activities. These mapped in segment 3. Mutants also occurred which were defective in steps 2 and 10, and which complemented only with those defective in step 3. These mapped in segment 2. Mutants in these two classes were interpreted as polar mutants, translation being in the sequence $3 \rightarrow 2 \rightarrow 10$. In support of this conclusion, no mutants were found which were defective in steps 2 and 3 but which were not defective in 10. Polarity was confirmed for the *hist*-4 region in *S. cerevisiae* by the suppression of many of the supposedly polar mutants, but not other kinds of mutants, by an amber suppressor. Ahmed *et al.* and Fink favoured an operon to explain their results. Catcheside (1965), however, found that mutants defective for all the *hist*-3 activities in *N. crassa* mapped in various positions, and those defective for one function overlapped in map position those defective for another. He favoured a single polypeptide as the *hist*-3 product with several enzymic functions. Polarity can arise on the one gene hypothesis from the polarized defect in a single polypeptide and from polarized complementation of allelic mutants. On the operon hypothesis, polarity results from the effects of a stop mutant on the translation of a neighbouring gene with a common messenger. Minson and Creaser (1969) purified the *hist*-3 product and obtained evidence that it was a single trifunctional protein. The one gene hypothesis therefore seems likely to be true.

Giles, Case, Partridge and Ahmed (1967) studied mutants of *N. crassa* having a requirement for aromatic amino-acids. There are thought to be 7 steps in the *arom* pathway: of these, steps 2–6 were found to be controlled by a region in the right arm of linkage group II. By contrast, Gollub, Zalkin and Sprinson (1967) found that these steps in *Salmonella typhimurium* were controlled by genes that were not closely linked. Giles *et al.* found that many of the mutants in the *arom*, steps 2–6, genetic region of *Neurospora* were deficient for only one of the 5 enzyme activities and mapped in a particular part of the region. This allowed the map shown in Fig. 15.3 to be constructed. Some of the *arom* 2–6 mutants, however, were deficient for all five activities and seemed to be polar, because those that mapped in the *arom*-2 segment at the right-hand end of the region showed no complementation with any of the single-deficiency mutants, and those that mapped in internal segments of the region mostly showed complementation only with single-deficiency mutants mapping to their right. Transcription and translation were inferred to proceed from right to left on the map. Polarity was confirmed by the suppression of some of the supposedly polar mutants, but not the single-deficiency mutants, by a probable suppressor of one class of stop mutants.

Density-gradient centrifugation revealed that the enzyme activities coded by the *arom* region form an aggregate with a molecular weight of about 200,000. The polar mutants, unlike the single-defect mutants, gave much smaller molecular weights: evidently stop mutants prevent the formation of the normal aggregate and so block all 5 steps, even though, as shown by the complementation data, normal enzyme molecules are usually formed by genetic segments situated to the right of that containing the polar mutation.

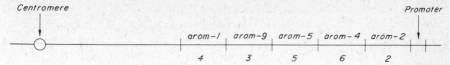

FIGURE 15.3   The diagram is a map of the *arom* region in the right arm of linkage
group II of *Neurospora crassa* from the work of Giles, Case, Partridge
and Ahmed (1967) and Rines, Case and Giles (1969). The names
of the segments are given above the line, and the sequence in which
their products act in the biosynthetic pathway leading to the aromatic amino-acids is indicated by the numerals below the line.

Initially no mutants were found which were deficient only for the step 3
enzyme. This step is the conversion of 5-dehydroquinic acid to 5-dehydro-
shikimic acid by dehydroquinase. All mutants with reduced levels of this
enzyme had marked reductions of the other four as well. From biochemical
studies Giles, Partridge, Ahmed and Case (1967) concluded that *N. crassa*
has two dehydroquinases—a constitutive enzyme in the synthetic pathway
coded by a segment in the *arom* region, and an inducible enzyme in a degradative pathway coded by a gene elsewhere. The two enzymes were
separated by sucrose gradient density centrifugation and found to differ in
thermolability. The discovery of a mutant non-inducible for the catabolic
dehydroquinase enabled Rines, Case and Giles (1969) successfully to select
for mutants in the predicted segment (*arom*-9) for the synthetic (constitutive)
enzyme.

Giles, Case *et al.* favoured a gene cluster as the basis of the *arom* region.
They pointed out that the presence of an aggregate of enzyme activities
distinguishes the *arom* region from most bacterial operons. Moreover,
there is no evidence for regulatory genes or an operator in connection with
the *arom* region. An operon, however, is probably best defined in terms of
one messenger for more than one polypeptide. The polar mutants of the
*arom* cluster indicate a single messenger. So, according to Giles's hypothesis,
the *arom* region consists of 5 genes forming an operon. Another possibility
is that the product of the region is a single polypeptide with multiple functions, in which case *arom* is only one gene. Until it has been possible to
establish whether the enzyme aggregate arises from a single polypeptide or
from an association of up to 5 different ones, the question of whether the
*arom* region is a gene or an operon remains open.

Giles *et al.* suggested that the *arom* aggregate of enzyme activities provides
a channelling mechanism for separating two potentially competing pathways
—one synthetic and one degradative—which have one step in common. The
occurrence of the multifunctional *arom* genetic region may therefore be
related to the existence of these two pathways. Aggregates of enzyme activities, however, are not necessarily associated with a single genetic region.
DeMoss, Jackson and Chalmers (1967) showed that mutations in either of
the genes *tryp-1* and *tryp-2* of *N. crassa*, which are in linkage groups III and
VI respectively, can reduce the size and activity of an enzyme aggregate
concerned with steps 1, 3 and 4 of the tryptophan pathway. The *tryp-2* gene

codes for anthranilate synthetase, which catalyzes the first step, but interaction with the *tryp-1* gene product is required for expression of this enzyme activity. On the other hand, such interaction is not required for expression of the two enzyme activities (steps 3 and 4) of the *tryp-1* product.

Another fungal example of a possible operon is provided by spore colour mutants mapping in segment 29 of *Podospora anserina* studied by Picard (1971). This fungus has binucleate ascospores, so complementation between spore colour mutants is easily tested. Picard compared the complementation and genetic maps for segment 29. She found polarized complementation and distal localization of non-complementing mutants consistent with transcription and translation beginning at the distal end of the segment with respect to the centromere. Two hypotheses were considered: one gene with interallelic complementation, or an operon of 5 genes. In the absence of evidence as to whether the product of segment 29 is one polypeptide or more than one, it is not possible to decide between these alternatives.

An example of genes of related function showing linkage but not forming an operon is provided by the work of Cove and Pateman (1969) on nitrate reduction in *Aspergillus nidulans*. Nitrate is used by this fungus as a source of nitrogen, following reduction first to nitrite and then via hydroxylamine to ammonium ions. On the basis of their results, Cove and Pateman suggested that a regulatory gene produces an activator (inducer) necessary for the synthesis of nitrate reductase and nitrite reductase from their respective structural genes. The nitrate reductase protein is believed to convert the activator into a repressor of the synthesis of both nitrate reductase and nitrite reductase. Nitrate acts as an inducer, and is believed to do so by combining with the nitrate reductase protein and preventing it from functioning as a co-repressor. The combination of positive and negative control is similar to that proposed by Englesberg *et al.* (1965) for the arabinose enzymes in *E. coli* (see § 15.7). Unlike that system, however, the regulatory gene for the nitrate pathway in *A. nidulans* is only loosely linked to the structural genes, which also show 10% recombination with one another. Clearly there is no operon, each structural gene evidently having its own regulatory sites.

Some authors have suggested that the human $\beta$ and $\delta$ haemoglobin polypeptides are the products of an operon. In the normal adult, 97·5% of the haemoglobin contains two $\alpha$ chains and two $\beta$ chains and 2·5% two $\alpha$ and two $\delta$ chains (Ingram and Stretton, 1961). Mutant forms of the $\alpha$ chain show no linkage in inheritance with mutant forms of $\beta$ (see § 13.7), but studies of the inheritance of mutant forms of $\beta$ and $\delta$, when these occurred in the same family, showed that the $\beta$ and $\delta$ genes are closely linked. The $\beta$ and $\delta$ polypeptides differ in only 10 of their 146 amino-acids: at positions 9, 12, 22, 50, 86, 87, 116, 117, 124 or 125, and 126. Moreover, Baglioni (1962) showed that in haemoglobin Lepore recombination had occurred within the $\beta$ and $\delta$ genes: a single hybrid polypeptide was formed having the normal number of amino-acids but with the $\delta$ sequence at the amino end and the $\beta$ sequence at the carboxyl end. The forms of Lepore haemoglobin known from people of New Guinean, West African and Mediterranean descent differ from one another, and are called Hollandia,

Baltimore and Boston, respectively, after the towns where they were first discovered. In the New Guinea form of the disease the recombination has occurred in the gene segments corresponding to the region between amino-acids 22 and 50, in the West African form between 50 and 86 (Ostertag and Smith, 1969) and in the Mediterranean form between 87 and 116 (see Labie, Schroeder and Huisman, 1966). The three forms have evidently arisen independently by separate recombination events (Fig. 15.4). The occurrence of the δ segment at the amino end of the polypeptide and hence at the 5' end of the gene (expressed in terms of the messenger RNA) suggests that β and δ may be contiguous genes orientated the same way and with δ before β. A

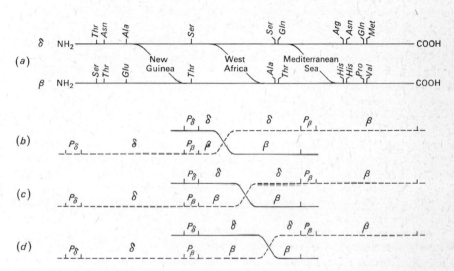

FIGURE 15.4   Diagram to illustrate the structure and proposed mechanism of origin of the hybrid δβ polypeptide of Lepore haemoglobin in people of New Guinean, West African and Mediterranean descent.

(a) The annotations on the horizontal lines show the nature and positions of the 10 amino-acid differences between the δ and β polypeptides of normal haemoglobin, each of which contain 146 amino-acid residues. The abbreviations of amino-acids correspond to Fig. 13.2. The oblique lines show the positions of the switches from the δ to the β sequence in the 3 forms of Lepore haemoglobin.

(b–d). The diagrams show the proposed explanation for the δβ polypeptides in the New Guinean (b), West African (c) and Mediterranean (d) Lepore haemoglobins. The lines indicate chromatids at the pachytene stage of meiosis. The δ gene precedes the β gene, displaced pairing has occurred, and a crossover has taken place between δ in one chromatid and β in the other. The position of the crossover determines which kind of δβ hybrid shall be generated. The δβ crossover chromatid is shown with a continuous line and the other crossover chromatid, which has the triple gene sequence δ – βδ – β, with a broken line. Its occurrence in human populations has not been demonstrated. $P_\beta$ and $P_\delta$ denote the promoters of the β and δ genes, respectively.

condition called hereditary persistence of foetal haemoglobin shows depression of both $\beta$- and $\delta$-chain production: in the form of this condition which occurs in Greece both $\beta$ and $\delta$ production are moderately reduced, while in its African form both $\beta$ and $\delta$ polypeptides are completely absent. The mutations responsible show close linkage to the $\beta$ and $\delta$ structural genes. This has led to the proposal that $\delta$ and $\beta$ might form an operon.

Other hereditary conditions, however, give contrary evidence. In the disease called thalassaemia the synthesis of either the $\alpha$ or the $\beta$ polypeptide is lowered or suppressed. The disease is usually lethal when homozygous, but nevertheless is of widespread occurrence in the Mediterranean countries and the tropics, apparently because like sickle-cell anaemia (§ 2.6) the heterozygote has greater than normal resistance to malaria. In $\beta$-thalassaemia, besides the reduction in synthesis of $\beta$-chains, there is an increased synthesis of $\delta$ chains, both in cis and trans. This means that if the $\delta$ and $\beta$ genes form an operon, there must be a control mechanism at translation rather than transcription, in order to allow increased synthesis of the $\beta$ polypeptide relative to $\delta$. The mutation responsible for $\beta$-thalassaemia shows close linkage to the $\beta$ and $\delta$ genes, although recombination has been detected with $\delta$ (see review by Motulsky, 1965). One possibility, assuming $\beta$ and $\delta$ are not part of an operon, is that $\beta$-thalassaemia is due to a mutation in the promoter of the $\beta$ structural gene, thereby reducing its affinity for RNA polymerase. Support for this hypothesis has been obtained by Clegg *et al.* (1968) who found by pulse-labelling that the rate of $\beta$ polypeptide synthesis is normal in $\beta$-thalassaemics. By labelling polypeptides with $^{35}$S-methionine in reticulocytes from a $\beta$-thalassaemia homozygote, Nathan *et al.* (1971) were able to show that the $\beta$ polypeptides were initiated, as well as elongated, at the normal rate. They concluded that $\beta$-thalassaemia is caused by a decrease in the amount of $\beta$ messenger RNA, either through decreased synthesis or decreased stability.

Haemoglobin Lepore is associated with the clinical and haematological features of thalassaemia and White *et al.* (1972) discovered an explanation for this. They found that in a Boston heterozygote the $\delta\beta$ hybrid polypeptide was not synthesized in the reticulocytes. This is similar to the normal behaviour of the $\delta$ polypeptide, synthesis of which stops at an earlier stage of red cell maturation than does the $\beta$ polypeptide, with the result that little or no $\delta$ is synthesized by the time the reticulocyte stage is reached. With the $\delta$ gene situated before the $\beta$ gene, as inferred above, the $\delta\beta$ hybrid polypeptide will have the $\delta$ and not the $\beta$ regulatory sites (Fig. 15.4). The level of $\delta\beta$ hybrid polypeptide in Lepore haemoglobin would therefore be expected to be low, like that of $\delta$ in normal haemoglobin. This low level of synthesis would account for the clinical features of thalassaemia. The amount of $\delta\beta$ synthesized was, in fact, ten-fold greater than that of $\delta$ in normal haemoglobin. White *et al.* suggested that, owing to the $\beta$ nucleotide sequence at its 3′ end, the $\delta\beta$ messenger RNA might be more stable than $\delta$ messenger.

There are two other human haemoglobin polypeptides: $\gamma$ formed in foetal life and $\varepsilon$ in embryonic life. Each is thought to be determined by a different gene, but it is not known whether the $\gamma$ and $\varepsilon$ genes show linkage to any of the other haemoglobin genes. Like $\beta$ and $\delta$, the $\gamma$ and $\varepsilon$ chains normally function in pairs associated with a pair of $\alpha$ chains. Baglioni and Campana (1967) believe that $\beta$ chains are not released from polyribosomes until an $\alpha$ chain has combined with the $\beta$. This would provide a mechanism for co-ordinating $\alpha$- and $\beta$-chain production. The mechanism of compensation, however, between $\beta$, $\gamma$ and $\delta$, by which reduced output of one of these polypeptides is often associated with increased output of another, is not understood.

There is evidence (references below and in H. R. Adams *et al.*, 1969) from at least 11 species of mammals (*Oryctolagus cuniculus, Mus musculus, Equus caballus, Bos buffelus, Odocoilus virginianus, Capra hircus, Ammotragus lervia, Macaca irus, Gorilla gorilla, Pan troglodytes* and *Homo sapiens*) for 2 genes for the $\alpha$ polypeptides of haemoglobin, with up to 9 differences in their amino-acid residues. In *E. caballus* (Clegg, 1970), *G. gorilla* and *P. troglodytes* (Boyer *et al.*, 1971) and *H. sapiens* (Ostertag, von Ehrenstein and Charache, 1972) the 2 $\alpha$ genes are believed to be closely linked, possibly adjacent. As with $\beta$ and $\delta$, however, there is no good reason to regard the pair of genes as forming an operon. In man (Beaven *et al.*, 1972) and the apes (Boyer *et al.*, 1971) some individuals seem to lack the second $\alpha$ polypeptide. Recombination following displaced pairing (§ 9.10 and Fig. 15.4) might account for this, as it would generate variable numbers of similar genes, and so could give rise to populations polymorphic for gene number. Adjacent genes, differing only slightly in nucleotide sequence are also inferred for at least one of the polypeptides of the human antibody molecule, where recombination following displaced pairing has also been demonstrated (Kunkel, Natvig and Joslin, 1969. See § 18.12). Two genes are also known for human haemoglobin $\gamma$ polypeptides (Schroeder *et al.*, 1968) and for pro-insulin* in *Rattus norvegicus* (Clark and Steiner, 1969). In these last two instances the pair of polypeptides differ in only one amino-acid residue.

Three main conclusions may be drawn about the possible occurrence of operons in chromosomal organisms. First, no example of an operon has been conclusively established. Secondly, several examples are known of a single polypeptide with more than one enzymic function, so there is a real possibility that this will also be the explanation of situations where the gene *versus* operon alternatives have not yet been resolved. Thirdly, examples are known of neighbouring and very similar genes concerned with the same function, and yet varying, in at least one instance, in their relative activity. There thus seems little doubt that operons, if they occur at all in chromosomal organisms, must be quite rare.

Although in chromosomal organisms the genes concerned with each biochemical pathway are usually widely scattered through the chromosomes, nevertheless, functionally related genes are not distributed completely at random. Elston and Glassman (1967) considered whether genes which control morphological structure are clustered on chromosomes. Some 850 genes of *Drosophila melanogaster*, when classified by chromosome and by body

* The two polypeptides of the insulin molecule are formed by enzymic removal of the middle part of the pro-insulin polypeptide.

part affected, including the nature of the change, revealed a tendency, significant at the 1% level, for genes of a kind to occur in the same chromosome. An analysis for 3 classes of genes affecting the eye (colour, shape, texture) and 5 affecting the wing (curvature, length, margin, position, veins) in terms of their positions on the linkage maps showed a significant clustering of genes of similar effect, but when gene sequence was considered without reference to recombination frequencies the genes affecting the same organ were randomly distributed. Thus the clustering of functionally related genes is due entirely to the clustering of all genes. This does not explain, however, the tendency for genes which affect the same structure to occur in a particular chromosome.

Genes affecting the same pathway or the same organ might be expected to be closely linked on functional grounds, in relation to such processes as co-ordinate control of activity or enzyme aggregation. The rarity of such linkage in chromosomal organisms is remarkable and suggests that an opposing force, such as the retention in evolution of particularly favourable combinations of genes of diverse function, may somehow have been of paramount importance.

# 16. The hybrid DNA theory of genetic recombination

## § 16.1 Introduction

When recombination between alleles was first discovered (in *Drosophila* and *Aspergillus*), it was assumed that its mechanism was the same as between different genes. The recombinant individuals, selected because they were wild-type for the character shown by the pair of alleles, usually had one specific new combination of outside marker genes (see § 9.11 and § 13.2), such as is expected with crossing-over*. Moreover, reciprocal recombinant genotypes were observed. In *Drosophila* these were from different meioses, but in diploid strains of *Aspergillus*, Pritchard (1955) found several instances of complementary recombinants in the two homologous chromosomes of the diploid, following mitotic recombination between alleles (see § 13.3.) These observations appeared to establish that crossing-over could occur between alleles.

This conclusion was supported by theoretical considerations. The discovery that the linkage map of a series of alleles was collinear with the genetic map of the genes in the linkage group (§ 13.2) suggested that there might be no physical discontinuity between successive genes in the chromosome. If cistrons were segments of a DNA molecule, recognizable from one another merely by their characteristic nucleotide sequence, then it might be expected that the recombination process, whatever its mechanism, would bear no relation to the points of junction of one cistron with another and would be the same irrespective of whether the mutant sites happened to be within the same cistron or not. However, later work, notably with fungi belonging to the Ascomycetes, revealed that recombination between alleles can occur by mechanisms other than crossing-over.

## § 16.2 Recombination in Ascomycetes

As described in § 8.6 with reference to *Neurospora*, the products of meiosis in Ascomycetes are held together in a large cell, the ascus. By isolating the spores from individual asci, detailed information about the process of recombination can be obtained, if there is segregation for linked character-differences.

---

* The term 'crossing-over' is used in genetics in two different senses: (1) for the process of reciprocal exchange between homologous chromatids or DNA molecules, and (2) as a synonym of recombination, irrespective of whether the exchange process is reciprocal or not. In this book 'crossing-over' is used exclusively in the first sense.

It was by this means that Lindegren (1933) showed that crossing-over occurs between chromatids one from each homologue and that the centromeres segregate at the first division of meiosis (§ 8.6). Subsequently (Lindegren, 1953), he obtained evidence that in *Saccharomyces cerevisiae* a mutant character, such as inability to ferment α-methyl glucoside, sometimes segregated from its normal allele at meiosis so as to give 3 normal and 1 mutant spores in the ascus, or 1 normal and 3 mutant, instead of the expected 2 of each kind. Lindegren followed Winkler (1930) in calling this phenomenon gene *conversion*. This discovery was so unexpected that it did not gain general acceptance until Mitchell (1955) demonstrated its occurrence

PLATE 16.1   Photographs of asci of *Sordaria brevicollis* from a cross between the wild-type, which has black spores, and a yellow-spored mutant. (*a*) Cluster of asci from one perithecium. Immature asci have pale spores. A majority of the mature asci have four wild-type (black) and four paler (yellow) spores, but one (marked C) shows conversion to wild-type, with six black and two yellow spores. (*b*) An ascus showing conversion to the mutant genotype, that is, with two black and six yellow spores. (*c*) An ascus showing postmeiotic segregation, with three black and five yellow spores.

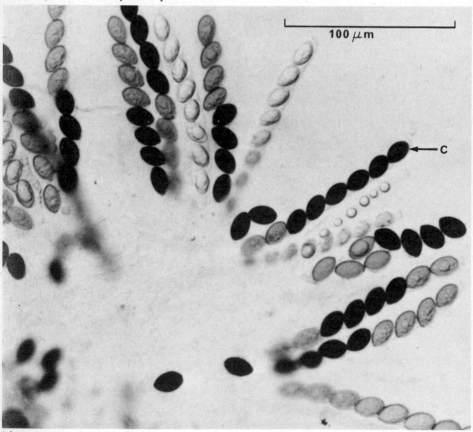

(*a*)

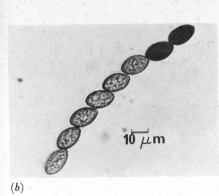

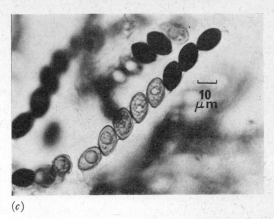

(b)                                             (c)

with pyridoxin-requiring mutants of *Neurospora crassa*. She made use of
linked marker genes on either side of the locus under study, such as were not
available in *Saccharomyces*, and so was able to show that the 3:1 ratios at
the pyridoxin locus were associated with normal 2:2 ratios for both of the
linked character-differences. This established that the 3:1 ratios were due to
abnormal segregation during meiosis at a particular mutant site in the
chromosome, rather than to abnormal behaviour of whole chromosomes or
nuclei.

Many examples of conversion have since been discovered. It is most
readily detected with spore characters since the abnormal segregation can
then be directly observed in the ascus (Pl. 8.1(*d*) & 16.1(*a*), (*b*)). Indeed, it
had been observed by Zickler (1934) using spore-colour mutants of *Bombardia
lunata*, but his observations had been neglected. Olive (1959) examined 2700
asci of *Sordaria fimicola* heterozygous for hyaline *versus* normal black ascospore
colour and found 6 asci with 6 black and 2 hyaline spores, and 5 asci with
2 black and 6 hyaline spores. A mutant (*g*) with grey spores showed not
only 6:2 and 2:6 ratios for spore colour in heterozygous asci but also
5:3 and 3:5 ratios. This was a remarkable discovery, as it implied that
segregation of the character-difference had occurred at the mitosis after
meiosis (*postmeiotic segregation*), (Pl. 16.1(*c*)).

Kitani, Olive and El-Ani (1962) discovered another class of asci showing
postmeiotic segregation. These asci had 4 black and 4 grey spores but the
sequence was abnormal, with half the spores not in pairs. Such arrangements
in the ascus can arise from the overlapping of two of the spindles at the third
division such that the nuclei pass one another (see Table 8.2, p. 106). However,
Kitani *et al.* made use of linked marker genes on either side of the grey spore
locus, and this enabled them to distinguish between abnormal behaviour of
whole nuclei (spindle overlap) and abnormal behaviour at the site of the
spore mutant. The method of distinction is shown in Fig. 16.1 by plus and
minus signs. These represent the behaviour of a linked character-difference
of the mycelium which could be recognized after dissecting the asci and
isolating and germinating the spores. With spindle overlap, the linked

character will show the same aberrant sequence as the spore character, but with postmeiotic segregation at the mutant spore site it will not. Moreover, the postmeiotic segregation may give spore sequences such as 1 grey, 4 black, 3 grey, which are unlikely to arise from spindle overlap. Kitani *et al.* examined over 200,000 asci segregating for black and grey spores and found that the 5 classes of aberrant asci resulting from abnormal segregation at the *g* site occurred with widely different frequencies. The numbers in each class expected in a sample of $10^5$ asci, according to their data, are shown below the drawings in Fig. 16.1. These numbers include other spore sequences in addition to those shown. The other sequences differ in spindle orientations or in the division of meiosis at which segregation occurred (see § 8.6). The number of alternative sequences which could arise in these ways is shown above the drawings.

A notable feature of the data of Kitani *et al.* is that about 36% of the aberrant asci (classes *c–g* of Fig. 16.1) showed crossing-over between the

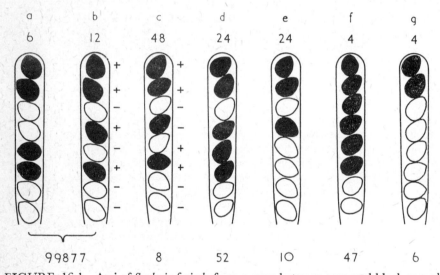

FIGURE 16.1    Asci of *Sordaria fimicola* from a cross between a normal black-spored strain and a mutant with grey spores. The grey spores are shown white in the diagram. The figures below the drawings of asci show the numbers of each kind of ascus expected in a random sample of 100,000 asci, according to counts made by Kitani *et al.* (1962). These numbers refer not only to the spore sequences shown but also to any permutations of them which may arise from variation in the orientation of the appropriate pair of chromosomes on the spindle at any of the nuclear divisions in the ascus, or from the presence or absence of crossing-over in the interval of the chromosome between the gene for spore colour and the centromere. The number of such permutations is shown above the drawings. The plus and minus signs in (*b*) and (*c*) show how a character-difference of the mycelium has segregated. This character-difference becomes evident after the spores germinate, and is due to a gene linked to that for spore colour.

marker genes on either side of the *g* site. Since these genes were closely linked, with only about 4% of recombination, it is evident that the aberrant segregation at the *g* site shows a highly significant correlation with the occurrence of crossing-over in the immediate neighbourhood. Furthermore, in 42 instances out of 46 the two chromatids involved in the crossing-over were the same as those involved in the aberrant segregation. Nevertheless, a majority of the aberrant asci (64%) showed the parental combinations of the outside marker genes. The less extensive data obtained by others, using the more laborious methods needed to study conversion of nutritional instead of spore characters, have given similar results. This is exemplified by the data of Stadler and Towe (1963) with allelic cysteine-requiring mutants of *Neurospora crassa*.

Some further remarkable discoveries concerning recombination were made by Lissouba and Rizet (1960) using *Ascobolus immersus*. This fungus normally has dark brown ascospores, but mutants are known with colourless spores. The spores are forcibly discharged from the asci when mature, and can be collected on a sheet of glass in clusters of 8, corresponding to individual asci. When mutants were crossed pairwise, some were found to be due to mutation at closely-linked sites, since almost all the resulting asci had pale spores only. Five mutants of spontaneous origin identified as *W*, and nos. 46, 63, 137, and 188, respectively, formed one such group, which was called series 46. These mutants were presumed to be allelic. The spores from a total of about 84,000 asci were examined from the 10 pairwise crosses between the mutants and 440 of the octads (groups of 8 spores from individual asci) were found to contain wild-type (dark) spores. Two of these octads had 4 dark and 4 pale spores and the remainder had 2 dark and 6 pale. It was evident that the dark spores were arising through recombination, because a total of nearly $10^6$ asci homozygous for individual mutants showed none. The frequencies of recombinants from each of the 10 crosses between the different mutants are shown in Fig. 16.2. The recombination frequencies fit a linear map satisfactorily when the mutants are placed in the sequence shown in the diagram, but the larger frequencies are somewhat irregular and tend to be more than the sum of the appropriate smaller ones.

A total of 120 of those octads which contained 2 dark and 6 pale spores were studied further. The spores were germinated and their genotype determined by crossing each with the parental genotypes. The dark spores were found to be wild-type as expected. The pale ones invariably gave a few wild-type recombinants with one or other parent but not with both. Double mutant spores would have been revealed, if present, because they would have given no recombinants with either parent, but none was found. Thus all the recombinant (wild-type) spores appeared to have arisen by a non-reciprocal event and not by crossing-over.

Furthermore, in each of these octads 4 pale spores always had the genotype of the parent whose mutant site was to the left on the map, as drawn in Fig. 16.2, and the other 2 pale spores had the genotype of the right-hand parent. For example, in a cross between mutants 188 and *W*, each ascus with 2 wild-type spores was found to have 4 spores of the same genotype as

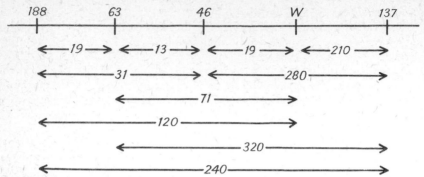

FIGURE 16.2   A recombination map of 5 pale-spored mutants belonging
to series 46 in *Ascobolus immersus*, according to the data of
Lissouba and Rizet (1960). The reference letters or numbers
of the mutants are given above the line. The figures below
the line show the frequencies of dark (wild-type) spores per
100,000 progeny ascospores from each pairwise cross.

188 and 2 like *W*, while if *W* was crossed with 137, there were 4 like *W* and
2 like 137 (see Fig. 16.3). It was as if in every tetrad, instead of 2 products of
meiosis with the genotype of the left-hand mutant and 2 like the right-hand
mutant, one of the latter had been transformed into wild-type without
affecting the others. The agreement with the recombination map implies
that a map of the mutant sites based on this difference between left- and
right-hand positions (polarity) would place the sites in the same sequence as
in Fig. 16.2. This polarity in recombination is distinct from the polarity in
gene expression discussed in § 15.3.

   Polarity in recombination was observed by Siddiqi (1962) at the *paba*-1
locus in *Aspergillus nidulans* and by Murray (1963) and Stadler and Towe
(1963) at the *me*-2 and *cys* loci, respectively, in *Neurospora crassa*. Murray
(1968) showed that the direction of polarity at the *me*-6 locus in *N. crassa*
was reversed relative to the centromere when the gene was transferred to a
paracentric inversion (§ 8.15). This showed that the direction of polarity
was a property of the *me*-6 region and was not imposed by the centromere.
Direct comparison between the data for these nutritional mutants and the
*Ascobolus* spore mutant data is not possible, since the techniques for observing
recombination differ, but it appears that there are a number of differences of
detail. Such variation is evident also between different genes in *Ascobolus*,
where Lissouba, Mousseau, Rizet and Rossignol (1962) found that the
pattern of recombination within two other series of spore colour mutants
(nos. 19 and 75) differed in a number of respects from series 46. Diverse
proportions of reciprocal and non-reciprocal recombination were found with
different mutants in the same series. This confirmed the discovery by Case
and Giles (1958) that both reciprocal and non-reciprocal recombination
could occur between a pair of alleles. Their example was at the pantothenic
acid 2 (*pan*-2) locus in *N. crassa*.

   A feature of recombination which any hypothesis of its mechanism must account for is the phenomenon of *negative interference*. A majority of the adenine-independent progeny obtained by Pritchard (1955) from crosses between allelic adenine-requiring mutants of *Aspergillus nidulans* had one specific crossover genotype for linked marker genes on either side of the adenine locus, for example, $y + + bi$ from the heterozygote $\dfrac{y + \; 8 +}{+ 11 + bi}$ (see § 13.2). However, a significant minority showed other combinations of the outside markers ($y + + +$, $+ + + bi$, and $+ + + +$). Giles (1952) had first described this phenomenon with inositol-requiring alleles of *N. crassa*, and it was observed with alleles at the pyridoxin-1 (*pdx-1*) and nicotinic acid 1 (*nic-1*) loci in this species by Mitchell (1956) and St Lawrence (1956), respectively. Among the recombinant progeny which did not require

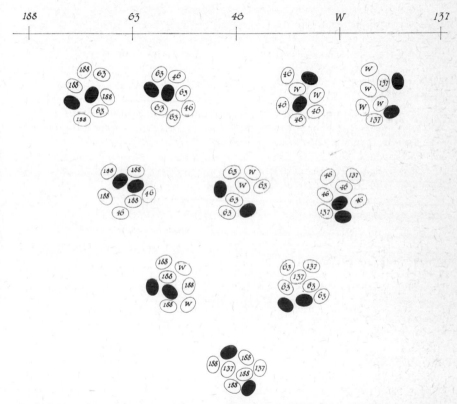

FIGURE 16.3   Diagram to illustrate the genotype of the spores of recombinant asci of *Ascobolus immersus* found by Lissouba and Rizet (1960). At the top is shown the linkage map of 5 pale-spored mutants belonging to series 46, from the data given in Fig. 16.2, and below the map are drawn sets of 8 spores from individual recombinant asci, one from each of the 10 pairwise crosses between the mutants. The black spores are wild-type. The genotype of each pale spore, established from backcrosses to both parents, is shown.

pyridoxin and nicotinic acid, respectively, for growth, not only were all 4 of the genotypes with respect to linked marker genes represented, but one crossover genotype was not markedly the most frequent class such as it would be expected to be with simple crossing-over between alleles. Thus, Mitchell found that 111 pyridoxin-independent progeny consisted of 43 and 28 with one or other parental genotype, respectively, for the outside markers, and 31 and 9 with one or other crossover genotype, respectively. Thus, less than one third of the recombinants could be attributed to a simple crossover between the alleles. St Lawrence found an even more extreme instance: 72 nicotinic acid-independent progeny from a cross between allelic mutants nos. 1 and 2 consisted of 25 and 14, respectively, with the two parental combinations of the outside marker genes, and 18 and 15, respectively, with the two crossover combinations.

Recombination between alleles thus appeared frequently to be associated with recombination in the neighbouring regions demarcated by the linked marker genes. Later work has confirmed the general occurrence of this negative interference between recombination in closely linked regions. This is in direct contrast to the positive interference found between recombination in longer intervals of the linkage map, where it has been well-established ever since the pioneer work of Muller (1916) that crossing-over in one region frequently reduces the likelihood of its simultaneous occurrence in a neighbouring region (see § 8.7). Pritchard (1955) found that negative interference was strictly localized in extent. However, Calef (1957) showed that it was not confined to alleles. In a cross between adenine-requiring mutant no. 15 and para-aminobenzoic acid requiring mutant no. 1 at the closely-linked *ad*-9 and *paba*-1 loci in *Aspergillus nidulans*, he found that out of 105 recombinants which did not require adenine or *p*-aminobenzoic acid for growth, 49 had additional recombination in one or both neighbouring regions.

## § 16.3   Recombination in bacteria

As indicated in § 10.2, the occurrence of recombination in bacteria was discovered by Lederberg and Tatum (1946) using biochemically deficient mutants of *Escherichia coli*. Two multiple mutants, one having requirements for threonine, leucine and thiamin, and the other for biotin, phenylalanine and cystine, had been obtained by successive treatment with X-rays. The two triple mutants were grown in mixed culture in complete medium containing yeast extract, peptone and glucose, and then suspensions of the washed cells were plated on minimal medium. Wild-type (prototrophic) colonies appeared with a frequency of about 1 per $10^6$ cells. This is a much greater frequency than could be attributed to mutation since simultaneous reverse mutation of all three mutants in one or other parent would be required to produce a wild-type colony. Control experiments showed no wild-types when the parents were cultured separately. It was also established by experiment that the prototrophs were not mere mixtures or heterokaryons.

By an extension of the original technique, the recombination of unselected character-differences such as resistance to streptomycin and to particular viruses was studied. Nutritional mutants could be included in the un-selected category by plating on a medium with the appropriate supplements. The colonies which developed on this medium were then scored for the unselected character-differences. In this way data on the relative frequencies of recombination between the various mutants were obtained. The re-combination frequencies were found to be non-random and to fit a single circular linkage map (Jacob and Wollman, 1958a). A similar linkage map has been obtained from recombination experiments with *Salmonella typhimurium*, and with *Streptomyces coelicolor* (Hopwood, 1967).

The mechanism of transfer of genetic material from cell to cell in *E. coli* has been investigated by a number of authors (cf. Hayes, 1966a, b, 1968). The details of the process are remarkable. It appears that there is sexual differentiation of cells, the two sexes differing in a number of respects. In particular, they are thought to differ in the charge on the cell surface such that cells of opposite sex may stay in contact if they happen to meet. The male cells are inferred to contain a 'sex factor' which evidently carries the genes for the male characters: it is absent from female cells. This sex factor is capable of replication independently of and more rapidly than the chromoneme*. The sex factor appears to consist of enough DNA to carry 100 or more genes, so it could be regarded as a second chromoneme of about $\frac{1}{50}$ the length of the other. The male cells have one or more special hair-like appendages or pili, with an axial hole 2–2·5 nm wide which is thought to function as a conjugation-tube. Through this tube one of the editions of the sex factor may be transferred to the female cell, whereupon the latter changes sex. Alternatively, the circular chromoneme of the male cell may be broken open at a particular point, with the sex factor joined to one specific end of it. One chain of the DNA is then transferred to the female cell, with the free end of the chromoneme leading the way. Very frequently transfer is interrupted before completion as a result of random breakage of the polynucleotide during transfer, with the result that the sex factor is rarely transferred since it is at the tail. Jacob and Wollman (1958a) separated conjugating cells by shaking the culture in a high-speed mixer, and they did this at various time intervals after the start of conjugation, which takes about 100 minutes at 37°C. On the basis of the characters of the male parent which appeared in the progeny, it was possible to find the time at which each gene was transferred. The time sequence map of the genes confirmed the map based on recombination frequencies. Strains of *E. coli* were found to arise, apparently by mutation, in which the sex factor was attached at the other end of the chromoneme, the orientation of the latter during conjugation being reversed so that the sex factor was again at the tail. In other mutants, the circular chromoneme was opened at different points in the linkage group, but the point of opening was constant for any particular mutant.

Jacob and Wollman (1958b) coined the term *episome* for structures

* Chromoneme = bacterial (or viral) chromosome—see § 11.1.

such as the *E. coli* sex factor and phages, which may exist attached to the chromoneme, whereupon they replicate only when it does, or they may exist unattached, when they replicate independently of and usually more rapidly than the chromoneme. The more general term *plasmid* includes episomes and also permanently unattached factors of a similar kind. Campbell (1962) proposed that the mechanism of integration of episomes into the host DNA was by crossing-over, the episome having first taken the form of a closed circle of DNA. Excision of the virus from the host DNA would also occur by crossing-over. This hypothesis has now been established, both for the *E. coli* sex factor (Hayes, 1966*a*, *b*) and for phages λ and φ80 (Signer, 1968; Echols, 1971). With λ and φ80 the evidence includes the demonstration that the linear order of genes in the prophage is a circular permutation of the order in the free virus, and the finding that bacterial genes on opposite sides of the prophage integration-site are farther apart (as measured by recombination) when the provirus is present (see Fig. 16.4). Integrative recombination is discussed in § 16.5.

Two other methods of transfer of hereditary material from cell to cell are known in bacteria, besides conjugation. In genetic transformation, which was discovered by Griffith (1928) in *Diplococcus pneumoniae* (see § 11.2), and has since been shown to occur in *Haemophilus influenzae* and a number of other bacteria, segments of DNA derived from the bacterial cells and equivalent to about $\frac{1}{200}$ of the total hereditary complement, are taken up by other cells of the same or a closely related species. These cells seem to absorb the DNA only during a particular phase of the cycle of cell division.

In *transduction*, which was discovered by Lederberg, Lederberg, Zinder and Lively (1952) in *Salmonella typhimurium*, a virus is responsible for transferring some of the bacterial DNA to another cell. In *generalized transduction*, exemplified by virus *P22* of *S. typhimurium*, any part of the *Salmonella* linkage group may be transferred. It seems that the viral coat can be filled with an amount of host DNA equal to that of the virus, and any part of the chromoneme may become inserted into the viral coat and so transferred to another cell (see Ozeki and Ikeda, 1968). In *specialized transduction*, exemplified by virus λ of *Escherichia coli*, only host genes in the immediate vicinity of the viral integration site can be transferred. This comes about through the crossover which releases the virus from the host chromoneme occurring in an abnormal position. For example, in Fig. 16.4(*c*), if a crossover to release the provirus took place between J in λ and the region of host chromoneme immediately below J in the drawing, instead of at the region with thick lines, the circular molecule released would include a segment of host DNA in place of the segment of λ between J and the integration site.

It is evident that in bacteria the process of transfer of hereditary material from one cell to another, whether it is by conjugation, by transformation, or by transduction, differs fundamentally from the nuclear fusion at fertilization characteristic of chromosomal organisms, since in bacteria only a part—and usually quite a small part—of the total hereditary complement of DNA is transferred.

The behaviour of the donor's DNA fragment on arrival in the acceptor

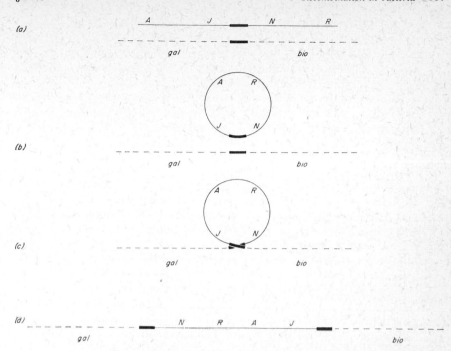

FIGURE 16.4    Linkage maps are shown of virus λ (unbroken line) and part of
that of its host, *Escherichia coli* (broken line). The letters *A*, *J*, *N*
and *R* identify particular genes in λ, and *gal* and *bio* refer to
galactose and biotin genes in *E. coli*. The thick line shows the λ
integration-site. (*a*) Free virus, with its linear map. (*b*) Circular
form of virus, following fusion of the cohesive ends of the DNA (see
§ 18.11). (*c*) Virus integrated with host DNA, as a result of crossing-
over at the integration-site. (*d*) The provirus as part of the same
linear structure as the host DNA. The steps in excision of the virus
would be represented by the same diagrams in the reverse sequence,
that is, (*d*) to (*a*).

cell appears to be similar whatever the means of transfer. Observations
which any hypothesis of the mechanism of recombination must explain
include the following (see Hayes (1968) for fuller discussion and references):

(*a*) A transferred gene from the donor may function at once in the
acceptor cell, for example, in conferring resistance to streptomycin on a
previously sensitive cell, but it does not follow from this that the progeny
of this cell will necessarily inherit the donor character. This phenomenon
is particularly well-known with transduction in *Salmonella* and is called
*abortive transduction*. It appears that the donor DNA fragment fails to
replicate and is therefore inherited by one daughter-cell only. When a
donor character does become fixed in the progeny of the acceptor cell,
it is evident that recombination has occurred such that the donor allele

has become incorporated into the chromoneme of the acceptor cell in place of the acceptor allele.

(*b*) By labelling the donor DNA of conjugating *E. coli* with radioactive phosphorus ($^{32}$P), Jacob and Wollman (1958*a*) showed that, probably before chromonemal replication and certainly less than an hour after transfer, genetic information in the donor fragment has begun to be conveyed to material free of $^{32}$P. This was shown by a slowing down in the rate of loss of recombinants through radioactive breakdown of the $^{32}$P.

(*c*) Isolation of the individual cells formed at each division after conjugation has shown that segregation of a parental character-difference may occur as late as the ninth division.

(*d*) Reciprocal differences in recombination frequency, depending on which parent is the donor, have been reported both with transformation and transduction.

(*e*) Isotope labelling experiments have established that both with conjugation (Siddiqi, 1963) and transformation (Fox and Allen, 1964), recombination can involve physical incorporation of part of the donor chromoneme into that of the acceptor cell.

(*f*) The work of Yanofsky and associates discussed in § 14.13 has shown that recombination in *E. coli* can occur between individual nucleotides of DNA. Some further features of their work are discussed in § 16.7.

Other experimental results from studies on recombination in bacteria are discussed in §§ 16.5–7.

## § 16.4   Recombination in viruses

As already mentioned (§ 13.4), the occurrence of genetic recombination in bacterial viruses was discovered by Delbrück and Bailey (1947) and by Hershey (1947) with the *T*-even viruses of *Escherichia coli*. Evidence was given in § 11.3 for the belief that only the DNA of the virus enters the host cell, and that it there undergoes replication a number of times prior to the formation of new virus particles, 100 or more of which may be formed per bacterial cell. Opportunity for recombination following mixed infection would appear to be confined to the period of DNA replication. The early data on recombination frequencies obtained with these *T*-even viruses from crosses (mixed infections) such as those described in § 13.4, were found to fit linear linkage maps, and more recently these maps have been shown to be part of a single circular linkage map (Epstein *et al.*, 1964—see Chapter 18). This does not necessarily mean that the DNA is in a circular molecule: it seems in fact to be linear with terminal redundancies, that is, the two ends have the same nucleotide sequence; and different individuals have the ends in different places on the map, so that they are circular permutations of one another (see Streisinger, Emrich and Stahl, 1967).

Outstanding discoveries which provide information about the mechanism of recombination in bacterial viruses include the following. Hershey and Rotman (1949) found that if the host is simultaneously infected with 3 different strains of virus $T2$, some of the progeny may have characters from all 3 parents. This would be accounted for if opportunities for recombination occurred repeatedly during the successive replications of the DNA. Hershey and Chase (1952*a*) discovered that about 2% of the virus progeny, following mixed infection by strains with clear and with turbid plaques, had mottled plaques. They established that the mottled plaques were initiated by virus heterozygous for the pair of alleles. It appeared that the heterozygous region was quite short. For every locus studied, a similar frequency (about 2%) of such heterozygotes was found. Using several genetic markers simultaneously, Levinthal (1954) found that heterozygosity at a particular locus was usually associated with recombination between markers situated at a distance on either side, suggesting that the origin of the heterozygosity is related to that of the recombinants. Chase and Doermann (1958) found that with closely-linked mutants of virus $T4$, recombination in one interval appeared to favour recombination in a neighbouring one, or in other words, there was 'negative interference' over short intervals of the linkage map, comparable to that known in chromosomal organisms (see § 16.2).

Meselson and Weigle (1961), by an ingenious experiment, showed that recombination in virus $\lambda$ of *Escherichia coli* could occur independently of replication of the chromoneme. A mixed infection of the host was made using two strains of the virus which differed from one another in two linked mutant characters concerning the morphology and the clarity of the plaque. As a result of multiplication in the host when cultured on appropriate media, the parental viral strains also differed in the atomic weight of their carbon and nitrogen ($^{13}C$ $^{15}N$, and the normal $^{12}C$ $^{14}N$, respectively). The progeny virus particles, resulting from inoculation of normal $^{12}C$ $^{14}N$ *E. coli* with the mixture, were centrifuged in caesium chloride solution to give a density gradient. The contents of the centrifuge tube were then removed drop by drop from the end, and the number of virus particles in each drop estimated by diluting and plating on the host. The genotype with respect to the two parental character-differences was observed for a random sample of the progeny particles in each drop. It was found that the progeny, whether of parental or recombinant genotypes, were concentrated at three positions in the centrifuge tube corresponding to heavy ($^{13}C$ $^{15}N/^{13}C$ $^{15}N$), half-heavy ($^{13}C$ $^{15}N/^{12}C$ $^{14}N$), and light ($^{12}C$ $^{14}N/^{12}C$ $^{14}N$) DNA, indicating 0, 1 (or more), and 2 (or more) replications of originally heavy DNA, respectively (cf. Fig. 12.4). The significant discovery was the occurrence of recombinants with an unreplicated ($^{13}C$ $^{15}N/^{13}C$ $^{15}N$) chromoneme.

Kellenberger, Zichichi and Weigle (1961) obtained similar results using a slightly different technique which did not involve culturing on 'light' or 'heavy' bacteria. As genetic markers they used mutants with abnormal DNA contents, which could be separated by density-gradient centrifugation, and as nucleotide-chain markers they used radiophosphorus ($^{32}P$) and followed the distribution of its radioactivity.

Meselson (1964), in further experiments with virus $\lambda$ labelled with heavy isotopes, has found some indication that a small amount of DNA is removed and resynthesized in the formation of recombinant molecules.

Further studies on recombination in $\lambda$ are described in § 16.5 and § 16.7.

## § 16.5   Hypotheses to explain recombination

Several hypotheses have been put forward to explain the remarkable series of discoveries concerning recombination described in the previous sections. Giles (1952) and Pritchard (1955) suggested that multiple crossover events might be responsible for the phenomena they had observed. If crossovers occurred in clusters, this would explain the negative interference within short regions. Pritchard attributed the positive interference over longer intervals to spacing of the regions where pairing of homologous chromosomes was sufficiently effective to allow crossovers to occur. The effective pairing segments would be variable in position, but rarely close together.

By itself, this hypothesis was clearly inadequate to explain non-reciprocal recombination. Lederberg (1955) suggested that recombination in bacteria could be related to the process of replication of the hereditary material, which might be copied first from one parent and then, further along the chain-molecule, from the other parent. Freese (1957) applied this model to chromosomes. It was suggested that if a daughter-chromatid were formed by a 'conservative' replication process, that is, by the copying of the hereditary material of a parent chromosome so as to leave the latter intact, then a process of copy-choice or switch-synthesis might operate when two homologous chromosomes were closely associated with one another. If the synthesis of the two daughter-chromatids, one from each parental chromosome, was not quite synchronized, both might be copied over a short interval from the same parent, thereby giving rise to a 3:1 ratio within a tetrad at a mutant site. This explanation of non-reciprocal recombination was combined with the multiple crossover idea by postulating that the switching back and forth between the parental templates during the synthesis of the daughter-chromatids occurred several times. This hypothesis explained a number of the observations on recombination, but it suffered from several serious defects:

(1) The conservative method of replication of chromosomes was in conflict with the evidence that both DNA and chromosomes replicate semi-conservatively (see Chapter 12). It was therefore incorrect to speak of parent- and daughter-chromatids, since the DNA of each chromatid is half old and half new.

(2) The copy-choice hypothesis predicts that successive crossovers along a chromosome arm will involve the same two chromatids (the daughters, not the parents), but the pair of chromatids involved in each crossover appear to be chosen at random, or nearly so (see § 8.8).

(3) Replication of both the DNA and the protein of the chromosome appears to take place before the homologous chromosomes start to associate at the zygotene stage of meiosis (see § 11.5 and Pl. 6.2(*a*)). Rossen and Westergaard (1966) showed by Feulgen micro-spectro-photometry that in the fungus *Neottiella rutilans* the premeiotic DNA synthesis actually took place before nuclear fusion. Henderson (1966), working with males of the locust *Schistocerca gregaria*, found by combined tritium labelling (to observe DNA synthesis) and high temperature treatment (to modify chiasma frequency), that chiasma formation occurred several days after the premeiotic DNA synthesis.

(4) Isotope labelling experiments have shown that recombinants, both in bacteria and their viruses, may contain parental atoms (§ 16.3 and § 16.4).

(5) Postmeiotic segregation (cf. Pl. 16.1(*c*)) was unexplained.

(6) The inequality in the frequency of 6:2 and 2:6 ratios of black:grey spores found by Kitani *et al.* with the *g* mutant of *Sordaria fimicola* was unexplained.

(7) Recombination must occur with molecular precision, such that nucleo-tides are rarely duplicated or omitted*, but it is not clear how this would be achieved with a switching of copying from one template to another.

Most of these defects in the copy-choice hypothesis can be avoided by making suitable additional postulates, but none of the variants of the hypothesis will explain all the observations.

Taylor (1965) obtained the first direct evidence that crossing-over at meiosis takes place by breakage and exchange. He used $^3$H-thymidine label-ling of the chromosomes in males of the grasshopper *Romalea microptera*. The label was introduced at the penultimate DNA replication before meiosis, with the result that at meiosis half the chromatids were labelled and half not. It was then possible to observe whether the chiasma frequency and distribu-tion at diplotene corresponded to the frequency and distribution of exchanges between labelled and unlabelled chromatids. The labelling pattern was observed at anaphase I, metaphase II or anaphase II, when the daughter-chromosomes were well separated from one another. Taylor's results for chiasma and label-exchange frequencies are given in Table 16.1. Since crossovers occur between either chromatid of one chromosome and either chromatid of the homologous chromosome (§ 8.6), and since one chromatid of each pair was labelled and the other not, only half the crossovers would be expected to be between a labelled and an unlabelled chromatid. As is evident from the table, there was excellent agreement between the label exchange frequency and half the chiasma frequency. A possible source of error, however, is the occurrence of sister-chromatid exchanges, as at mitosis (§ 12.2), though Taylor inferred that these were of rare occurrence. As pointed out by Douglas and Kroes (1969), Taylor's data also show good agreement between chiasmata and label-exchanges as regards their distribu-tion in the chromosomes. Taylor's results establish for the first time that crossing-over takes place by breakage and rejoining, and provide the first

* Evidence for occasional duplication or omission of nucleotides is discussed in § 16.6.

TABLE 16.1   Data of Taylor (1965) for *Romalea microptera*

| Chromosome identification numbers | Diplotene | | | Anaphase I or II or metaphase II | | |
|---|---|---|---|---|---|---|
| | Number of chromosome pairs | Number of chiasmata | Frequency of chiasmata per chromosome ($\times \frac{1}{2}$) | Number of chromosome pairs | Number of label exchanges | Frequency of exchanges per chromosome |
| 2 and 3 | 88 | 323 | 1·84 | 32 | 57 | 1·78 |
| 4, 5 and 6 | 74 | 194 | 1·31 | 38 | 49 | 1·29 |

direct experimental evidence of the truth of the chiasmatype theory, includ-ing a direct demonstration from the labelling pattern that homologous centromere regions separate from one another at the first division of meiosis (cf. § 8.6).

Peacock (1970) eliminated the effects of sister-chromatid exchange in an ingenious way. One crossover in a chromosome arm will cause non-sister label exchange in the chromosome segment distal to the crossover in 50% of cases. Sister-chromatid exchange will not affect this proportion for segments distal to both the exchange and the crossover. In *Goniaea australasiae*, Peacock observed distal non-sister label exchange in 20 out of 46 metaphase I chromosome pairs with a single chiasma, in excellent agreement with expectation on the assumptions that chiasmata correspond to crossovers and crossovers result from breakage and rejoining. Furthermore, Peacock varied the chiasma frequency by means of a temperature shock and found a corresponding change in distal non-sister label exchange. At meiosis in *Stethophyma grossum* each chromosome pair has a single proximally localized chiasma. Out of 60 chromosome pairs scored at anaphase I, metaphase II or anaphase II, Jones (1971) found 32 with a non-sister label exchange, and all the exchanges were in the chiasma region.

These studies by Taylor, Peacock and Jones have established the truth of the chiasmatype theory, some 60 years after Janssens proposed it, and have shown that crossing-over takes place by breakage and rejoining.

I proposed (Whitehouse, 1963) a model based on breakage and rejoining of chromatids. It was assumed that replication of chromosomes is completed prior to the occurrence of recombination, and that during the zygotene or pachytene stages of meiosis, when the homologous chromosomes are asso-ciated in pairs, recombination occurs at a number of points along the length of the chromosome, and that one chromatid from each chromosome, selected at random, is involved each time. It was also assumed that the reconstruction of DNA molecules in new ways, following breakage, took place not by an end-to-end association of broken chains but by a lateral association of complementary segments from homologous regions, to give *hybrid DNA*. End-to-end joining would appear readily to allow deficiencies or duplications of nucleotides to arise, whereas lateral association would automatically give great precision to the recombination process. Furthermore, it has been shown by Marmur and Lane (1960) that complementary nucleotide chains of DNA will associate spontaneously *in vitro* under suitable conditions to give the duplex molecule. It therefore seemed likely that re-assembly of broken molecules might occur *in vivo* in a similar way. If, in the process of crossing-over, the two DNA molecules assumed to be involved were broken one chain at a time, re-association in new ways could occur before the second chains were broken. This might have a selective advantage by reducing the likelihood of complete breakage without rejoining. Failure of rejoining would lead ultimately to loss of the acentric part of the chromosome, with lethal consequences for that cell lineage (cf. § 9.8). If the initial breakage was in homologous regions of complementary chains, one from each molecule, crossing-over could occur by a succession of steps involving chain separation, synthesis of new chain segments, and base pairing of complementary segments from the two molecules.

Holliday (1964) independently proposed a different hybrid DNA model. He suggested that the initial breakage was in chains of the same polarity. Crossing-over could then take place merely by chain separation followed by annealing between the complementary chains from the two molecules, without the necessity for any DNA synthesis. Breakage and exchange of the other two nucleotide-chains would complete the process (Fig. 16.5).

Without specifying the sequence of events in detail, Meselson (1965) has also suggested that recombination involves pairing between complementary DNA chains of different parentage. He favours a model in which recombination may be associated with the excision of chain segments and their resynthesis along the complementary chain derived from the other parent.

In order to account for polarity in recombination, it appears necessary to postulate the existence of discontinuities at intervals along the chromosome in the pattern of recombination. Hastings and Whitehouse (1964) suggested that the initial points of breakage of the DNA chains could occur only at fixed positions which an enzyme could recognize. The steps in the process of crossing-over according to this model are illustrated in Fig. 16.6. Only 2 of the 4 chromatids are shown, as the other 2 do not take part (although they may participate in another crossover further along the chromosome). No attempt has been made to show the spiral coiling. In Fig. 16.6(ii), breakage of complementary chains has occurred at corresponding positions, and the chains have uncoiled. In (iii), synthesis of nucleotide-chains has occurred alongside both of the unbroken chains and complementary to them. The newly-synthesized chains are shown by broken lines, and need not be of equal length. In (iv), the newly-synthesized chains have dissociated from their templates, and in (v) they have associated with the complementary broken chains from the other molecule. The final stages of crossing-over involve the filling of any gaps in the duplex crossover molecules with nucleotides complementary to the other chain (Fig. 16.6(vi)), and the breakdown

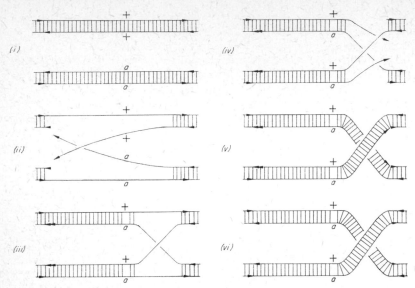

FIGURE 16.5    The mechanism of crossing-over according to the hybrid
DNA model of Holliday (1964). The lines with arrow-heads
represent nucleotide-chains, with the arrows showing the
direction of the phosphate-sugar backbones. Transverse lines
indicate hydrogen bonding between complementary bases.
The letter *a* indicates the site of a mutation, and the plus sign
its normal allele.

of the unpaired non-crossover chains (vii) to give the completed crossover
(viii). It will be noticed that there is no net synthesis of DNA.

There is experimental evidence for the occurrence of a limited amount of
DNA synthesis at the pachytene stage of meiosis, such as this hypothesis
demands. Hotta, Ito and Stern (1966) found from studies of the uptake of
radioactive phosphorus ($^{32}P$) that about 0·3% of the total DNA of *Lilium
longiflorum* and *Trillium erectum* was synthesized at the zygotene and pachytene
stages of meiosis in the anthers and that this DNA synthesis was necessary
for chromosome pairing and chiasma formation (see §§ 18.10 and 11).
Hotta, Parchman and Stern (1968) showed that protein synthesis also
took place at this stage of meiosis, and that this was necessary for the DNA
synthesis to occur. The newly-synthesized DNA and protein are physically
associated and evidently functionally related. Howell and Stern (1971)
demonstrated the presence in *Lilium* pollen-mother-cells at zygotene and
pachytene, but not at other times, of three enzymes: (i) an endonuclease
which breaks single chains of duplex DNA to give 5'-hydroxyl and 3'-
phosphoryl ends, (ii) a polynucleotide kinase which phosphorylates 5'-
hydroxyl ends, and (iii) a polynucleotide ligase which joins adjacent
5'-phosphoryl and 3'-hydroxyl ends if the other chain is intact. Such enzymes
are just those required for recombination by breakage and rejoining via
hybrid DNA formation, and Howell and Stern suggested that they function

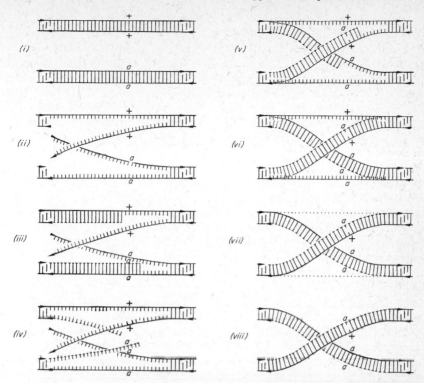

FIGURE 16.6   The mechanism of crossing-over according to the hybrid
DNA model proposed by Whitehouse and Hastings (1965).
The lines with arrow-heads represent nucleotide-chains, with
the arrows showing the direction of the phosphate-sugar backbones. Broken lines represent newly-synthesised chains, and
dotted lines chains which are breaking down. The arrow-
heads are placed at the points where breakage of the chains
can occur enzymatically. Transverse lines extending from one
nucleotide-chain to another indicate hydrogen bonding
between complementary bases. The letter *a* indicates the
site of a mutation, and the plus sign its normal allele.

in the sequence in which they are referred to above, a non-specific phosphatase, such as is present throughout prophase, dephosphorylating the 3′
ends before the ligase can act.

Hotta and Stern (1971*b*, *c*) made a further biochemical discovery about
meiotic prophase of exceptional interest. They demonstrated in *Lilium*,
*Rattus*, *Bos* and *Homo* the occurrence of a protein in zygotene and pachytene
nuclei with properties similar to those of the product of gene 32 of virus
*T*4 of *Escherichia coli* which was isolated by Alberts and Frey (1970). In *T*4
the 32-protein is required for replication and recombination, and it functions
structurally rather than catalytically, the quantity of 32-protein synthesized
directly limiting the number of progeny. The protein is not used, however,

in the construction of the virus particle, and seems to play a structural part in separating the chains of duplex DNA, by binding to the single chains, and holding the otherwise highly folded chains in an extended conformation and thus facilitating matching of complementary chains. Hotta and Stern found that the *Lilium* meiotic protein likewise had a high affinity for single chains of DNA, and that it promoted duplex formation. These properties of the 32 and meiotic proteins are precisely those required for recombination by hybrid DNA formation, because it is known that complementary nucleotide sequences will find one another spontaneously. Thomas (1966) discussed the pairing of single chains and concluded that a sequence of 10 nucleotides in each chain, if complementary, might be sufficient to initiate duplex formation, that is, provide a minimum recognition length, and that once a segment had annealed the process would extend rapidly along the molecule.

If crossing-over is associated with hybrid DNA formation, the site of a mutation will sometimes happen to lie within the hybrid region, giving rise to *heterozygous DNA*. The consequences of such heterozygosity in the molecule are discussed in § 16.7.

Hybrid (heteroduplex) DNA formation as a result of base-pairing between complementary nucleotide-chains derived from two different molecules, would also appear to fit many of the observations on recombination in bacteria and viruses described in § 16.3 and § 16.4. Hypotheses to this effect were put forward by Szybalski (1964), Meselson (1964) and Thomas (1966). Indeed, the idea of heteroduplex DNA, originally suggested by Watson and Crick (1953*b*), was proposed by Levinthal (1954) as one of the possible explanations for the heterozygosity, already referred to, which he and others observed in virus *T*2. Much support for heteroduplex models in bacterial and viral recombination has since been obtained. Fox and Allen (1964) working with *Diplococcus pneumoniae* and Bodmer and Ganesan (1964) with *Bacillus subtilis* showed by experiments using heavy isotopes that genetic transformation involved the physical incorporation of segments of donor DNA into the DNA of recipient cells, and that single-chain segments of donor DNA were covalently bound within recipient DNA molecules. Gurney and Fox (1968), in further experiments on transformation in *D pneumoniae* using heavy isotopes, obtained DNA that was both physically and genetically hybrid, and they showed that either chain of the duplex could participate in the breakage and rejoining mechanism which effects transformation. Anraku and Tomizawa (1965) obtained evidence, by caesium chloride density-gradient centrifugation of labelled DNA from mutants defective in particular genes, that in recombination in virus *T*4 of *E. coli* the heteroduplex is held together at first only by hydrogen-bonding between the parental chains, but that after single-chain gaps have been filled by DNA synthesis the heteroduplex becomes covalently bonded. Weiss and Richardson (1967) purified an enzyme, DNA ligase, from *E. coli* infected with *T*4. This enzyme catalyzes the covalent joining of two segments of a broken chain of a DNA duplex to give a phosphodiester bond, and is coded by gene 30 (Fig. 18.5) in *T*4 (Fareed and Richardson, 1967). Similarly, Oppenheim and Riley (1967) showed that, following conjugation in *E. coli*, the covalent joining of donor and recipient DNA segments takes place after non-covalent bonding.

The occurrence of recombination in bacteria and viruses by breakage followed by rejoining by hybrid DNA formation has thus been established. There is little evidence, however, for polarity in recombination in these organisms, such as is known in fungi; this probably means that, in general, bacterial and viral recombination can be initiated at any point in the DNA rather than at specific ones. This conclusion is in agreement with Kozinski, Kozinski and James's (1968) belief that recombination in *T*4 is initiated by random single-chain cuts in the DNA caused by an enzyme coded by the virus, and Altman and Meselson (1970) have purified such an endonuclease.

Crossing-over at a specific site is required, however, for the insertion (integration) and removal (excision) of an episome from its specific attachment-site on the host chromoneme (§ 16.3). With virus λ, a λ gene called *int* is involved in the process of integration, and both it and another λ gene called *xis* function in the excision process for the removal of λ DNA from the host chromoneme (see Signer, 1968, and Echols, 1971). The exact functions of these genes are not yet clear, but a remarkable feature of the integration and excision processes is the apparent lack of homology between the integration-sites in the *E. coli* and λ DNA molecules. This has been established by Davis and Parkinson (1971) from electron microscopy of heteroduplex DNA molecules using deletion mutants and mutants defective in the integration site. The heteroduplex molecules were obtained by adding alkali to a mixed suspension of wild-type and mutant virus. The alkali extracts the DNA from the particles and separates the two chains. Addition of formamide allows them to reanneal and this will often involve complementary chains from the two strains, giving heteroduplex DNA. Regions lacking homology, that is, where the two chains have failed to pair because of a difference in base sequence, are directly visible under the electron microscope as a pair of fine contorted threads, while the duplex regions appear as a single stouter and straighter thread (Pl. 16.2 and 16.3). No homology was found between the host and viral attachment regions. Davis and Parkinson inferred that there could not be more than 12 bases in the same sequence in the two regions. Thus, the specificity of the integration and excision processes seems to reside in the *int* and *xis* gene products, which can evidently recognize one particular nucleotide sequence in the host DNA and a different one in the virus. The mechanism of recombination involved in λ integration and excision therefore appears to be a highly specialized one, and probably has little in common with other recombination mechanisms.

The general recombination process in λ is quite independent of the integration mechanism. It has been shown that *red* mutations (*recombination deficient*) which prevent general recombination involve the structural gene for λ exonuclease and its associated β protein. This exonuclease digests 5′ ends of duplex DNA, and Cassuto and Radding (1971) showed that *in vitro* it acts at the site where a duplicate length of 3′-terminal chain leaves a stretch of duplex DNA. The enzyme excises nucleotides from the end of the paired chain, and allows the unpaired chain to take its place. The

PLATE 16.2   Electron micrograph of the heteroduplex DNA molecule formed by
annealing one nucleotide chain of virus $\lambda$ with the complementary
chain of an abnormal strain of $\lambda$ in which part of the genome has been
replaced by host genes. Where the two chains are homologous a duplex
molecule is formed and this appears as a rigid rod-like structure.
Where the nucleotide sequences of the two chains do not correspond
the chains remain separate. These single chains are more flexible and
less distinct than the duplex regions. The ends of the non-homologous
region are indicated by arrows. For fuller discussion, see Schleif,
Greenblatt and Davis (1971). The DNA molecule of $\lambda$, about half of
which is shown in the photograph, is about 17·2 $\mu$m long.

*(Courtesy of R. W. Davis)*

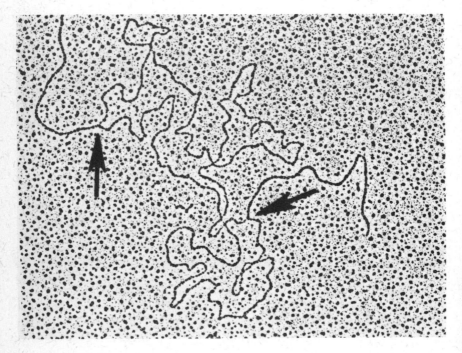

excision stops immediately the 3′ terminus has fallen into place, with the
result that there is no gap and polynucleotide ligase can seal the junction
(Cassuto *et al.*, 1971). This action of $\lambda$ exonuclease is shown in Fig. 16.7,
together with the likely steps in recombination mediated by the enzyme.

Mutants of genes in *E. coli* called *recB* and *recC* have a much reduced level
of recombination and lack a deoxyribonuclease activity which is dependent
on adenosine triphosphate (ATP). This enzyme is of exceptional interest
because it seems to act exonucleolytically on duplex DNA (Wright and
Buttin, 1969) and endonucleolytically on single-chain DNA (Goldmark
and Linn, 1970). These properties of the *recBC* DNase have been confirmed
by Goldmark and Linn (1972), who have purified it and shown that it
consists of two non-identical polypeptides, as expected with two structural

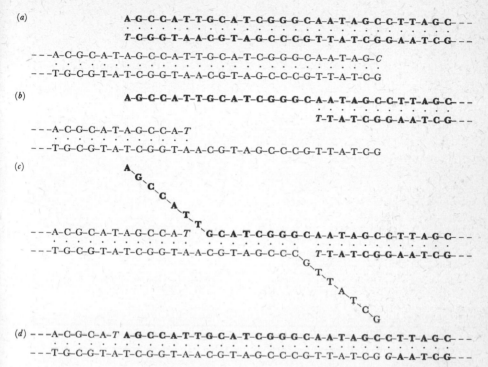

FIGURE 16.7   Diagram to illustrate the action of λ exonuclease in recombination
in virus λ of *Escherichia coli* according to Cassuto and Radding (1971).
The letters indicate the nucleotide sequence in the DNA, with
dashes for the covalent joining along the chains and dots for the
hydrogen bonding between chains. The bold type distinguishes the
nucleotides from one parent molecule. The sites of action of λ
exonuclease are the 5′ chain ends, and these are indicated by the
letters in italics. (*a*) Two partly homologous DNA molecules. (*b*)
The 5′ ends have been eroded by λ exonuclease. (*c*) The complementary nucleotide sequences exposed by the action of the enzyme
have paired to give a segment of hybrid DNA with a redundant
3′-terminal chain at each end of it. (*d*) The exonuclease molecules
have continued to erode the 5′ ends, allowing the redundant
chains to pair instead, but the action of the enzyme has then
stopped without a gap forming. The missing phosphodiester bond
can then be synthesized by polynucleotide ligase.

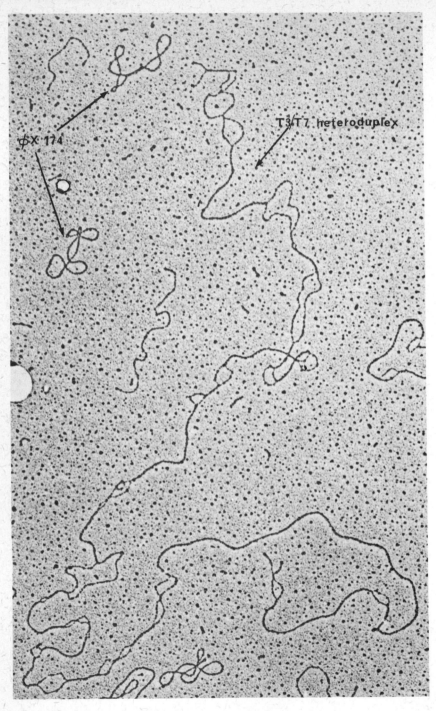

PLATE 16.3   Electron micrograph of the heteroduplex DNA molecule obtained by annealing one chain of the DNA of virus *T*3 of *Escherichia coli* with the complementary chain of the DNA of virus *T*7. As can be seen there are numerous places where the bases do not match and so the two chains remain separate. For discussion, see Davis and Hyman (1971). The DNA of virus *T*3 has a length of about 11·6 $\mu$m and that of *T*7 of about 12·5 $\mu$m. The small circular molecules are DNA of virus $\phi X174$ included as a scale object. The DNA of this virus has a length of about 1·77 $\mu$m.                                    (*Courtesy of R. W. Davis*)

genes coding for it. They found that the enzyme can degrade single-chain DNA exonucleolytically as well as endonucleolytically. Barbour *et al.* (1970) found that some suppressors (*sbcA* mutants) of *recB* and *recC* mutants have a high level of an exonuclease which does not require ATP but can presumably take over the exonucleolytic function of the *recB* and *C* genes. Kushner *et al.* (1971) obtained evidence that other suppressors (*sbcB*) of *recB* and *recC* mutants were mutants of the structural gene for exonuclease I, an enzyme which *in vitro* cleaves nucleoside 5′-phosphates stepwise from the 3′-hydroxyl end of single-chain DNA. Kushner *et al.* (1972) suggested that the residual recombination activity shown by *recB* and *C* mutants resulted from an alternative pathway for recombination which is normally kept at a low level by the action of exonuclease I, not in removing 3′-terminal single chains, but through some other unidentified activity of the enzyme.

Broker and Lehman (1971) made multiple infections of *E. coli* cells with virus *T*4 defective in both DNA polymerase (gene 43) and polynucleotide ligase (gene 30) and obtained evidence that branched duplex DNA molecules seen under the electron microscope about 20 minutes after infection were intermediates in the recombination pathway. The evidence included the finding that mutants which altered the recombination frequency changed the frequency of the branched molecules to the same extent. The branches often arose from H-shaped junctions, with the cross-bar of the H estimated to be 200–2000 base pairs long. To account for these structures, Broker and Lehman suggested that, following endonuclease action to cut single chains, exonucleolytic breakdown from these nicks exposed the other chain. If a complementary nucleotide sequence was similarly exposed in another molecule, pairing would take place, promoted by the 32-protein (see above), and would generate an H-shaped duplex molecule such as were observed. The top half of the H would then derive from one parent molecule and the bottom half from the other, the cross-bar being hybrid DNA. The H-shaped molecules were seen sometimes to have the cross-bar extending a short distance beyond one of the uprights. In order to account for this, Broker and Lehman proposed that after a hybrid DNA segment has formed it may extend at the expense of the adjoining parental duplex segments. This process would involve complementary nucleotides separating in these homoduplex segments and taking new—heteroduplex—partners. The consequence would be that the cross-bar of the H would increase in length between the uprights and also extend beyond one of them (Fig. 16.8). The propagation of the hybrid DNA would continue until a nick was reached. Finally, an endonuclease would remove the branches, and polymerase and ligase would close gaps and nicks.

Sobell (1972) applied to recombination in chromosomal organisms the Broker-Lehman idea of hybrid DNA extension. From his study of the mechanism of binding of actinomycin to DNA (§ 15.2), Sobell inferred that nucleotide sequences with twofold symmetry had important functions. In particular, he suggested that Gierer's idea for the structure of the operator, that is, secondary loops arising from reversed duplications (see § 15.2) could also explain chromosome pairing at meiosis: for example, the G-A-G sequence in one loop apex on p. 307 could pair with the C-T-C- sequence of one of the corresponding loops in a homologous chromatid, the loops having first been cut by an endonuclease. Hybrid DNA, initiated in this way, could then spread in both directions along the pair of DNA molecules to give an H-shaped junction with projecting cross-bar, as in Fig. 16.8(*b*). Sobell pointed out that, if the original cuts were then resealed by ligase, the pair of hetero-duplex segments would be trapped within the two DNA molecules, but it could migrate along them in either direction for any distance. Implications of hybrid DNA migration are discussed in § 16.7.

Boon and Zinder (1971) studied recombination in virus $F1$ of *E. coli*. The chromoneme of this virus consists of one circular polynucleotide of DNA long enough to contain about 6 genes. An amber mutant of viral gene II prevents replication proceeding beyond one chain complementary to the original one, but such mutants can recombine. When the host is infected with two different amber mutants of this gene, progeny virus particles are produced only if recombination has occurred to generate a wild-type chromoneme. All the virus particles from individual host cells will consist exclusively of the products of this recombination event. In addition to the pair of alleles, the parents differed by two unselected mutations which mapped at widely spaced positions relative to gene II and one another. Study of the progeny revealed that most of the recombination events generated one recombinant and one parental genotype. Denoting the parental genotypes by Al +B and a +2b, where 1 and 2 are the alleles, + means wild-type, and A/a and B/b are the distant markers, the wild-type recombinant progeny had the genotypes A + +B, a + +b and a + +B about equally often, and each could be associated with either parental genotype, though A + +B with Al +B, and a + +b with a +2b were only about half as frequent as the converse combinations (A + +B with a +2b, and a + +b with Al +B). To explain these results, Boon and Zinder favoured a break-and-copy model in which a heteroduplex segment linking the two parental molecules, P and Q, initiates a replication fork on Q. Subsequent breakage at the fork could lead to the insertion of a segment of Q DNA into the homo-logous region of P, while Q remains intact. The results can also be explained as a direct result of heteroduplex formation, if it is assumed that in addition to the recombinant polynucleotide, one of the other nucleotide chains, but not more than one, survives.

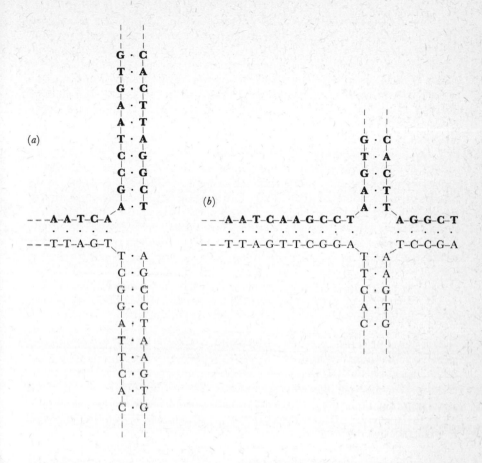

FIGURE 16.8   Diagram to show the process of hybrid DNA extension proposed by
Broker and Lehman (1971) to explain the origin of projecting
cross-bars to the H-shaped junctions seen in electron micrographs
of virus *T*4 DNA molecules at an intermediate stage in recombination. The letters indicate the nucleotide sequence in the DNA, with
dashes for the covalent joining along the chains and dots for the
hydrogen bonding between chains. The bold type distinguishes the
nucleotides from one parent molecule. (*a*) Part of an H-shaped
configuration, the cross-bar being of hybrid DNA. (*b*) Extension of
the cross-bar and formation of a new heteroduplex segment to its
right, the parental (homoduplex) upright of the H being correspondingly shortened.

## § 16.6   Repair of radiation damage to DNA

Kelner (1949) showed that the lethal effect of ultraviolet light on the conidia of *Streptomyces griseus* could be reversed by subsequent treatment with visible light. This process of photo-reactivation has since been found to apply to a wide range of organisms, although not to all, and to diverse effects of ultraviolet light on them. Numerous observations pointed to an effect of the ultraviolet light on DNA as the primary source of the changes induced. *Haemophilus influenzae* shows no photoreactivation, but Rupert, Goodgal, and Herriott (1958) demonstrated that a cell-free extract from *Escherichia coli*, which is a photoreactivable species, will reactivate the DNA of *H. influenzae* in the presence of visible light. This was shown by an increased ability of DNA from streptomycin-resistant bacteria to transform strepto-mycin-sensitive cells. The ultraviolet treatment reduced the transforming ability to 1% of the original, and illumination with visible light in the presence of the *E. coli* extract raised it again to between 10 and 50% of the original value. Rupert (1960) isolated an enzyme from *Saccharomyces cerevisiae* which had a similar effect to the *E. coli* extract. It is evident that a number of organisms possess an enzyme system capable of bringing about 'repair' of 'damage' to DNA caused by ultraviolet light. Beukers and Berends (1960) showed that UV irradiation of thymine in a frozen aqueous solution leads to the formation of thymine dimers. These have covalent bonds between carbon atom no. 4 (Fig. 11.3) in adjacent thymine residues in the poly-nucleotide, and likewise between carbon atom no. 5 in each. Wacker, Dellweg and Weinblum (1960) studied the effect of ultraviolet irradiation of the DNA from *Streptococcus faecalis* labelled with $^{14}$C in the thymine, and concluded that the mutagenic and lethal effects of ultraviolet light on cells were likely to be due to its chemical effect on thymine in DNA. Wulff and Rupert (1962) showed that the enzymic repair in the presence of light occurs by specific cleavage of the thymine dimers *in situ*, that is, by a reversal of the ultraviolet-induced process. Setlow, Boling and Bollum (1965) discovered that ultraviolet light caused the formation of cytosine dimers in DNA as well as thymine dimers, and that the photoreactivating enzyme splits both kinds. Pyrimidine dimers involving one thymine and one cytosine residue can also be generated by UV and split by the enzyme.

Setlow and Carrier (1964) and Boyce and Howard-Flanders (1964*a*) showed that an ultraviolet-resistant strain of *Escherichia coli* had the ability to remove thymine dimers from its DNA, whereas a particular ultraviolet-sensitive strain lacked this capacity. This reaction did not require visible light. It appeared that a specific gene was responsible. Boyce and Howard-Flanders showed that the dimers were extracted as a whole, including the intervening phosphodiester bond, and that nucleotides were then inserted into the excised region by complementary pairing with the intact opposite chain, and the broken phosphodiester backbone was rejoined. It is of particular significance that the repair mechanism appears to make use of the base sequence of the complementary chain. Extending these studies, Boyce and Howard-Flanders (1964*b*) found that in *E. coli* the repair in the dark of ultraviolet-irradiated DNA is controlled by 3 genes (*uvrA*, *uvrB* and *uvrC* for ultraviolet repair) which are widely spaced in the chromoneme. Mutation of any one of these genes produced an ultraviolet-sensitive mutant unable to excise thymine dimers.

Pettijohn and Hanawalt (1964) discovered that repair-replication of DNA took place after *E. coli* had been irradiated with ultraviolet light. From caesium chloride density-gradient studies they showed that 5-bromouracil (a density label) was incorporated into short segments of single chains. They concluded that damaged regions of single chains were excised and replaced by normal nucleotides using the undamaged chain as template. Kelly, Atkinson, Huberman and Kornberg (1969) showed that both the excision of a segment of polynucleotide containing thymine dimers and the DNA synthesis to replace it can be brought about by DNA polymerase I. A thymine dimer causes distortion of the DNA duplex and it is believed that an ultraviolet-specific endonuclease recognizes the dimer or the distortion and cuts the chain on the 5′ side of it. Mutants of the three *uvr* genes all have the same phenotype and appear to be defective in the endonuclease (Howard-Flanders, Boyce and Theriot, 1966). Following the endonucleolytic action, DNA polymerase I evidently recognizes the site and excises nucleotides from the 5′ chain-end a few at a time, and hence including the dimer, and inserts 5′-nucleotides at the adjoining 3′-hydroxyl chain-end, using the complementary chain as template (§ 12.4). The outcome is that the enzyme and the associated nick move along the chain in the 5′ to 3′ direction and will continue to do so until ligase intervenes to close the nick.

Recombination-deficient mutants of *E. coli* were first isolated by Clark and Margulies (1965). The mutants allowed normal conjugation, either as donor or recipient, but had lost the ability to integrate the donor DNA into that of the recipient. *RecA* mutants were devoid of recombination and after exposure to UV showed extensive breakdown of the DNA and were readily killed (Clark, Chamberlin, Boyce and Howard-Flanders, 1966), while *recB* and *C* mutants (discussed in § 16.5) had some residual recombination, less UV-induced DNA breakdown, and a lower, though still abnormally high, UV-sensitivity.

Rupp and Howard-Flanders (1968) obtained evidence from sedimentation studies that when *E. coli* DNA containing pyrimidine dimers replicates, a gap is formed in the newly-synthesized chain opposite each dimer. Such gaps would preclude excision of the dimer and repair by DNA polymerase I, since this requires an intact complementary chain as template. Rupp and Howard-Flanders suggested that, after replication of a duplex containing dimers, recombination took place between sister duplexes and this allowed an intact polynucleotide to be reconstructed, thus enabling the cell to survive. The UV-sensitivity of *recA* mutants was attributed to the failure of this recombination repair.

The recombination hypothesis for the repair of damaged and then replicated DNA was confirmed by Rupp, Wilde, Reno and Howard-Flanders (1971) from density-labelling experiments. Cells were grown for several generations in a culture medium containing $^{13}C$ and $^{15}N$, then exposed to ultraviolet light and grown for less than a generation in normal medium ($^{12}$C and $^{14}$N). Excision repair and photoreactivation were prevented by using a *uvrA* mutant kept in the dark. After extraction of the DNA and separation of the chains by heating, density-gradient centrifugation in caesium chloride revealed chains of intermediate density. This implied recombination to associate heavy and light atoms in a single polynucleotide. Breakage by shearing gave a partial separation into heavy and light components and suggested that segments more than 150 nucleotides long had been exchanged. Rupp *et al.* favoured the hypothesis outlined in Fig. 16.9 in which the exchange, presumably triggered by the internal free ends at the gap opposite each dimer, leads to the filling of the gap by a segment of the sister chain, followed by repair of the latter.

*(a)*          *(b)*          *(c)*          *(d)*

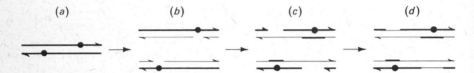

FIGURE 16.9   Diagram to illustrate the hypothesis proposed by Rupp, Wilde, Reno and Howard-Flanders (1971) for recombination repair of DNA containing pyrimidine dimers in *Escherichia coli*. The lines represent nucleotide chains and the dots pyrimidine dimers. The thick lines indicate the DNA present at the time of ultraviolet irradiation, and the thin lines that synthesized subsequently. *(a)* DNA molecule immediately after UV irradiation, with two pyrimidine dimers. *(b)* Gaps formed opposite the dimers in the newly-synthesized chains. *(c)* A segment of one chain transferred from the sister molecule to fill each gap. *(d)* Newly-formed gaps filled by DNA synthesis using the complementary chain as template.

The nature of the product of the *recA* gene and how it functions in this recombination repair are not known. Boyle, Paterson and Setlow (1970) made the surprising discovery that pyrimidine dimer excision is normal in a DNA polymerase I (*polA*) mutant, implying that there was an alternative mechanism for excision repair besides that mediated by the polymerase. Gross, Grunstein and Witkin (1971) found that the *recA polA* double mutant was inviable. They suggested that this lethality could result from inability to repair single-chain gaps in the DNA, the *recA* gene product being needed for this repair in the absence of DNA polymerase I.

Miura and Tomizawa (1968) and Witkin (1969) independently discovered that ultraviolet light will not induce mutations in *recA* strains. It is inferred that the process of recombination repair is prone to errors and these are the source of most UV-induced mutations in *E. coli*.

Radiation-sensitive mutants showing abnormal frequencies of recombination were isolated by Holliday (1967) in *Ustilago maydis*. Mitotic recombination in fungi is normally much increased by ultraviolet light and ionizing radiation, but two of these mutants showed little response to UV. This suggests that the recombination normally observed after irradiation is part of a repair process, which is blocked in the mutants. Holliday (1971) found that a *U. maydis* diploid from a cross between two different mutants in the structural gene (*nar*-1) for nitrate reductase began to produce nitrite 4 hours after a low dose of $\gamma$-rays. The nitrite production indicated that recombination had occurred and had generated a wild-type *nar*-1 gene. Nitrite levels with different doses of $\gamma$-rays were used as an assay of recombination and compared with the number of colony-forming cells surviving each dose. After allowing for enzyme inactivation by the radiation, it was evident that few of the non-colony-forming cells had undergone recombination. Holliday suggested that a recombination repair mechanism is induced in cells containing damaged DNA, and that the non-colony-forming cells failed to show recombination because the structural gene for a recombination enzyme had been damaged. The cell inviability would then be a direct consequence of recombination failure.

Hunnable and Cox (1971) used UV-sensitive strains of *Saccharomyces cerevisiae* to study the effects on survival and on recombination of exposing the cells to UV light and then applying various régimes of storage in saline in the dark and of exposure to photoreactivating light. They concluded that there were pathways for the repair of damage to yeast DNA similar to those known in *E. coli*, namely, photoreactivation, excision repair, and postreplication recombination repair, and that in addition there was another repair pathway involving recombination. Rodarte-Ramón and Mortimer (1972) isolated recombination-deficient mutants of *S. cerevisiae* using a disomic strain, that is, haploid but with one chromosome represented twice. The two copies of this chromosome (no. VIII) carried different mutations of the *arginine*-4 gene. Recombination-deficient mutants were detected because arginine-independent cells arose less often than normal. Such cells, which result from mitotic recombination, are revealed by their ability to grow on minimal medium. The mutants mapped at 4 different genes (*rec*-1 to 4). Some of the mutants were killed no more readily than wild-type by X-rays or ultraviolet light, while others were sensitive to X-rays and UV or only to X-rays. These results suggested that some of the *rec* genes were involved in recombination repair and others in recombination pathways not involving repair of radiation damage.

In *Aspergillus nidulans*, Shanfield and Käfer (1969) isolated UV-sensitive mutants which affected recombination, and further such mutants have been studied by Jansen (1970) and Fortuin (1971). The mutants map in 5 genes which are believed to be concerned in the recombination repair of UV damage to the DNA.

Rasmussen and Painter (1964) found that tritiated thymidine is incorporated into the chromosomes of human cells in tissue culture following ultraviolet irradiation, and the effect is much enhanced if the thymine of the DNA has been partially replaced by 5-bromouracil. Control experiments without ultraviolet treatment showed no uptake of tritium, except in the fraction of cells undergoing normal DNA replication. It was inferred that a mechanism exists for the repair of ultraviolet damage to DNA, and that thymine is inserted into the DNA in the process. This conclusion has been confirmed from study of skin cell cultures derived from patients with xeroderma pigmentosum. In this disease, which is caused by a recessive gene, the skin is extremely sensitive to ultraviolet light. From density gradient and autoradiographic studies using 5-bromodeoxyuridine and tritium as labels, Cleaver (1968) discovered that, after irradiation with UV or X-rays, xeroderma cells showed no repair replication of DNA such as occurs in normal cells. Setlow, Regan, German and Carrier (1969) found that xeroderma cells had a much reduced ability to excise pyrimidine dimers compared with normal, and sedimentation patterns revealed an absence of single-chain breaks following the UV-irradiation, such as appear temporarily in normal cells. Setlow *et al.* concluded that xeroderma pigment-osum is a deficiency for an ultraviolet-specific endonuclease necessary to initiate repair of UV damage to DNA in man.

Lehmann (1972), using *Mus* lymphoma cells in culture, showed that the gaps which formed opposite pyrimidine dimers when the DNA of ultraviolet-irradiated cells replicated, were filled with newly-synthesized DNA and not with material introduced from the sister molecule by recombination. His technique was to label the chains synthesized after the UV treatment with ³H-thymidine and then to incubate in bromodeoxyuridine at the time when the gaps opposite the dimers were being filled. Exposure to light, which breaks the regions containing bromine atoms, was found to lower the molecular weight of the ³H-labelled material, implying that gaps had been filled with newly-synthesized DNA.

Riley and Miller (1966) found that a mutant of *Hordeum vulgare* (Barley), when homozygous, lowered the mean chiasma frequency per pollen mother cell from 15 to 6, and caused a two- to fourfold increase in abnormal joining of chromosomes in root tip mitoses following X-irradiation of seed, when compared with normal plants given the same dose. Pearson, Ellis and Evans (1970) reported on a man with much reduced chiasma frequency and whose cells, after irradiation, showed a lower than normal yield of aberrations at mitosis. In spite of the contrary effect of radiation, these discoveries in barley and man suggest a link, such as an enzyme in common, between meiotic recombination and radiation repair.

## § 16.7   Correction of mispairing in hybrid DNA

From study of mutation caused by ultraviolet light in *Ustilago maydis*, Holliday (1962) came to the conclusion that both of the daughter cells arising from the first mitosis after irradiation were often mutant. In order to account for this, he suggested that the radiation modified a nucleotide (or nucleotides) in one chain, and that an enzymatically-controlled repair mechanism, in restoring normal hydrogen bonding, was liable to replace the complementary nucleotide in the other chain, so as to match the mutated one.

The overall effect of ultraviolet light and 'repairing' enzyme would then be either to remove the effect of the ultraviolet light or to extend it to both chains. In other words, if ultraviolet light causes mispairing, that is, heterozygosity in the DNA, the repairing enzyme in restoring homozygosity may make the molecule homozygous mutant instead of homozygous normal. Holliday suggested that a similar process could be responsible for gene conversion if the DNA had become heterozygous at the mutant site through the pairing of complementary homologous chains.

I pointed out (Whitehouse, 1965) that on a hybrid DNA model of recombination, the 5 kinds of aberrant asci found by Kitani *et al.* (1962) with the grey-spored mutant of *Sordaria fimicola* (Fig. 16.1(*c*)–(*g*)) are what are expected if correction of heterozygosity in the DNA at the mutant site does not always happen, but may occur in either direction, that is, to homozygous normal (black) and to homozygous mutant (grey). If hybrid DNA occurred at the *g* site in both of the chromatids participating in crossing-over, and if there was no correction of the mispairing, asci with 4 black and 4 grey spores and with half of them not in pairs (Fig. 16.1(*c*)) would result. The 5:3 and 3:5 ratios would be the result of correction in one chromatid but not in the other, while the 6:2 and 2:6 ratios would be the consequence of correction in the same direction in both chromatids (see Fig. 16.10). Postmeiotic segregation, implying the persistence of mispairing through meiosis, might be due to failure of the postulated 'correcting' enzyme to reach its substrate during a critical period of time. One of the implications of the correction of heterozygosity in DNA as a source of recombination is that it can give rise to reciprocal as well as to non-reciprocal recombination. This is evident from Fig. 16.10, where correction in one direction in one chromatid and in the opposite direction in the other one will lead to normal reciprocal recombination, with asci containing 2 pairs of black and 2 pairs of grey spores.

Direct evidence that conversion is associated with the interior of a crossover, as the hybrid DNA models predict, was obtained by Rizet and Rossignol (1966) from study of *Ascobolus* asci segregating simultaneously for three closely-linked spore-colour mutants. Reciprocal recombination of the outer two was frequently associated with conversion at the intervening site and, moreover, the conversion involved the crossover chromatids.

The correction of mispairing in DNA at a mutant site would cause recombination simultaneously in two intervals of the chromosome, that is, between the mutant and neighbouring markers on both sides. Moreover, the hybrid DNA segments, where heterozygosity in the DNA arises, can occur only within crossovers and so there will be an association between crossing-over and the correction of mispairing. Whitehouse and Hastings (1965) assumed that there may be dissociation of the chains of the DNA molecules on one side of an opening-point, as illustrated in Fig. 16.6, or on both sides. The latter would give simultaneous crossovers in the same two chromatids in adjacent segments, with the result that outside markers would remain in parental combinations. This would account for the aberrant asci for the grey-spored mutant of *Sordaria fimicola* with parental combinations of

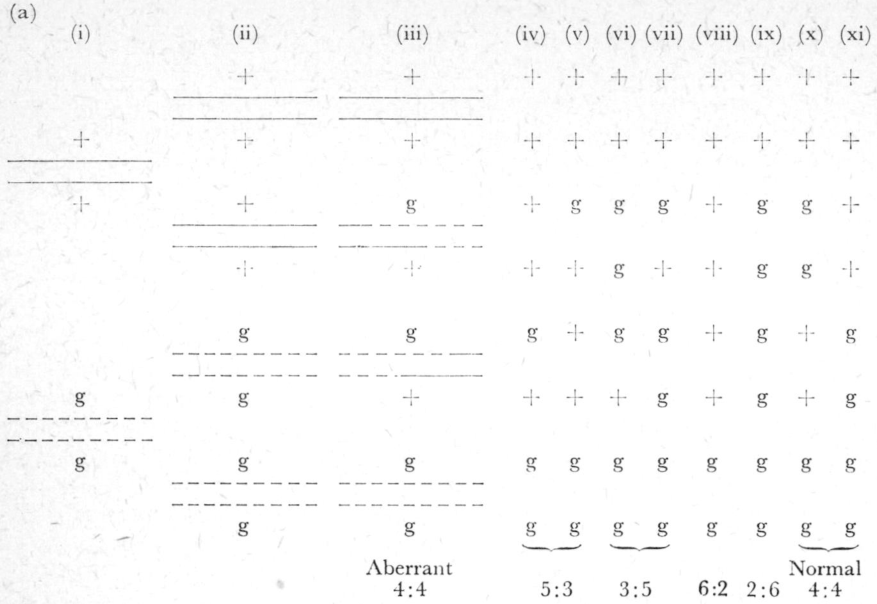

(a)

| (i) | (ii) | (iii) | (iv) | (v) | (vi) | (vii) | (viii) | (ix) | (x) | (xi) |
|---|---|---|---|---|---|---|---|---|---|---|

(The diagram shows pairs of horizontal lines representing DNA molecules, with + and g symbols corresponding to the eight spores.)

|  |  |  | Aberrant 4:4 |  | 5:3 | 3:5 | 6:2 | 2:6 | Normal 4:4 |
|---|---|---|---|---|---|---|---|---|---|

FIGURE 16.10   The diagram shows the origin of aberrant asci of *Sordaria fimicola* heterozygous for black (+) and grey (g) ascospore colour, according to hybrid DNA models of crossing-over. The pairs of horizontal lines represent DNA molecules (chromosomes or chromatids), the individual lines corresponding to nucleotide chains, and the full or broken lines distinguishing their parental origin.

The diagrams show (i) the parental chromosome-pair, (ii) the chromatids derived from them, (iii) the chromatids after a crossover has occurred such that there is hybrid DNA in the region of *g* in two of the chromatids, and (iv)—(xi) the consequences of correction of the heterozygosity in one or both chromatids. In (ii)—(xi), the 8 nucleotide chains, each carrying either + or *g*, correspond to the 8 spores in the mature ascus. The numbers of spores of each colour in the asci are shown below the 8 symbols, with the number of black (+) spores given first. Asci of classes (iii), (v), (vii), (viii) and (ix) correspond to (*c*)–(*g*), respectively, in Fig. 16.1. The various permutations of spore sequence due to the presence or absence of a proximal crossover, or to the chromosome orientation at the nuclear divisions in the ascus, are not shown.

outside markers (see § 16.2), and the similar results for genes in *Neurospora* and *Aspergillus*. Apart from these 'double crossovers', which are essentially single events in origin, there appears to be no evidence for the existence of negative interference between crossovers. It appears that recombination by correction of heterozygosity in DNA, together with the association between this correction and crossing-over, will account for negative interference. Pontecorvo (1958) pointed out that negative interference could be the result of the proximity to one another of the mutants necessary to detect it, if heterozygosis at one point along a chromosome favours recombination in its immediate vicinity, and in essence this is the explanation of negative interference provided by the hybrid DNA models.

As already mentioned in § 16.5, according to the model of recombination proposed by Holliday (1964), the primary breakage is in identical rather than in complementary chains in the two molecules, and there is no DNA synthesis. The secondary breakage, necessary to complete the crossover, may then occur in the same two chains, giving parental combinations for outside markers, or it may occur in the complementary chains, giving a crossover.

A hypothesis of recombination based on the correction of heterozygosity in DNA has to account for the fact that recombination frequencies between closely-linked mutants fit a linear map (cf. Fig. 16.2). The fit is often rather an approximate one, but nevertheless is well established. If recombination frequency depended merely on correction frequency, it would bear no relationship to the distance between sites. However, another factor to be considered is the frequency with which heterozygosity arises in the first place. If there is variability in the length of the hybrid DNA segments from a fixed origin, it is evident that mutant sites near the origin will occur in hybrid DNA more often than those further away. There will then be a gradient of decreasing conversion frequency with distance from the origin. It is evident that the unit of measurement of distance on a map of closely-linked mutants may not bear a constant relationship to the physical distance between sites. Where hybrid DNA is most frequent, the map may be relatively extended.

A further factor to be considered is the possibility that the heterozygosity itself interferes with the formation of hybrid DNA. This is particularly likely with deletions, to judge from the results obtained by Drake (1966) in experiments with virus *T*4 of *E. coli*. He studied the frequency with which *r*II mutants of known extent could be recovered from heterozygotes, and he found that the longer the addition or deletion the more often it failed to form a heteroduplex.

Howard-Flanders and Boyce (1966) obtained evidence that several hundred nucleotides are excised when ultraviolet-induced thymine dimers are removed from the DNA of *Escherichia coli*. If a similar length of chain were removed in the correction of mispairing in hybrid DNA, a second site of heterozygosity might be included in the excised segment. Both sites would then necessarily be corrected to the same parental genotype, and there would be no recombination between the two alleles. Two mutant sites close together in the DNA would be expected to show such linked correction more often than two sites further apart, and so the closer the mutants the lower their recombination frequency. This would explain the additivity of recombination frequencies of alleles. It would also explain why Lissouba and Rizet found that conversion did not occur at left-hand sites in series 46 of *Ascobolus* (Fig. 16.3): it is assumed that hybrid DNA forms from the right-hand end of the gene and when it extends to both mutant sites correction of mispairing of both occurs together and so does not cause recombinant asci. Only those asci in which hybrid DNA reached the first (right-hand) site but not the second would then be detected. Direct support for linked correction was obtained by Case and Giles (1964) who dissected *Neurospora* asci segregating for 3 alleles at the *pan*-2 locus and found several examples of simultaneous conversion of all 3 to the same parent. Paszewski (1967) found that simultaneous conversion of two closely-linked spore colour mutants in *Ascobolus* was 7 times more frequent than would be expected if the conversion at the two sites had been independent.

Fogel and Mortimer (1969) showed that the frequency of linked or co-conversion decreases with increasing distance between the sites, as expected if the linkage depends on both mutant sites lying in a single excised segment of polynucleotide. They analysed between 500 and 700 unselected asci of *Saccharomyces cerevisiae* from each of 3 crosses between mutants of the *arginine*-4 gene. From each cross about 45 asci showed conversion, and the proportion of these with co-conversion fell from 75% with the closest pair of alleles to 46% with the intermediate pair and to 6% with the most distant pair. It was concluded that the modal length of nucleotide chain excised in the correction of mispairing was to be measured in hundreds of nucleotides.

Gutz (1971) obtained similar results with *Schizosaccharomyces pombe*. Two unequal intervals in the *adenine*-6 gene were studied, and from a total of over 2500 unselected asci nearly 100 showed conversion. For the shorter interval (one intervening mutant site out of 20 alleles mapped) the proportion of linked conversion was 100% while with the longer interval (7 intervening sites) the proportion was 54%.

Holliday (1964) pointed out that in pairwise crosses between allelic mutants, where linkage maps have been prepared from the frequency of wild-type recombinants, the longer intervals on the map are often greater than the sum of the intervening ones (see Fig. 16.2). He suggested (Holliday, 1968) that this map expansion with distance can be accounted for by extensive excision in the correction of mismatched bases in hybrid DNA. With mutations that are closer together than the mean excision length, many recombination events will not be detected because linked conversion involving both sites will not generate a wild-type polynucleotide. Short intervals on the map will therefore be disproportionately reduced. Fincham and Holliday (1970) worked out the consequences of this explanation of map expansion. Assuming the *tryp*-3 gene of *Neurospora crassa*, which codes for tryptophan synthetase, is about 900 nucleotides long, they estimated that the minimum length of the DNA segment involved in correction was about 130 nucleotides and the average length not less than 500 nucleotides.

For *Ascobolus immersus* there is clear evidence that conversion is influenced by the molecular nature of the mutation. Rossignol (1969) discovered homogeneous categories of spore colour mutants exhibiting various conversion patterns, and Leblon (1972a) showed the existence of a correlation between mutagen and conversion pattern. When crossed with wild-type, mutants induced by nitrosoguanidine gave numerous asci with postmeiotic segregation and there was an excess of conversion to wild-type over conversion to mutant. On the other hand, mutants induced by the acridine mustard ICR170 showed almost no postmeiotic segregation and a marked excess of conversion to mutant. These results can be accounted for if a correcting enzyme recognizes different kinds of mismatching in hybrid DNA with different efficiencies, and if the nature of the mispairing influences from which of the two polynucleotides a segment shall be excised. Study of revertants of some of these spore colour mutants, obtained with particular mutagens, led Leblon (1972*b*) to conclude that *Ascobolus* mutants with low and with high frequencies of postmeiotic segregation were frameshift mutants and base-substitution mutants, respectively, and that in hybrid DNA involving a frameshift mutant and wild-type it was the shorter chain which was excised. In other words, mutants with a nucleotide added show conversion predominantly to mutant, and those with a nucleotide deleted convert chiefly to wild-type.

In contrast to *Ascobolus*, conversion frequencies to wild-type and to mutant in *Saccharomyces cerevisiae* have always proved equal (Fogel and Mortimer, 1969), suggesting that the yeast enzyme does not recognize the molecular nature of the mispairing. On the other hand, Gutz (1971) obtained a mutant, M26, in the *adenine*-6 gene of *Schizosaccharomyces pombe* which showed conversion to wild-type more than 10 times as often as to mutant. Furthermore, this bias extended to allelic mutants when they were crossed with M26: they converted predominantly to mutant in trans crosses and to wild-type in cis crosses, whereas when crossed with wild-type they showed no bias and a much lower conversion frequency. These results can be explained if mispairing caused by M26 in hybrid DNA is recognized by a correcting enzyme much more often than the mispairing of the other *ad*-6 mutants studied, and if the excision is predominantly of the mutant chain and is extensive, with the result that allelic mutants are affected by it. The excision must occur in both directions along the DNA from the site of M26, because Gutz found that the high frequency and the bias in conversion affected mutants on both sides of M26 in asci heterozygous at 3 sites within *ad*-6. Fincham and Holliday (1970) found it necessary to postulate such bidirectional degradation from mismatched base pairs in order to explain the fact that recombination frequencies seem to depend on the distance between the sites even for very short map intervals.

Yanofsky *et al.* (1964) found that certain mutants affecting the *A* protein of tryptophan synthetase in *E. coli* showed consistently higher recombination frequencies than others, even though their sites were almost identical in position and within the same codon (see § 13.8). Moreover, the recombination behaviour was found to be associated with the amino-acid substitution which the mutation induced. Thus, mutants nos. 23, 27, 28, and 36, which led to the replacement of glycine by arginine at position no. 210 in the polypeptide, gave higher frequencies of recombination with allelic mutants than mutants nos. 46, 95, and 178, which caused the substitution of glutamic acid at the same position. It is inferred that the heterozygosity in the DNA in these two groups of mutants differ in their influence on hybrid DNA formation or on a correction mechanism. Drapeau, Brammar and Yanofsky (1968) found that two mutants which replace GAG (for glutamic acid) at codon 48 by the stop signal UAG increased recombination with close alleles up to 20-fold compared with three mutants which substitute another amino-acid at position 48. In spite of this influence of individual mutants on recombination, it is also evident from the data of Yanofsky and associates that allelic recombination frequencies reflect the physical distance apart of mutant sites, as can be seen from Fig. 13.8.

Emerson (1966) reported that 571 *Ascobolus* asci showing aberrant segregation for mutant *w*-62 comprised 45·0% 6:2, 21·0% 2:6, 10·2% 5:3 and 23·8% 3:5 ratios of wild-type: mutant spores. (The frequency of the aberrant 4:4 class was not known.) This favouring of correction to wild-type in the even ratios (6:2 > 2:6) and to mutant in the odd ratios (5:3 < 3:5), which was not shown by the data of Kitani *et al.* for the *g* mutant in *Sordaria fimicola* (Fig. 16.1), implies that the pattern of correction is different in the two chromatids involved. Emerson suggested that this difference might be caused by a different kind of mispairing in each chromatid, such as Holliday's model predicts (for example, A·C and G·T with a transition). Another possibility is that the correction of mispairing is influenced by the presence of a break in a nucleotide-chain in the neighbourhood of the mispaired site. This idea is supported by the work on transformation in *Diplococcus pneumoniae* by Ephrussi-Taylor and Gray (1966), who proposed that correction of mispairing in the heteroduplex DNA took place with certain mutants which gave integration with low efficiency, and that this correction was strongly biassed in favour of excision of the donor chain. Such a bias implies enzymic recognition not only of a site of mispairing but also of some feature of the DNA which distinguishes donor and recipient, such as a free end to a nucleotide chain. Both the bacterial and fungal data could be explained if an enzyme complex, after recognizing a site of mispairing, 'scrutinizes' neighbouring regions of the DNA and, if a break is found, excises the broken chain. In biochemical terms this might mean that an endonuclease acts only if an exonuclease has failed to find its substrate. Selection might have favoured such a mechanism as it would prevent breaks near together in both chains.

Support for the idea that *D. pneumoniae* has an enzyme system capable of correcting mispairing in heteroduplex DNA has been obtained by Roger (1972). She prepared heteroduplex molecules by annealing complementary chains of different genotype *in vitro*, and then used the heteroduplex DNA to transform recipient cells. She found that with two mutants initially in the trans configuration, that is, one in one polynucleotide and the other in the complementary one, doubly transformed progeny were as frequent as with both mutants in the same chain. Since only one chain of donor DNA is integrated into the recipient DNA molecule in transformation, she inferred that a correction mechanism acts on the heteroduplex DNA immediately after it enters the cell, and at each site of mispairing excises a segment of one chain and replaces it by nucleotides complementary to the other one. This would often generate polynucleotides carrying both mutants.

Repair of mispaired bases in virus λ was demonstrated in an ingenious experiment by Hogness *et al.* (1967). Complementary nucleotide chains of $\lambda$ differing by a mutation in gene $N$ (Fig. 16.4) were annealed. Whether the mutation was in the heavy or the light chain (Fig. 14.6) made no difference to the activity of the wild-type allele, evidently because repair could transfer the wild-type base sequence to the chain that was transcribed. In confirmation of this conclusion, pre-irradiation of the host with a heavy dose of ultraviolet light evidently saturated the repair system, because the $\lambda$ duplex having the $N$ mutant in the heavy chain was then inactive. This experiment suggests that repair in either direction may occur equally often. The conclusions reached from this experiment would be invalid if sufficient gene $N$ activity had occurred to allow a wild-type transcribed chain to be formed where it was previously lacking but, as Doerfler and Hogness (1968) pointed out, the sensitivity to UV would then be unaccounted for.

Spatz and Trautner (1970) experimented with heteroduplex DNA of virus *SPP*1 of *Bacillus subtilis*. As with $\lambda$, the complementary nucleotide chains of this virus can be separated by centrifuging in a caesium chloride density gradient. Reciprocal heteroduplex molecules were prepared by mixing complementary polynucleotides derived from wild-type and mutant virus, respectively, and incubating at 70°C to allow them to anneal. Bacteria were then infected with the viral DNA: this transfection, as it is called, is possible at a certain stage of the cell division cycle, as with transformation. The frequencies of cells yielding wild-type virus only, mutant only, or both, were measured. For example, with mutant 161 induced by hydroxylamine, over 800 cells were investigated for each kind of heteroduplex: when the heavy chain of the virus was wild-type and the light mutant, 69% of the cells gave wild-type particles only and 8% mutant only, while with the heavy chain mutant and light wild-type, 80% of the cells gave mutant particles only and 5% wild-type only. In each experiment, the remaining cells gave both wild-type and mutant particles. It was evident that correction of mispairing was occurring in the heteroduplex DNA and that with mutant 161 it favoured the heavy chain. Other mutants behaved similarly except that the chain preference was less pronounced and with some the light chain was favoured.

Hydroxylamine is known to change cytosine to thymine (§ 14.2), so with such mutants one heteroduplex will have guanine mispaired with thymine and the other adenine with cytosine, with the purines in the heavy chain and the pyrimidines in the light one, or *vice versa*. By inducing mutations in the separated chains of the virus, the purine-pyrimidine orientation relative to heavy and light was known. It was found that the correction mechanism favoured purine excision with some of the mutants and pyrimidine excision with others. Spatz and Trautner concluded that with *SPP*1 the specificity of conversion is influenced by the neighbouring base sequence, and it is now evident (Trautner *et al.*, 1973) that the heavy *versus* light preference depends on the position of the mutation on the linkage map. With mutants in the left-hand part of the map, correction generally favours the light chain, while for mutants in the right-hand part, and also on the extreme left, it favours the heavy chain. Trautner *et al.* suggested that this conversion pattern might be related to that of transcription, which also shows such patterns of specificity (cf. $\lambda$, Fig. 14.6), though the transcription pattern of *SPP*1 is unknown.

Trautner *et al.* also discovered that the majority of heteroduplex molecules with deletions, and some with amber mutants, showed preferential correction to the mutant genotype. Although single chain excision and repair occurring independently of the sites of mispairing has not been ruled out as an explanation of the *SPP*1 data, the specific conversion patterns of mutants point to correction triggered by the mismatched bases. The direction of conversion seems usually to be influenced by the neighbouring base sequence, but with some mutants, particularly deletions, the direction is evidently determined by the molecular nature of the mismatch. This variation provides an interesting parallel with conversion in fungi where, in general, some mutants show conversion to wild-type and to mutant equally often, and others favour one parent.

Evidence for correction of mispairing in *B. subtilis* has also been obtained from studies on transformation. Bresler, Kreneva and Kushev (1968) found that transformed cells, when transferred to non-selective medium before they had replicated, often gave rise to cells all of only one kind, that is, all with the donor character or all with the recipient one, implying that the initial mispairing in hybrid DNA had been corrected.

Zimmermann (1968) studied the phenotype resulting from mitotic gene conversion at the isoleucine-1 (*is*-1) locus in *Saccharomyces cerevisiae*. Conversion from mutant to wild-type was induced by N-methyl-N-nitroso-urethane in a diploid heterozygous for two *is*-1 mutants *a* and *b*; this led to the production of the enzyme threonine dehydrase. Zimmermann found that 23 convertants were all identical with the wild-type for each of 9 properties of the enzyme that were examined. He concluded that conversion is an accurate process of information transfer, as predicted by hybrid DNA models. Fogel and Mortimer (1970) reached similar conclusions for conversion at meiosis in *S. cerevisiae* from study of a large sample of unselected asci heterozygous for ochre mutants in the *arginine*-4 gene or for an ochre-specific suppressor. Altogether, 189 asci showing conversion to the mutant character were analysed and all showed complete fidelity of information transfer, even though single base changes could have been detected, either in the ochre codons or in the anticodon region of the gene for the tyrosine transfer RNA believed to be responsible for the ochre suppression.

In spite of this precision in conversion, the idea that crossing-over occurs with such molecular precision that no nucleotide is lost or gained must be qualified in view of the discovery by Magni and von Borstel (1962) that mutation at a number of nutritional loci in *Saccharomyces cerevisiae* is 6 to 20 times as frequent at meiosis as at mitosis and, moreover (Magni 1963), is often associated with crossing-over at or near the site of mutation. The strains used were initially homozygous at the loci where mutation occurred. Demerec (1962) observed a similar phenomenon in *Salmonella typhimurium*. Magni (1964) found that nutritional mutants in *Saccharomyces* thought to be due to base substitution (for example, because they were induced by base analogues) show at meiosis the same low rate of reverse mutation to wild-type as at mitosis. He inferred that the high mutation-rate at crossing-over is exclusively with mutations whose molecular basis is the addition or deletion of one or more nucleotides. Paszewski and Surzycki (1964) studied what appears to be the same phenomenon using a spore-colour mutant in *Ascobolus immersus*, and found that back-mutations affect 2 of the 6 spores in the ascus and that, in the one ascus tested, the 6 mutant spores all behaved alike as regards further mutation. It is tentatively concluded that the phenomenon affects only one of the two chromatids taking part in crossing-over.

On hybrid DNA theories of crossing-over, it is possible that one or more nucleotides might be gained or lost if a nucleotide chain ended at the position of a run of identical bases, and if the tip of the chain initiated the contact with the complementary chain from the other molecule. Displaced pairing might then arise. For example, if the chain ended with a run of thymines, it might begin to associate with the complementary chain displaced by one nucleotide, thus:

$$--- C C A C C T T T G G A G_{G\,T_{T\,T\,T}}$$
$$G G T G G A A A C C T C C A A A A T T A C T G A ---$$

This might lead to a distorted molecule:

$$\phantom{--- C C A C C T T T G G A G G}\; T$$
$$--- C C A C C T T T G G A G G\; T T T$$
$$G G T G G A A A C C T C C\; A A A A T T A C T G A ---$$

with the possibility of correction of the distortion by the addition of a nucleotide complementary to the unpaired one. Conversely, loss of a nucleotide could arise if the pairing was displaced in the other direction, and if the resulting distortion in the molecule were corrected by removal of the unpaired nucleotide. It is possible that the high spontaneous mutability at certain sites ('hot spots') in the *r*II region of virus *T*4 of *Escherichia coli* (see § 14.2) might be due to such displaced pairing. Streisinger, Okada *et al.* (1967) made the same suggestion to explain hot spots and obtained experimental evidence for it: the neighbourhood of mutant eJ42 in the lysozyme gene of *T*4 is a hot spot and is thought to have 5 adenines in a row (cf. Fig. 14.14(*a*)).

## § 16.8  Crossing-over as a gene-specific process

Whitehouse and Hastings (1965) concluded from study of all the relevant data that polarity in recombination was of general occurrence in Ascomycetes. From the pattern of this polarity it was inferred that the opening-points in the DNA where crossing-over is initiated are situated between the genes (or at their ends) and not within them. Most of the genes that have been studied show a reversal of polarity towards one end, as though hybrid DNA could be initiated at each end of every gene, but more extensively from one end than the other. I suggested (Whitehouse, 1966) that the DNA could be opened for recombination (that is, one chain broken and the pair dissociated) only at one specific end of each gene—the 'recombination-operator' end, but that the hybrid DNA from this starting-point might extend through this gene and into another. The predominant direction of polarity within a gene would then indicate at which end it was opened, while the weaker reversed polarity at the other end would be the overshoot of another gene.

This hypothesis is supported by the remarkable discoveries made by Catcheside and associates (Jessop and Catcheside, 1965; Catcheside, 1966; Smith, 1966; Angel, Austin and Catcheside, 1970; Catcheside and Austin, 1971) of dominant repressors of intragenic recombination in *Neurospora*. Each recombination-repressor seems to be specific for particular sites at scattered positions in the chromosomes, and to be determined by a specific *rec* (recombination) gene which is not linked (or only loosely linked) to the sites on which it acts. The repressors reduce recombination 3- to 15-fold within genes near their sites of action. Moreover, in every instance investigated the pattern of polarity in recombination within these genes is altered in the manner expected if the repressors act to close an opening point for recombination to one side of the gene, without affecting opening points on the other side (Whitehouse, 1966; Thomas and Catcheside, 1969; Angel, Austin and Catcheside, 1970; Smyth, 1971). Three recombination repressors have so far been detected and each affects recombination in one or two genes out of about 10 tested. Angel *et al.* suggested that each repressor controls the synthesis of an endonuclease which initiates recombination at specific sites. They favoured this hypothesis because a mutant (*cog*) at one of these sites behaved as a dominant promoter of recombination in a neighbouring gene (*his*-3) when recombination was derepressed. They argued that the *cog* mutation was associated with a difference in the ease of recognition of the site by the endonuclease.

In bacteria, site-specific endonucleases are known which recognize particular nucleotide sequences and thereby restrict the action of foreign DNA by enabling the organism to destroy the DNA of any infecting virus which possesses the sequence. Self-destruction of the host DNA is believed to be prevented by modifications to the sequence such as methylation of the bases. Kelly and Smith (1970) determined the base sequence of the recognition site of such a restriction enzyme from *Haemophilus influenzae* and found it to be:

$$5'-G-T-Y-R-A-C-3'$$
$$\cdot \quad \cdot \quad \cdot \quad \cdot \quad \cdot \quad \cdot$$
$$3'-C-A-R-Y-T-G-5'$$

The symmetry suggests that the enzyme is a dimer and that it cuts both chains at once. The phosphodiester links are cleaved between the purine (R) and pyrimidine (Y) in each chain. If site-specific endonucleases initiate recombination in chromosomal organisms, as Angel *et al.* believe, they are likely to cut only one chain of the DNA, because severing both chains at the same position precludes rejoining by hybrid DNA formation.

Another possibility to account for the specificity of action of the recombination repressors is that they block hybrid DNA formation from unsealed replicon junctions (Whitehouse, 1972; see § 18.11). D. E. A. Catcheside (1968) found that the repressor which reduces recombination within the *am-1* gene of *Neurospora* did not repress production of the glutamate dehydrogenase specified by *am-1*, so there is no reason to think that the repression of recombination is related to that of transcription.

Simchen and Stamberg (1969) found an organization in *Schizophyllum commune* similar to that in *Neurospora*, with dominant repressors of recombination affecting specific regions of the chromosomes.

Murray (1970) obtained genetic evidence from *Neurospora* that hybrid DNA can extend from one opening point through another potential opening point. This inference was based on studies of recombination in two neighbouring genes concerned with the methionine pathway, *me-7* and *me-9*. Evidence that hybrid DNA can extend from one gene to the other was obtained by selecting for wild-type recombinants in crosses between a mutant in each gene, and also by such selection within one of the genes, followed by scoring for an unselected mutant in the other gene. The pattern of polarity in recombination, however, indicated a discontinuity between the genes, with much of the hybrid DNA that entered *me-7* originating between them. It is evident that the hybrid DNA entering each end of a gene may arise from more than one source, and that to prevent all recombination in a chromosome region might require closure of two or more opening points on each side of it.

The origin of hybrid DNA between genes rather than within them is accounted for in the recombination model proposed by Sobell (1972) and discussed in § 16.5, because he suggests that the hybrid DNA is initiated in secondary loops arising from reversed duplications of nucleotide sequence.

If crossing-over can be initiated only at a gene end and not within it, this represents a step back towards the old bead hypothesis of the gene, but with a significant difference. The old idea (§ 9.9) was that crossing-over was confined to the links between genes, but it is now suggested that crossing-over is merely initiated at the end of a gene and then the molecular processes involved extend into it. Moreover, there is a secondary process of recombination, namely, the correction of heterozygosity, which occurs only when a heterozygous gene lies within a crossover, and is responsible for the phenomenon of conversion.

## § 16.9   Conclusions

It is now generally accepted that recombination occurs, at least predominantly, by breakage and rejoining through hybrid DNA formation. There are two main exceptions (see § 16.5) to this conclusion: (1) the special recombination process involved in the integration of viral DNA into the bacterial chromoneme may not require hybrid DNA, and (2) the importance of break and copy models, such as that proposed by Boon and Zinder, cannot yet be judged.

The migratory hybrid DNA structure which Sobell (1972) has proposed (see § 16.5) will account for several hitherto puzzling features of fungal recombination data:

(1) The M26 mutant in the *adenine*-6 gene of *Schizosaccharomyces pombe* studied by Gutz (1971), and discussed in § 16.7, shows a much higher conversion frequency than alleles, although it is not situated at one end of the gene. If, unlike M26, most of the *ad*-6 mutants frequently fail to trigger conversion, a zone of hybrid DNA might often pass through the gene and not be detected when they are present.

(2) It is well documented that reciprocal recombination of alleles may be of frequent occurrence with particular mutants and that, unlike conversion, it is almost always associated with coincident recombination of genes on either side (see Whitehouse, 1970). As with M26, particular mutants, on entering a migratory hybrid DNA structure, may trigger endonuclease attack. This could fix the starting point at which the hybrid DNA begins in the final recombinant molecule, and could explain how a hybrid DNA segment can ultimately be confined to mid-gene although initiated outside it.

(3) Recombination of many allelic mutants is almost entirely non-reciprocal, even though correction of mispairing in opposite directions in the two chromatids would be expected often to lead to reciprocal recombination with a mutant outside the region of hybrid DNA (§ 16.7). Sobell suggested that a mutant, on entering hybrid DNA, triggers the enzymic nicking of one chain in each duplex and that the enzyme also has exonuclease activity (like the *recBC* DNase of *E. coli*—see § 16.5) and degrades one of the two nicked chains. This would mean that the hybrid DNA was largely confined to one molecule, with the consequence that recombination between closely linked mutants was non-reciprocal. The idea that in individual recombination events hybrid DNA is of greater extent in one chromatid than the other is directly supported by data for asci with 5:3 and 3:5 ratios of wild-type to mutant spores. Such data have been obtained by Kitani and Olive (1967) with the *grey* spore colour gene in *Sordaria fimicola* (see Whitehouse, 1970, 1972) and by Stadler and Towe (1971) with the *w*17 spore colour gene in *Ascobolus immersus*.

There is good reason to believe that the recombination phenomena discovered in fungi are likely to apply generally in chromosomal organisms, because essentially similar results have now been obtained with *Drosophila melanogaster*. Carlson (1971) with *rudimentary* wing mutants and Chovnick, Ballantyne and Holm (1971) with *rosy* eye mutants (which are mutants of the structural gene for xanthine dehydrogenase) have demonstrated (1) conversion with and without reciprocal exchange of flanking genes, (2) polarity, (3) crossing-over between alleles, and (4) probable examples of postmeiotic segregation. Carlson's extensive data, involving over 3000 intragenic recombinants scored for flanking marker genes, also revealed two other fungal recombination phenomena: map expansion, and a lack of interference between a crossover and a conversion event when this is associated with a parental arrangement of flanking genes. This is unlike two crossovers, which usually show interference with one another (§§ 8.7 and 8.9).

The development of means for selecting fungal mutants deficient in recombination opens the possibility of substantial progress in understanding the biochemistry of the process. Roth and Fogel (1971) used an otherwise haploid strain of *Saccharomyces cerevisiae* carrying two copies of chromosome III which differed in the mating-type alleles and in mutations of the *leucine*-2 gene. Numerous mutants deficient in meiotic recombinants were detected by their reduced yield of leucine-independent ascospores. The similar technique applied by Rodarte-Ramón and Mortimer (1972) for isolating mutants defective in mitotic intragenic recombination, using a different disomic strain, was described in § 16.6. Rodarte-Ramón (1972) found that one of these mutants, *rec*-4, was also deficient in meiotic intragenic recombination but not in crossing-over. It seems likely, therefore, that the *rec*-4 gene plays an essential part in conversion, whether meiotic or mitotic, but not in the basic recombination mechanism of reciprocal exchange. This conclusion is in keeping with hybrid DNA models, in which correction of mispairing in heteroduplex DNA is regarded as a secondary non-essential attribute of the basic breakage and rejoining process.

The possibility is discussed in § 18.11 that the initial step in crossing-over is not breakage of nucleotide-chains but failure to join them at replicon junctions after the premeiotic DNA replication.

# *17.* The theory of the single-stranded chromosome

## § 17.1  Single-stranded *versus* multistranded chromosomes

Ever since the establishment of the DNA theory of the molecular basis of heredity (Chapter 11), there has been controversy over the question of whether the germ-line chromosome is single-stranded or multistranded. According to the single-strand hypothesis, the chromosome axis consists essentially of only a single DNA duplex (or a number joined end-to-end), while the multistrand hypothesis supposes that the chromosome is built like a cable with the genetic information repeated in two or more DNA molecules lying parallel to one another. The single-strand model is favoured by most geneticists because of the difficulty otherwise of explaining mutation and recombination. The multistranded structure has been favoured by many cytologists. The evidence relating to this controversy is reviewed in this chapter.

## § 17.2  Light microscopy of chromosomes

Claims that telophase chromosomes are multistranded have been made on many occasions on the basis of their appearance under the light microscope. References to such reports are given by Bajer (1965) who has found from phase-contrast time-lapse photography of mitosis in living endosperm cells of *Haemanthus katherinae* that the chromosomes are sometimes clearly double at anaphase and telophase. Trosko and Wolff (1965) observed four strands in the chromosomes of *Vicia faba* root-tip cells after trypsin digestion and concluded that the chromosomes are multistranded. Maguire (1966) observed double-stranded chromosomes in *Zea mays*. It is manifest, however, that light microscope observations, with a resolving-power of 0·2–0·3 $\mu$m (Bajer 1965), cannot distinguish between a multistranded condition and one very long thread less than 0·2 $\mu$m wide and folded into aggregates.

## § 17.3  Electron microscopy of chromosomes

From study of electron micrographs of chromosomes from a number of different plants and animals at various stages of mitosis and meiosis, Ris (1957) came to the conclusion that a chromosome is multistranded in the sense that it is built like a cable with numerous identical sub-units. Kaufmann and McDonald (1957) reached similar conclusions. However, the possibility that the chromosome is single-stranded does not appear to be excluded by these observations because the multiple strands observed with the electron microscope might represent a single strand folded back and forth upon itself a number of times.

Gall (1966) used the technique of spreading material at an air-water interface, first developed by Kleinschmidt *et al.* (1962) (see § 11.3), and applied it to chromosomes. By electron microscopy after such treatment he observed fibres of diameter 25–30 nm in chromosomes from *Triturus* erythrocytes and human cells in culture. He saw no evidence for a multi-stranded condition within these fibres and from their size and their ability to stretch he believed they contained one coiled or folded nucleohistone molecule. Wolfe (1965) and DuPraw (1965) reached a similar conclusion using Gall's technique applied both to interphase nuclei and to metaphase chromosomes of *Triturus*, *Bos* and *Apis mellifica*. The problem of the number of strands per chromosome then resolves into how many of these fibres make up a chromosome. In this connection, Wolfe and Hewitt (1966) made a significant observation. They found that the number of 25 nm fibres in the chromosomes of the hemipterous insect *Oncopeltus fasciatus* increases during prophase of meiosis from 15–40 at zygotene to 25–50 at pachytene and to 30–65 at diakinesis; all counts refer to bivalents, that is, four chromatids. Since it is well known that chromosomes show thickening and contraction in length during prophase, Wolfe and Hewitt concluded that the increase in the number of fibres seen under the electron microscope is caused by parallel folding. They argue from this that it is quite possible that the multistranded appearance seen at leptotene and zygotene is also caused by such folding.

Miller (1965) reported that lampbrush chromosomes (§ 17.6) of *Triturus viridescens*, after digestion with proteases and ribonuclease, revealed under the electron microscope a thread 5 nm wide between chromomeres and 3 nm wide in the lateral loops. The interchromomeric region in *Triturus* lampbrush chromosomes is thought to correspond to two chromatids in close association, and the lateral loops to one chromatid. Since the DNA molecule is 2 nm wide, Miller's observations, if confirmed, would establish that the chromosome is single-stranded as regards its DNA.

## § 17.4   Tritiated thymidine labelling

If chromosomes replicate once in the presence of tritiated thymidine and then once in an unlabelled solution, each pair of sister chromatids is usually labelled in one chromatid and not in the other, though sister-chromatid exchanges may occur (§ 12.2). Corresponding segments of both chromatids, or even the whole of them, however, are sometimes found to be labelled at this second mitosis (La Cour and Pelc, 1958; Peacock, 1963). This iso-labelling, as it has been called, is usually interpreted as evidence that the chromosome is multistranded. The iso-labelling can be explained, however, with a single-stranded chromosome. Comings (1971) pointed out that longitudinal folding of the DNA axis, in conjunction with a sister-chromatid exchange, would account for it (see Figs. 12.2 (iv) and 12.3 (iv)). According to this hypothesis, iso-labelling is merely an exchange that has become hidden by the folding of the chromatid axis in the much contracted metaphase chromosome.

As mentioned in § 12.2, strong support for the single-strand hypothesis of the structure of the chromosome is given by Herreros and Giannelli's (1967) confirmation of the 2 : 1 ratio of single : twin exchanges, implying that the two longitudinally-arranged sub-units of the chromosome, which replicate semi-conservatively, are structurally different from one another. This argument has been strengthened still further by the demonstration by Brewen and Peacock (1969) that this structural difference is polarized. Cells of *Cricetulus griseus* in culture were treated with $^3$H-thymidine for 6 hours and then exposed to 200 r of X-rays. Colcemide was added 1 hour later to inhibit anaphase. Beginning 24 hours later cells were fixed and stained, and the positions of tetraploid metaphase cells with mirror-image dicentrics marked. These symmetrical chromosomes with two centromeres had evidently arisen as a consequence of X-ray induced sister-chromatid exchange but, unlike a normal sister-chromatid exchange, the rejoining had given U-shaped configurations (Fig. 17.1 (*d*), (*e*)) instead of an X-shaped one. One of the U-shaped chromatids would lack a centromere and the other would have two. The cells were autoradiographed and the labelling pattern in the mirror-image dicentrics recorded. Out of 137 such chromosomes, 104 had the label confined to one chromatid in the region between the two centromeres (Fig. 17.1 (*l*)), and the remaining 33 had a label switch in this region, 6 at the mid-point (Fig. 17.1 (*m*)) and 27 not (Fig. 17.1 (*k*)). Label exchanges outside the intercentromere region were ignored and are not shown in the diagram. With the observed frequency of sister-chromatid exchanges (Fig. 17.1 (*i*)), which evidently caused the asymmetrical label switches, it was estimated that between 3 and 8 would have been expected to occur sufficiently near the mid-point (Fig. 17.1 (*j*)) to have been counted as such. Brewen and Peacock concluded that all the joining to give the dicentrics was labelled sub-unit to labelled sub-unit and unlabelled to unlabelled (Fig. 17.1 (*d*)), and none labelled to unlabelled (Fig. 17.1 (*e*)). This is the opposite rejoining pattern to that observed with an X-shaped exchange (Figs. 12.2 (ii) and 12.3 (ii)) and implies that the structural difference between the two sub-units of a chromatid is one of polarity, such that reversing the direction of a sub-unit removes the difference. This is expected with a single-stranded chromosome, because the two nucleotide-chains of the DNA molecule differ in polarity.

## § 17.5   X-ray-induced aberrations

It has been known for some time that doses of X-rays applied to cells in G1 (the first gap) between telophase and the period of DNA synthesis (S period) give rise in general to chromosome-type aberrations at the following metaphase, whereas irradiation in G2 (after the S period) causes chromatid-type aberrations such as those discussed in § 9.8. These results are expected if the division of the chromosome into two chromatids coincides

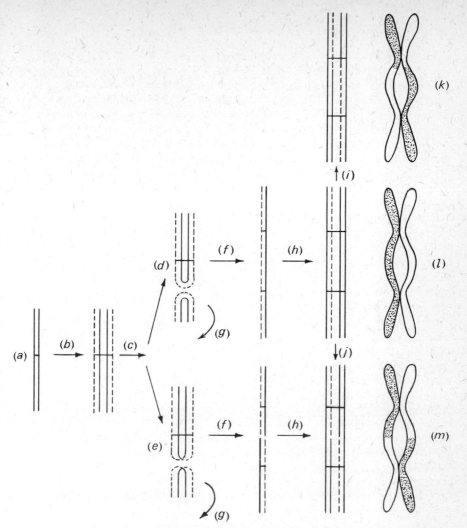

FIGURE 17.1    Diagram to illustrate the experiment of Brewen and Peacock (1969)
labelling X-ray induced mirror-image dicentric chromosomes of
*Cricetulus griseus* with ³H-thymidine. Drawings (*k-m*) show the
labelling patterns observed, and diagrams (*a-j*) the suggested
explanation. The vertical lines represent the sub-units of the
chromatids, broken lines indicating sub-units synthesized in the
presence of ³H-thymidine. Horizontal lines indicate centromeres.
(*a*) Chromosome before replication. (*b*) Replication in the presence
of ³H-thymidine. (*c*) X-irradiation. (*d*), (*e*) X-ray-induced sister-
chromatid exchanges with rejoining to give U-shaped dicentric and
acentric chromatids; (*d*) joining of labelled sub-unit to labelled
sub-unit and unlabelled to unlabelled; (*e*) joining of labelled
sub-units to unlabelled. (*f*) Division of centromere. (*g*) Behaviour
of acentric chromatid not followed further. (*h*) Replication of
dicentric chromatid in the absence of ³H-thymidine. (*i*) (*j*) Sister-
chromatid exchange in the region between the centromeres;
(*i*) not near the mid-point; (*j*) near the mid-point. (*k*) (*l*) (*m*)
Labelling patterns observed in the region between the centromeres;
(*k*) exchange not at mid-point; (*l*) no exchange; (*m*) exchange at
mid-point.

with the period of DNA synthesis because chromosome-type aberrations have both chromatids involved in similar structural changes, as though the aberration occurred before chromosome duplication. Evans and Savage (1963) and Wolff and Luippold (1964), however, independently discovered, by the use of $^3$H-thymidine labelling in conjunction with X-irradiation, that the chromosomes of *Vicia faba* become double to X-rays in late G1 about one hour at room temperature before DNA synthesis begins. This discovery has been used as an argument for the multistrandedness of the chromosome. A single-strand model, however, does not seem to be excluded when the probable steps in the formation of aberrations are considered, using such a model.

Damage to chromosomes by X-rays is likely to involve breakage of single nucleotide-chains, either as a direct effect, or indirectly through the enzymic excision of damaged bases (§ 16.6). The formation of structural aberrations is probably caused by chance encounters between unpaired chains of non-homologous origin exposed in this way. Annealing would occur if there were segments in antiparallel juxtaposition with a sufficient length of a mutually complementary sequence of nucleotides. Subsequently, on the basis of what is known from *Escherichia coli*, repair of DNA would be expected to occur, and to require the successive action of an exonuclease to erode one chain in order to enlarge a gap, a polymerase to fill the gap, and a ligase to seal it. An endonuclease would also seem to be needed at some stage of repair in order to cut one chain where Y-shaped configurations occurred; depending on which chain was cut, a potential aberration would either be fixed or removed. Evans (1967) proposed a similar 'misrepair' model for the formation of aberrations. He suggested that in *Vicia faba* root-tips grown at room temperature the repair of DNA following X-ray damage occurred 5 to 6 hours after irradiation, because he found a fractionated X-ray dose with this time interval caused as many aberrations as a single dose of the same total size given at the earlier time. He suggested that during repair the chromosomes were temporarily more vulnerable to the formation of aberrations. In molecular terms, the single chains exposed through breakage by the first half-dose are exposed again during repair, with the possibility of chance contact with chains exposed through breakage by the second half-dose.

If aberrations are not fully defined until DNA repair occurs, it is possible that the single-strandedness or double-strandedness of chromosomes to X-rays, that is, whether they form chromosome-type or chromatid-type aberrations, is a reflection of their condition, not at the time of irradiation, but at the time of the subsequent repair of the DNA. It seems significant that the time of transition from singleness to doubleness to X-rays (late G1) corresponds to a time for DNA repair (allowing a 5–6 hour interval) in the middle of the S period, since this lasts about 8 hours at room temperature. On this interpretation, therefore, the X-ray aberration data are in keeping with a single-strand model for the chromosome.

## § 17.6   Lampbrush chromosomes

The chromosomes at diplotene of meiosis in female *Triturus* (newts) show a remarkable feature which has earned them the name of lampbrush chromosomes (review: Callan, 1963; see § 8.13 and Pl. 17.1). The chromosomes at this stage are very long (up to 800 $\mu$m) and have hundreds of loops projecting laterally for 10–15 $\mu$m or more, giving a fuzzy or lampbrush appearance. The loops arise in pairs from chromomeres (Fig. 17.2(*d*)), and are thought to be associated with the functioning of some of the genes, since RNA and protein synthesis have been demonstrated autoradiographically as occurring alongside them (see § 18.7). This synthesis is no doubt related to the considerable growth of the oocytes which occurs during the period of at least 200 days

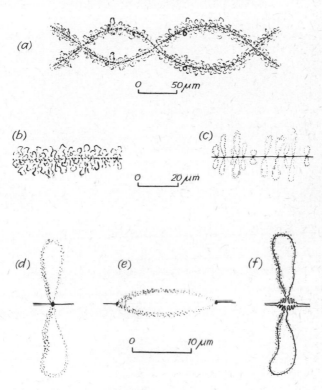

FIGURE 17.2   Drawings to illustrate the appearance of the chromosomes at the diplotene stage of meiosis in oocytes of *Triturus*.

(*a*) One chromosome-pair with three chiasmata.

(*b*) Part of one chromosome, to show the lampbrush appearance in more detail.

(*c*) Diagrammatic representation of (*b*) to illustrate how the loops arise in pairs from the chromomeres.

(*d*) One such pair of loops in more detail.

(*e*) The appearance of such a loop-pair after stretching (from Callan).

(*f*) The structure of a loop-pair, according to Gall (1956).

PLATE 17.1 Lampbrush chromosomes from oocytes at the diplotene stage of
meiosis in subspecies *carnifex* of the newt *Triturus cristatus*. The
highly refractive spherical objects are free nucleoli (compare § 18.11).
(*Courtesy of H. G. Callan*)

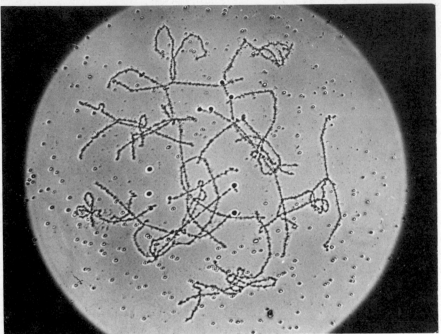

(*a*) Phase-contrast photograph of the entire chromosome complement
consisting of 12 bivalents (×60). Each has 2 or 3 chiasmata.

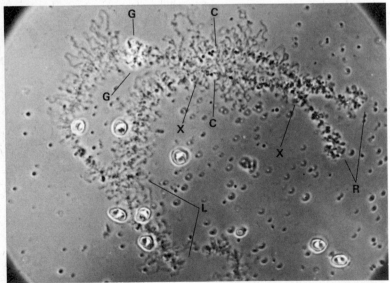

(*b*) Bivalent XII showing the lateral loops in more detail
(×300). C = Centromere. G = Giant fusing loop. L = Left-
hand end. R = Right-hand end. X = Chiasma. The individual
from which this chromosome-pair was obtained was heterozygous
for the presence and absence of loop G. In spite of a proximal
chiasma, the two chromatids with this loop remain paired, and
likewise with the two without the loop (see § 8.13).

PLATE 17.1 (continued) Lateral loops of lampbrush chromosomes from oocytes of subspecies *cristatus* of the newt *Triturus cristatus*, showing the polarity in loop organization (see § 18.7) (× 700).                    (*Courtesy of H. G. Callan*)

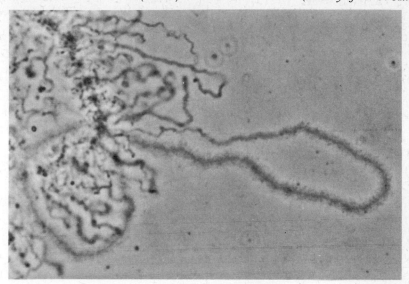

(*c*) A large normal loop on chromosome I showing the greater thickness of matrix at one end of the loop compared with the other. The partner loop encircles the end of the chromosome.

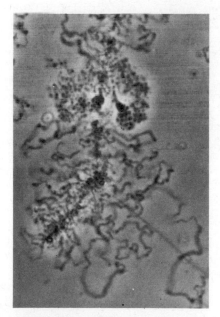

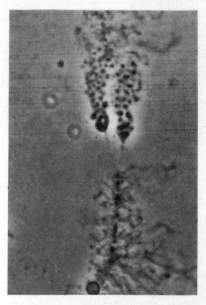

(*d*) The pair of giant granular loops which occur on the left arm of chromosome XII. These loops showed progressive labelling starting at the thin end when exposed to ³H-uridine.

(*e*) The pair of loops seen in (*d*) forms a double bridge when the chromosome is stretched (cf. Fig. 17.2). These particular loops always have their thin ends directed away from the centromere.

that the nucleus remains at the diplotene stage. The nucleus itself increases greatly in size during this period. Callan found that when a lampbrush chromosome is stretched, the chromomeres break transversely and a pair of lateral loops spans the break (Fig. 17.2 and Pl. 17.1(*d*) and (*e*)). He inferred that the loops have an axis which is continuous with the fine thread which joins the chromomeres, and forms the main axis of the chromosome. From electron micrographs of lampbrush chromosomes after pepsin digestion, Gall (1956) confirmed that there is an axis to the lateral loops, and moreover showed that it was resistant to the enzyme. His interpretation of the structure of a lampbrush chromosome is shown in Fig. 17.2(*f*). Callan and Macgregor (1958) transferred unfixed *Triturus* lampbrush chromosomes to physiological saline, and studied the effect of enzymes on them. They found that pepsin and trypsin, which split peptide linkages near free acidic and basic groups respectively, and ribonuclease, which breaks down RNA, did not destroy the linear integrity of the chromosomes, whereas a deoxyribonuclease (DNase) which breaks down DNA, had a dramatically different action: the lateral loops and the thread joining the chromomeres rapidly disintegrated. They inferred that DNA runs throughout the length of the chromosome, including the axis of the lateral loops. Dutt and Kaufmann (1959) observed no disruption of meiotic chromosomes in males of the grasshoppers *Trimerotropis maritima* and *Melanoplus femur-rubrum* when unfixed material was treated with DNase, but Macgregor and Callan (1962) attribute this to the relatively compact condition of the chromosomes at the time of treatment, compared with the remarkably extended diplotene chromosomes in female *Triturus*. Macgregor and Callan have found that much contracted parts of these amphibian chromosomes, such as the chromomeres, also fail to show breakage by DNase.

Gall (1963) studied the kinetics of the DNase breakage of *Triturus* lampbrush chromosomes. Unfixed chromosomes were photographed at intervals over a period of one hour after placing in the enzyme solution. From the photographs the number of breaks was counted and plotted against the time. Breaks in the lateral loops were scored separately from breaks in the main axis. It was assumed that the chromosome consists of $n$ longitudinal sub-units, that each is attacked independently and with equal probability by the enzyme, and that all of them have to be digested at approximately corresponding positions in order to cause a break in the chromosome. The number of breaks will then be directly proportional to the time if there is one sub-unit, proportional to the square of the time if there are two sub-units, to the time cubed if there are three, and so on. Gall obtained the values $n = 2.6 \pm 0.2$ for the loops, and $n = 4.8 \pm 0.4$ for the main axis. He considered the true sub-unit numbers to be 2 and 4, respectively, because delay in penetration of the enzyme through the nuclear sap, and delay in observing a break after it had occurred, would both tend to raise the observed values of $n$ above the true one. That there should be twice as many sub-units in the main axis as in the loops is in keeping with his earlier interpretation of the structure of lampbrush chromosomes (Fig. 17.2(*f*)), according to which the loops are parts of the axis of individual chromatids, while the main axis consists of a pair of sister chromatids in close association with one another. From *in vitro* studies, Thomas (1956) found evidence that DNase cleaves the phospho-ester links in the two nucleotide chains of DNA independently and at random, and that breakage of the molecule occurs only when the two chains are severed less than about 3 nucleotide-pairs apart. Gall inferred that the two sub-units of the chromatid were the two nucleotide chains, and that the chromatid contains either one very long DNA molecule or a series of DNA molecules connected end-to-end. Gall's results thus provide strong support for the single-strand hypothesis.

Lampbrush loops are not confined to the chromosomes of amphibian oocytes. They have been seen at diplotene in the oocytes of a wide range of animals (see Callan and Lloyd, 1960) and in the spermatocytes of insects: 7 loop-pairs have been observed by Hess and Meyer (1968) in the Y-chromosome of *Drosophila hydei* and numerous loops were seen by Henderson (1964) in the autosomes of orthopteran insects (see Plate 6.2($g$) and ($h$)). Lampbrush loops were also observed at diplotene in *Allium cernuum* by Grun (1958), who gave reasons for believing that these loops are of general occurrence at this stage. Henderson considered it likely that all prophase chromosomes have the lampbrush organization.

## § 17.7   Induced subchromatid exchanges

Exchanges which appear to be between sub-units of chromatids were first clearly described by Swanson (1947). These aberrations were seen at anaphase in the pollen-tube mitosis of *Tradescantia*, following X-ray treatment at the preceding prophase. The anaphase bridges with two side arms caused by these subchromatid exchanges have since been observed in many organisms, and at meiosis as well as mitosis. If they were due to subchromatid exchange, as most observers have inferred, this would imply that the chromatid was at least two-stranded. It is possible, however, that these exchanges, when induced by irradiation or chemicals, occur between lampbrush loops, either between those of sister chromatids or of non-sisters, or between non-homologous loops. Assuming that the axis of the loop is continuous with the chromatid axis (Fig. 17.2($f$)) such an exchange would really be between whole chromatids. The subchromatid appearance would be due to a failure of transverse breakage of the chromomeres whose loops underwent exchange; such breakage would not be expected in contracted chromosomes. Subchromatid exchanges are normally induced only at mitotic prophase or late prophase I of meiosis, which are the stages when chromosomes are known or believed to have the lampbrush configuration (§ 17.6).

Irradiation or chemical treatment in interphase does not produce exchanges of subchromatid appearance at the next mitosis. In interphase the chromosome or chromatid is not aggregated into chromomeres, and consequently, with an exchange at positions corresponding to lampbrush loops, the double association by loop axis and by chromomere would not be expected. An interphase exchange will thus not be expected to give rise at mitosis to the subchromatid appearance characteristic of exchanges induced at a time when the chromosomes are strongly contracted. Taylor (1958) has obtained direct experimental evidence by autoradiography with $^3$H-thymidine that subchromatid exchanges are not induced in the S or G2 periods of interphase.

Induced subchromatid exchanges, if between half-chromatids, would be expected to give rise to chromatid-type aberrations at the next mitosis. On the other hand, if subchromatid exchanges are really between lampbrush loops and hence whole chromatids, they would be expected to give rise at the next mitosis to chromosome-type aberrations. One of the characteristic features of induced subchromatid exchanges is the rarity of chromatid aberrations at the next division, in agreement with the present hypothesis. Moreover, Östergren and Wakonig (1954) and Kihlman and Hartley (1967) found numerous chromosome-type aberrations in *Allium cepa* and *Vicia faba* root-tips at the mitosis following that at which subchromatid exchanges had been induced.

## § 17.8   DNA content of nuclei

One of the main arguments for the multistranded chromosome has been based on the variation in DNA content of the chromosomes of related organisms. Both animals and plants show this variation to an extraordinary degree. Mirsky and Ris (1951) measured the DNA content of erythrocytes of various animals and found great variation. Expressed in picograms (see § 11.4) per diploid cell, the values for 31 teleost fish ranged from 0·94 to 3·49, with figures of 5·46 and 6·67 for two elasmobranch fish (*Carcharias* spp.) and 100 for a dipnoan fish (*Protopterus* sp.). Two anuran amphibians had values of 7·33 and 15·0, and two urodelan species 48·4 and 168. McLeish and Sunderland (1961), using root-tip cells, found similar variation in flowering plants: 5·5 to 60·5 in 4 dicotyledons and 15·5 to 107·7 in 5 mono-cotyledons; these figures do not include polyploid species. Rothfels *et al.* (1966) obtained figures ranging from 1·3 to 52·5 for 22 diploid members of the Ranunculaceae. It is evident that within some groups of organisms such as fish, amphibians and flowering plants, DNA contents per cell can vary 20- to 100-fold.

Associated with the variation in the amount of DNA is a variation in chromosome size: Rothfels *et al.* found that in the Ranunculaceae the total length of the metaphase chromosomes increased approximately as the square root of the DNA content, ranging from 20 to 220 $\mu$m in the species in which the DNA was measured.

The great variation in DNA content is found even between species of the same genus. Schrader and Hughes-Schrader (1956) measured the density of Feulgen staining at prophase I of meiosis in males of two species of hemi-pterous insect of the genus *Thyanta* and found in this way that one had 25 % more DNA than the other, although they had the same chromosome number. Rothfels *et al.* found a 3·5-fold range of variation in diploid species of *Anemone*, and Martin and Shanks (1966) and Rees *et al.* (1966) a sixfold range in diploid *Vicia* spp. Rees *et al.* also found a 2·3-fold range in *Lathyrus* species with 14 chromosomes ($2n$). Callan (1967) quoted the finding of a threefold variation within the crustacean genus *Gammarus* and 11-fold in the flatworm genus *Mesostoma*, again between species with the same chromosome number.

Moreover, the specific differences in DNA content are an intrinsic property of the particular chromosomes. This was shown by Ullerich (1966). He found that the toad *Bufo bufo* had 40 % more DNA per cell than *B. viridis*. Both species have the same chromosome number: $2n = 22$. He measured the relative DNA content of individual chromosomes determined from the density of Feulgen stain. He found that *bufo* chromosomes nos. 5, 7 and 8 in order of decreasing size had about 53 % more DNA than their homologues in *viridis*, and the other 8 had about 36 % more. The same differences were found in a hybrid of the two species.

Recent studies with the giant salivary gland chromosomes of dipteran larvae (§ 9.7) have solved the problem of the source of these large variations in DNA content. This was an unexpected development because these chromosomes are clearly abnormal, and unquestionably multistranded (*polytene*), and might not have been expected to provide information relevant to germ-line chromosomes. The correspondence, however, between chromomeres and the transverse bands of these chromosomes has been established. Beermann and Pelling (1965) used ³H-thymidine to label the DNA of fertilized eggs of *Chironomus tentans*. Autoradiographic examination of the chromosomes of the salivary glands of the larvae showed that some had a radioactive thread resembling a leptotene chromosome extending the length of an otherwise unlabelled giant chromosome (see Pelling, 1966).

Keyl (1965a, b) studied two subspecies, *thummi* and *piger*, of *Chironomus thummi* and found that the type subspecies had 27 % more DNA per cell than *piger*. In genera such as *Drosophila* and *Chironomus* it has been possible to establish evolutionary relationships between species from study of the salivary gland chromosomes because structural changes in these chromosomes can be recognized by identifying the sequence of transverse bands (§ 9.7). From comparison of the salivary gland chromosomes of the two subspecies with those of other species of *Chironomus*, Keyl concluded that *piger* is the nearer of the two to the ancestral stock. He inferred that an increase in the amount of DNA in the chromosomes had occurred in the evolution of subspecies *thummi*, since *piger* has the lower value.

Keyl measured the relative DNA content of individual transverse bands by microspectrophotometry after Feulgen staining. When he compared the DNA contents of corresponding bands in the hybrid of the two subspecies he made a remarkable and quite unexpected discovery. The DNA content of each band of *thummi*, regardless of the absolute amount, always differed from that of its counterpart in *piger* by a power of 2: either 1, 2, 4, 8 or 16 times as much DNA in the one as the other. Different individuals of *thummi*, particularly when derived from a different population, sometimes differed in the DNA content of certain bands (see Table 17.1). Moreover, this variation was not caused by more rounds of replication in some bands than others during the development of the salivary gland chromosomes, because the total

DNA content of the *thummi* cell was 1·27 times that of *piger* whether measured in the salivary glands or the spermatocytes. Evidently the germ-line chromosomes of subspecies *thummi* also showed the geometric differences from *piger* in the DNA content of individual chromomeres collectively responsible for the 1·27:1 ratio.

TABLE 17.1   Data of Keyl (1965*b*) for the DNA content of particular transverse bands of the salivary gland chromosomes of *Chironomus thummi* subspecies *thummi* relative to subspecies *piger*.

| Chromosome | Band | DNA Content | | | | | Chromosome | Band | DNA Content | | |
|---|---|---|---|---|---|---|---|---|---|---|---|
| IR | c3/3 | | ×2 | | | | | b5/28 | ×1 | ×2 | ×4 |
| | 4 | ×1 | | | | | | c1/7 | ×1 | ×2 | |
| | 5 | ×1 | | | | | | 8 | ×1 | ×2 | |
| | 6 | ×1 | ×2 | | | | | 9 | ×1 | ×2 | |
| | c4/1 | | | ×4 | | | | 11 | | ×2 | ×4 |
| IL | d3/8 | ×1 | ×2 | | | | IIR | c2/1 | ×1 | | |
| | e1/2 | ×1 | ×2 | | | | | 2 | ×1 | | |
| | 3 | ×1 | | | | | | 3 | ×1 | ×2 | |
| | 4 | | ×2 | | | | | 4 | | ×2 | |
| | 6 | ×1 | | | | | | 11 | | ×2 | |
| | 7 | | ×2 | | | | | 12 | | | ×4 |
| | e2/3–5 | ×1 | | | | | | c4/3 | | ×2 | |
| III | b1/13 | | | | ×8 | ×16 | IIL | 6 | | | ×4 |
| | b3/11 | | ×2 | ×4 | ×8 | ×16 | | 7 | | ×2 | ×4 |
| | 17 | | ×2 | | | | | 12 | | ×2 | |

The source of the variation in DNA content of related organisms was thus revealed as a successive doubling of the amount of DNA in individual chromomeres. Moreover, Keyl had reason to believe that this doubling was in the length and not the number of strands within each chromomere because Keyl and Pelling (1963) had found by [3]H-thymidine autoradiography that the larger bands take longer to replicate, as though they contained a greater length of DNA. Further aspects of Keyl's discovery, which is clearly of fundamental importance for an understanding of chromosome organization, are discussed in Chapter 18.

## § 17.9   Conclusions about chromosome strands

From the discussion of the cytological evidence for the multistranded chromosome in §§ 17.2–8, it is evident that none of the data provides good evidence for multistrandedness, and in no instance is the single-strand alternative eliminated. Moreover, the results obtained by Taylor, by Herreros and Giannelli, and by Brewen and Peacock from [3]H-thymidine labelling, and by Callan, Gall, and Miller from study of lampbrush chromosomes, provide strong arguments in favour of single-strandedness.

It has been apparent for many years that mutation and recombination are phenomena that are normally shown by chromatids rather than by sub-units within the chromatid. The occurrence of mutation at the level of the whole chromosome, rather than a sub-unit of it, is well illustrated by the work of Abel (1967*a*) using the same experimental material as Knapp and Schreiber had employed in their classic demonstration that DNA is the molecular basis of heredity (§ 11.1). Sperm of the liverwort *Sphaerocarpos donnellii* were irradiated with ultraviolet light and then used to fertilize untreated eggs. Of 253 sporophytes obtained, 41 showed segregation for mutations and every tetrad of spores analysed in these 41 sporophytes showed segregation, with 2 wild-type and 2 mutant spores.

That recombination occurs at the level of the whole chromatid is manifest from the results of tetrad analysis (cf. § 8.6), although postmeiotic segregation (see § 16.2) constitutes an exception. With the acceptance of the DNA theory of the molecular basis of heredity, it appeared most improbable that several DNA molecules carried the same genetic information within one chromatid. If a mutation can represent a change in only one nucleotide-pair, and if, as would appear, recombination normally occurs with such precision that no nucleotide is lost, added or changed*, the simultaneous occurrence of any particular gene in more than one DNA molecule in a chromatid is most unlikely. This argument applies only to the germ-line, and it is evident that in certain differentiated cells the chromosomes are multistranded (see § 9.7 and § 17.8). As indicated in Chapter 16, data on recombination are in agreement with hybrid DNA theories of crossing-over based on the assumption that a chromatid can be looked upon as one DNA molecule (or as a number joined end-to-end). The recombination data contradict hypotheses of chromatid structure which postulate a multistranded condition, since these would require exceedingly elaborate and implausible models to account for the data on conversion, postmeiotic segregation and crossing-over. It is evident that the genetic data confirm the cytological observations of Taylor with *Vicia* and *Bellevalia* root-tips, and of Callan and Gall with *Triturus* oocytes, and establish that a chromatid is a single-stranded structure in the sense that each gene is represented in only one DNA molecule.

The simplest hypothesis of chromosome structure which fits the data is that the whole of the DNA of a chromosome is contained in one giant DNA molecule. Gall (1956) has estimated that the 12 chromosomes in the haploid set of *Rana pipiens* contain an average of 20 cm of DNA if it were straightened out, and the 11 chromosomes of *Triturus viridescens* an average of 90 cm. It is evident from labelling experiments that the chromosome replicates in discrete segments in a controlled sequence, and not progressively from one end (§ 12.7). This means that during replication a number of separate DNA molecules will exist, but as indicated in § 12.7, the simplest explanation of the data is to suppose that when replication is completed these are joined end-to-end into a single giant DNA molecule. Direct support for the idea that a chromosome contains only one DNA molecule has been obtained by Petes and Fangman (1972). Following gentle extraction of the DNA from

---

* The rare occurrence of mutation in association with recombination is discussed in § 16.7.

the nuclei of *Saccharomyces cerevisiae*, the sedimentation rates of the DNA molecules were determined in a caesium chloride density gradient. From this sedimentation analysis it was inferred that the average molecular weight, $\times 10^{-8}$, was 6.2, with a range from 0.5 to 14. *S. cerevisiae* is believed to have a haploid chromosome number of 17, so these figures are reasonable for whole chromosomes, since estimates of the molecular weight, $\times 10^{-8}$, of the total DNA of the haploid nucleus have ranged from 84 to 120, or about 14 to 20 million nucleotide pairs.

If the central axis of a chromosome is a single DNA molecule, how does the protein fit into its construction? The unequal spacing of the two nucleotide chains about the DNA helix axis leaves two spiral grooves. Atoms nos. 2 and 3 of the purines and pyrimidines lie at the bottom of the smaller groove, and atoms nos. 5 and 6 at the bottom of the deep groove (cf. Fig. 11.5). Watson and Crick (1953*b*) pointed out that the molecular dimensions of DNA were such that a fully extended polypeptide chain could be wound round the same helix axis as the nucleotide chains. Feughelman *et al.* (1955) confirmed this for nucleoprotamine, the nucleoprotein found in the sperm of some animals. Protamines appear to consist of one polypeptide chain in which about two-thirds of the amino-acid residues are arginine. Feughelman and associates showed from X-ray crystallographic evidence that the polypeptide chain lies in the smaller groove. The spacing is such that the basic end-groups of the arginine side-chains of the polypeptide can combine with the phosphate groups of the DNA, while non-basic amino-acid residues probably occur in pairs at folds in the polypeptide chain. Basic proteins have been shown to be absent from the chromosomes of the dinoflagellates *Amphidinium elegans* (Ris, 1962) and *Prorocentrum micans* (Dodge, 1964) and from the eu-fungi *Microsporum gypseum*, *Neurospora crassa*, *N. tetrasperma* and *Phycomyces blakesleeanus* (Leighton *et al.*, 1971).

Wilkins (1957) tentatively suggested that in nucleohistone there might be a polypeptide chain in both grooves of the DNA. However, the X-ray diagrams were much less clear than for nucleoprotamine, and the structure of nucleohistone remains uncertain. Pardon, Wilkins and Richards (1967), from X-ray diffraction studies, obtained strong evidence that nucleohistone is folded in some regular way with the DNA inclined to the fibre axis. They suggested that a nucleohistone thread of 3 nm diameter might be coiled to give a helix of pitch 12 nm and of diameter 10 nm measured from the axial centre of the thread, that is, 13 nm outside diameter. The discrepancy between this fibre thickness inferred from X-ray diffraction and that observed by DuPraw, Wolfe and Gall in electron micrographs (20–25 nm) is unresolved. Bram and Ris (1971), however, studied nucleohistone from *Bos taurus* by both electron microscopy and small-angle X-ray scattering in solution, and found that the nucleohistone fibres varied from 8 to 12 nm in diameter with numerous protuberances 10 nm wide randomly spaced and projecting in random directions for 8 to 20 nm. These projections were estimated to make up $\frac{1}{4}$ to $\frac{1}{3}$ of the total mass of the fibre. By comparing the X-ray scattering data with the expectations from precise models, the best agreement was obtained from a nucleohistone fibre containing one DNA double helix irregularly supercoiled or folded with an average pitch of 4.5 nm. Pardon and Wilkins (1972) obtained further evidence for their super-coil model for nucleohistone and discussed the relation between it and structures observed in the electron microscope.

Kellenberger (1961) has discussed the process of condensation of DNA in the formation of the head in bacterial viruses. The DNA of the head occupies only about one-fifteenth of its previous volume, through removal of water molecules. The condensation process does not occur in the presence of chloramphenicol, which is an inhibitor of protein synthesis. When virus *T*2 of *Escherichia coli* is ruptured by osmotic shock, small quantities of proteins and polyamines are released, in addition to the DNA. It is inferred that protein plays a part in the condensation process. It appears that the polyamines may have a similar function, since Kaiser, Tabor and Tabor (1963) found that the polyamine spermine protects DNA from breakage by hydrodynamic shear. Spermine has the chemical formula:

$$NH_2 . (CH_2)_3 . NH . (CH_2)_4 . NH . (CH_2)_3 . NH_2$$

These authors found that if an aqueous suspension of DNA obtained from virus $\lambda$ of *E. coli* is stirred, genes placed well apart on the linkage map are inherited independently, indicating that the DNA has been broken. The marker genes were found often to stay together in the progeny if spermine was added before stirring the DNA. Only a minute quantity of spermine was required. It was suggested that the spermine might make the DNA more resistant to shear, either by alignment of the aliphatic chains along the helix, or by pulling together phosphate groups from different parts of the molecule to give a more compact configuration.

It is possible that the basic proteins (protamines or histones) in chromosomes have a similar function to the polyamines and protein in the heads of bacterial viruses. The protein part of the chromosome may be responsible for its characteristic morphological features, such as chromomeres and spiral coiling, since bacterial chromonemes, which lack these features, also lack protein in association with their DNA. That protamines should occur in association with the chromosomes in the sperm of some animals, instead of the histones found in other nuclei, may be related to the smaller size of the protamine molecule allowing the chromosomes to become more compact (see Ling, Jergil and Dixon, 1971). For sperm, small size means greater mobility and hence has a selective advantage.

The histones of the chromosomes are known to show species specificity, and this suggests that their function is not merely to protect the DNA from breakage, and to allow it to fold into condensed configurations. It is possible that they are concerned in the control of gene action in cell-differentiation. This is discussed in § 18.13.

# 18.   The theory of the chromomere as the basic unit of the chromosome

## § 18.1   Introduction

The chromomere is a characteristic feature of chromosomes, particularly in early prophase I of meiosis and, as transverse bands, in the giant salivary gland chromosomes of Diptera (Fig. 9.1). Suggestions such as that of Belling (1928*b*) (see § 9.9) that chromomeres correspond to genes have received little support because the chromomere was much too large to correspond to 1000 or 1500 nucleotide-pairs of DNA and associated proteins. Recent work, however, suggests that the chromomere may, nevertheless, be the basic unit of chromosome organization. The control of gene action is discussed in this chapter with particular reference to the chromomere. One way in which information about this control has been obtained has been from study of the causes of instability in gene expression, such as give rise to mottled colouring or other kinds of variegation.

## § 18.2   Unstable phenotypes

Emerson (1914) was the first to show that a variable or unstable phenotype may be caused by mutations occurring with abnormally high frequency. He studied the variegated pigmentation of the pericarp found in some strains of *Zea mays* and established that the coloured stripes were due to mutation to a dominant allele of a gene affecting the pigment, for he found that the coloured condition was sometimes inherited. This was evidently when the mutation had occurred in the germ-line.

An outstanding example of a variegated phenotype resulting from mutation is due to the work of Rhoades (1936, 1938), also with *Zea*. The production of red or purple anthocyanin pigment in the aleurone layer of the endosperm of the maize grain requires the simultaneous presence of four dominant genes called $A_1$, $A_2$, $C$, and $R$, and situated in chromosomes nos. 3, 5, 9, and 10, respectively. After self-pollination of a maize plant homozygous for all these dominant factors, Rhoades (1936) found an ear with grains in the proportion of 12 with coloured aleurone: 3 dotted: 1 with colourless aleurone, instead of all the grains coloured as expected. The dotted aleurone character, which had not been observed previously, consisted of small spots of colour of fairly uniform size distributed apparently at random over the aleurone layer (see Fig. 18.1(*a*)). The 12:3:1 ratio was evidently a modified 9:3:3:1 ratio, and implied that the dotted character was due to a dominant gene inherited independently of that responsible for the otherwise colourless aleurone. Further breeding work confirmed this conclusion. The dominant gene responsible was called Dotted, $Dt$. It was found that a mutation had occurred at the $A_1$ locus to the recessive allele $a_1$ giving colourless aleurone, and that $Dt$ interacted specifically with $a_1$ to give the dots, provided the genes $A_2$, $C$ and

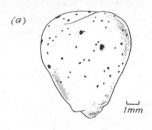

(a)

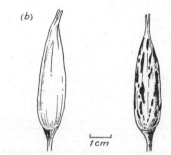

(b)

FIGURE 18.1    (a) A *Dotted* grain of *Zea mays*. The spots represent regions of purple anthocyanin pigment in the otherwise colourless aleurone layer of the endosperm.
(b) Flower-buds of *Oenothera blandina* showing the uniformly coloured calyx of a normal plant (left), and the mottled appearance (right) of an individual heterozygous for a pair of alleles affecting sepal pigmentation and with a structural change in this particular chromosome in the neighbourhood of that gene (from Catcheside, 1947).

$R$ were also present. $Dt$ was found to have no effect on recessive alleles at these other loci. It was found that $Dt$ had been segregating previously in the strain of maize in which the mutation from $A_1$ to $a_1$ took place, but it had not been detected owing to the homozygosity for $A_1$.

The $A_1$ gene causes anthocyanin pigment to form in various other parts of the plant, provided the genotype is appropriate. These parts are the leaf-sheaths, the bracts surrounding the female inflorescence ('husks'), the stems ('culms'), the male inflorescence ('tassel'), the anthers and the pericarp. Rhoades (1938) found that $Dt$ in the presence of $a_1$ and also of whatever other genes were necessary, caused small and narrow longitudinal stripes on the husks and culms; numerous small spots on the anthers or occasionally an entire anther coloured; and fine streaks or sectors on the pericarp. By making crosses with pollen from the rare wholly purple anthers, he showed that the purple character was sometimes inherited and that it was quite stable and was due to the presence of the $A_1$ gene. It was evident that the dots or stripes in potentially coloured parts of $Dt$ $a_1$ maize were arising as a result of mutation from $a_1$ to $A_1$. Rhoades found that $Dt$ also stepped up the rate of mutation of $a_1$ to another allele called $a_1{}^p$ which gives a paler anthocyanin than $A_1$. However, the rate of mutation from colourless to pale was much lower than from colourless to the normal dark pigmentation. Out of 80,955 aleurone dots on 2426 grains, 80 were pale and the remainder dark, implying that about 1 mutation in 1000 was to the pale allele.

A striking feature of Rhoades' observations was that the mutations of $a_1$ induced by $Dt$ apparently occurred only at the time when the $A_1$ gene might be expected to be functioning, that is, at a late stage of the development of each tissue. This was indicated by the small and relatively uniform size of the dots in the anthers and aleurone and of the stripes in the husks, culms and pericarp. Mutation at an earlier stage would be expected to give rise to whole regions or sectors pigmented, but these were rare or absent.

From studies of its linkage with other character-differences, Rhoades (1945) established that $Dt$ was situated in the region of the knob at the end of the short arm of chromosome 9 (see Fig. 7.5). This knob is composed of heterochromatin (see § 18.3).

Muller (1930a) found that a new dominant mutation at the notched-wing locus in the $X$-chromosome of *Drosophila* had appeared following X-ray treatment. In order to test whether it represented a deficiency of a small chromosome segment, like some previous Notch mutants, he crossed it with white-eyed flies, since the gene for white eye-colour was known to be only about 1.5 map units from that for notched wings. Most unexpectedly, the notched-winged offspring had mottled eyes, with lighter and darker areas of diverse size and ranging in colour over the whole series of colours known for alleles of white eye. Muller found that the mottled character was inherited as an allele of white. He also found that the notching of the wings was irregular in its manifestation, both within and between individuals, and that the behaviour of the wing and eye characters was to some extent correlated, as though there was some peculiarity in a chromosome segment rather than in just one gene. Moreover, Muller found from study of linkage with other mutants that the notched-winged mottled-eyed flies had a structural change in the $X$-chromosome, such that the part of it containing the notched-wing and white-eye loci had been removed and attached to chromosome no. III. Four other independent occurrences of alleles at the white-eye locus giving mottled eyes were found following irradiation of sperm. Although these had normal wings, linkage studies revealed that in every instance there was a structural change involving the $X$-chromosome. These observations suggested that there might be a causal connection between the mottling and the structural change.

Such a connection was subsequently demonstrated with the white-eye alleles and a number of other genes in *Drosophila*. One of the clearest demonstrations was by Catcheside (1939, 1947) in *Oenothera blandina*. A plant

homozygous for a gene giving light red sepals was self-pollinated using pollen
treated with X-rays. One of the progeny was found to have irregular patches
of light red on otherwise green sepals. By examination of the chromosomes at
meiosis, it was found that in this plant an interchange had occurred between
segments of two of the chromosomes, and that one of these chromosomes was
that which carried the gene for sepal colour. Through crossing-over in the
interval between gene and point of interchange (recombination frequency
about 1.7%), the sepal colour gene was subsequently transferred to a plant
with normal chromosomes. This plant was found to show no variegation. It
was inferred that the mottling of the calyx was due to a position effect and not
to mutation.

Catcheside (1947) confirmed the position effect when he showed that an
allele giving a uniformly deep red sepal colour also gave rise to a variegated
calyx, this time of dark red patches on a green background, when transferred
through crossing-over to plants carrying the interchange (see Fig. 18.1(*b*)).
Moreover, a gene giving paler petal-colour than normal, which was linked
to the sepal-colour gene (recombination frequency 8·5%), gave rise to variegated petals consisting of a mosaic of patches of the two shades of yellow when
similarly transferred to plants with the interchange. Both sepal and petal
colours became uniform again in progeny in which the genes concerned were
restored to the normal chromosome configuration.

These experimental results, and similar data obtained by several authors
from studies of variegated phenotypes in *Drosophila*, have usually been
interpreted as evidence that the variegation was due to instability in the
functioning of the gene and not to a mutational change of the gene itself. An
alternative explanation is that mutations occur with high frequency in the
tissues where the gene produces its specific effect, but that in other tissues,
including the reproductive cell-lineage, the gene is stable.

There is good evidence for a connection between the occurrence of
variegated phenotypes in *Drosophila* and variability in the extent of regions
of the chromosome which stain differently from normal and consequently
are called heterochromatic.

## § 18.3   Heterochromatin

One of the most striking features of chromosomes in many organisms is that
certain parts of them stain differently from other parts. Heitz (1928), after
studying the mitotic chromosomes of the liverwort *Pellia endiviifolia*, introduced the term *heterochromatin* to describe segments which stained more deeply
than the majority (*euchromatin*). Subsequently, the use of the term heterochromatin was extended to include regions which stain differently from the
rest, since it was found that the same region may stain more deeply at one
period of the nuclear cycle and less deeply at another. Heterochromatic
regions lack chromomeres, such as the euchromatic parts of a chromosome
show in early prophase I of meiosis. Corresponding to this, the heterochromatic segments of the salivary gland chromosomes of *Drosophila* show an

amorphous appearance without transverse bands. Small segments of hetero-chromatin occur at numerous points in the chromosomes, and are revealed by cold treatment prior to fixation. In addition, there are usually extensive heterochromatic regions near the centromere of each chromosome. These centromere regions form an aggregate in the *Drosophila* salivary gland chromosomes (see Fig. 9.1), apparently as a result of a general attraction between the heterochromatin of the different chromosomes. Non-homologous pairing between heterochromatic segments has been observed at the pachy-tene stage of meiosis in *Zea*, and may be responsible for an association in inheritance between genes in different linkage groups observed by Longley (1945) in *Zea* (female side only) and by Michie and Wallace (1953) in *Mus*. Camenzind and Nicklas (1968) found that, unlike Fig. 6.4, in spermatocytes of *Gryllotalpa hexadactyla* the $X$-chromosome regularly moves to the same pole as the larger component of an unequal bivalent; micromanipulation showed that the $X$ is the active partner. Another feature of heterochromatin, dis-covered by Lima-de-Faria (1959), is that it appears to replicate after euchromatin (see Lima-de-Faria and Jaworska, 1968). Thus, Schmid (1963) has shown by tritium labelling that the late replicating regions of human chromosomes in tissue culture correspond to secondary constrictions. Further aspects of heterochromatin are discussed in §§ 18.9 and 18.13.

In many organisms, chromosomes composed wholly of heterochromatin occur (review: Battaglia, 1964). They have been called supernumerary or $B$-chromosomes, and they usually vary in number in different individuals or populations of the species.

The heterochromatic regions appear to correspond to parts of the linkage maps largely without genes. Examples of this in *Drosophila* are the $Y$-chromosome and the centromere regions of the other chromosomes. The viability of individuals lacking a $Y$-chromosome, or lacking $B$-chromosomes, contrasts with the lethality of even small deficiencies from the normal or $A$-chromosomes. On the other hand, the heterochromatic regions are evidently not devoid of function even though they appear to lack normal genes. This is shown by the necessity for a $Y$-chromosome if male *Drosophila* is to be fertile, and by the existence of adaptations in the haploid generation of flowering plants favouring an increase in the number of $B$-chromosomes. The means by which such increase is brought about include failure of $B$-chromosomes to segregate normally at mitoses in the pollen or embryo sac, such that daughter nuclei in the germ-line receive more of them than do other nuclei. Selective fertilization by gamete nuclei with the larger numbers may also occur (for references, see Battaglia, 1964). Likewise, Rhoades (1942) found that in *Zea* heterozygous for a terminal heterochromatic knob in chromosome no. 10 the homologue with the knob reached the functional embryo-sac in 70% of meioses instead of the 50% expected if the orientation of the chromosome-pair on the meiotic spindle were at random.

From study of the number of sternopleural chaetae in stocks of *Drosophila melanogaster* differing only in their $Y$-chromosome, Mather (1944) found evidence for the occurrence in this chromosome of a number of genes having small and more or less additive effects on bristle number. It was inferred that heterochromatin is not inert but contains genes which are qualitatively different from those occurring in euchromatin.

Ris (1957) obtained electron micrographs of thin sections through heterochromatin in the nuclei of amphibians and found that its structure did not differ from that of euchromatin, except in the degree of packing of the fine threads seen. He also found that the chromomeres in leptotene chromosomes of *Lilium longiflorum* differ from the interchromomeric segments merely in the density of the folded and coiled threads. That heterochromatin should differ from euchromatin merely in the way its material is arranged is in keeping with some observations which suggest that under certain circumstances euchromatin can become heterochromatic.

Schultz (1936) studied the salivary gland chromosomes of a number of mutants of *Drosophila* showing mottled phenotypes, and found that in every instance the euchromatic region containing the genes in question had been transferred, as a result of a structural change, to the neighbourhood of heterochromatin. Moreover, Caspersson and Schultz (1938) found that this euchromatic region bordering on heterochromatin tended to acquire the appearance of heterochromatin, the transverse bands being lost. According to Schultz (1948) it appeared as if in some cells the heterochromatin modified the neighbouring euchromatin so that it resembled heterochromatin both in appearance and in lack of genetical activity, while in other cells there was no such modification, thus accounting for the mottled phenotype. The cytological observations thus appear to confirm the inference from the genetic data and suggest that the mottling results from the genes in question having been rendered inactive in some of the cells, the appropriate part of the chromosome having at the same time become heterochromatic in appearance.

These variegated phenotypes resulting from a position effect are reviewed by Baker (1968). Further aspects of them are discussed in §§ 18.4, 6 and 13.

## § 18.4   Controlling elements

McClintock (1950) discovered an interaction in *Zea* between mutable loci and the variegated phenotypes caused by position effect. A dominant gene which she had called Activator (*Ac*) appeared to cause mutation of certain other genes not linked to it. McClintock had established that *Ac* acted through an unlinked gene called Dissociation (*Ds*). In the presence of *Ac*, genes situated within about 5 map units on either side of *Ds* gave rise to a variegated phenotype similar to the effect of structural changes in *Drosophila* (described above) which bring euchromatin into proximity to heterochromatin. Moreover, in the presence of *Ac*, strains arose with different patterns of variegation, apparently as a result of mutation of *Ds* to new alleles.

A further remarkable feature of these genes such as *Dt*, *Ac*, and *Ds* which cause variability in the manifestation of other genes is that they occasionally become transferred to new positions either in the same or a different linkage group. This was shown for *Ac* and *Ds* by McClintock (1950), and for *Dt* by Nuffer (1955), who found *Dt* was situated in chromosomes nos. 6 and 7, respectively, in two South American strains of *Zea*, instead of in no. 9. With *Ac* and *Dt* such transposition does not affect the phenotype, but this is not true of *Ds*, since it is genes in proximity to *Ds* which give rise to variegation. The regions which *Ac*, *Ds*, and *Dt* occupy in the chromosomes appear to be heterochromatic.

Through the transfer of *Ds* to new positions, McClintock was able to confirm that on removal of *Ds*, the original phenotype was restored. As with the mottled phenotypes associated with transfer of euchromatin to the neighbourhood of heterochromatin described above, the effect of *Ds* on neighbouring genes is therefore usually interpreted as influencing gene action rather than as causing mutation. However, McClintock (1950) has put forward the view that these position-effect variegations and those due to mutable loci are essentially one phenomenon, and she has given a striking series of parallels between them (McClintock, 1952). However, it is not clear how far the resemblances are due to the fact that both phenomena are associated with heterochromatin.

Another feature of *Ds* is that it is frequently associated with chromosome breakage or other structural changes at its locus. Alleles of *Ds* differ in the frequency of occurrence of such aberrations. Many of these structural changes could be the consequence of non-homologous crossing-over between different chromosomes or different parts of the same chromosome (cf. § 9.8).

In view of the evidence that *Ac*, *Ds*, *Dt*, and other factors like them (see reviews by Brink *et al.*, 1968, and McClintock, 1965, 1968) are not normal genes, but appear to be abnormal states of a system affecting the manifestation of particular characters at a certain stage of development, McClintock (1957) has proposed the name *controlling element* for them. The characteristic features of such an element are that it does not occupy a fixed position in the chromosome complement, and that it appears to act only within the nucleus, serving as a gene mutator, or modifying or suppressing gene action.

## § 18.5   Controlling episomes

Dawson and Smith-Keary (1963) have studied instability in the suppression of leucine requirement in *Salmonella typhimurium* and found similarities to the mutable loci of chromosomal organisms. They have postulated that *controlling episomes*, having many of the properties of McClintock's controlling elements, are responsible for the genic instability which they have observed. As mentioned in § 16.3, the term *episome* was introduced by Jacob and Wollman (1958b) for genetic particles that can exist either free in the cytoplasm or integrated into the chromoneme. When free, such particles replicate independently of and usually more rapidly than the chromoneme. Examples of episomes include many bacterial viruses, and the sex factor in *Escherichia coli*. Dawson and Smith-Keary found from transduction experiments that certain mutants having a requirement for leucine could become leucine independent as a result of mutation at a neighbouring locus. Some of these mutants which suppressed the leucine requirement were found to give colonies on minimal medium showing great diversity in size. Evidence was found that this instability was due to genetic factors capable of undergoing transduction, and that in any particular unstable mutant, the instability was probably confined to a particular site within the suppressor locus. They suggested that a controlling episome became attached at this site within the gene and was retained there through successive backward and forward mutations which it caused. Loss of the controlling episome would restore stability. This hypothesis was supported by the discovery of a correlation of instability at the leucine-suppressor locus with stability of suppressors for proline requirement, and *vice versa*. It was suggested that the leucine and proline suppressors often competed for a single controlling episome.

A two-element system, such as *Ac* and *Ds*, or *Dt* and $a_1$, in *Zea*, has not been found in *Salmonella*, nor is there any evidence from *Salmonella* of a modification of gene expression, either of the gene to which the controlling episome becomes attached, or of neighbouring genes such as *Ds* affects in *Zea*. The mutability in the presence of *Dt* of genes such as $a_1$ when they occur in specific tissues, and the stability of the same genes when in other cells, would be accounted for on Dawson and Smith-Keary's hypothesis if in the development of the tissues where the mutation occurs, a controlling episome was transferred from *Dt* to $a_1$, while in other tissues there was no such transfer. The association between gene and episome might be through complementary base pairing. If there was a difference between them in nucleotide sequence, this would lead to mispairing, and if there was correction of heterozygosity in the DNA, a mutation might arise (cf. § 16.7). The high mutability would then be a consequence of a failure of the base sequence in the episome precisely to match that of the gene (or a part of it).

Green (1969) studied transpositions of a controlling element in the *w* (white eye) genetic region of *Drosophila melanogaster*. The controlling element causes mutation of *w* with high frequency. In the transpositions it had moved from within *w* to various positions in chromosome 3 and apparently carried with it part of the *w* region. Green concluded that the controlling element was an episome.

Both the hypothesis of mutation caused by controlling episomes, and of gene inactivation caused by proximity to heterochromatin (or to a controlling element within the heterochromatin) are discussed further in § 18.6, since they appear to have relevance to the control of differentiation.

## § 18.6   The control of differentiation

One of the fundamental problems in biology is how a diversity of types of cell can develop from a single cell, such as a fertilized egg. At one time it was thought that cell differentiation must be controlled from the cytoplasm, since mitosis ensures that daughter-nuclei are identical. However, this inference was based on the assumption that all the genes were active in every cell. This assumption is clearly no longer justified in view of the demonstration by Monod, Jacob, and collaborators that the synthesis of $\beta$-galactosidase in *Escherichia coli* is under the control of a regulator gene (see Chapter 15). Differentiation could therefore reflect the control of gene action.

Accepting that the specificity of the enzymes of an organism is inherited in the structural genes in its nuclei, enzyme activity might be controlled at any of 3 stages:

1. Control of enzyme synthesis at transcription, that is, messenger RNA formation from DNA.
2. Control of enzyme synthesis at translation, that is, polypeptide formation from messenger RNA.
3. Direct control of enzyme activity after synthesis.

Regulation at all these levels is known in bacteria. Control at transcription may be positive or negative. Three examples of positive control have already been discussed: $\sigma$ factors which combine with RNA polymerase (§ 14.9); activators such as the $C$ gene product for the arabinose operon in *Escherichia coli* (§ 15.7); and the cyclic AMP receptor protein for activating catabolite-sensitive genes (§ 15.7). Examples of negative control are the repressors discussed in §§ 15.1 and 2. The action of a repressor may occur only in the absence of another molecule (an inducer) as with the lactose operon in *E. coli*, or only in the presence of another molecule (a co-repressor) as with the histidine operon in *Salmonella typhimurium*. As already discussed in § 15.2, induction and co-repression may be closely related in evolution.

Control at translation is also known to be brought about in several different ways: by initiation factors acting in conjunction with the ribosomes (§ 14.16); by the secondary structure of the messenger preventing the access of ribosomes to the starting-point of a gene (§ 15.5); and by repression. Weber *et al.* (1972) found that the replicase of RNA virus $Q\beta$ of *E. coli* functions as a repressor and binds to the initiation site of the gene for the coat protein. The significance of this repression is believed to be to clear the viral RNA of ribosomes in preparation for replication, which begins at the 3' end of the molecule and therefore proceeds in the opposite direction to that of ribosome movement. Weber *et al.* determined the nucleotide sequence of the replicase binding site after removing the remainder of the RNA with a ribonuclease. The binding site, which was protected from RNase attack by the replicase, was 100 nucleotides long, overlapped the ribosome binding site and ended with the AUG start signal of the coat gene. Part of the nucleotide sequence is shown in Table 14.10.

Control of enzyme activity, as distinct from synthesis, may be brought about by transfer RNA molecules or by their activating enzymes (see p. 398). The first enzyme of a sequence, or if the biosynthetic pathway is branched, the first enzyme of each branch, may be inhibited by the end-product of the reaction chain (cf. Umbarger 1962). Monod and Jacob (1962) proposed the name *allosteric inhibition* for this phenomenon, because the inhibitor is not a steric analogue of the substrate, and because, theoretically at least, there is no reason why substances other than the end-product of the reaction chain should not act as inhibitors. They pointed out that if the end-product of each of two independent biosynthetic pathways acted as an inhibitor of the enzyme activity for the first step in the other pathway, this would provide a basis for alternative stable states. However, they considered that a steady state would be insufficient to account for differentiation, which appears to require controlled changes in the capacity of individual cells to synthesize specific proteins.

In chromosomal organisms, little is known in biochemical terms of the control of enzyme activity. If, as Monod and Jacob suppose, differentiation implies differential enzyme synthesis, there is the possibility, as in bacteria, of control at transcription or at translation. Control at transcription might be organized differently in chromosomal organisms compared with bacteria and their viruses, owing to the almost general occurrence of histones in association with the DNA (see § 18.13).

A controlling mechanism implies a signal, and a transmitter and receiver. Control of gene action at the level of the transcription of DNA into messenger RNA requires the receiver of a signal to be in the nucleus. On the other hand, much of the protein synthesis of the cell occurs in the cytoplasm, so that control at the level of the translation of messenger RNA into polypeptide might occur in the cytoplasm. Equally, the transmitter might be either nuclear or cytoplasmic. Four possibilities are evident: intranuclear signals; extranuclear signals; and signals from nucleus to cytoplasm, and *vice versa*.

The occurrence of intranuclear signals in the control of gene action is exemplified by the regulator gene for $\beta$-galactosidase synthesis in *Escherichia coli* (see § 15.2), and by the evidence for controlling episomes in *Salmonella* and controlling elements in *Zea* (see §§ 18.4 and 5). In particular, Dawson and Smith-Keary's hypothesis of controlling episomes not only offers a possible explanation for mutable genes (see § 18.2), but it also suggests a mechanism for the control of differentiation. It appears significant that mutable genes often show mutation only at the time of development when the gene may be expected to be functioning (cf. Peterson, 1966). This suggests that the normal function of the episome might be to act as a signal for the formation of messenger RNA from that particular gene. If there is any truth in the suggestion that the mutability is due to the correction of heterozygosity in DNA (see § 18.5), in normal development the nucleotide sequences of gene and episome would presumably correspond, with the result that the gene would be stable.

One fungal control system was discussed in § 15.10. Many others have been studied (see Fincham and Day, 1971). They reveal a diversity of mechanisms, with the rather frequent occurrence of gene products with dual functions, either more than one regulatory activity or a combination of enzymic and regulatory actions. Proteins with more than one function, however, are not peculiar to chromosomal organisms: the translation repression by $Q\beta$ replicase illustrates this, and other examples are mentioned below.

Control of gene action at translation has been suggested by Ohno (1971) to explain the differentiation of secondary sex organs in *Mus musculus*. It is known that a gene called *Tfm* in the X-chromosome is involved in this process and also the hormone testosterone. Ohno suggested that the *Tfm* protein binds to specific messenger RNAs and inhibits their translation, and that testosterone is an inducer which removes the repressor from the messengers. He further suggested that this repressor-inducer complex stimulates the activity of RNA polymerase.

As already mentioned, one of the methods of control of enzyme activity, as distinct from synthesis, is believed to be through the action of transfer RNAs and of their activating enzymes. Inhibition of an enzyme in *Drosophila* by a transfer RNA was discussed in § 14.15. Many examples are known in bacteria of the apparent repression by an amino-acyl *t*RNA synthetase of the enzymes catalyzing the synthesis of the corresponding amino-acid. Calvo and Fink (1971), in reviewing data on such repression, pointed out that *t*RNA activating enzymes might readily acquire a repressor function in evolution because they already possess a binding site for a potential co-repressor, namely, the amino-acid which forms the end-product of the biosynthetic pathway. Similar dual functions for these enzymes seem to occur in chromosomal organisms. McLaughlin, Magee and Hartwell (1969) found that mutants of the isoleucyl-*t*RNA synthetase gene in *Saccharomyces cerevisiae* showed twofold derepression of two of the enzymes of the isoleucine pathway, and Nazario, Kinsey and Ahmad (1971) made a similar discovery with *Neurospora crassa*: a tryptophan-requiring mutant, *tryp*-5, was found to be deficient in tryptophanyl-*t*RNA synthetase and to show lowered repression of the tryptophan biosynthetic enzymes.

Schultz (1948), using strains of *Drosophila* having chromosomal structural changes which brought genes into proximity to heterochromatin, found that the extent of the resulting variegated phenotypes associated with position effect (see § 18.3) was easily modified by genetic factors. In the presence of an extra *Y*-chromosome the development of a variegated phenotype did not occur and the chromosome region brought into proximity to heterochromatin by the structural change remained euchromatic. The influence of the relative numbers of the different chromosomes on development was referred to in § 6.8 with reference to the abnormality of trisomics. Some genes, for example, *Dt* in *Zea* (Rhoades 1936) and *w* in *Drosophila* (see § 9.1), show a dosage effect, that is, their action is dependent on the number of copies present in the nucleus, and so a trisomic individual will be expected to have a modified phenotype. However, the mechanism by which an extra *Y*-chromosome, which is more or less entirely heterochromatic, can modify the phenotype, is unknown, although progress in understanding this phenomenon has been made by Hess (1970).

A number of further observations indicate the receipt within the nucleus of signals affecting gene action, although their source (whether intra- or extra-nuclear) is usually not known. King and Briggs (1955) transplanted nuclei from *Rana pipiens* embryos at various stages of development into enucleated eggs. They found that the nuclei from certain tissues had lost certain potentialities of expression possessed by the original nucleus of the egg. It appeared that progressive and stable nuclear changes occurred during differentiation, even though, as later studies showed, there were no visible changes in the chromosomes. In similar experiments with *Xenopus laevis*, Gurdon (1970) found that nuclei from differentiated somatic cells quickly came to resemble in structure and metabolism the zygote nuclei of fertilized eggs. At the same time as changes in gene activity are induced by the egg cytoplasm, labelling experiments have shown that particular recently synthesized proteins are transferred from the cytoplasm to the nucleus. Johnson and Harris (1969) formed heterokaryons by introducing nuclei of erythrocytes of *Gallus domesticus* into the cytoplasm of human cells in culture. The cell fusion was induced by Sendai virus which had been inactivated by ultraviolet light. It was found that the *Gallus* nuclei responded to signals from the human cytoplasm and underwent DNA synthesis synchronized with that of the human nuclei.

## § 18.7   The master and slaves hypothesis

Gall and Callan (1962) studied the incorporation of tritiated uridine (the ribonucleoside of uracil containing $^3$H) into RNA, and of tritiated phenylalanine into protein, in association with *Triturus* lampbrush chromosomes (the diplotene stage of meiosis in oocytes) and they found that RNA and protein synthesis occur on the lateral loops but not on the chromomeres. There was marked variation in intensity of labelling from one pair of loops to another, each of which has a characteristic morphology. These observations were in agreement with the hypothesis of Callan and Lloyd (1960) that each loop-pair is associated with the activity of one specific gene. Callan and Lloyd had reached this conclusion from their detailed studies of the lampbrush chromosomes in the oocytes of four subspecies of *Triturus cristatus*. They found that the presence or absence of particular loops of recognizable morphology showed Mendelian inheritance. Some individuals were heterozygous for particular differences (see § 8.13) and the relative frequencies in 22 individuals of the 9 combinations of homozygotes and heterozygotes for the presence or absence of a particular loop on chromosome X and another on chromosome XII (Pl. 17.1(*b*)) agreed with the Hardy-Weinberg equilibrium (§ 2.6). Support for the idea that a chromomere, or its counterpart the transverse band in salivary gland chromosomes, corresponds to a gene, has been obtained from studies of mutations in the white-notch region of the *X*-chromosome of *Drosophila melanogaster*, summarized by Beermann (1967).

In order to explain the discrepancy between the size of a lampbrush loop and the size of a gene as a nucleotide sequence coding for a polypeptide, Callan and Lloyd proposed that the lampbrush loop consisted of a series of duplicate copies of one gene in tandem array. Moreover, they suggested that at each chromomere there was a 'master copy', mutations in which were somehow transferred to the other copies, from which messenger RNA was synthesized. Callan (1967) proposed a mechanism for this transfer of information from master to slaves. He suggested that the complementary chains of each slave in turn are separated from one another and paired with their respective complements in the master copy. If slave and master differ in nucleotide sequence as a result of mutation there will be mispairing in the master/slave hybrid DNA. Correction of mispairing is then assumed to occur, and moreover to take place in a defined direction, namely, from master chain to slave chain. In other words, it is postulated that the mispaired nucleotides of the slave chain are excised and replaced by nucleotides synthesized with the master chain as template. As explained in § 16.6, correction of mispairing arising from recombination is thought to occur widely, and moreover the direction of correction may be determined by features of the DNA other than the mispairing itself. Such external control of direction is an essential requirement for the matching of the slave nucleotide sequence to that of the master. Following the correction of the slave chains to correspond with the master, it is assumed that the slave chains separate from the master chains and anneal with one another. The whole process of matching slave to master would then be repeated with the next slave copy of the gene, and so on through the whole series of them.

Callan's hypothesis of matching slave nucleotide sequences to a master sequence makes the master and slaves hypothesis compatible with the results of studies on mutation and recombination, because a single master copy with means of transferring its nucleotide sequence to the slaves is equivalent in genetical terms to the single copy of the gene which the genetical data require.

Callan's matching hypothesis also explains an extraordinary polarity in the appearance of lampbrush loops, which he and Gall had observed. Callan and Lloyd (1960) reported that, irrespective of the characteristic texture of the protein matrix of particular loops, lateral loops had one end thin and the other much thicker at the points of insertion in the chromomere (Fig. 17.1($d$) and Pl. 17.1($c$)). They observed that when a chromomere was stretched until it broke (Fig. 17.1($e$)), the polarity of any particular loop was always constant in direction with respect to the centromere of the chromosome (Pl. 17.1($e$)), but either orientation was possible: 2 particular loops always had the thin end towards the centromere, and for 3 other loops the converse was true. To explain the loop asymmetry, Gall (1955) and Callan and Lloyd (1960) suggested that the loop axis was spun out from the chromomere at the thin end and returned to it at the thick end, for example, clockwise movement in the lower loop of Fig. 17.1($f$). Such movement would explain the greater accumulation of gene product on the one side of the loop than the other. Support for this idea of polarized loop extension and retraction was obtained by Gall and Callan (1962) who found from their RNA labelling with $^3$H-uridine that some loops (cf. Pl. 17.1($d$)) showed progressive labelling starting at the thin end but not reaching the thick end for 10 days. The matching hypothesis explains this movement of the loops by supposing that the master copy is at one end of the tandem array and that the thin end of the loop is the point where slaves emerge from the chromomere after they have been matched against the master copy.

Keyl's discovery of the successive doubling in evolution of the DNA content of individual chromomeres, probably as a result of doubling the length of the DNA molecule or molecules which they contain (§ 17.8), implies on the master and slaves hypothesis a doubling of the number of slave copies of particular genes. The number of copies of a gene in a large chromomere might be very high. Allowing for the feeding back of the lampbrush loop into the chromomere, Callan (1963) estimated that one of the largest loops in *Triturus cristatus*, together with its chromomere, might contain 1–2 mm of DNA. A gene of 1000 to 1500 nucleotide-pairs would occupy 0·34 to 0·5 $\mu$m of DNA, so the largest *Triturus* chromomere might contain $2^{12}$ or more copies of the gene.

Edström and Beermann (1962) have shown that the RNA associated with diffuse areas or puffs at specific points in the giant chromosomes of the salivary glands of *Chironomus tentans* has a base composition which is characteristic for each puff. Beermann (1961) has shown that in *C. pallidivittatus* granules are secreted from particular cells in the salivary gland, and that in these cells but not in others, puffing occurs at the position in the appropriate chromosome of a gene concerned with production of the granules (see Fig. 9.1(*b*) and (*c*)). In hybrids with *C. tentans*, which lacks the granules, the *C. pallidivittatus* chromosome alone shows the puff. The inference from these experiments is that loop-formation in lampbrush chromosomes, and puffing in salivary gland chromosomes are associated with the activity of specific genes (see Beermann, 1964). The evidence that a chromomere, and therefore a lampbrush loop or salivary puff, corresponds to one gene has been reviewed by Thomas (1971). Strong support for this hypothesis has been obtained by Judd, Shen and Kaufman (1972) from a combined cytological and genetical study of mutants located in a particular region of the *X*-chromosome of *Drosophila melanogaster*.

Since there is evidence for characteristic patterns of puffing which vary with the stage of development and with the tissue, Beermann's findings constitute good evidence that differentiation is associated with differential gene activity. Moreover, Clever (1961) has shown with *Chironomus* larvae that injection of ecdysone, which causes premature pupation, initiates within a few minutes puffing at a specific locus in one of the salivary gland chromosomes, to be followed by a whole series of puffs corresponding in position and time sequence to those occurring at normal pupation. There is an obvious resemblance between lampbrush loops and the puffs of salivary gland chromosomes, and it seems that both represent the activity of specific chromomeres. According to Callan's master and slaves hypothesis the puff would represent a set of copies of one gene in each of the many parallel strands making up the giant chromosome.

Beerman (1965, 1967) put forward an alternative explanation for puffs and lampbrush loops. He suggested that the newly-made RNA must be packaged in protein before it can leave the chromosome, and that a function of the DNA axis of the loop was to provide a substrate for binding and stabilizing the messenger, in readiness for its association with transport protein. On Beermann's hypothesis the master segment at the base of a loop is active in transcription and the remainder of the loop is concerned primarily, at least, with packaging, whereas on Callan's hypothesis the master copy does not take part in transcription, which is confined to the slaves in the loop. For a discussion of Beermann's hypothesis, see Hess and Meyer (1968).

Attardi, Parnas, Hwang and Attardi (1966) and a number of other authors have discovered giant-sized RNA molecules in the cells of vertebrates, having sedimentation coefficients of 30$S$ to 80$S$ or more. This RNA has a composition similar to that of the total DNA of the organism, and may possibly be derived from a set of copies of one gene. The significance of the duplication of genes within chromosomes is presumably to allow a higher rate of messenger synthesis.

Miller and Beatty (1969$b$) obtained beautiful electron micrographs of parts of lampbrush loops of *Triturus viridescens*. Numerous closely spaced ribonucleoprotein fibrils could be seen arising laterally from the DNA loop axis and increasing progressively in length round the loop. A particle on the loop axis at the base of each fibril was presumed to be an RNA polymerase molecule.

Fincham (1967) suggested that the master and slaves hypothesis may account for some of the features of mutable genes (§§ 18.2 and 4). Certain mutations or structural changes in the master copy might interfere with the correction of mispairing in the process of matching the slaves to the master, and thus also affect transcription. The gene could then be inactive, although the slaves were non-mutant. The mutability would then lie in the possibility, which might arise in various ways, of one of the slaves taking over the function of the master-copy.

By DNA–RNA hybridization, Kedes and Birnstiel (1971) found evidence for clustered repetitive DNA sequences complementary to histone messenger RNA in the echinoderm *Psammechinus milaris*, but in similar experiments Bishop, Pemberton and Baglioni (1972) found no evidence for multiple copies of the haemoglobin genes in the duck *Anas platyrhynchos*.

Thomas *et al.* (1970) purified the DNA of *Salmo*, *Necturus* and *Bos*, fragmented it by shearing and then treated it with an exonuclease to expose single chains at the ends of the fragments. After maintenance at 60°C for several hours, it was found that 20% of the *Salmo* fragments and 35% of the *Necturus* ones were circular, as seen under the electron microscope. It was inferred that the exonuclease treatment frequently exposes nucleotide chains at opposite ends of a fragment that are complementary to one another and so can anneal to form a circular molecule. No circles were formed by the DNA of *Bacillus subtilis*, *Escherichia coli* or its *T*7 virus when given similar treatment. Thomas *et al.* concluded that, in agreement with Callan's hypothesis, there is much tandem duplication of nucleotide sequence in chromosomal organisms.

## § 18.8　Genes for ribosomal and transfer RNA

The base composition of the RNA of the nucleolus agrees with that of the ribosomes and led to the idea that the nucleolus is the site of ribosome formation. It has now been established that the nucleolus represents visible evidence of the activity of a set of copies of a gene for part of the ribosomal RNA. Elsdale, Fischberg and Smith (1958) discovered a female *Xenopus laevis* (Clawed Toad) in which half the offspring possessed only one nucleolus in each nucleus instead of the normal two. These heterozygotes for the

nucleolar deficiency were of normal phenotype and on intercrossing gave a
1:2:1 ratio of individuals with 2, 1, and 0 nucleoli, respectively, per nucleus.
The homozygotes lacking nucleoli died as young larvae. Brown and Gurdon
(1964) showed that the embryos lacking nucleoli were unable to synthesize
either of the major RNA molecules which occur in ribosomes. These have
sedimentation rates of 18$S$ and 28$S$. The survival of the homozygotes to the
tadpole stage appeared to depend on the use made of ribosomes inherited
cytoplasmically from the mother. Wallace and Birnstiel (1966) showed from
DNA–RNA hybridization studies that the genetic defect in the nucleolus-
deficient *Xenopus* is a deletion of at least 95% and possibly the whole of the
DNA which is complementary to ribosomal RNA. They found that in
normal *Xenopus* 0·07% of its DNA hybridizes with 28$S$ ribosomal RNA and
0·04% with 18$S$. With 6 pg of DNA per erythrocyte nucleus, they estimated
there were about 800 copies per haploid set of chromosomes of the genes for
the ribosomal RNA components. Brown and Weber (1968), using an improved technique involving density-gradient centrifugation in caesium
chloride to fractionate the DNA, followed by hybridization of RNA to DNA
immobilized on nitrocellulose filters, concluded that each nucleolar organizer
contains an alternating sequence of '18$S$' and '28$S$' DNA segments with
intervening nucleotide sequences not corresponding to either. Similar
results were obtained by Birnstiel *et al.* (1968). The 18$S$ and 28$S$ ribosomal
RNA molecules are formed from a single precursor molecule, an extra
segment of polynucleotide being excised during maturation (see Birnstiel,
Chipchase and Spelrs, 1971).

Direct examination under the electron microscope by Miller and Beatty
(1969*a*, *b*) of active ribosomal DNA of *Triturus viridescens* has confirmed the
existence of a tandem array of genes alternating with inactive DNA (Frontispiece and Pl. 18.1).

Brown, Wensink and Jordan (1972) used DNA–RNA hybridization to
compare the ribosomal DNA of *Xenopus laevis* and *X. mulleri*. They found that
the 18$S$ and 28$S$ sequences were identical throughout, while the spacer
sequences were identical within species but differed between them. Brown
*et al.* considered various mechanisms, including Callan's hypothesis, by which
the uniformity of the repeated sequence might be maintained and yet change
as a whole in evolution.

Ritossa and Spiegelman (1965) hybridized *Drosophila* ribosomal RNA
with the DNA of strains of *Drosophila* having from 1 to 4 doses of the nucleolar-
organizer region, which is in the *X*- and *Y*-chromosomes. The abnormal
doses were from stocks which had either a duplication or a deletion of the
nucleolar-organizer region of the *X*-chromosome. The hybridization was
achieved by incubating the denatured (single chain) DNA with the RNA
for 7–10 hours at 65°C. It was found that the amount of DNA which would
hybridize with ribosomal RNA was proportional to the number of nucleolar
organizers present and that no less than 0.27% of the DNA of a normal fly
could be hybridized with ribosomal RNA. Ritossa and Spiegelman favoured
the idea that the genes for ribosomal RNA are repeated consecutively along
the chromosome 100 or more times. Quagliarotti and Ritossa (1968) obtained

PLATE 18.1   Nucleolar genes isolated from an oocyte of the spotted newt, *Triturus viridescens*. The electron micrograph shows alternating active and inactive regions of the DNA in a free nucleolus (cf. Plate 17.1(*a*)). The active regions are about 2·5 μm in length and have an RNA and protein matrix in the form of about 100 fibrils extending laterally from the DNA axis and increasing progressively in length along each active segment. Each fibril is believed to be a ribosomal RNA molecule undergoing synthesis as an RNA polymerase molecule moves along the DNA of the nucleolar gene. The inactive regions are of variable length. For fuller discussion, see Miller and Beatty (1969*a*). F Fibrils surrounding active region of DNA. I Inactive region of DNA.
   (*Courtesy of O. L. Miller, Jr. and Barbara R. Beatty*, Biology Division, Oak Ridge National Laboratory, U.S.A.)

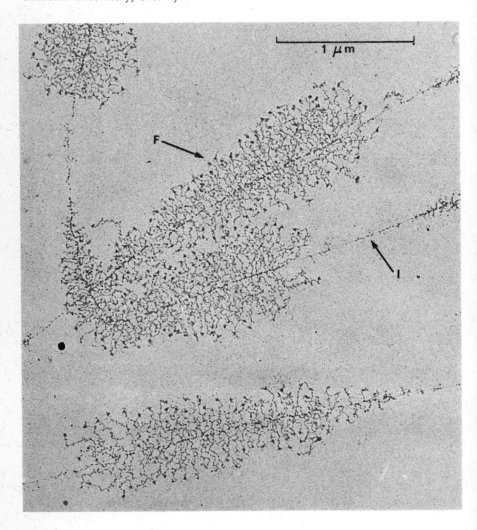

PLATE 18.2   Phase-contrast photographs of the loop which extends into the
nucleolus from the nucleolar chromosome at the nucleolar organizer.
In these examples the nuclei contain only one nucleolus and each
copy of the nucleolar chromosome is associated with it.

*(Courtesy of L. F. La Cour)*

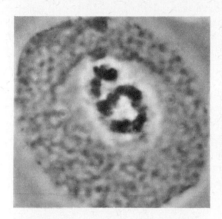

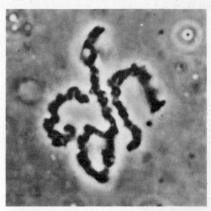

(a) Two loops within the nucleolus of
a diploid nucleus from a root-tip of
*Ipheion uniflorum* ( × 1650).

(b) Three filaments (broken loops)
within the nucleolus of a triploid
nucleus from endosperm of *Scilla
sibirica* ( × 1100).

evidence that there are equal numbers of the 18*S* and 28*S* segments and that,
as in *Xenopus*, each 18*S* segment alternates with a 28*S*. Ritossa, Atwood and
Spiegelman (1966*b*) made the surprising discovery that the gene in *D. melanogaster* for bobbed bristles (*bb*), which is situated in heterochromatin near the
centromere (see Fig. 8.10), is none other than the nucleolar organizer.
Different *bb* mutants showed from 30% to 55% of the normal amount of
DNA complementary to ribosomal RNA.

In all chromosomal organisms in which hybridization of DNA and
ribosomal RNA has been carried out, the evidence has pointed to many
copies of the 18*S cum* 28*S* gene, though with considerable diversity of number
(see Birnstiel, Chipchase and Speirs, 1971).

The duplication of the gene for 18*S cum* 28*S* ribosomal RNA is in agreement
with the experimental results obtained by McClintock (1934) with *Zea mays*
and Beermann (1960) with hybrids of *Chironomus tentans* and *C. pallidivittatus*.
These authors discovered that when the nucleolar organizer is broken into
two parts by X-irradiation, each part can fulfil the functions of a complete
organizer and sustain normal development.

The occurrence in chromosomal organisms of duplicate copies of the gene
at the nucleolar organizer for 18*S* and 28*S* ribosomal RNA gives support to
the master and slaves hypothesis of chromomere organization, because Callan
(1966) has found that the nucleolar organizer locus in the lampbrush
chromosomes of *Ambystoma mexicanum* is organized in the same way as other
chromomeres. Moreover, La Cour and Wells (1967) found that in the

nucleoli of *Ipheion uniflorum* a loop about 20 μm long, and with a DNA axis, extends from the nucleolar organizer (see Pl. 18.2(*a*) and (*b*)).

As indicated in § 14.5, in addition to the two large RNA molecules, ribosomes also contain a small one, with a sedimentation rate of 5*S* and containing a single chain of 120 nucleotides. Surprisingly, the 5*S* gene is separate from the 18 *cum* 28*S*. Brown and Weber (1968) estimated that there were 24,000 genes for 5*S* RNA per haploid chromosome set in *Xenopus laevis*, that is, 25–30 times as many as for the 18 *cum* 28*S*. Furthermore, the 5*S* genes had not been lost in the nucleolus-deficient mutant. Brown, Wensink and Jordan (1971) purified the '5*S*' DNA and showed that it was repetitive, with 83% of each repeating unit consisting of spacer DNA that did not hybridize with the RNA. Wimber and Steffensen (1970) located the 5*S* genes in *Drosophila melanogaster* by hybridizing labelled 5*S* RNA with the salivary gland chromosomes and preparing autoradiographs. The 5*S* genes, of which there are believed to be about 200, were found to be in a transverse band in the right arm of chromosome 2. On the other hand, Aloni, Hatlen and Attardi (1971), after separating human chromosomes into size groups by glycerol-sucrose density gradient centrifugation, found that 5*S* RNA hybridized with DNA from each group, implying that the 5*S* genes occurred in several different chromosomes.

There seem also to be duplicate copies of the genes for transfer RNAs. Ritossa, Atwood and Spiegelman (1966*a*) estimated that in *Drosophila melanogaster* 0·015% of the haploid DNA content of 0·2 pg hybridized with transfer RNA, or enough for about 750 *t*RNA genes. The minimum number of different transfer RNAs is 32, but a larger number is probable (cf. § 14.15). It is evident, however, that there is considerable duplication of the 40 or 50 *t*RNA genes likely to be present. Goodman *et al.* (1968) found two identical tyrosine *t*RNA genes in *E. coli* (see § 14.17) and Gilmore *et al.* (1968) found 8 genes for tyrosine *t*RNA in *Saccharomyces cerevisiae*; these genes were not closely linked, however.

The identification of genes for ribosomal RNA and transfer RNAs implies that some genes are concerned in specifying nucleotide sequence in RNA for its own sake, rather than as coded information for polypeptide synthesis.

## § 18.9    Satellite DNA

Satellite DNA was discovered by Sueoka (1961*a*) who found that 30% of the DNA of the crab *Cancer borealis* formed a separate peak in a caesium chloride density gradient, implying a guanine-cytosine content of only 20% compared with 42% in the remainder. Swartz, Trautner and Kornberg (1962) showed that the base sequence of the *Cancer* satellite consisted largely of an alternating sequence of adenine and thymine. *Drosophila melanogaster* has a similar component comprising about 4% of its DNA, but here the adenine and thymine do not always alternate. Blumenfeld and Forrest (1971) measured the amount of this DNA in flies of different genotypes and concluded that it was mostly in the *Y*-chromosome, including both arms of it.

Kit (1961) discovered satellite DNA in the mouse *Mus musculus* and the guinea pig *Cavia porcella*, with G + C contents less and more, respectively, than the bulk of the DNA. Waring and Britten (1966) found that the separated chains of the *Mus* satellite, which comprises about 10% of its DNA, re-annealed rapidly, implying a short sequence repeated many times. Hennig and Walker (1970) studied the rapidly re-annealing component of the DNA of various species of rodents and found large differences even within a genus in the amount and in the G + C content. Southern (1970) determined the nucleotide sequence of the *Cavia* satellite, which forms 5% of its DNA, and found it to be:

$$5'—C—C—C—T—A—A—3'$$
$$3'—G—G—G—A—T—T—5'$$

repeated millions of times. He concluded, however, that a substantial number, perhaps 20%, of the copies of this sequence contained mutations. He believed the *Mus* satellite to have a repeating sequence of between 8 and 13 base pairs unrelated to the *Cavia* sequence.

Caspersson *et al.* (1968) discovered that certain regions of chromosomes showed intense fluorescence when stained with quinacrine mustard. Subsequently, it was found that quinacrine hydrochloride also causes fluorescence. By means of this technique it has become possible for the first time to distinguish all the human chromosomes. The regions of these chromosomes showing quinacrine fluorescence were found by Ganner and Evans (1971) to correspond, in general, with late-replicating heterochromatic regions, though not all such regions were fluorescent. Ellison and Barr (1972) discovered that quinacrine fluorescence occurs in chromosome regions rich in adenine and thymine. Labelling of chromosomes of the fly *Samoaia leonensis* with ³H-deoxycytidine, followed by autoradiography and quinacrine staining, showed that the regions of intense fluorescence, which were close to the centromeres and late replicating, contained little or no cytosine. All parts of the chromosomes, however, including the fluorescent regions, would take up ³H-thymidine. The Y-chromosome of *Drosophila melanogaster* fluoresces brightly with quinacrine, in agreement with Blumenfeld and Forrest's conclusion that it contains much of the A·T-rich satellite. Ellison and Barr's discovery was confirmed by Weisblum and de Haseth (1972), who found that quinacrine fluorescence *in vitro* required duplex DNA rich in adenine and thymine. It did not matter whether the adenine was in one chain and the thymine in the other, or whether the bases alternated along each chain. Single chains did not fluoresce.

John, Birnstiel and Jones (1969) and Pardue and Gall (1969) used heat or alkali treatment, respectively, to separate the nucleotide chains of DNA *in situ* in cytological preparations of chromosomes. This made possible hybridization with radioactive DNA, the position of which could then be revealed by autoradiography. Using this technique, Jones (1970) and Pardue and Gall (1970) discovered that the *Mus* satellite DNA was localized in the centromere regions of all the chromosomes except the Y-chromosome. A variant of this technique was to transcribe satellite DNA into radioactive

RNA, using a bacterial DNA-dependent RNA polymerase, and use the RNA for *in situ* hybridization. This method was used by Pardue and Gall (1970) and Jones and Robertson (1970) and confirmed the centromeric location of satellite DNA in *Mus musculus*, and also showed that highly reiterated sequences occurred near the centromeres in *Drosophila melanogaster*, because RNA transcribed from *Drosophila* DNA selectively annealed with the chromocentre in salivary gland nuclei. The centromeric heterochromatin of the chromocentre evidently contained DNA capable of rapidly annealing with complementary RNA.

Walker (1971) suggested that satellite DNA might facilitate the pairing of the centromere regions of chromosomes at meiosis, and hence favour their regular separation at anaphase I. Ellison and Barr (1972) found that the A·T-rich quinacrine-fluorescent regions of chromosomes tend to be physically associated with one another at interphase, and they suggested that this association might explain why these regions replicate late.

## § 18.10    The cycloid model for the chromosome

Callan (1967) drew attention to a feature of chromosome organization which seems to be of fundamental importance, namely, that lampbrush loops do not take part in crossing-over. There are three lines of evidence which point to this conclusion. First, he had never observed chiasmata in the loops. Secondly, the chromomeres are conspicuous in early prophase of meiosis when crossing-over takes place; they are too densely folded for the occurrence of homologous pairing at the molecular level, such as crossing-over requires. It is significant that the greater part of the material of homologous chromosomes lies outside the synapton at pachytene (see § 8.21 and Plate 8.2). Thirdly, accepting the master and slaves hypothesis, if pairing took place at meiosis between slaves, it would be expected often to occur with the homologues relatively displaced, for example, slave no. 5 from one parent with slave no. 6 from the other. The behaviour of the Bar-eye duplicated segment demonstrates this (§ 9.10). Crossing-over between slaves after such displaced pairing, however, would break down the geometrical

---

FIGURE 18.2    Diagrams to show a structure for the chromosome in keeping with recombination data and with the work of Callan and Gall on lampbrush chromosomes and of Keyl on dipteran salivary gland chromosomes. In (*a*) is shown the normal state of the chromosome with sets of duplicate genes inserted in the DNA axis to give the form of an irregular cycloid, while (*b*) represents the situation at pachytene of meiosis with the duplicates removed as rings of DNA. Eight master-genes are shown, each with 2, 4, 8 or 16 slave-copies. Much larger numbers of copies would be present in organisms with high DNA contents to their chromosomes. Transverse lines and arrows indicate the ends of each gene copy and its orientation, with double lines separating unlike genes. The letter O indicates the hypothetical position of the operator of each gene. The loops or rings of DNA would correspond to chromomeres.

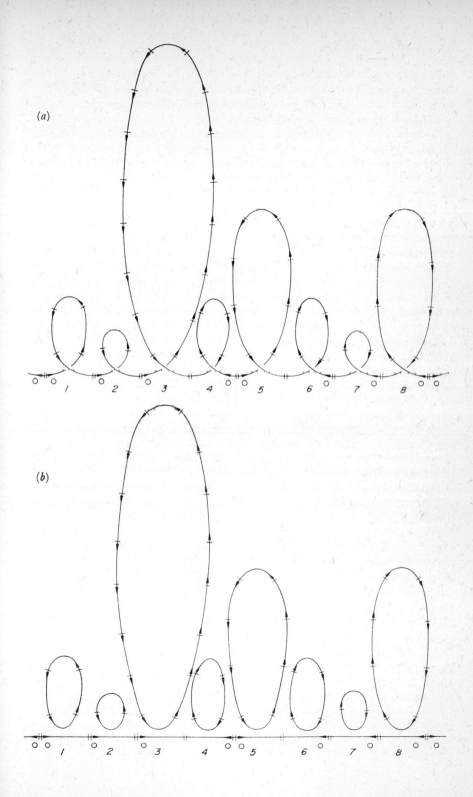

differences in DNA content which Keyl (1965*a*, *b*) observed (§ 17.8). The inference is that crossing-over does not occur between slaves. Callan therefore suggested that crossing-over was confined to the master-copies, which might be situated in the inter-chromomeric regions.

If the master-copies of neighbouring genes in a chromosome are separated by slaves which do not take part in crossing-over, the hybrid DNA of a crossover in one gene cannot overshoot into its neighbour (§ 16.8). I therefore suggested (Whitehouse, 1967) that the slaves were detached from the chromosome before crossing-over took place. If crossing-over occurred between the first and last copies of the gene, all but one of them would be detached in the form of a circle of DNA. The slaves in the ring could subsequently be integrated once more with the chromosome by crossing-over between one of them and the master copy of the gene which had remained in the chromosome. The sets of duplicate genes inserted in the DNA axis of the chromosome would give it the form of an irregular cycloid (Fig. 18.2(*a*)), except at pachytene of meiosis when the duplicates would be removed as rings (Fig. 18.2(*b*)). In keeping with Callan and Keyl's observations, the diagram shows some genes orientated the opposite way to others, and with geometric variation in the number of their slaves.

Wolstenholme (1971) could find only linear DNA molecules in electron micrographs of material extracted from pollen mother cells of *Lilium longiflorum* at various stages of meiosis. There was no evidence for a change in molecular form or size at the zygotene or pachytene stages when crossing-over takes place.

Ito, Hotta and Stern (1967) found that the DNA synthesized in early prophase I of meiosis in *Lilium longiflorum* and *Trillium erectum* (§ 16.5) was of two kinds. At early zygotene DNA of a higher than normal guanine–cytosine content was formed. This was thought to be a structural element of the chromosome (§ 18.9). In late zygotene and pachytene DNA of the same density as the total DNA of the species was synthesized. It seems likely that this DNA synthesis is associated with crossing-over. The total amount of such synthesis in conjunction with the crossovers between homologous chromosomes is expected to be slight, however, and it is possible that the greater part derives from the crossovers postulated for chromomere excision and re-insertion. Hotta and Stern (1971*a*) extended these studies and concluded that the pachytene synthesis, unlike that at zygotene, was of the nature of repair synthesis as it was not inhibited by hydroxyurea. The total DNA content of the pollen mother cell nucleus is 110 pg and the pachytene synthesis was estimated to be 0.1% of this or $10^8$ nucleotide pairs. When the DNA synthesized at pachytene carried 5-bromo-uracil as a heavy label, shearing the total DNA into fragments of about 800 nucleotide pairs did not affect the position on a caesium chloride gradient of the heavy pachytene-synthesized material. Howell and Stern (1971) inferred that the pachytene synthesis occurred at not less than $6.6 \times 10^5$ sites in the nucleus. This figure is puzzlingly high if it represents sites of recombination between homologous chromosomes, but is understandable if it includes the number of chromomeres.

## § 18.11   The chromomere as the replicon

Keyl and Pelling (1963) found by ³H- and ¹⁴C-thymidine labelling of the salivary gland chromosomes of larvae of *Chironomus thummi* that the transverse bands all start to replicate within 30 minutes but that each takes its own specific time and they finish replicating one after another in a well-defined time sequence. Moreover there was a correlation between the length of the replicating period of a transverse band and its DNA content. These discoveries, in conjunction with Keyl's finding that duplication in evolution of the DNA content of a chromomere occurs independently of that of its neighbours, led Keyl (1965*a*, *b*) and Pelling (1966) to the important conclusion that the chromomere corresponds to the replicon.

Support for this idea has been obtained from the study of nucleoli. The nuclei of amphibian oocytes contain several hundred free nucleoli (Pl. 17.1). Miller (1966) found from enzymic treatment with pepsin, trypsin, RNase and DNase, and from electron microscopy that each nucleolus of *Triturus pyrrhogaster* contains a circular DNA molecule surrounded by matrix. This construction resembles that of lampbrush loops and led him to suggest that a single chromosomal locus may replicate independently of the rest of the chromosome. Callan (1966) found that the lampbrush loop at the locus of the nucleolar organizer in *Ambystoma mexicanum* resembles ring-shaped nucleoli which occur both attached to the chromosome at this point and free in the nucleus. Kunz (1967) made similar findings with the free nucleoli in the oocytes of some orthopteran insects. It appears from these studies that the nucleolar locus can generate circular DNA molecules, presumably consisting of a tandem array of duplicate copies of the genes for ribosomal RNA. This organization has been confirmed by Miller and Beatty (1969*a*, *b*) from electron micrographs of detached nucleoli of *T. viridescens*: active regions of DNA about 2·5 μm long alternated with inactive regions devoid of lateral fibrils (Frontispiece and Pl. 18.1). These varied from 0·8 to 25 μm in length. Dawid *et al.* (1970) found that 4·5% of the bases in the DNA at the nucleolar organizer in *Xenopus laevis* were 5-methylcytosine, whereas the DNA of the detached nucleoli contained no 5-methylcytosine. They could find no other differences between them. Macgregor (1968) and Bird and Birnstiel (1971), from autoradiographic studies, favoured cascade amplification, that is, successive duplication of the detached copies, rather than all synthesized directly from the nucleolar organizer.

Keyl and Pelling's hypothesis of the chromomere as the replicon, taken in conjunction with Callan's master and slaves hypothesis, leads to the conclusion that the replicon consists of the master copy of a gene and its slaves. As pointed out in § 16.8, there is reason to believe that crossing-over is initiated at the ends of genes, possibly at one specific end. From the argument of § 18.10 this would be one end of the master-copy and might therefore coincide with the end of the replicon. Thus, it is possible that the initial step in crossing-over is not breakage of nucleotide-chains but failure of DNA ligase to join chains at the end of a replicon after the pre-meiotic round of DNA synthesis. This would explain one of the earliest discoveries about crossing-over, namely, that it always occurs at the four-strand stage (§ 8.6), and this is true also of mitotic crossing-over (§ 13.3).

Direct support for this idea of a failure to complete DNA synthesis before meiosis has been obtained by Ito, Hotta and Stern (1967) from their studies on pollen-mother-cells of *Lilium longiflorum* (§§ 16.5 and 18.10) using $^{32}$P-phosphate labelling, density-gradient centrifugation and DNA hybridization. They found that the addition of deoxyadenosine, which inhibits DNA synthesis, stopped meiosis when added in early zygotene and caused chromosome fragmentation when added in mid-zygotene. They concluded that the *GC*-rich DNA synthesized in early zygotene (see § 18.10) forms an essential part of the chromosome and consists of comparatively short regions interspersed along the entire length of the chromosomes. These regions would be synthesized in mitotic interphase, but before meiosis their synthesis would be delayed until zygotene. In support of this hypothesis, Stern and Hotta (1967) found that pollen-mother-cells transferred to culture medium when in premeiotic G2 were diverted to mitosis instead of meiosis, and a small amount of *GC*-rich DNA synthesis then occurred before the mitosis; such synthesis was not found before a normal mitosis. It is evident that the delayed synthesis of segments of the DNA is peculiar to meiosis. Further study by Hotta and Stern (1971a) of the DNA synthesized at zygotene supported the earlier conclusions.

Assuming that the chromosome is single-stranded and with a continuous DNA axis, it is likely that the replicons making up the chromosome have staggered ends, that is, the breaks in the chains needed for replication do not occur opposite one another but some nucleotides apart. This would prevent breakage of the whole molecule and would be comparable to the cohesive ends in phage $\lambda$ of *Escherichia coli*, where the 5′-terminal chain has been found by Wu and Taylor (1971) to extend 12 nucleotides beyond the 3′-terminal chain at both ends. The protruding chains at each end of $\lambda$ have complementary nucleotide sequences:

$$- - - -C\text{-}C\text{-}C\text{-}G\text{-}C\text{-}C\text{-}G\text{-}C\text{-}T\text{-}G\text{-}G\text{-}A5'p$$
$$\cdot \quad \cdot \quad \cdot \quad \cdot \quad \cdot \quad \cdot \quad \cdot \quad \cdot \quad \cdot \quad \cdot$$
$$p5'G\text{-}G\text{-}G\text{-}C\text{-}G\text{-}G\text{-}C\text{-}G\text{-}A\text{-}C\text{-}C\text{-}T\text{-} - - -$$

thus allowing the ends to join to give first a hydrogen-bonded and then a covalently-closed circle of DNA (Gellert, 1967; and see Figs. 14.6 and 16.4). The protruding ends are believed to arise through the action of a specific endonuclease which recognizes the nucleotide sequence and cuts the two chains 12 nucleotides apart (Mousset and Thomas, 1969). The cohesive ends of virus 186 of *E. coli* have a different sequence from those of $\lambda$ and are 19 nucleotides long (Padmanabhan and Wu, 1972):

$$- - - -C\text{-}C\text{-}G\text{-}C\text{-}A\text{-}C\text{-}C\text{-}G\text{-}C\text{-}C\text{-}C\text{-}C\text{-}T\text{-}T\text{-}T\text{-}C\text{-}G\text{-}T\text{-}A5'p$$
$$\cdot \quad \cdot \quad \cdot \quad \cdot \quad \cdot \quad \cdot \quad \cdot \quad \cdot \quad \cdot \quad \cdot \quad \cdot \quad \cdot \quad \cdot \quad \cdot \quad \cdot$$
$$p5'G\text{-}G\text{-}C\text{-}G\text{-}T\text{-}G\text{-}G\text{-}C\text{-}G\text{-}G\text{-}G\text{-}G\text{-}A\text{-}A\text{-}A\text{-}G\text{-}C\text{-}A\text{-}T\text{-} - - -$$

In the chromosomes of higher organisms, the protruding chain at the right-hand end of each replicon would need to be complementary to the protruding chain at the left-hand end of the next replicon to the right. The *GC*-rich DNA synthesized in early zygotene might represent delayed formation of the protruding ends of the chains, the remainder of which had been synthesized at the previous S period.

Daughter-chromatids sometimes fail to separate at mitosis and this process, called *endoreduplication*, can be induced by desacetyl-methyl-colchicine. It gives rise to associations of 4 chromatids called *diplochromosomes*. Walen (1965) with cultures of kidney cells of the marsupial *Potorous tridactylus*, and Schwarzacher and Schnedl (1966) with human blood cells in culture, independently discovered that diplochromosomes regularly have the outer chromatids labelled and the inner not, after one replication in the presence of $^3$H-thymidine followed by another without the label (Pl. 18.3 and Fig. 18.3). The distinction between inner and outer chromatids is possible because the four lie in one plane. The labelling pattern would be accounted for if at each replication the 2 new sub-units were formed on the outside of the 2 old

PLATE 18.3    Autoradiograph of metaphase of an endoreduplicated cell from a culture of human peripheral blood cells, following treatment with $^3$H-thymidine during the first of the two rounds of DNA synthesis ($\times 1650$). The label is restricted to the two outer chromatids of each group of four that make up a diplochromosome, except where exchanges have occurred between chromatids.

*(Courtesy of H. G. Schwarzacher)*

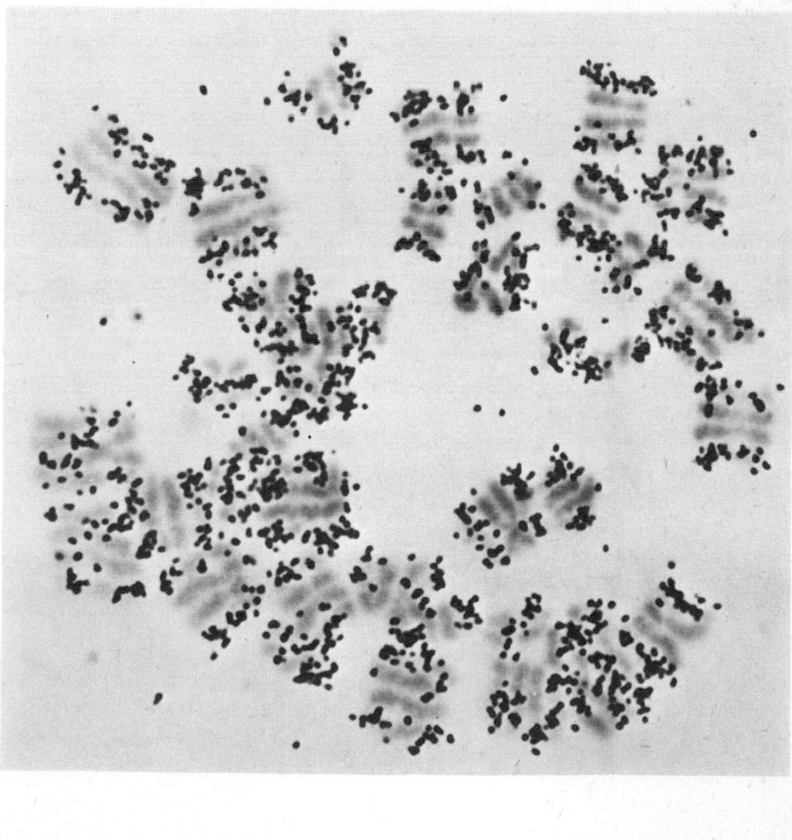

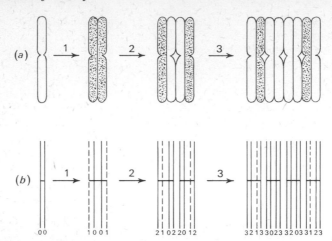

FIGURE 18.3 Drawing to illustrate the labelling pattern when chromosomes undergo one replication (no. 1) in the presence of ³H-thymidine, followed by one or two replications (nos. 2 and 3) in its absence, and the daughter chromatids fail to separate. (*a*) Appearance of chromosomes at metaphase, the dots indicating silver grains in the autoradiograph. (*b*) Proposed explanation, the vertical lines indicating the two sub-units of each chromatid, with broken lines for radioactive sub-units. The numbers show the replication at which each sub-unit was synthesized. Horizontal lines indicate undivided centromeres.

ones. The appearance of chromosomes which had undergone a third replication, and in which the 8 chromatids had failed to separate, was in keeping with this hypothesis (Fig. 18.3). According to the single-strand theory of the chromosome, the sub-units are individual nucleotide-chains of DNA. A new chain can be on the outside of an old one only at specific points, however, because old and new form one double helix. These specific points are evidently not the centromeres because Herreros and Giannelli (1967) found the characteristic outer-chromatid labelling pattern in acentric diplochromosomes. It seems likely that the points where the new pair of chains regularly lie on the outside of the old pair are at the junctions of the replicons because, as indicated in § 12.8, it already seemed likely that a mechanism existed for joining new chain to new, and old to old, at these points.

The discovery by Huberman and Riggs (1968) of pairs of diverging replication-forks in *Cricetulus* chromosomes (§ 12.8) suggests that replication of the stretches of DNA corresponding to two neighbouring chromomeres may be initiated simultaneously at a common starting-point between them. Two kinds of replicon junctions would then be predicted, corresponding to the origin and the terminus, respectively, of the replication of the adjacent chromomeres. It may be significant that Haut and Taylor (1967), from study of DNA replication in *Vicia faba* root-tips, using 5-bromouracil as a density label, obtained evidence for short primer segments of two kinds, with different thymine contents.

As pointed out in § 12.8, Callan (1972) prepared autoradiographs of DNA extracted from cells of *Triturus cristatus* following tritiated thymidine labelling. He found that the intervals between the initiation-points of replication at the premeiotic period of DNA synthesis were of the same order of magnitude as the length of DNA in a chromomere. On the other hand, in somatic cells the interval between replication origins was considerably shorter, and in embryonic cells it was much shorter still. It is evident that, although replication before meiosis may be initiated only between the chromomeres, at other times it can start at many sites within a chromomere as well.

## § 18.12   Antibody diversity

The antibody molecules of mammals present a major riddle for genetics. The main discoveries which any hypothesis of the origin of antibody diversity must explain are the following (see Hood and Talmage, 1970; Gally and Edelman, 1970; Milstein and Pink, 1970; Milstein and Munro, 1970; and Smith, Hood and Fitch, 1971, for references to original sources).

Mammalian antibodies (immunoglobulins) are composed of polypeptide chains of two kinds: light chains with about 215 amino-acid residues and heavy chains with about 450. Variation in antibody molecules is of three kinds called isotypic, idiotypic and allotypic. All the variation is caused by differences in the primary structure of the light and heavy chains. Isotypic variation occurs within individuals, can be recognized serologically and chemically, and arises from the occurrence of two primary classes of light chains called kappa and lambda (relative frequencies in human serum 70% and 30% respectively) and three primary classes of heavy chains called gamma, alpha and mu (relative frequencies 70–80%, 15–20% and 5–10%, respectively). These heavy chains define the three major classes of antibodies in a normal human serum, which are called immunoglobulins G, A, and M, respectively. Two identical light chains of either class combine, through the formation of disulphide bridges between cysteine residues, with two identical heavy chains of any of the three classes. In immunoglobulin A, the unit comprising two light and two alpha chains combines with another identical four-chain unit through a disulphide bridge between cysteine residues near the carboxyl end of the alpha chains. In immunoglobulin M, five identical four-chain units, each consisting of two light and two mu chains, are similarly combined. Two other classes of immunoglobulins, D and E, contain heavy chains called delta and epsilon, respectively. These account for less than 1% of the heavy chains in human serum.

Although every individual can produce more than $10^5$ different antibody molecules, single cells normally produce antibodies with only a single specificity, having either kappa or lambda light chains of one specific kind and heavy chains also of only one specific kind.

Idiotypic variation occurs within individuals and is responsible for the immense diversity of antibody structure. The source of this diversity has been established primarily from study of the antibody molecules excreted by myeloma patients. Such patients normally excrete antibody molecules that are all alike and are believed to be derived from a clone of cells all capable of producing only the one kind of antibody molecule. There is good evidence that the individual antibodies excreted by myeloma patients are representative of the normal population of antibodies. Idiotypic variation is caused by much diversity in the half of the light chains (amino-acids 1–108 approximately) and the quarter of the heavy chains (amino-acids 1–114 approximately) nearest to the amino end of these polypeptides. The variability does not involve all the residues in these parts of the chains: some are constant, while at other positions as many as 10 alternative amino-acids are known. It is notable that, despite the great diversity, the basic structure of the molecule is constant with, for instance, disulphide bridges always present between cysteine residues at approximate positions 23 and 88 in the variable region of each light chain, and likewise between cysteine residues in the constant region at approximate positions 134 and 194 (see Fig. 18.4).

Isotypic variation arises from the many differences between kappa and lambda chains, both in the constant and variable regions, and from the numerous differences between the constant parts of the gamma, alpha and mu polypeptides.

Allotypic variation occurs between individuals and shows Mendelian inheritance. Antigenic specificities of human antibodies called *Inv* have been found in immunoglobulins G, A and M. Two of the variants of the *Inv* character in man, which can be detected serologically and show simple Mendelian inheritance, have been shown to correspond to the substitution of leucine for valine at position 191 in kappa chains. This position is in the so-called constant part of the molecule. In addition to kappa-chain genetic markers, allotypic characters determined by the lambda chains are known in *Oryctolagus cuniculus*, with preliminary evidence that the kappa and lambda characters are inherited independently. Human antigenic specificities called *Gm* and *Am* have been found only in immunoglobulins G and A respectively. The variants of the *Gm* character have been shown to reside in the 'constant' part of gamma chains, and the *Am* variants in the 'constant' part of alpha chains. The *Gm* and *Am* characters show close linkage in inheritance, and this is also true of gamma and alpha characters in *Mus musculus*. In man and other mammals, characters carried by the light chains are inherited independently of characters carried by the heavy chains.

Study of the *Gm* characters of different human populations has shown that they are carried by three closely linked genes. Investigation of the antigenic and other characteristics of human gamma chains has shown that they fall into four subclasses numbered (in decreasing frequency) 1–4: each is present in every normal human serum. Combined study of *Gm* characters and gamma subclasses has revealed a precise correspondence between the three genes and three of the four subclasses: no genetic marker is yet known for subclass no. 4. Support for this genetic organization is provided by the discovery by Kunkel, Natvig and Joslin (1969) of a human family which seems to lack gamma genes 1 and 3 and to contain instead a single gene consisting of the

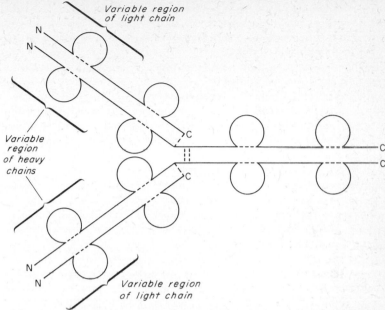

FIGURE 18.4   Diagram to show the structure of the antibody molecule. The
diagram refers specifically to subclass 1 of immunoglobulin G, and
is based on Milstein and Pink (1970). The unbroken lines show the
two light and the two heavy polypeptide chains, N indicating the
amino and C the carboxy terminus of each. Broken lines indicate
disulphide bridges between cysteine residues: these cause the
formation of loops containing about 60 amino-acid residues, spaced
at regular intervals along each polypeptide. Disulphide bridges
also unite the four chains, though the number and positions of these
links depend on the class or subclass of immunoglobulin. In
immunoglobulins A and M four-chain units, similar to that shown,
combine with one another through disulphide bridges near the
carboxyl end of the heavy chains.

amino-terminal part of 3 and the carboxyl-terminal part of 1. This is
comparable to Lepore haemoglobins (§ 15.10) and has evidently arisen
through crossing-over at meiosis between gene 3 from one parent and gene 1
from the other, following displaced pairing. A further implication of this
finding is that these genes have the same orientation in the chromosome
and are adjacent to one another, with the transcription terminus (that is,
the carboxy end of the polypeptide) of gene 3 next to the transcription
origin (amino end of the polypeptide) of gene 1 (cf. Fig. 15.4.) Studies of
gamma chain subclasses and genetic markers in *Mus musculus* likewise indicate
four closely linked genes. The four human gamma chain subclasses differ
in the numbers and positions of the cysteine residues which provide the
disulphide bridges between chains. Depending on the subclass, from two
to five bridges occur close together linking the two heavy chains near their
mid-point. The bridge between heavy and light chains occurs at position

220 in gamma chain 1 (see Fig. 18.4) but at 131 in the others. The subclasses known in the gamma polypeptides of other mammals do not correspond to the human subclasses. Two subclasses are known in human alpha chains and also in human mu chains.

Allotypic variation is not always caused by variability of amino-acid sequence in the so-called constant parts of the polypeptides: in *O. cuniculus* the allotypes *Aa*1, *Aa*2 and *Aa*3, which are inherited as 3 alleles, have been correlated with the variable region of the heavy chains. These allotypes differ from one another by amino-acid substitutions at numerous positions extending over at least 30 amino-acids in the variable region. The genetic markers in the variable region (*Aa*1, *Aa*2, *Aa*3) show linkage in inheritance with allotypes *A*11 and *A*12 which correlate with the substitution of methionine for threonine in the 'constant' region of gamma chains.

The amino-acid sequences of the variable regions of the kappa, lambda and heavy polypeptides fall into subgroups. In human kappa chains three subgroups are known, each with characteristic amino-acid residues at particular positions at least as far as no. 94, that is, 88% of the variable region. In human lambda polypeptides five subgroups have been recognized but they are not as clearly defined as in kappa. In human heavy chains four subgroups are known and their characteristic residues extend over the first 100 residues or 87% of the variable region (Köhler *et al.*, 1970). Polypeptides within subgroups have about 75–85% of the amino-acid sequence of the variable region in common, whereas between subgroups either within kappa, or within lambda, or within the heavy chains, the identity is only about 50–70% or even lower. Some of the subgroups have additions or deletions of one or a few amino-acids at particular positions in the variable regions of their polypeptide chains. Additions or deletions may also occur within subgroups. There is evidence which suggests that all the subgroups of the human light and heavy chains occur in each individual. Moreover, highly inbred mice (*M. musculus*) show as much variation as occurs in the human population in the amino-acid sequence of the polypeptides of their antibody molecules. The variation in the amino-acids within human subgroups is similar to that observed within haemoglobin or cytochrome *c* of different individuals or species. This similarity extends to the class and frequency of base changes needed in the DNA coding for these polypeptides. The difference between subgroups is also similar to that seen between evolutionarily related sets of proteins. Within subgroups, particular amino-acid substitutions occur repeatedly and most of the variation is restricted to rather few positions. Both the constant and variable regions of mouse (*M. musculus*) kappa chains differ from human kappa chains by about 40% of their amino-acids. Although kappa and lambda light chains have been recognized in many mammals, they do not correspond to any of the human subgroups.

Not all variation in the so-called constant regions of the polypeptides is allotypic. Two human serological subtypes (*Oz* + and −) have been correlated with the substitution of arginine for lysine at position 190 in lambda chains. Every individual can synthesize both of these forms of the constant part of lambda, and both occur in conjunction with each of the subgroups of the variable part of the chain.

The same subgroups of the heavy chain variable regions can be recognized in human gamma, alpha and mu chains (Köhler *et al.*, 1970). In *O. cuniculus* the allotypes *Aa*1, *Aa*2 and *Aa*3, which are determined by the variable region of heavy chains, have been recognized serologically in gamma, alpha, mu and epsilon chains, and furthermore, gamma-chain and mu-chain antibodies with the same idiotype, that is, with the same distinctive antigenic determinants, are known. A human myeloma patient has been found to excrete two kinds of antibodies, one consisting of gamma and kappa chains and the other of mu and kappa (Wang *et al.*, 1970). The G and M immuno-globulins are synthesized in different cells, yet the gamma and mu chains of this patient were identical for at least the first 27 amino-acids of the variable region, and the kappa chains associated with gamma were identical with those associated with mu.

The heavy chains in the antibodies excreted by three patients with heavy chain disease were found all to have long but different deletions which, in at least two instances, spanned the junction of the variable and constant regions and ended at identical positions (see Franklin and Frangione, 1971).

There has been much speculation about the genetic basis of antibody polypeptides. It may be tentatively concluded that three unlinked chromosome regions are involved, corresponding to the kappa, lambda and heavy polypeptides, respectively. Contrary to the basic postulate of one gene: one polypeptide (§ 13.7), separate genes are favoured for the variable and constant regions of each antibody polypeptide. If this is true, does the joining occur at the DNA, RNA or polypeptide levels, and how is it possible to have deletions spanning the junction? The problem is further complicated by the existence of more than one gene for each class of constant regions, for example, four genes, probably arranged consecutively, for the constant regions of the human gamma polypeptides. Underlying all these questions is the basic one: what is the source of the immense diversity of amino-acid sequence in the variable regions? Does each subclass of these regions correspond to a different gene, with the variability arising by somatic mutation? And if so, how is it that the particular amino-acid sequence of a favoured antibody molecule is then maintained in a huge clone of cells. Is there a germ-line gene for every variable region amino-acid sequence? Or, is there a smaller number of germ-line genes, with every permutation of sequence derivable, potentially at least, by somatic recombination? Does each variable-region gene consist of a large number of copies in tandem array which, unlike Callan's master and slaves, are non-identical? In spite of the rapid advance in knowledge of the molecular basis of antibody diversity, the underlying genetic mechanism remains quite obscure.

## § 18.13 Chromosome inactivation

Lyon (1961, 1962) has postulated that in female mammals one of the two X-chromosomes is inactive. She has suggested that in the human embryo inactivation occurs at about the 12th day after fertilization, and that the choice is made independently in each cell, but once made is maintained indefinitely not only in these nuclei but also in all nuclei derived from them during later development. Cytological studies have shown that one *X*-chromosome is more tightly coiled than the other and replicates later. As a consequence of the random imposition of inactivity, female mammals will be mosaics for any sex-linked characters with local gene action for which they are heterozygous. This is exemplified by a number of mottled phenotypes known in mice (*Mus musculus*), and by the tortoiseshell cat (*Felis catus*).

Strong support for Lyon's hypothesis was obtained by Davidson, Nitowsky and Childs (1963). They found that skin fibroblasts in women heterozygous for alleles A and B of the sex-linked gene for glucose-6-phosphate dehydrogenase (G-6-PD) were of two kinds, some showing the electrophoretic band of allele A and some of allele B. This discovery was made by growing cultures from individual fibroblasts. All the cells of a colony were alike, as expected if the inactivation of a particular *X*-chromosome is maintained in the daughter cells. Goldstein, Marks and Gartler (1971) confirmed the hypothesis from a study of hair follicles in two women heterozygous not only for the A and B variants of G-6-PD but also for another sex-linked character difference, the presence (+) and absence (−) of hypoxanthine: guanine phosphoribosyltransferase (HGPRT). Both women inherited an A− chromosome from their mothers and a B+ chromosome from their fathers. It was found that the hair follicles of both women were either A− like the maternal chromosome or B+ like the paternal one, or thirdly, showed both the A and B electrophoretic bands of G-6-PD and had an intermediate level of HGPRT activity, as expected for follicles which arose from a mixture of the two kinds of cells. The lack of A+ and B− cells was striking evidence that the inactivation is at the level of the chromosome rather than the gene, because the two genes investigated are situated well apart in the *X*-chromosome. Similar evidence was found by Gartler *et al.* (1972), who examined 56 skin cells from a woman heterozygous for the G-6-PD variants and for types 1 and 2 of the enzyme phosphoglycerate kinase (PGK). They found 33 A1 cells and 23 B2. There were none with the other two combinations. PGK is determined by a sex-linked gene which is not closely linked to either the G-6-PD or the HGPRT genes.

A feature of Lyon's hypothesis of great interest is that the inactive *X*-chromosome shows all the peculiarities of heterochromatin, that is, differential staining properties, late replication, and genetical inertness. Furthermore, Cattanach and associates have demonstrated that this heterochromatin can show a spreading phenomenon comparable to that described in § 18.3. Cattanach and Isaacson (1967) studied the behaviour of autosomal genes in *Mus* when these were transferred to the *X*-chromosome as a result of a translocation. Mottled phenotypes were observed for genes in the transferred segment of the autosome and the greater the distance of an autosomal gene from the point where the autosome was joined to the *X*-chromosome the less often that gene was found to be inactivated. The extent to which the inactivation spread from the *X*-chromosome to the autosome appeared to be determined by a controlling element situated in the *X*-chromosome. This gene existed as a number of different alleles giving inactivation to different distances. Mutations of these gene were frequent. McClintock found similar mutations with the controlling elements which she studied in *Zea* (see § 18.4). Two distinct processes may be at work in the *X*–autosome translocation: first, inactivation of an *X*-chromosome, and secondly, extension of this inactivation into the autosomal segment. It seems more probable, however, that the controlling element is responsible for both processes.

In kangaroo hybrids (*Macropus robustus erubescens* × *M. r. robustus* and *M. r. erubescens* × *Megaleia rufa*) the pair of *X*-chromosomes in the female differ in morphology. By $^3$H-thymidine labelling Sharman (1971) discovered that the late replicating *X*-chromosome was always the paternal one. Confirmation of this difference from the random inactivation of placental mammals was obtained by Richardson, Czuppon and Sharman (1971) who found that the hybrid females did not express the G-6-PD variant (a sex-linked character as in man) of their male parents.

In considering the mechanism of *X*-chromosome inactivation, two observations are important. First, in abnormal individuals with more than two *X*-chromosomes all but one of these are inactivated if the individual is diploid, but Schindler and Mikamo (1970) found that in a human XXY triploid the two *X*-chromosomes remained active. This suggests that inactivation results from a balance between the number of *X*-chromosomes and the number of sets of autosomes,* as in sex-determination in *Drosophila* (see § 6.13). Secondly, Comings (1967), using the technique of polyacrylamide-gel electrophoresis, failed to detect any differences between the histones of euchromatin and heterochromatin. It may be tentatively concluded, therefore, that gene inactivation results either from the binding of specific acidic or neutral proteins to operators in the DNA, as with the *lac*, $\lambda$ and 434 repressors in *Escherichia coli* (§ 15.2) or, alternatively, that non-basic proteins which act as specific activators of *X*-chromosome genes, are absent, perhaps maintained so through masking of the DNA by histones (see p. 422). The latter alternative is attractive because the spreading effect of heterochromatin into euchromatin might then result from blanketing by histones, following the removal of the specific gene activators.

Brown (1966) recognized two kinds of heterochromatin which he called *constitutive* and *facultative*, respectively. Constitutive heterochromatin remains in this state in all types of cell and developmental stages studied. Facultative heterochromatin is material which can be euchromatic or heterochromatic. It is constitutive heterochromatin that contains satellite DNA. Ganner and Evans (1971) found that the facultative heterochromatin of an inactivated human *X*-chromosome stained like its active partner and did not show quinacrine fluorescence. The genetical activity of the two kinds of heterochromatin, with particular reference to the effects of various numbers of *X*- and *Y*-chromosomes in man, have been discussed by Polani (1969). He

* Brown and Chandra (1973) believe that in placental mammals the number of *X*-chromosomes remaining active corresponds to the number of sets of autosomes of maternal origin. They suggested that a single episome having the ability to keep an *X*-chromosome active is released by each maternal copy of a specific autosome. In marsupials the episome donor and receptor sites would be on *X*-chromosomes of maternal origin.

drew attention to the remarkable correlation between the number of such chromosomes and the number of dermal ridges on the finger tips. Individuals with one $X$-chromosome and no $Y$-chromosome have about 168 ridges. Additional $X$- and $Y$-chromosomes reduce the number of ridges by about 30 and 20, respectively, per chromosome added—an excellent example of the quantitative, rather than directly qualitative, effects of heterochromatin (cf. § 18.3).

Henderson (1964), using tritiated uridine, studied RNA synthesis in association with the meiotic chromosomes of males of the locusts *Schistocerca gregaria* and *Cyrtacanthacris tartarica* and the grasshopper *Chorthippus brunneus*. He found that with the autosomes RNA synthesis occurs at all stages except when the chromosomes are strongly contracted (metaphase and anaphase, I and II). The single $X$-chromosome was found to remain tightly coiled and heavily staining throughout meiosis (see Plate 6.2), and to show no RNA synthesis, indicating that the activity of its genes is under generalized control (see Pl. 6.2($f$)).

Stedman and Stedman (1951) discovered that the erythrocyte nuclei of *Gallus domesticus* contained a different histone from that of the spleen and thymus. This cell specificity of histones led them to advance the hypothesis that the basic proteins of cell nuclei are gene inhibitors, each suppressing the activity of specific groups of genes. That histones play a part in the control of gene action is suggested by the discovery of Izawa, Allfrey and Mirsky (1963) that the addition of arginine-rich histones to *Triturus* lampbrush chromosomes stops the formation of RNA in association with the lateral loops, and leads to their retraction into the chromomeres. Bonner, Huang and Gilden (1963) found that nucleoprotein extracted from *Pisum sativum* will bring about the synthesis of proteins *in vitro*. They used the protein synthesizing system derived from *Escherichia coli* and developed by Nirenberg and Matthaei (see § 14.12). Bonner *et al.* discovered that the proteins synthesized with this system, using nucleoprotein from the developing cotyledons as primer, include a globulin characteristic of the cotyledons. The nucleoprotein from pea buds, which do not produce this globulin *in vivo*, failed to lead to its synthesis *in vitro* unless the histone had first been removed from the DNA. Barr and Butler (1963) made similar studies using DNA and histones from *Bos* and inferred that the effect of histone in inhibiting the priming activity of DNA was unspecific.

Paul and Gilmour (1968) also obtained evidence that the effect of histones in masking the DNA is non-specific. They confirmed the discovery of Bonner and associates that organ-specific masking is preserved in isolated chromatin: the RNA of the thymus of the rabbit, *Oryctolagus cuniculus*, hybridised with 5–10% of rabbit DNA, and was synthesized *in vitro* when thymus chromatin was used as template; similar results were obtained with bone marrow RNA, part of which hybridized with a different fraction of the DNA. The specificity in the masking of the DNA appeared to reside, however, not in the histones but in a non-histone protein fraction. This was shown by the behaviour of the DNA as an RNA template when it was combined with each protein fraction separately. Thymus chromatin from *Bos*, when treated with dilute hydrochloric acid to remove the histones but leave other proteins associated with the DNA, had more template activity than the untreated chromatin but less than half that of pure DNA. Addition of the histones to the acid-treated chromatin restored the organ-specific template activity of the original chromatin. This is in contrast to the behaviour of the histones when added to the DNA alone, without the non-histone fraction: no RNA was then synthesized, evidently because the histones masked the DNA completely. Paul and Gilmour concluded that non-histone proteins are responsible for the unmasking of the genes which function in a particular tissue and that these proteins also prevent the non-specific masking of these genes by histones. It may be significant, as they point out, that the repressor molecules isolated from bacteria and shown to bind to a specific nucleotide sequence in DNA are acidic or neutral proteins (§ 15.2), such as would be included in the non-histone fraction in their experiments, if similar proteins occurred in mammalian chromosomes.

Histones can be divided (see DeLange and Smith, 1971) into 5 major classes of which no. I is very rich in lysine, II slightly so, III and IV are rich in arginine, and V is a slightly lysine-rich histone characteristic of nucleated erythrocytes (fish, amphibia and birds). Some of these classes contain molecules of more than one kind, for example, Greenaway and Murray (1971) reported heterogeneity in histone V between different individuals of *Gallus domesticus*. The complete amino-acid sequence has been established for histone IV from 4 mammals and 1 flowering plant, and for histones II*b* and III from *Bos taurus* (see DeLange, Hooper and Smith, 1972). These proteins are all single polypeptide chains containing 102, 125 and 135 amino-acid residues, respectively. DeLange, Fambrough, Smith and Bonner (1969) made the remarkable discovery that histone IV of *Pisum sativum* is almost identical with that of *Bos taurus*. The only differences were isoleucine and arginine in *Pisum* in place of valine and lysine in *Bos* at positions 60 and 77, respectively, apart from methylation of lysine at position 20 in *Bos* but not in *Pisum*. This unparalleled conservation for thousands of millions of years led them to suggest that histone IV interacts with the helical backbone of DNA and not with the base sequence. It may be questioned, however, why so strict an amino-acid sequence should be maintained in histones if they are concerned with packaging the DNA. DeLange and Smith (1971) concluded that histones, besides protecting the DNA, regulate its replication and transcription, but are non-specific repressors which can interact with any region of DNA to reduce its template activity, though non-histone proteins or RNA may confer specificity to this repression. The amino-terminal halves of histones III and IV contain many more basic residues than the carboxy-terminal halves, suggesting to DeLange *et al.* that the first half is the primary site of interaction with DNA, and the second with other proteins. Johns (1972) found that the amino half of histone II*b* bound to DNA *in vitro* much more strongly than the carboxy half. He also found that the carboxy half, unlike the amino, did not repress template activity. He concluded that DNA is not necessarily repressed by protein binding, and that transcription may not require removal of histones. He pointed out that one histone molecule would cover only about 30 nucleotide pairs of DNA, so it was pointless to consider half a molecule as repressor and half not. He concluded, like DeLange *et al.*, that part of the molecule attaches to DNA and the rest is available for interactions with other molecules.

Hamilton (1968) and Arnott *et al.* (1968) suggested that for transcription DNA may have to change its form so that the bases are tilted 20° to the helix axis, similar to double-helical RNA ( § 14.15) ; these two forms of DNA were recognized *in vitro* from X-ray crystallography by Franklin and Gosling (1953). Failure of base-tilting might thus provide a mechanism of gene inactivation.

## § 18.14    The cytoplasm in differentiation

In spite of these diverse pieces of evidence, which point to differential gene activity in the chromosomes as the basis of differentiation, there are cogent reasons for believing that such action forms only one aspect of the process. Thus, Spencer and Harris (1964), working with *Acetabularia crenulata*, found that synthesis of enzymes took place in enucleated cells during regeneration of the cap, and moreover that synthesis of one of the enzymes was specifically associated with the formation of the cap. The whole process of regeneration took 6–8 weeks. It was evident that enzyme synthesis can be regulated over long periods in the absence of the nucleus. A number of other observations support the belief that wholly cytoplasmic control of enzyme synthesis is of widespread occurrence in chromosomal organisms. Thus, Reich *et al.* (1962) showed that actinomycin $D$ inhibited the formation of RNA in fibroblasts of *Mus* in tissue culture, but that protein synthesis, revealed by incorporation of tritiated leucine, continued for many hours, and Sokoloff *et al.* (1964) found that the addition of thyroxine to cell-free preparations of the liver of *Rattus* stimulated the incorporation of amino-acids into protein in the absence of any synthesis of RNA. Actinomycin $D$ or deoxyribonuclease were used to prevent RNA synthesis.

That differential activity at the level of transcription cannot form more than a part of the mechanism of differentiation is also evident from study of the development of unicellular (or acellular) organisms such as *Acetabularia* or *Paramecium*, and from the embryology of multicellular organisms. From such studies it is apparent that differentiation occurs within cells as well as between one cell and another. Intracellular differentiation implies a spatial organization within the cell, and this presumably arises through interaction between molecules in the cytoplasm and physical or chemical factors in the external environment. In multicellular structures, this external environment would include other cells or tissues of the organism. There is little evidence, at present, about the means by which the spatial organization within a cell is determined, but the activity of an enzyme is known sometimes to be affected by genes other than its structural gene. The necessity for enzyme aggregates in the *arom* and *tryp* pathways in *Neurospora crassa* was discussed in § 15.10. Woodward and Munkres (1966) found that the activity of malate dehydrogenase in *Neurospora* is affected by mutations of the structural protein of the mitochondrion, in association with which the enzyme functions. The mitochondrial protein is coded by a cytoplasmically-inherited gene (see below), while the structural genes for the two polypeptides of the enzyme are each in a different chromosome. Ganschow and Paigen (1967) identified a gene in the mouse, *Mus musculus*, mutation of which caused the enzyme glucuronidase to be absent from the ergastoplasm of liver cells, though still present in the lysosomes. This gene was not linked to the structural gene for the enzyme.

Differentiation within a cell includes the differentiation of the organelles which it contains. Clues as to how such differentiation may take place are provided by the development of virus *T*4 of *Escherichia coli*. Epstein *et al.* (1964) discovered two classes of conditional lethal mutants in this virus: temperature-sensitive mutants which formed plaques at 25°C but not at 42°C, and ambivalent (amber) mutants which formed plaques on a mutant (amber suppressor) strain of the host but not on the wild-type host (see § 13.8 and § 14.17). Both classes of mutations can occur in numerous genes, and have enabled the map positions and function of the genes to be established. In this way 77 genes, possibly representing about $\frac{3}{4}$ of the total number of genes in the *T*4 chromoneme, have been mapped (Fig. 18.5). Epstein *et al.* found that mutants which were defective in DNA synthesis failed to show synthesis and assembly either of the head or the tail components. It thus appeared that the mutants affecting DNA synthesis also prevented development proceeding past an early stage. This functional interaction defined what they called early- and late-acting genes, which (with a few exceptions) were also grouped together physically in the chromoneme (Fig. 18.5). A further point of great interest was that it appeared unlikely that there were more than 20 different proteins in the virus and so Epstein *et al.* imagined that many of the genes were 'morphogenetic' in function, that is, were involved in the assembly of the components rather than in their synthesis.

Study of the *in vitro* assembly of the virus particles by Edgar and Lielausis (1968) showed that there were three separate pathways concerned, respectively, with the head, the tail and the tail-fibres (Fig. 11.1). Within each pathway gene products interact in a fixed sequence. There is a tendency for genes concerned with each pathway to be clustered together on the linkage map (Fig. 18.5). King (1968) studied the formation of the tail of the virus and showed that the base-plate is completed first (*t*1 genes in Fig. 18.5) and then the core is formed on the base-plate (*t*2), followed by the sheath (*t*3–5). Unlike these steps, Edgar and Lielausis found that completed heads and tails combined spontaneously without gene control. The tail-fibres, products of the third independent assembly-line, were attached to the tail after it had combined with the head. Terzaghi (1971) obtained evidence that the distal half of each tail-fibre is first loosely bound to the head, and then the proximal half is attached to the base-plate and to the distal half, which is then detached from the head. This use of the head as an assembly jig for the tail-fibres would explain the otherwise puzzling need to assemble head and tail before attaching the fibres.

These studies on *T*4 assembly have not yet revealed how many of the genes code for minor sub-units of the structure—nuts and bolts, as it were—and

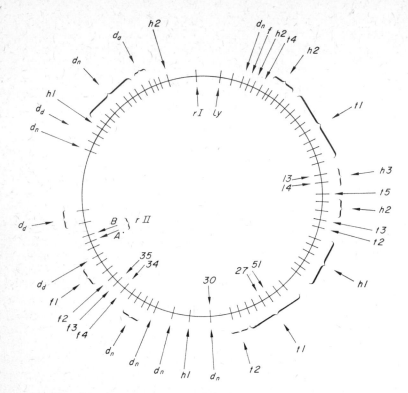

FIGURE 18.5   The diagram shows a genetic map of virus *T*4 based
on the work of Edgar, Epstein and associates (see King,
1968). The virus has a linear, not circular, DNA molecule, but the map is circular because individual virus
particles contain circular permutations of the nucleotide sequence, with some duplication of sequence at
each end (see § 16.4). Symbols outside the circle indicate the function of the gene: *d*, early-acting genes—
relating to DNA synthesis: $d_a$, DNA synthesis arrested
after a short time; $d_d$, delay in starting DNA synthesis;
$d_n$, no DNA synthesis (DNA negative); *f*, *h*, *t*, late-
acting genes—*f*, fibres; *h*, head; *t*, tail. (See Fig. 11.1).
Numerals associated with these letters indicate the
sequence, as far as this is known, in which the products
of these genes interact with one another: *f*1–3, formation of tail-fibre segments; *f*4, union of segments; *h*1,
formation of protein coat of head; *h*2, formation of
active head; *h*3, completion of head; *t*1, formation of
active base-plate; *t*2, formation of core on base-plate;
*t*3–5, formation and stabilization of sheath. Symbols
inside the circle identify some of the genes mentioned in
the text: *ly*, gene for lysozyme.

how many for enzymes which catalyze the joining of different parts, that is, act as temporary scaffolding. The uncertainty whether some gene products function catalytically or structurally is illustrated by the work of Meezan and Wood (1971) and King (1971) on genes 48, 54 and 19 of the *t*2 group, the products of which have been shown from complementation studies to function in the order quoted. The products of genes 48 and 54 act on the base-plate to modify it somehow so that the core of the tail can be attached to it. Study of the base-plate by electron microscopy revealed no detectable difference before and after those products had interacted with it. Gene 19 specifies the major structural sub-unit of the core, but whether the products of genes 48 and 54 are enzymes or minor structural components is not known. The product of gene 38 of the *f*1 group, on the other hand, is believed to be a morphogenetic factor functioning catalytically in tail-fibre assembly. The evidence includes the inability of King and Laemmli (1971) to detect the gene 38 product in mature *T*4, although the electrophoretic band corresponding to this protein could be recognized at the time of tail-fibre assembly.

The differentiation of cytoplasmic organelles in chromosomal organisms may perhaps follow a similar pattern to the assembly of *T*4, with genes for the formation of the essential components, and morphogenetic genes, giving products which interact in a fixed sequence. The flagellar apparatus of *Chlamydomonas reinhardii* promises to be a suitable organelle in which to study this possibility, as numerous flagellar mutants of various kinds are known, mapping in many different genes scattered over the 15 linkage groups (Randall, 1969).

## § 18.15  Cytoplasmic inheritance

It is inferred from Spencer and Harris's results described in § 18.14 that messenger RNA can persist for long periods in the cytoplasm of *Acetabularia*, and such persistence could also account for the examples of delayed Mendelian inheritance described in Chapter 5. On the other hand, the occurrence of cytoplasmic inheritance, also discussed in Chapter 5, indicates that some features of the cytoplasm are self-replicating rather than merely persistent. Many RNA viruses are known to be self-replicating (§ 14.4). So one possible explanation of cytoplasmic inheritance would be to postulate an RNA genetic system in the cytoplasm supplementary to the DNA system in the nucleus. There appears to be no evidence for the existence of such a system. However, it has been discovered that DNA occurs in the chloroplasts and mitochondria, and that it replicates at a different time from nuclear DNA. Chloroplasts and mitochondria resemble chromonemal organisms in many respects (see below) and for this reason the term chromoneme will be used for their DNA.

There has been rapid progress in understanding the organization of chloroplasts and mitochondria. The main findings are as follows. For references to original sources, see Boulter, Ellis and Yarwood (1972) (plants) and Wolstenholme, Koike and Renger (1971) (animals).

The DNA contents of chloroplasts of *Euglena gracilis, Chlamydomonas*

*reinhardii, Nicotiana tabacum* and *Lactuca sativa* are in the range 3·3–9·6 fg*.
From the rate at which the separated chains re-anneal it has been estimated
that the unique sequence of nucleotides involves 0·15–0·33 fg, implying
about 20 copies of the chromoneme per chloroplast. In *Euglena* a circular
DNA molecule of circumference 40 μm has been observed under the electron
microscope, in good agreement with the estimate of chromoneme size from
the annealing rate.

On the basis of electron microscopy and of re-annealing rates, the chromoneme lengths of mitochondrial DNA are believed to be about 25 μm in
fungi, about 50 μm in flowering plants and only about 5 μm in metazoan
animals. The mitochondrial DNA of animals is circular. In *Mus musculus*,
bursting of mitochondria by osmotic shock indicated 2 to 6 chromonemes in
each. Piperno, Fonty and Bernardi (1972) fragmented the mitochondrial
DNA of *Saccharomyces cerevisiae* with a nuclease and found that the DNA
consists of gene-length stretches of normal composition alternating with
stretches of similar length in which over 80% of the bases were adenine and
thymine. Wolstenholme, Koike and Renger (1971) studied the mitochondrial
DNA of *Rattus norvegicus* by denaturation mapping. This technique was
described by Inman (1966) and depends on the fact that the nucleotide
chains of regions of DNA rich in adenine and thymine separate at a lower
temperature (about 50°C) than those rich in guanine and cytosine. Formaldehyde prevents reannealing on cooling and so allows the A·T-rich segments to be observed under the electron microscope. Wolstenholme *et al.*
found three such regions all in one half of the *Rattus* mitochondrial DNA.

As pointed out in § 14.5, chloroplasts and mitochondria have their own
ribosomes and these are of chromonemal type with a sedimentation rate of
70$S$ compared with 80$S$ elsewhere in chromosomal organisms. The main
RNA components are correspondingly smaller: 16$S$ and 23$S$, as in bacteria,
instead of 18$S$ and 28$S$. DNA–RNA hybridization studies have indicated
that the chloroplast ribosomal RNA is coded by the chloroplast DNA, and
the mitochondrial ribosomal RNA by the mitochondrial DNA. The source
of the proteins of the chloroplast and mitochondrial ribosomes is less certain.
Chloramphenicol and cycloheximide are examples of inhibitors of protein
synthesis which are specific to chromonemal and chromosomal organisms,
respectively. The cause of the differential response is no doubt related to the
difference in ribosome size, and to associated differences in the process of
recognition by ribosomes of initiation sites for translation. It is not surprising,
therefore, that protein synthesis in chloroplasts and mitochondria responds
to the chromonemal inhibitors but not to the chromosomal ones. From studies

---

* 1 fg $\equiv \dfrac{606\cdot2}{m} \times 10^6$ nucleotide-pairs, where $m$ is the mean molecular weight of one
nucleotide-pair of the DNA. This formula is derived from Avogadro's constant ($6\cdot062 \times 10^{23}$)
which is the number of molecules in one gram molecule (for example, in 18 g of water).
The molecular weights of nucleotide-pairs in the duplex polynucleotide are as follows:
when the bases are adenine and thymine, 617·3; guanine and cytosine, 618·3; guanine and
5-methylcytosine, 632·3; guanine and 5-hydroxymethylcytosine, 648·3. It follows that 1 fg
of DNA will usually be equivalent to about 1600 genes of 600 nucleotide-pairs. It would have
a length of about 310 μm.

with these inhibitors, evidence has been obtained that the proteins of fungal mitochondrial ribosomes are synthesized outside the mitochondrion and are therefore most probably coded in the nucleus. For chloroplasts, however, the evidence points to synthesis in the organelle.

Chloroplasts and mitochondria have at least some of their own transfer RNA molecules and amino-acyl *t*RNA synthetases, and polypeptide initiation is of chromonemal type with formylation of the methionine of the initiating methionyl transfer RNA (§ 14.16). DNA-RNA hybridization studies have shown that at least some of the *t*RNAs are coded in the DNA of the organelles. Whether the genes for the activating enzymes are also in the respective organelles is uncertain.

It is also not known where the DNA and RNA polymerases found in these organelles are coded. Küntzel and Schäfer (1971), however, isolated the mitochondrial RNA polymerase from *Neurospora crassa* and found it to be a relatively simple molecule of chromonemal type, unlike the RNA polymerase of the *Neurospora* nucleus.

Attempts have been made to identify the proteins synthesized in chloroplasts and mitochondria from study *in vitro* of the isolated organelles, and from the effects *in vivo* of inhibitors of chromonemal—and hence organelle—protein synthesis. There is much evidence that many of the enzymes found in chloroplasts and mitochondria are synthesized outside these structures on ribosomes of chromosomal type. Surprisingly, the larger sub-unit of ribulose-diphosphate carboxylase seems to be synthesized within chloroplasts and the smaller outside (see Boulter *et al.*, 1972). Likewise, cytochrome oxidase seems to require the combination of components synthesized inside and outside the *Saccharomyces* mitochondrion, respectively, and similar conclusions have been reached from study of the *mi*-1 mutant of *Neurospora crassa* (Woodward, Edwards and Flavell, 1970). There is evidence which suggests that some membrane proteins of chloroplast and mitochondrion may be synthesized within these organelles.

The many points of resemblance between chromonemal organisms and chloroplasts and mitochondria support the idea that these organelles are derived from such organisms which have adopted a symbiotic existence within the cells of chromosomal organisms. Progressive transfer of organelle genes to the nucleus may then have taken place. This might explain why animals have relatively small mitochondrial chromonemes. An alternative explanation of the evolutionary origin of chloroplasts and mitochondria is to suppose that these organelles are remnants of the chromonemal ancestor of chromosomal organisms.

It was shown in Chapter 5 that many examples of cytoplasmic inheritance are known, and that these are concerned predominantly with characters of the plastids and mitochondria. How far can cytoplasmic inheritance be attributed to genes in the DNA of these organelles? Woodward and Munkres (1966) discovered that two cytoplasmically-inherited respiratory-deficient mutants of *Neurospora crassa* were defective in the structural protein of the mitochondrion. One of the mutants (*mi*-1) had one less tryptophan and one more cysteine residue per molecule of the protein, a change which would be accounted for by a single base substitution (see Table 14.7). There is no evidence where in the cytoplasm this gene is situated, but it is natural to attribute it to the mitochondrial DNA. Rifkin and Luck (1971), however, inferred that the defect in the *mi-1* mutant was in the synthesis or assembly of small sub-units of the mitochondrial ribosome. Mounolou *et al.* (1966) found that the mitochondrial DNA of cytoplasmic petite mutants of *Saccharomyces cerevisiae* (§ 5.15) was of abnormal density. Wintersberger and Viehhauser (1968) showed by DNA–RNA hybridization that the mitochondrial DNA of a cytoplasmic petite mutant lacking the normal RNA components of mitochondria also lacked DNA complementary to mitochondrial ribosomal and transfer RNA. Mutants of *S. cerevisiae* resistant to erythromycin, chloramphenicol and other drugs have been studied by Linnane, Slonimski, Wilkie and their respective associates (see Bunn *et al.*, 1970; Coen *et al.*, 1970; Wilkie, 1970). These mutants show non-Mendelian inheritance and are believed to be caused by mutations in the mitochondrial DNA affecting the mitochondrial ribosomes or membranes. Evidence for linkage and recombination of such mutants has been obtained by Wilkie and by Coen *et al.*, implying a sexual life-cycle for mitochondria with genetic continuity and exchange. The results are discussed by Preer (1971). Of particular interest is the discovery by Coen *et al.* that the contributions of the two parents are unequal and in many ways comparable to those found with conjugation in *Escherichia coli* (§ 16.3). The polarized transfer of part of the chromoneme of a donor mitochondrion into a recipient one is a possible explanation.

Results similar in many ways to those derived from mitochondrial genetics have been obtained from study of characters believed to be inherited in the DNA of the chloroplasts. Loss of the plastids of *Euglena* (§ 5.14) can be induced by irradiation of the cytoplasm, but not the nucleus, with ultraviolet light. A wavelength of 260 nm gave the maximum effect, which could be reduced by exposure to visible light (see Kirk and Tilney-Bassett, 1967). Sensitivity to ultraviolet light of this wavelength, and photoreactivation, are characteristics of DNA. Similarly, Sager and Ramanis (1967) found that the behaviour of the cytoplasmic streptomycin and acetate mutants of *Chlamydomonas reinhardii* (see § 5.14) showed ultraviolet-sensitivity and photoreactivation. Moreover, they had found (Sager and Ramanis, 1965) that allelic mutants at the acetate locus showed linkage and recombination, and they recovered reciprocal recombinants in the same clone. Linkage and recombination were also found at the streptomycin locus. Sager and Ramanis argued that these results require a process of close intermolecular pairing and exchange such as is known only in nucleic acid. They concluded that their results supported the hypothesis that the cytoplasmic genes are composed of nucleic acid. Later studies (Sager and Ramanis, 1970) indicated a single cytoplasmic linkage group for all the mutants studied and suggested that it was situated in the chloroplast, though the evidence for this was indirect. Cavalier-Smith (1970) obtained evidence from electron micrographs for fusion of the two chloroplasts in the zygotes of *C. reinhardii*. Such fusion would provide an opportunity for recombination of plastid-inherited characters. Gillham, Boynton and Burkholder (1970) found that certain streptomycin-resistant and streptomycin-dependent non-Mendelian mutants of *C. reinhardii* had abnormal chloroplast ribosomes and were presumably defective in genes of the chloroplast DNA.* Surzycki and Gillham (1971) used rifampicin, which specifically inhibits transcription of chloroplast DNA, to distinguish between chloroplast and mitochondrial mutants. Preer (1971) discussed their results, some of which are puzzling.

In flowering plants, studies by Michaelis (1967) with *Epilobium* and by Schötz (1970) and Diers (1970) with *Oenothera* have supported the hypothesis that non-Mendelian inheritance of plastid characters results from determinants carried by the plastids and hence presumably by their DNA. Diers (1970) reviewed the evidence for and against genetic continuity of chloroplasts, and concluded that they arise by division of pre-existing chloroplasts. Herrmann and Bauer-Stäb (1969) discovered by polyacrylamide-gel electrophoresis that the lamellar protein of the chloroplasts of a non-Mendelian *albomaculata* mutant of *Antirrhinum majus* lacked certain components. Dolzmann (1968), by electron microscopy, likewise found abnormality of the lamellar membrane in the plastids of cytoplasmically-inherited mutants of *Oenothera* with pale plastids. Hallier (1968) and Fork and Heber (1968) found that the *Oenothera* mutants were defective in the electron transport system of photosynthesis, but this could be a consequence of the membrane defects. The work of Schötz (1970) on the lamellar membranes of the chloroplasts of *Oenothera* hybrids has emphasized how much nuclear and plastid factors interact in plastid development. These studies reinforce the early discoveries (Chapter 5) that some mutations affecting the plastids show Mendelian inheritance and others are non-Mendelian.

Although not conclusively established, it thus seems probable that many of the known examples of cytoplasmic inheritance are caused by genes in the DNA of plastids and mitochondria.

It has long been realized that cytoplasmic inheritance does not necessarily imply the existence of self-perpetuating particles such as DNA or RNA might provide. Cytoplasmic mutant phenotypes might result from alteration of a self-perpetuating metabolic system from one steady state to another. This is exemplified by Monod and Jacob's hypothetical system described in § 18.6 where each of two alternative reaction-chains is inhibited by the end-product of the other. It would be necessary, in addition, to postulate that each chain contained features which made it self-perpetuating when once established. However, on present evidence the steps in the reaction-chains would be expected to depend on nuclear gene activity. An example of a condition that may be caused by cytoplasmic inheritance of this metabolic kind is the disease of sheep, *Ovis aries*, called scrapie. This disease develops slowly in the nervous system and was thought to be of viral origin until the causative agent was found to be abnormally insensitive to ultraviolet light and other treatments to which nucleic acids are vulnerable. Latarjet *et al.* (1970) found that, unlike its effect on nucleic acids, ultraviolet light of wavelength 237 nm

---

* Labelling experiments by Sager and Lane (1972) indicated that the chloroplast DNA of the male parent disappears soon after zygote formation. This agrees with the predominantly maternal inheritance of characters believed to be determined by the chloroplast DNA.

inactivated the scrapie agent much more than UV of 250 nm. Pattison and Jones (1968) detected the scrapie agent in normal mice, *Mus musculus*, and suggested that the disease may be caused, not by self-replication of an infecting particle, but by unmasking of some cell constituent which is normally inhibited. Other diseases of the nervous system, such as multiple sclerosis and kuru in man, may have a similar cause.

It seems likely that the greater part of the cytoplasmic organization of a cell at any particular time is a reflection of the previous nuclear activity. This is supported by studies of the development of, for example, echinoderms and *Acetabularia* (see Chapter 5). On the other hand, the occurrence of cytoplasmic inheritance demonstrates that the organization cannot all be so explained, and that parts of the cytoplasm constitute self-perpetuating particles or reaction-chains. Until more is known biochemically about the nature of cytoplasmic inheritance, and about how species differ from one another cytoplasmically, the relative importance of Mendelian and non-Mendelian heredity cannot be judged with any precision, and likewise of nucleus and cytoplasm in the control of differentiation. Does positional information in membrane structure, or a gradient in a cell, ultimately derive from nucleotide sequence in DNA or RNA? Or, are these factors in differentiation inherited independently of nucleic acids? These are the kinds of question which it has not yet been possible to answer and which, in consequence, leave open some fundamental questions about how organisms develop.

## § 18.16   The environment and gene activity

An important aspect of differentiation is the influence of the external environment, which can frequently modify profoundly the development of an organism. This is particularly evident in plants, where individuals growing in different habitats may differ greatly and yet when cultivated under similar conditions appear alike. An example of environmental influence is provided by sun red *Zea mays*, which has the genotype *A B pl pl*, that is, with the dominant genes *A* (anthocyanin pigment) and *B* (booster of 'plant colour'), and homozygous for the recessive gene *pl* (plant colour). In sun red maize, anthocyanin is formed only in those parts of the plant which are exposed to sunlight. Another example of the influence of the environment is provided by the number of facets to the eye in Bar-eyed *Drosophila melanogaster*. As indicated in § 9.10, the facet number is greatly influenced by the temperature during development: there are 4 or 5 times as many facets at 15°C than at 30°C. In general, to give precision to an account of the effect of a particular gene on an organism, the genotype and the environmental conditions must be specified. Many genes show *incomplete penetrance*, that is to say, they are not expressed in some individuals. Incomplete penetrance reflects variability of genotype or environment. If a gene acts at an early stage of development, it may affect many different characters of the organism, but the various parts may differ in their sensitivity to modification by other genetic factors and by the external environment. This gives rise to *variable expressivity* of the gene, that is, variation in the way in which a gene is expressed in different individuals. The plants of *Antirrhinum majus* found by Darwin to be intermediate in appearance between the normal and peloric forms (see Fig. 2.4) are likely to have arisen in this way.

How the environment influences gene activity is not understood. However, it appears significant that the variegated phenotypes associated both with position effect and with mutable genes show great variability of expression depending on environmental factors. Harrison and Fincham (1964) have shown that the frequency of mutation to a magenta colour in the stems and petals of *A. majus*, which occurs with high frequency in strains of a particular genotype, is much influenced by the temperature when the tissue in question is developing. Many similar instances are known of sensitivity of mutable genes to the temperature at the stage of development when the gene is acting. The various ways in which mutation in flowering plants can be influenced, including environmental and developmental controls, are discussed by Fincham (1970), who also considers the possible relevance of high mutability to normal control of gene activity.

## § 18.17   Conclusions

From the experimental results described in this chapter the following tentative conclusions may be drawn.

The functional unit of the chromosome is the chromomere, which is a linear series of copies of a structural gene within one DNA molecule. Control of its activity seems to be possible in at least four ways. The primary control may involve a particular chemical, such as the end-product of a preceding step in a biosynthetic pathway or morphological assembly-line, interacting with an acidic or neutral protein which can recognize a specific nucleotide sequence in the DNA of the chromomere (or interchromomeric region alongside it). This protein would be coded by a regulatory gene.

A second control mechanism can cause the inhibition of gene activity in a segment of a chromosome or a whole chromosome as a result of the activation of a controlling element situated in heterochromatin. Such activation can lead to euchromatin becoming heterochromatic for a certain distance along the chromosome from the controlling element. Heterochromatin lacks chromomeres, as though the masking of their activity also masked their morphology. Histones seem to be a likely candidate for the molecules which cause this spreading effect, since they are believed to mask gene activity in a non-specific way, once non-basic proteins have been removed from the chromosome.

Thirdly, controlling elements can cause frequent mutation in specific unlinked genes, possibly acting by way of episomes. The episome would be expected to be a circular DNA molecule. The mutability, which may be an abnormal feature, could arise from differences in nucleotide sequence between gene and episome. Mutability unrelated to a controlling element could arise from a failure of the matching process by which the slaves would have been rendered identical to the master copy of the gene.

A fourth control mechanism may occur at the translation of messenger into polypeptide. Ribosomes, transfer RNAs or the conformation of the messenger might provide the control. Ribosomal RNA and transfer RNAs are themselves the products of special genes which determine RNA nucleotide sequences for their own sake.

Each of these control mechanisms involves a different molecule (non-basic protein, histone, DNA, RNA) and a different process. Genetic and environmental factors would be expected to influence all of them, but not in a uniform way in view of their diversity. A further complication is the existence of subsidiary hereditary systems in plastids and mitochondria, with a chromonemal type of organization.

The basic unit of the chromosome from the point of view of its replication also seems to be the chromomere, as the largest segment of DNA that replicates as a unit. A control mechanism evidently exists which determines the sequence in which replication is initiated in the various replicons. That heterochromatin replicates late suggests that whatever masks the chromomeres morphologically and functionally in heterochromatic segments also temporarily masks them from the signals which trigger the replication of the euchromatic replicons.

The basic unit of the chromosome from the point of view of recombination also seems to be the chromomere. Here, too, there is evidence for specific control in each chromomere. Moreover, crossing-over is normally absent from heterochromatin. This was shown, for example, by Baker (1958) with *Drosophila*. Woollam and Ford (1964) found that the chromosome material lateral to the synapton was much denser in heterochromatic than in euchromatic regions of *Mus* chromosomes. Assuming that crossing-over occurs in the synapton, there is evidently much less opportunity for crossing-over in heterochromatin than euchromatin.

The similarity in the organization of all three primary functions of the hereditary material—its specific action, its replication and its recombination—is remarkable. All seem to depend on the chromomere as the basic unit, to be under specific control at the level of the individual chromomere, and to be subject to delay or inactivation when the chromomere becomes heterochromatic. The existence of a common unit for all three functions raises the question whether they act through a common mechanism: a transcriptase, a replicase and a recombinase might each be capable of functioning in every unit, but be active only when combined with the key—a specific activator or derepressor—for unlocking that unit.

# Appendix 1. *Organisms mentioned in the text*

*(In order to make the classification more meaningful, a few species and phyla have been included which are not referred to in the text.)*

## I. CHROMONEMAL KINGDOM

The term 'chromoneme' is used for 'bacterial chromosome' (see § 11.1). Chromonemal organisms differ from chromosomal in having no basic proteins (histones or protamines) associated with their DNA, no mitotic apparatus, no nuclear membrane, no mitochondria, and no plastids and in having ribosomes with a sedimentation coefficient of $70S$ instead of $80S$. The cell-walls have a distinctive composition containing mucopeptides, and the flagella, where present, are of a different construction from those of chromosomal organisms. Lysine synthesis is by a pathway involving diamino-pimelic acid (Vogel, 1965). For a detailed account of the characteristic features of chromonemal organisms, see Stanier (1964). The relationship between bacteria and blue-green algae has been discussed by Echlin and Morris (1965).

## A. BACTERIA

Photosynthesis absent, or, if present, without evolution of oxygen.

(i) EUBACTERIALES

a Pseudomonadaceae
*Pseudomonas aeruginosa*

b Azotobacteriaceae
*Azotobacter vinelandii*

c Micrococcaceae
*Micrococcus pyogenes*

d Parvobacteriaceae
*Haemophilus influenzae*

e Lactobacteriaceae
*Diplococcus pneumoniae* (Pneumococcus)
*Streptococcus faecalis*

f Achromobacteriaceae

g Enterobacteriaceae
*Aerobacter aerogenes*
*Escherichia coli* (Colon bacillus)
*Salmonella typhimurium* (Mouse typhoid bacillus)

    *h*  Bacillaceae
       *Bacillus cereus*
       —*subtilis*

(ii) ACTINOMYCETALES
    *a*  Mycobacteriaceae
       *Mycobacterium phlei*
       —*tuberculosis avium* (Avian tubercle bacillus)
    *b*  Streptomycetaceae
       *Streptomyces coelicolor*
       —*griseus*

## B. CYANOPHYTA

(Myxophyta, Blue-green Algae)
Photosynthesis with evolution of oxygen
    *Phormidium luridum*

## II. CHROMOSOMAL KINGDOM

The classification of fungi with a motile stage follows Sparrow (1959), whose division of them into the truly fungal Chytridiomycetes and the algal Phycomycetes has received strong support from the remarkable discovery by Vogel (1965) that the Eu-fungi have a different pathway for lysine synthesis from other plants, and so have probably evolved from animals. *Euglena* has the lysine pathway of Eu-fungi and is therefore presumably of fungal origin, having perhaps become photosynthetic secondarily by symbiosis with a chloroplast from another organism. Some—perhaps all—of the Phycofungi are diploid organisms with meiosis in the gametangia (Sansome, 1961).

## A. PLANT SUBKINGDOM

Lysine synthesis by the diaminopimelic acid pathway.

  (i)  CHLOROPHYTA (Green Algae)
    *a* Volvocales
      1 Green forms
        *Chlamydomonas reinhardii*
      2 Colourless forms
        *Polytoma* sp.
    *b* Siphonales
      *Acetabularia crenulata*
      —*mediterranea*
 (ii)  XANTHOPHYTA (Yellow-green Algae)
      *Vaucheria* sp.
(iii)  PHYCOFUNGI (Zoospores biflagellate or anteriorly uniflagellate)
      *Pythium* sp.
(iv)  DINOPHYTA (Dinoflagellates)
      *Amphidinium elegans*
      *Prorocentrum micans*

(v) PHAEOPHYTA (Brown Seaweeds)

(vi) RHODOPHYTA (Red Seaweeds)

(vii) BACILLARIOPHYTA (Diatoms)

(viii) BRYOPHYTA
  *a*  Musci (Mosses)
  *b*  Hepaticae (Liverworts)
      *Pellia endiviifolia*
      *Sphaerocarpos donnellii*

(ix) PTERIDOPHYTA (Ferns, etc.)
      *Ophioglossum petiolaris* (Adder's-tongue Fern)

(x) GYMNOSPERMAE (Conifers, Cycads, etc.)

(xi) ANGIOSPERMAE (Flowering Plants)
  *a*  Monocotyledones
      1  Gramineae (Grasses)
          *Avena sativa* (Oats)
          *Hordeum vulgare* (Barley)
          *Oryza sativa* (Rice)
          *Triticum aestivum* (Wheat)
          *Zea mays* (Maize)
      2  Commelinaceae
          *Tradescantia paludosa*
      3  Juncaceae (Rushes)
          *Luzula* sp. (Wood-rush)
      4  Liliaceae (Lilies, etc.)
          *Allium cepa* (Onion)
          —*cernuum*
          —*fistulosum*
          *Aloe purpurascens*
          *Bellevalia romana*
          *Disporum sessile*
          *Endymion nonscriptus* (Bluebell)
          *Fritillaria* sp.
          *Hyacinthus orientalis* (Hyacinth)
          *Ipheion uniflorum*
          *Lilium formosanum*
          —*henryi*
          —*japonicum*
          —*longiflorum*
          —*maximowiczii*
          —*regale*
          *Paris verticillata*
          *Scilla scilloides*
          —*sibirica*
          *Trillium erectum*
          *Tulipa* sp. (Tulip)

    5 Amaryllidaceae
      *Clivia cytranthiflora*
      *Haemanthus katherinae*
    6 Iridaceae
      *Crocus balansae*
    7 Orchidaceae (Orchids)
*b*  Dicotyledones
    1 Urticaceae
      *Urtica pilulifera* (Roman Nettle)
    2 Nyctaginaceae
      *Mirabilis jalapa* (Marvel of Peru)
    3 Caryophyllaceae
      *Agrostemma githago* (Corn Cockle)
      *Silene alba* (White Campion)
      —*dioica* (Red Campion)
    4 Ranunculaceae
      *Anemone* spp.
      *Helleborus foetidus*
    5 Papaveraceae
      *Chelidonium majus* (Greater Celandine)
      *Papaver somniferum* (Opium Poppy)
    6 Cruciferae
      *Lunaria annua* (Honesty)
    7 Leguminosae
      *Lathyrus odoratus* (Sweet Pea)
      *Phaseolus vulgaris* (Dwarf Bean)
      *Pisum sativum* (Edible Pea)
      *Trifolium pratense* (Red Clover)
      *Vicia faba* (Broad Bean)
    8 Geraniaceae
      *Pelargonium zonale* (Geranium)
    9 Linaceae
      *Linum usitatissimum* (Flax)
  10 Violaceae
      *Viola cornuta*
  11 Caricaceae
      *Carica papaya* (Papaw)
  12 Onagraceae
      *Epilobium hirsutum* (Hairy Willow-herb)
      —*luteum*
      *Oenothera blandina*
      —*hookeri*
      —*lamarckiana* (Evening Primrose)
      —*muricata*
  13 Primulaceae
      *Primula sinensis*

14 Labiatae
  *Salvia horminum*
15 Solanaceae
  *Datura tatula*
  *—stramonium* (Thorn Apple)
  *Hyoscyamus niger* (Henbane)
  *Nicotiana tabacum* (Tobacco)
  *Solanum nigrum* (Black Nightshade)
16 Scrophulariaceae
  *Antirrhinum glutinosum*
  *—majus* (Snapdragon)
  *Veronica longifolia*
17 Plantaginaceae
  *Plantago lanceolata* (Ribwort Plantain)
18 Compositae
  *Aster tripolium* (Sea Aster)
  *Chrysanthemum roxburghi*
  *Coreopsis tinctoria*
  *Haplopappus gracilis*
  *Lactuca sativa* (Lettuce)

## B. FUNGAL SUBKINGDOM

Lysine synthesis by a pathway involving α-aminoadipic acid.
  (i) CHYTRIDIOMYCETES (Zoospores posteriorly uniflagellate)
    *Allomyces* sp.
  (ii) EUGLENIDA
    *Euglena gracilis* (including *Astasia longa*)
  (iii) ZYGOMYCETES
    Mucorales
      *Phycomyces blakesleeanus*
  (iv) ASCOMYCETES
    *a* Endomycetales (Yeasts, etc.)
      *Saccharomyces cerevisiae*
      *Schizosaccharomyces pombe*
      *Torulopsis utilis**
    *b* Eu-Ascomycetes
      *Microsporum gypseum**
      1 Plectomycetes
        *Aspergillus nidulans*
        *—niger**
        *Penicillium notatum**
      2 Discomycetes
        *Ascobolus immersus*
        *Neottiella rutilans*

* Fungi Imperfecti: nuclear fusion and meiosis unknown.

       3 Pyrenomycetes
          *Bombardia lunata*
          *Neurospora crassa*
          *—sitophila* (Red Bread Mould)
          *—tetrasperma*
          *Podospora anserina*
          *Sordaria brevicollis*
          *—fimicola*

(v) BASIDIOMYCETES
    *a* Uredinales (Rust Fungi)
       *Puccinia glumarum* (Yellow Rust)
    *b* Ustilaginales (Smut Fungi)
       *Ustilago maydis* (Maize Smut)
    *c* Hymenomycetes (Mushrooms and Toadstools)
       *Schizophyllum commune*
    *d* Gasteromycetes (Puffballs)


## C. ANIMAL SUBKINGDOM

Lysine not synthesized.
(i) PROTOZOA
    *a* Mycetozoa (Slime Moulds)
       1 Myxomycetales (Plasmodial Slime Moulds)
          *Physarum* sp.
       2 Acrasiales (Cellular Slime Moulds)
          *Dictyostelium* sp.
    *b* Sarcodina
       *Amoeba* sp.
    *c* Flagellata (Zoomastigina)
       *Trypanosoma* sp. (Sleeping sickness parasite)
    *d* Sporozoa
       *Plasmodium falciparum* (Falciparum malarial parasite)
    *e* Ciliata
       *Paramecium aurelia*

(ii) PORIFERA (Sponges, etc.)
(iii) COELENTERATA (Jelly-fish, etc.)
(iv) PLATYHELMINTHES (Flatworms)
       *Mesostoma* spp.
(v) NEMATODA (Threadworms)
       *Parascaris equorum* (*Ascaris megalocephala*) (Horse Threadworm)
(vi) ANNELIDA (Worms)
(vii) ARTHROPODA
    *a* Crustacea (Shrimps, Crabs, etc.)
       *Cancer borealis*

       *Gammarus* spp.
       *Procambarus clarkii* (Crayfish)
*b*  Myriapoda (Centipedes)
*c*  Insecta (Insects)
    1  Orthoptera (Grasshoppers)
       *Acrida lata*
       *Arphia simplex*
       *Brachystola magna*
       *Calliptamus palaestinensis*
       *Chorthippus brunneus*
       *—parallelus*
       *Cyrtacanthacris tartarica* (Locust)
       *Goniaea australasiae*
       *Gryllotalpa hexadactyla* (Mole Cricket)
       *Gryllus argentinus*
       *Melanoplus femur-rubrum*
       *Phrynotettix tschivavensis* (*P. magnus*)
       *Romalea microptera*
       *Schistocerca gregaria* (Locust)
       **Stauroderus bicolor**
       *Stethophyma grossum*
       **Trimerotropis maritima**
    2  Hemiptera (Bugs)
       *Euschistus variolarius*
       *Lygaeus turcicus*
       *Oncopeltus fasciatus*
       *Protenor belfragei*
       *Pyrrhocoris apterus*
       *Thyanta* spp.
    3  Lepidoptera (Butterflies, Moths)
       *Abraxas grossulariata* (Magpie Moth)
       *Ephestia kuhniella*
    4  Coleoptera (Beetles)
       *Tenebrio molitor* (Meal worm)
    5  Hymenoptera (Ants, Bees, Wasps)
       *Apis mellifica* (Honey Bee)
    6  Diptera (Flies)
       *Chironomus pallidivittatus*
       *—tentans*
       *—thummi* subsp. *piger*
       *— —*subsp. *thummi*
       *Drosophila hydei*
       *—melanogaster* (Fruit Fly)
       *—pseudo-obscura*
       *—subobscura*
       *Samoaia leonensis*

    *d*  Arachnida (Scorpions, Spiders)

(viii)  MOLLUSCA
    *a*  Lamellibranchiata (Oysters, etc.)
    *b*  Gasteropoda (Snails, Slugs)
        *Helix hortensis* (Garden Snail)
        —*nemoralis*
        *Limnaea peregra* (Freshwater Snail)
    *c*  Cephalopoda (Octopus, etc.)

(ix)  ECHINODERMATA (Sea Urchins, etc.)
      *Lytechinus variegatus*
      *Paracentrotus lividus*
      *Psammechinus microtuberculatus*
      —*milaris*
      *Sphaerechinus granularis*
      *Strongylocentrotus purpuratus*

(x)  CHORDATA (Vertebrates, etc.)
    *a*  Pisces (Fish)
      1 Teleostei
        *Alosa* sp. (Shad)
        *Salmo gairdnerii* (Rainbow Trout)
        —sp. (Salmon)
      2 Elasmobranchii
        *Carcharias* spp. (Shark)
      3 Dipnoi
        *Protopterus* sp. (Lung-fish)
    *b*  Amphibia (Amphibians)

      1 Anura (Frogs, Toads)
        *Bufo bufo* (European Toad)
        —*viridis* (Green Toad)
        *Rana pipiens* (American Frog)
        —*temporaria* (European Frog)
        *Xenopus laevis* (Clawed Toad)
        —*mulleri*

      2 Urodela (Newts, Salamanders)
        *Ambystoma mexicanum* (Axolotl)
        *Batrachoseps attenuatus* (Slender Salamander)
        *Plethodon cinereus*
        *Salamandra maculosa* (Spotted Salamander)
        *Triturus cristatus* (European Crested Newt)
        —*pyrrhogaster* (Japanese Newt)
        —*viridescens* (American Newt)
    *c*  Reptilia (Reptiles)
      *Crotalus adamanteus* (Diamond-back Rattlesnake)

*d* Aves (Birds)
  *Anas platyrhynchos* (Duck)
  *Gallus domesticus* (Domestic Fowl)
*e* Mammalia (Mammals)
  1 Marsupialia
    *Macropus robustus* (Wallaroo)
    *Megaleia rufa* (Red Kangaroo)
    *Potorous tridactylus* (Rat Kangaroo)
  2 Lagomorpha (Rabbits, Hares)
    *Oryctolagus cuniculus* (Rabbit)
  3 Rodentia (Rodents)
    *Cavia porcella* (Guinea Pig)
    *Cricetulus griseus* (Chinese Hamster)
    *Mesocricetus auratus* (Golden Hamster)
    *Microtus agrestis* (Field Vole)
    *Mus musculus* (Mouse)
    *Rattus norvegicus* (Rat)
  4 Cetacea (Whales)
    *Balaenoptera borealis* (Sei Whale)
  5 Carnivora (Carnivores)
    *Canis familiaris* (Dog)
    *Felis catus* (Cat)
  6 Perissodactyla (Odd-toed Ungulates)
    *Equus caballus* (Horse)
  7 Artiodactyla (Even-toed Ungulates)
    *Ammotragus lervia* (Barbary Sheep)
    *Bos buffelus* (Buffalo)
    *—taurus* (Domestic Cattle)
    *Capra hircus* (Goat)
    *Odocoilus virginianus* (Virginian White-tailed Deer)
    *Ovis aries* (Sheep)
    *Sus scrofa* (Pig)
  8 Primates
    *Cercopithecus aethiops* (African Green Monkey)
    *Gorilla gorilla* (Gorilla)
    *Homo sapiens* (Man)
    *Macaca irus* (Irus Macaque)
    *—mulatta* (Rhesus Monkey)
    *Pan troglodytes* (Chimpanzee)

## III. VIRUSES

For the classification of viruses, see Lwoff and Tournier (1966).

### A. DEOXYVIRA (DNA Viruses)

   (i)  DEOXYHELICA (Nucleocapsid with helical symmetry)
       *Poxvirus variolae* (Small-pox virus)
  (ii)  DEOXYCUBICA (Nucleocapsid with cubical symmetry)
     *a*  Microviridae
       *Microvirus monocatena* ($\phi$X174)
     *b*  Papillomaviridae
       *Polyomavirus neoformans* (Polyoma virus)
       Simian virus 40 belongs to this genus.
     *c*  Inophagoviridae
       *Inophagovirus bacterii*
       F1 belongs to this genus.
 (iii)  DEOXYBINALA (With head and tail)
       *Phagovirus (Coli) T secundus* (T2)

       The other T phages and $\lambda$, 434, $\phi$80, P1, P22 and SPP1 belong
       to this class.

### B. RIBOVIRA (RNA Viruses)

   (i)  RIBOHELICA (Nucleocapsid with helical symmetry)
     *a*  Protoviridae
       *Protovirus tabaci* (Tobacco mosaic virus)
       Ribgrass virus belongs to this genus.
     *b*  Paramyxoviridae
       *Paramyxovirus primus* (Parainfluenza virus I)
       Sendai virus belongs to this genus.
  (ii)  RIBOCUBICA (Nucleocapsid with cubical symmetry)
     *a*  Napoviridae
       *Poliovirus primus* (Polio virus)
       Mengovirus belongs to this genus.
       *Androphagovirus bacterii* (RNA phage)
       F2, MS2, Q$\beta$ and R17 belong to this genus.
     *b*  Reoviridae
       *Reovirus (Mammalis) primus* (Reovirus I)
     *c*  Arboviridae
       The avian myeloblastosis and Rous sarcoma viruses, the
       murine mammary tumour, Moloney sarcoma and Rauscher
       leukaemia viruses, and the feline leukaemia virus belong to
       this family.

# Appendix 2. *Glossary of genetical terms used in the text*

**Acentric**   Applied to a chromatid or a chromosome when it lacks a centromere. This condition may arise in an inversion heterozygote as a result of crossing-over between a normal and an inverted segment that does not include the centromere.

**Allele (allelomorph)**   The term coined by Bateson and Saunders (1902) for characters which are alternative to one another in Mendelian inheritance (Gk. *allelon*, one another; *morphe*, form). The term is also applied to the wild-type and mutant forms of a gene responsible for such a character-difference.

**Anaphase**   Strasburger (1884) originally introduced this term for the stage of nuclear division when the contents of the nuclei were going back (Gk. *ana*) to their normal appearance, but from about 1905 he used the term in the now universally adopted sense of the stage of mitosis or of meiosis I or II when the daughter-chromosomes (or homologous chromosomes at meiosis I) move towards opposite poles of the spindle.

**Anticodon**   The triplet of nucleotides in transfer RNA which associates by complementary base pairing with a specific triplet (codon) in messenger RNA during protein synthesis.

**Autosomes**   A term coined by Montgomery (1906) for the chromosomes other than the $X$- and $Y$-chromosomes (which he called allosomes) by which the sexes differ.

**Back-cross**   A cross between a parental strain and its progeny from a previous cross.

**Centromere**   The term introduced by Darlington (1936a) for the point on a chromosome which reacts to the spindle at nuclear division, and to which a spindle-fibre appears to be attached at the times when the spindle is evident (metaphase and anaphase) (Gk. *centron*, centre; *meros*, part). The chromosome is usually constricted and often bent at the centromere. In a few organisms the centromeric properties are distributed over the entire length of the chromosome (diffuse centromere).

**Chiasma**   The term which Janssens (1909) introduced for the nodes (Gk. *chiasma*, cross) where the individual chromosomes making up each pair remain in contact during the diplotene and diakinetic stages of prophase I and during metaphase I of meiosis. It is now established that a chiasma arises as a consequence of crossing-over. In some organisms the chiasmata move towards the ends of the chromosome arms during late

445

diplotene and diakinesis (terminalization of chiasmata). The position of the chiasma will then no longer correspond to that of the cross-over. The expression 'terminal chiasma' was coined by Newton and Darlington (1929) for the end-to-end association of homologous chromosome arms as a result of chiasma movement to the end.

**Chromatid**    The term which McClung (1900) proposed for each of the four threads making up a chromosome-pair at meiosis (Gk. *chroma*, colour; for the derivation of *id*, see Diploid). The use of the term was subsequently extended to mitosis, and is now applied to the individual daughter-chromosomes into which each chromosome is divided in all nuclear divisions. The term chromatid is used so long as the daughter centromeres remain in contact with one another. As soon as they separate (anaphase of mitosis and anaphase II of meiosis), the expression daughter-chromosome is substituted for chromatid.

**Chromomere**    The term applied by Wilson (1896) to each of the serially aligned granules of a chromosome, best seen when the thread is relatively elongated as in early prophase I of meiosis (Gk. *chroma*, colour; *meros*, part).

**Chromoneme**    The term used in this book for the DNA thread of bacteria and their viruses (see § 11.1 and Appendix 1) (Gk. *chroma*, colour; *nema*, thread). It is appropriate also to use it for the DNA of plastids and mitochondria (see § 18.15).

**Chromosome**    The term proposed by Waldeyer (1888) for the individual threads within the cell nucleus (Gk. *chroma*, colour; *soma*, body). The expression **lampbrush** chromosome describes the appearance of the chromosomes at the diplotene stage of meiosis in the oocytes of some amphibia, when the chromatids extend from each chromomere to form lateral loops. Wilson (1909) proposed that a chromosome which is represented twice (the normal diploid condition) in the nuclei of one sex, but is present only once per nucleus in the other sex, should be called an *X*-chromosome, while the dissimilar chromosome which may also be present as a single chromosome in the latter sex, but is absent from the sex with two *X*-chromosomes, he called a *Y*-chromosome. The normal chromosomes were called by Randolph (1928) *A*-chromosomes to distinguish them from supernumerary or *B*-chromosomes which occur in some individuals or populations of certain species, and are usually composed wholly of heterochromatin.

**Cis**    The term proposed by Haldane (1941), by analogy with chemical isomerism, to describe a double heterozygote in coupling phase, that is, with the two dominant factors derived from one parent and the two recessives from the other. The term is used chiefly for closely linked mutants, to which it was first applied by Pontecorvo (1950). The cis configuration for two recessive mutants, 1 and 2, is $\dfrac{+\,+}{1\ 2}$, where $+$ stands for the wild-type, and the line separates the contributions from the two parents. (Cf. Trans.)

**Cistron**    The term which Benzer (1957) introduced, derived from cis and trans, for the functional unit of the hereditary material defined by the phenotype of the trans (repulsion) heterokaryon or heterozygote for two recessive mutations: if this phenotype is mutant, the mutations are said to be alleles and to belong to the same cistron; if the phenotype is normal (wild-type), the mutations are said to be non-allelic and to belong to different cistrons. Benzer's definition has since been modified to accommodate allelic complementation, and a cistron is now defined (Fincham 1959a) as a segment of the hereditary material within which pairs of mutations in the trans configuration are either deficient for a particular enzyme (or other protein), or give rise to that protein in a structurally abnormal form. There is a good case for equating Fincham's definition of a cistron with the term gene.

**Codon** The term proposed by Crick (1963) for the sequence of nucleotides in DNA or RNA which is responsible for determining that a specific amino-acid shall be inserted into a polyptptide chain. There is more than one codon for most amino-acids, and it has now been established that the codons are triplets of nucleotides.

**Complementation** The process by which two recessive mutants can supply each other's deficiency, such that a heterokaryon or diploid derived from them and having the trans (repulsion) configuration is phenotypically normal (wild-type) or nearly so. Allelic complementation appears to be due to the formation of hybrid protein molecules.

**Conversion** The term proposed by Winkler (1930) for a process of interaction between alleles at meiosis. The term was re-introduced by Lindegren (1953) to account for aberrant ratios in the products of meiosis, apparently arising from such interaction. Whitehouse and Hastings (1965) have suggested that if gene conversion is due to the correction of mispairing of bases in DNA, it may give rise to reciprocal as well as nonreciprocal recombination.

**Coupling** The term introduced by Bateson, Saunders and Punnett (1905) for the condition (phase) of a heterozygote for two linked characterdifferences in which the two dominant factors were derived from one parent and the two recessive factors from the other. This condition may be represented as $\dfrac{A\ B}{a\ b}$, where $A/a$ and $B/b$ represent the two characterdifferences and the capital letters indicate the dominants. The horizontal line separates the contribution from each parent. (Cf. Repulsion.)

**Crossing-over** The term coined by Morgan and Cattell (1912) for the occurrence of new combinations of linked characters. With the acceptance of the chromosome theory, the term is also applied to the process of exchange between homologous chromosomes which gives rise to the new character-combinations.

**Deficiency** ⎫
**Deletion** ⎬ Loss of a segment of a chromosome.

**Diakinesis** The term coined by Häcker (1897) for the stage of late prophase I of meiosis when the chromosomes are well separated from one another (Gk. *kinesis*, movement; *dia*, apart). This stage is recognized by the highly condensed condition of the chromosomes, the homologous pairs of which are held together by chiasmata. In some organisms, just before or during diakinesis the chiasmata move to the ends of the chromosome arms (terminalization of chiasmata). When there are interstitial chiasmata, the successive loops into which the chromosome-pair is divided are set in planes at right angles to one another.

**Dicentric**   Applied to a chromatid or chromosome with two centromeres, such as arises in an inversion heterozygote as a result of crossing-over between a normal and an inverted segment that does not include the centromere.

**Diploid**   With two homologous sets of chromosomes, such as arise at fertilization. The term was introduced by Strasburger (1905), and is derived from the Greek *diploos* meaning double, and *id*, which was Weismann's term for hypothetical structural units of the nucleus assumed to be represented by the chromomeres. The term *id* was derived from Nägeli's term idioplasm (Gk. *idion*, peculiar).

**Diplotene**   The term proposed by von Winiwarter (1900), originally as an adjective, to describe the nuclei at a particular stage of prophase I of meiosis when the two chromosomes making up each homologous pair have separated from one another except at nodes (chiasmata) distributed along their length. The successive loops between the chiasmata all lie in one plane. (Gk. *diploos*, double; *taenia*, ribbon.)

**Dominant**   The term which Mendel (1866) introduced for a character which is manifest in all the members of the first filial ($F_1$) generation from a cross between two pure-breeding strains differing in respect of this character, and which is evident in three quarters of the individuals of the second filial ($F_2$) generation. (Cf. Recessive.)

**Duplication**   The repetition of a sequence of chromomeres or other chromosome segment, usually in linear juxtaposition to the original.

**Episome**   The term introduced by Jacob and Wollman (1958*b*) for a particle which may at some time exist attached to and apparently integrated with the hereditary material of a cell, and which may at other times occur free from such attachment. When attached, episomes replicate along with the hereditary material, but when free they usually replicate independently of it and more rapidly. Examples of episomes are many bacterial viruses and the male sex factor in *Escherichia coli*. (Gk. *epi*, upon; *soma*, body.)

**Euchromatin**   Parts of chromosomes showing the normal cycle of condensation and normal staining properties at nuclear divisions. (Gk. *eu*, true.) (Cf. Heterochromatin.)

**Gene**   The term coined by Johannsen (1909) for the unit of heredity, that is, the hypothetical entity which determines the development of a particular character. The word gene was derived from De Vries's term pangen, itself a derivative of the word pangenesis which Darwin (1868) had coined. At one time it was assumed that the hereditary units defined by mutation, recombination, and function were the same, but with the discovery that they are not, the term gene is now applied to the unit of function, that is, the cistron.

**Genetics**  The term which Bateson (1907*a*) coined for the science of heredity and variation. (Gk. *genesis*, descent.)

**Genotype**  The term proposed by Johannsen (1909) for the hereditary constitution of an individual, or of particular nuclei within its cells.

**Haploid**  With a single set of chromosomes, such as occurs in gamete nuclei (Strasburger, 1905). (Gk. *haploos*, single; for the derivation of *id*, see Diploid.)

**Heterochromatin**  The term proposed by Heitz (1928) for parts of chromosomes with an abnormal degree of contraction or of staining properties at nuclear divisions. (Gk. *heteros*, different; chromatin was Flemming's term for the material of which the chromosomes are composed.) (Cf. Euchromatin.)

**Heteroduplex**  Applied to a nucleic acid molecule in which the two nucleotide chains are derived from different individuals and so may not be exactly complementary. The expression hybrid DNA is often used as a synonym of heteroduplex DNA.

**Heterokaryon**  A multinucleate cell containing nuclei of more than one genotype, such as are of widespread occurrence in fungi (Gk. *heteros*, different; *karyon*, nut, nucleus).

**Heterozygote**  The term coined by Bateson and Saunders (1902) for a zygote, or a diploid individual derived from it, which carries both members (Gk. *heteros*, different) of a pair of alleles. This might be as a result of fusion of dissimilar gametes, or from mutation.

**Homokaryon**  A multinucleate cell containing nuclei of only one genotype. (Gk. *homos*, alike; *karyon*, nut, nucleus.)

**Homozygote**  The term coined by Bateson and Saunders (1902) for a zygote, or a diploid individual derived from it, which carries only one member of a pair of alleles, that is, a zygote derived from the union of gametes identical (Gk. *homos*, alike) in respect of a particular gene.

**Interchange**  An exchange of segments between non-homologous chromosomes. In an interchange heterozygote, the chromosomes contributed by one parent have the normal configuration and those contributed by the other parent have the interchange of segments.

**Interference**  The term proposed by Muller (1916) for the interaction between crossovers such that the occurrence of one exchange between homologous chromosomes reduces the likelihood of another in its vicinity. Chromatid interference is the expression which Mather (1933*a*) introduced for a non-random distribution between the chromatids involved in successive crossovers.

**Inversion**  The term introduced by Sturtevant (1926) for the reversal of the linear sequence of the genes in a segment of a chromosome. In an inversion heterozygote one of the chromosomes contributed by one parent has an inverted segment while the homologous chromosome contributed by the other parent has the normal gene sequence.

**Leptotene**   The term proposed by von Winiwarter (1900), originally as an adjective, to describe the nuclei at the earliest stage of prophase I of meiosis, when the chromosomes first become visible as fine threads and have not yet associated in pairs. (Gk. *leptos*, thin; *taenia*, ribbon.)

**Linkage**   An association in inheritance between characters such that the parental character-combinations appear among the progeny more often than the non-parental. Thus, if the parental combinations were *AB* and *ab*, where *A/a* and *B/b* are two pairs of contrasting characters, then linkage is shown if *AB* and *ab* are significantly more frequent than *Ab* and *aB* among the progeny.

**Locus**   The position occupied by a gene in the chromosome. (Cf. Site.)

**Meiosis**   The term coined by Farmer and Moore (1905) to describe the special nuclear divisions associated with the halving (Gk. *meion*, smaller) of the chromosome number in compensation for the doubling at fertilization. Two successive nuclear divisions occur, with no intervening chromosome replication. Grégoire (1904) proposed that the two divisions be denoted by the appropriate Roman numerals I and II. At the first division of meiosis, homologous chromosomes, already divided into chromatids, associate, undergo crossing-over at certain points along their length, and subsequently separate to opposite poles of the spindle. At the second division, their chromatids separate, leading to four haploid nuclei.

**Metaphase**   Strasburger (1884) originally introduced this term for the stage of nuclear division after (Gk. *meta*) the chromosomes have divided into chromatids, but from about 1905, with the realization that the chromosomes are already double when nuclear division begins, he used the term in the now universally adopted sense of the stage of mitosis or of meiosis I or II when the chromosomes lie on the equatorial plate of the spindle.

**Mispairing**   The presence in one nucleotide chain of a DNA molecule of a nucleotide which is not the complement of that at the corresponding position in the other chain.

**Mitosis**   The term proposed by Flemming (1882) for the process of nuclear division, during which the chromosomes become evident as they shorten and thicken (prophase), and then take up position on the equatorial plate of the spindle (metaphase), to be followed by separation of the two daughter-chromosomes of which each is composed to opposite poles of the spindle (anaphase), and their gradual return to a slender elongated condition (telophase). (Gk. *mitos*, thread.)

**Mutation**   The term which De Vries introduced into biological literature for an abrupt change of phenotype which is inherited. Mutation may be due to a molecular change within a gene, or to the loss or duplication of one or more entire genes.

**Nucleolus**   The more or less spherical structure which occurs in association with a particular point (the nucleolar organizer) on a specific chromosome in the nucleus. At nuclear divisions, the nucleoli disappear in late prophase, are absent at metaphase and anaphase, and reappear during telophase.

**Nucleus**  The term proposed by Brown (1833) for the more or less spherical structure which occurs in cells and stains deeply with basic dyes.

**Operator**  The term introduced by Jacob, Perrin, Sanchez and Monod (1960) for the site at one end of an operon where a repressor molecule binds to the DNA and thereby inhibits transcription. The term is appropriate also for the corresponding site when a gene does not form part of an operon.

**Operon**  The term which Jacob, Perrin, Sanchez and Monod (1960) introduced for a group of closely-linked genes which appear to affect different steps in a single biosynthetic pathway and which appear to function as an integrated unit. With the discovery that the genes of an operon are transcribed in one messenger RNA molecule, an operon could be defined as an aggregate of genes with a single messenger.

**Pachytene**  The term proposed by von Winiwarter (1900), originally as an adjective, to describe the nuclei at a particular stage of prophase I of meiosis when homologous chromosomes are associated throughout their length. (Gk. *pachys*, thick; *taenia*, ribon.)

**Phenotype**  The term coined by Johannsen (1909) for the appearance (Gk. *phainein*, to appear) of an organism with respect to a particular character or group of characters, as a result of its genotype and its environment.

**Plasmid**  A more general term than episome for hereditary elements that may become attached to the hereditary material of the cell at certain times, or may remain permanently unattached (see Hayes, 1968).

**Polyploid**  Having three or more (Gk. *polys*, many) sets of homologous chromosomes. The term was introduced by Strasburger (1910) by analogy with his earlier terms haploid and diploid.

**Polytene**  The term proposed by Koller (1935) to describe the multi-stranded condition (Gk. *polys*, many; *taenia*, ribbon) of the chromosomes found in certain specialized tissues such as the salivary glands of *Drosophila*.

**Postmeiotic Segregation**  See Segregation.

**Promoter**  The term introduced by Jacob, Ullman and Monod (1964) for a site at one end of an operon necessary for its expression, and now identified as the site where RNA polymerase binds to the DNA. The term is appropriate also for the corresponding site when a gene does not form part of an operon.

**Prophase**  Strasburger (1884) originally introduced this term for the early stage of nuclear division before (Gk. *pro*) the chromosomes divide into two chromatids, but from about 1905, with the realization that the chromosomes are double from the beginning of nuclear division, he used the term in the now universally adopted sense of the stage of mitosis or of meiosis I or II before breakdown of the nuclear membrane.

**Recessive**   The term which Mendel (1866) proposed for a character which was not evident in the first filial generation $(F_1)$ of a cross between two pure-breeding strains differing in respect of this character, and which re-appeared in one quarter of the second filial generation $(F_2)$. (Cf. Dominant.)

**Recombination**   The process by which new combinations of parental characters may arise in the progeny. Recombination may come about through random orientation of non-homologous chromosome-pairs on the meiotic spindles, from crossing-over between homologous chromosomes, from gene conversion, or by other means.

**Replicon**   The term proposed by Jacob and Brenner (1963) for a unit of replication of the hereditary material, that is, a segment within which replication occurs in a unified way from a pre-determined origin in a fixed direction to a pre-determined end-point.

**Repulsion**   The term introduced by Punnett and Bateson (1908) for the condition (phase) of a heterozygote for two linked character-differences in which the two dominant factors were derived one from each parent of the individual, and likewise with the two recessive factors. This condition may be represented as $\dfrac{A\ b}{a\ B}$, where $A/a$ and $B/b$ represent two pairs of alleles, and the capital letters represent the dominants. The horizontal line separates the contribution from each parent. (Cf. Coupling.)

**Ribosome**   The term proposed by Roberts (1958) for the *ribo*nucleoprotein particles (Gk. *soma*, body) associated with protein synthesis.

**Segregation**   The separation of allelic differences from one another. Segregation may occur at the first or second division of meiosis, or at the first mitosis after meiosis (postmeiotic segregation).

**Semi-conservative**   The term proposed by Delbrück and Stent (1957) to describe the method of replication of DNA postulated by Watson and Crick (1953*b*) in which the molecule divides longitudinally, each half being conserved and acting as a template for the formation of a new half.

**Site**   The position occupied by a mutation within the gene. (Cf. Locus.)

**Strand**   It is customary in genetics to use strand for chromatid. It has therefore been used in that sense in this book, where it has also been used for one double-chain (that is, duplex) DNA molecule. Chain, but not strand, has been used for a single polynucleotide.

**Structural change**   A general term for a deletion, duplication, or inversion of a chromosome segment, or an interchange of segments which may be between non-homologous chromosomes.

**Synapton**   The term used in this book as a shortened version of synaptinemal complex. This expression was introduced by Moses (1958) for the structure which he discovered in electron micrographs of paired chromosomes at the pachytene stage of meiosis. It is derived from the word synapsis (Gk. union) which has been used by some authors for chromosome pairing.

**Telophase**   The last stage (Gk. *telos*, end) of mitosis, or of either division of meiosis, during which the chromosomes become progressively thinner and more elongated (Heidenhain, 1894). Telophase is said to begin with the formation of a nuclear membrane round each group of daughter-chromosomes.

**Test-cross**   A cross between a heterozygote and the corresponding homozygous recessive, for example, $\dfrac{A\,b\,C}{a\,B\,c} \times \dfrac{a\,b\,c}{a\,b\,c}$, where $A/a$, $B/b$ and $C/c$ represent three pairs of alleles, and the horizontal lines separate the contributions from the two parents of each individual.

**Tetrad**   The four haploid cells formed at the end of meiosis. The term was formerly used for the four chromatids making up a chromosome-pair at the first division of meiosis.

**Trans**   The term proposed by Haldane (1941), by analogy with chemical isomerism, to describe a double heterozygote in repulsion phase, that is, with the dominant factors derived from opposite parents. The term is used chiefly for closely-linked mutants, to which it was first applied by Pontecorvo (1950). The trans configuration for two recessive mutants, 1 and 2, is $\dfrac{+\,2}{1\,+}$, where $+$ stands for the wild-type, and the line separates the contributions from the two parents. (Cf. Cis.)

**Transcription**   The transfer of information in the form of nucleotide sequence from DNA to RNA.

**Transduction**   Gene transfer from one cell to another brought about by a virus. The phenomenon was first described by Lederberg, Lederberg, Zinder and Lively (1952) with reference to *Salmonella typhimurium*, fragments of the hereditary material of which may be transferred by virus *P22*.

**Transformation**   Gene transfer from one cell to another brought about by fragments of the hereditary material of a cell persisting after its breakdown and subsequently entering another cell. The phenomenon was first described by Griffith (1928) in *Diplococcus pneumoniae*.

**Transition**   The term proposed by Freese (1959) for a mutation caused by the substitution in DNA or RNA of one purine by the other, and similarly with the pyrimidines. (Cf. Transversion.)

**Translation**   The transfer of information in the form of nucleotide sequence in DNA or RNA into amino-acid sequence in a polypeptide.

**Translocation**   Transfer of a segment of a chromosome to a non-homologous chromosome. Translocations are usually reciprocal. (Cf. Interchange.)

**Transversion**   The term proposed by Freese (1959) for a mutation caused by the substitution of a purine for a pyrimidine, and *vice versa*, in DNA or RNA. (Cf. Transition.)

**Trisomic**   The term proposed by Blakeslee (1922) to describe an otherwise diploid individual which has one of the chromosomes represented three times instead of twice.

**Wild-type**    The normal condition, either with regard to a whole organism (wild-type strain), or with reference to a particular mutation (wild-type at that locus or site, denoted by a plus sign).

**Zygotene**    The term proposed by Grégoire (1907) to describe the nuclei at a particular stage of prophase I of meiosis when the homologous chromosomes are associating side by side. (Gk. *zygon*, yoke; *taenia*, ribbon.)

# References—Author index

*References within the text are indicated by italic figures in parentheses.*

ABEL, W. O. (1967*a*). Genetische Versuche zur Frage der Ein oder Mehrsträngigkeit der Chromatide. *Ber. dt. bot. Ges.*, **80**, 128–132. *(385)*

ABEL, W. O. (1967*b*). Analyse der Interferenz unter verschiedenen Temperatur-bedingungen bei *Sphaerocarpos.* I. Chromatidinterferenz. *Molec. gen. Genet.*, **99**, 49–61. *(114)*

ADAMS, H. R., WRIGHTSTONE, R. N., MILLER, A. and HUISMAN, T. H. J. (1969). Quantitation of hemoglobin α chains in adult and fetal goats; gene duplication and the production of polypeptide chains. *Archs Biochem. Biophys.*, **132**, 223–236. *(325)*

ADAMS, J. M. and CAPECCHI, M. R. (1966). N-formylmethionyl-sRNA as the initiator of protein synthesis. *Proc. natn. Acad. Sci. U.S.A.*, **55**, 147–155. *(287)*

ADAMS, J. M., JEPPESEN, P. G. N., SANGER, F. and BARRELL, B. G. (1969). Nucleotide sequence from the coat protein cistron of R17 bacteriophage DNA. *Nature, Lond.*, **223**, 1009–1014. *(279)*

ADLER, J., LEHMAN, I. R., BESSMAN, M. J., SIMMS, E. S. and KORNBERG, A. (1958). Enzymatic synthesis of deoxyribonucleic acid. IV. Linkage of single deoxy-nucleotides to the deoxyribonucleoside ends of deoxyribonucleic acid. *Proc. natn. Acad. Sci. U.S.A.*, **44**, 641–647. *(196)*

AHMED, A., CASE, M. E. and GILES, N. H. (1964). The nature of complementation among mutants in the *histidine-3* region of *Neurospora crassa. Brookhaven Symp. Biol.*, **17**, 53–65. *(319–20)*

ALBERTS, B. M. and FREY, L. (1970). T4 bacteriophage gene 32: a structural protein in the replication and recombination of DNA. *Nature, Lond.*, **227**, 1313–1318. *(345)*

ALLISON, A. C. (1954). Protection afforded by sickle cell trait against sub-tertian malarial infection. *Br. med. J.*, **1**, 290–292. *(14)*

ALONI, Y., HATLEN, L. E. and ATTARDI, G. (1971). Studies of fractionated HeLa cell metaphase chromosomes. II. Chromosomal distribution of site for transfer RNA and 5S RNA. *J. molec. Biol.*, **56**, 555–563. *(406)*

ALTMAN, S., BRENNER, S. and SMITH, J. D. (1971). Identification of an ochre-suppressing anticodon. *J. molec. Biol.*, **56**, 195–197. *(297)*

ALTMAN, S. and MESELSON, M. (1970). A T4-induced endonuclease which attacks T4 DNA. *Proc. natn. Acad. Sci. U.S.A.*, **66**, 716–721. *(347)*

ALTMAN, S. and SMITH, J. D. (1971). Tyrosine tRNA precursor molecule polynucleotide sequence. *Nature New Biol.*, **233**, 35–39. *(264)*

AMES, B. N. and GARRY, B. (1959). Coordinate repression of the synthesis of four histidine biosynthetic enzymes by histidine. *Proc. natn. Acad. Sci. U.S.A.*, **45**, 1453–1461. *(301)*

ANDERSON, E. G. (1923). Maternal inheritance of chlorophyll in maize. *Bot. Gaz.*, **76**, 411–418.                                                          (*53, 54*)

ANDERSON, E. G. (1925). Crossing over in a case of attached X chromosomes in *Drosophila melanogaster*. *Genetics, Princeton*, **10**, 403–417.   (*100–3, 108–14*)

ANDERSON, E. G. (1936). Induced chromosomal alterations in maize. In *Biological Effects of Radiation* edited by B. M. DUGGAR, New York, McGraw-Hill, pp. 1297–1310.                                                                    (*150–57*)

ANGEL, T., AUSTIN, B. and CATCHESIDE, D. G. (1970). Regulation of recombination at the *his-3* locus in *Neurospora crassa*. *Aust. J. biol. Sci.*, **23**, 1229–40.   (*368*)

ANRAKU, N. and TOMIZAWA, J. (1965). Molecular mechanisms of genetic recombination of bacteriophage. V. Two kinds of joining of parental DNA molecules. *J. molec. Biol.*, **12**, 805–815.                                              (*346*)

ARDITTI, R. R., SCAIFE, J. G. and BECKWITH, J. R. (1968). The nature of mutants in the *lac* promoter region. *J. molec. Biol.*, **38**, 421–426.              (*314*)

ARISTOTLE (ca. 350 B.C.) *The Reproduction of Animals* (*De Generatione Animalium*), especially book 1, chapters 17–23. (English translation by A. PLATT, Oxford, Clarendon Press, 1910.)                                                  (*1–2*)

ARLETT, C. F., GRINDLE, M. and JINKS, J. L. (1962). The "red" cytoplasmic variant of *Aspergillus nidulans*. *Heredity*, **17**, 197–209.                     (*55*)

ARNOTT, S., FULLER, W., HODGSON, A. and PRUTTON, I. (1968). Molecular conformations and structure transitions of RNA complementary helices and their possible biological significance. *Nature, Lond.*, **220**, 561–564.                     (*424*)

ARNOTT, S., HUTCHINSON, F., SPENCER, M., WILKINS, M. H. F., FULLER, W. and LANGRIDGE, R. (1966). X-ray diffraction studies of double helical ribonucleic acid. *Nature, Lond.*, **211**, 227–232.                                              (*286*)

ASTBURY, W. T. and BELL, F. O. (1938). X-ray study of thymonucleic acid. *Nature, Lond.*, **141**, 747–748.                                       (*182, 234, 235*)

ATTARDI, G., NAONO, S., ROUVIÈRE, J., JACOB, F. and GROS, F. (1964). Production of messenger RNA and regulation of protein synthesis. *Cold Spring Harb. Symp. quant. Biol.*, **28**, 363–372.                                          (*304, 309, 311*)

ATTARDI, G., PARNAS, H., HWANG, M.-I. H. and ATTARDI, B. (1966). Giant-size rapidly labeled nuclear ribonucleic acid and cytoplasmic messenger ribonucleic acid in immature duck erythrocytes. *J. molec. Biol.*, **20**, 145–182.          (*401*)

AUERBACH, C. and ROBSON, J. M. (1947). The production of mutations by chemical substances. *Proc. R. Soc. Edinb. B*, **62**, 271–283.            (*146, 149–50*)

AUGUST, J. T., COOPER, S., SHAPIRO, L. and ZINDER, N. D. (1964). RNA phage induced RNA polymerase. *Cold Spring Harb. Symp. quant. Biol.*, **28**, 95–97.   (*245*)

AVERY, O. T., MACLEOD, C. M. and MC CARTY, M. (1944). Studies on the chemical nature of the substance inducing transformation of pneumococcal types. Induction of transformation by a desoxyribonucleic acid fraction isolated from pneumococcus Type III.   *J. exp. Med.*, **79**, 137–158.        (*173, 176, 185*)

BAGLIONI, C. (1962). The fusion of two peptide chains in hemoglobin Lepore and its interpretation as a genetic deletion. *Proc. natn. Acad. Sci. U.S.A.*, **48**, 1880–1886.                                                                   (*322*)

BAGLIONI, C. and CAMPANA, T. (1967). α chain and globin: intermediates in the synthesis of rabbit hemoglobin. *Eur. J. Biochem.*, **2**, 480–492.        (*325*)

BAGLIONI, C. and INGRAM, V. M. (1961). Four adult haemoglobin types in one person. *Nature, Lond.*, **189**, 465–467. *(226)*

BAILEY, N. T. J. (1959). *Statistical Methods in Biology*. London, English Universities Press. *(23)*

BAJER, A. (1957–58). Ciné-micrographic studies on mitosis in endosperm. *Expl Cell Res.*, **13**, 493–502; **14**, 245–256; **15**, 370–383. *(59)*

BAJER, A. (1965). Subchromatid structure of chromosomes in the living state. *Chromosoma*, **17**, 291–302. *(372)*

BAKER, W. K. (1958). Crossing-over in heterochromatin. *Am. Nat.*, **92**, 59–60. *(434)*

BAKER, W. K. (1968). Position-effect variegation. *Adv. Genet.*, **14**, 133–169. *(393)*

BALTIMORE, D. (1970). RNA-dependent DNA polymerase in virions of RNA tumour viruses. *Nature, Lond.*, **226**, 1209–1211. *(245)*

BALTIMORE, D. and FRANKLIN, R. M. (1962). Preliminary data on a virus-specific enzyme system responsible for the synthesis of viral RNA. *Biochem. biophys. Res. Commun.*, **9**, 388–392. *(245)*

BARBOUR, S. D., NAGAISHI, H., TEMPLIN, A. and CLARK, A. J. (1970). Biochemical and genetic studies of recombination proficiency in *Escherichia coli*, II. Rec⁺ revertants caused by indirect suppression of rec⁻ mutations. *Proc. natn. Acad. Sci. U.S.A.*, **67**, 128–135. *(351)*

BARR, G. C. and BUTLER, J. A. V. (1963). Histones and gene function. *Nature, Lond.*, **199**, 1170–1172. *(422)*

BATESON, W. (1907*a*). The progress of genetic research. *R. hort. Soc. Rep. 3rd int. Conf. Genet.*, pp. 90–97. *(449)*

BATESON, W. (1907*b*). Facts limiting the theory of heredity. *Science, N.Y.*, **26**, 649–660. *(165)*

BATESON, W. (1909). *Mendel's Principles of Heredity*. Cambridge Univ. Press, 396 pp. *(5, 13, 45, 164)*

BATESON, W. and SAUNDERS, E. R. (1902). Experimental studies in the physiology of heredity. *Rep. Evol. Comm. R. Soc.*, **1**, 1–160. *(9, 11, 18, 31, 164, 445, 449)*

BATESON, W., SAUNDERS, E. R. and PUNNETT, R. C. (1905). Experimental studies in the physiology of heredity. *Rep. Evol. Comm. R. Soc.*, **2**, 1–55 and 80–99. *(12, 18–20, 44, 81, 86, 164–65, 447)*

BATTAGLIA, E. (1964). Cytogenetics of B-chromosomes. *Caryologia*, **17**, 245–299. *(392)*

BAUERLE, R. H. and MARGOLIN, P. (1967). Evidence for two sites for initiation of gene expression in the tryptophan operon of *Salmonella typhimurium*. *J. molec. Biol.*, **26**, 423–436. *(315)*

BAUR, E. (1909). Das Wesen und die Erblichkeitsverhältnisse der "varietates albomarginatae" von *Pelargonium zonale*. *Z. VererbLehre*, **1**, 330–351. *(51–4)*

BAUR, E. (1912). Ein Fall von geschlechtsbegrenzter Vererbung bei *Melandrium album*. *Z. VererbLehre*, **8**, 335–336. *(76)*

BAUTZ, E. K. F., BAUTZ, F. A. and DUNN, J. J. (1969). *E. coli* σ factor: a positive control element in phage T4 development. *Nature, Lond.*, **223**, 1022–1024. *(257)*

BAUTZ, E. K. F. and HALL, B. D. (1962). The isolation of T4-specific RNA on a DNA-cellulose column. *Proc. natn. Acad. Sci. U.S.A.*, **48**, 400–408. *(248, 251)*

BAYEV, A. A., VENKSTERN, T. V., MIRSABEKOV, A. D., KRUTILINA, A. I., AXELROD, L. L. and AXELROD, V. D. (1967). The primary structure of valine tRNA of baker's yeast. *Molec. Biol., USSR*, **1**, 754–766. *(263)*

BEADLE, G. W. (1945). Genetics and metabolism in *Neurospora*. *Physiol. Rev.*, **25,** 643–663. *(168)*

BEADLE, G. W. and COONRADT, V. L. (1944). Heterocaryosis in *Neurospora crassa*. *Genetics, Princeton*, **29,** 291–308. *(166)*

BEADLE, G. W. and TATUM, E. L. (1941). Genetic control of biochemical reactions in *Neurospora*. *Proc. natn. Acad. Sci. U.S.A.*, **27,** 499–506. *(165–66, 210)*

BEAVEN, G. H., HORNABROOK, R. W., FOX, R. H. and HUEHNS, E. R. (1972). Occurrence of heterozygotes and homozygotes for the α-chain haemoglobin variant Hb-J (Tongariki) in New Guinea. *Nature, Lond.*, **235,** 46–47. *(325)*

BECKWITH, J. R. (1964a). Restoration of operon activity by suppressors. *Abh. dt. Akad. Wiss. Berl., Kl. Med.*, 1964, (4), 119–124. *(309, 311)*

BECKWITH, J. R. (1964b). A deletion analysis of the *lac* operator region in *Escherichia coli*. *J. molec. Biol.*, **8,** 427–430. *(308)*

BEERMANN, W. (1960). Der Nukleolus als lebenswichtiger Bestandteil des Zellkernes. *Chromosoma*, **11,** 263–296. *(405)*

BEERMANN, W. (1961). Ein Balbiani-Ring als Locus einer Speicheldrüsenmutation. *Chromosoma*, **12,** 1–25. *(148, 401)*

BEERMANN, W. (1964). Control of differentiation at the chromosomal level. *J. exp. Zool.*, **157,** 49–62. *(401)*

BEERMANN, W. (1965). Operative Gliederung der Chromosomen. *Naturwissenschaften*, **52,** 365–375. *(401)*

BEERMANN, W. (1967). Gene action at the level of the chromosome. In *Heritage from Mendel* (ed. R. A. Brink), pp. 179–201. Madison. *(399, 401)*

BEERMANN, W. and PELLING, C. (1965). H³-Thymidin-markierung einzelner Chromatiden in Riesenchromosomen. *Chromosoma*, **16,** 1–21. *(383)*

BELLING, J. (1927). The attachments of chromosomes at the reduction division in flowering plants. *J. Genet.*, **18,** 177–205. *(151)*

BELLING, J. (1928a). A working hypothesis for segmental interchange between homologous chromosomes in flowering plants. *Univ. Calif. Publs Bot.*,**14,** 283–291. *(117–18, 136)*

BELLING, J. (1928b). The ultimate chrommomeres of *Lilium* and *Aloë* with regard to the numbers of genes. *Univ. Calif. Publs Bot.*, **14,** 307–318. *(158, 388)*

BELLING, J. (1928c). Contraction of chromosomes during maturation divisions in *Lilium* and other plants. *Univ. Calif. Publs Bot.*, **14,** 335–343. *(118, 119)*

BELLING, J. (1929). Nodes and internodes of trivalents of *Hyacinthus*. *Univ. Calif. Publs Bot.*, **14,** 379–388. *(63, 118, 124)*

BENEDEN, E. VAN (1875). La maturation de l'oeuf, la fécondation, et les premières phases du développement embryonnaire des mammifères d'après des recherches faites chez le Lapin. *Bull. Acad. r. Belg., Cl. Sci.*, 2me Sèr., **40,** 686–736. *(40)*

BENEDEN, E. VAN (1883). Recherches sur la maturation de l'oeuf et la fécondation. *Archs Biol., Paris*, **4,** 265–638. *(58, 62)*

BENZER, S. (1955). Fine structure of a genetic region in bacteriophage. *Proc. natn. Acad. Sci. U.S.A.*, **41,** 344–354. *(214–15)*

BENZER, S. (1957). The elementary units of heredity. In *The Chemical Basis of Heredity* edited by W. D. MCELROY and B. GLASS, Baltimore, Johns Hopkins Press, pp. 70–93. *(215–18, 224, 446–47)*

BENZER, S. (1961). On the topography of the genetic fine structure. *Proc. natn. Acad. Sci. U.S.A.*, **47**, 403–415. (*217, 236, 237*)

BENZER, S. and CHAMPE, S. P. (1962). A change from nonsense to sense in the genetic code. *Proc. natn. Acad. Sci. U.S.A.*, **48**, 1114–1121. (*294, 296*)

BENZER, S. and FREESE, E. (1958). Induction of specific mutations with 5-bromouracil. *Proc. natn. Acad. Sci. U.S.A.*, **44**, 112–119. (*235*)

BERNSTEIN, F. (1925). Zusammenfassende Betrachtungen über die erblichen Blut-strukturen des Menschen. *Z. VererbLehre*, **37**, 237–270. (*139*)

BESSMAN, M. J., LEHMAN, I. R., ADLER, J., ZIMMERMAN, S. B., SIMMS, E. S. and KORNBERG, A. (1958). Enzymatic synthesis of deoxyribonucleic acid. III. The incorporation of pyrimidine and purine analogues into deoxyribonucleic acid. *Proc. nat. Acad. Sci. U.S.A.* **44**, 633–640. (*196*)

BESSMAN, M. J., LEHMAN, I. R., SIMMS, E. S. and KORNBERG, A. (1958). Enzymatic synthesis of deoxyribonucleic acid. II. General properties of the reaction. *J. biol. Chem.*, **233**, 171–177. (*194*)

BEUKERS, R. and BERENDS, W. (1960). Isolation and identification of the irradiation product of thymine. *Biochim. biophys. Acta*, **41**, 550–551. (*354*)

BIRD, A. P. and BIRNSTIEL, M. L. (1971). A timing study of DNA amplification in *Xenopus laevis* oocytes. *Chromosoma*, **35**, 300–309. (*411*)

BIRNSTIEL, M. L., CHIPCHASE, M. and SPEIRS, J. (1971). The ribosomal RNA cistrons. *Progr. nucleic Acid Res. molec. Biol.*, **11**, 351–390. (*403, 405*)

BIRNSTIEL, M. L. SPEIRS, J., PURDOM, I., JONES, K. and LOENING, U. E. (1968). Properties and composition of the isolated ribosomal DNA satellite of *Xenopus laevis*. *Nature, Lond.*, **219**, 454–463. (*403*)

BISHOP, J. O., PEMBERTON, R. and BAGLIONI, C. (1972). Reiteration frequency of haemoglobin genes in the duck. *Nature New Biol.*, **235**, 231–234. (*402*)

BLAKE, R. D., FRESCO, J. R. and LANGRIDGE, R. (1970). High-resolution X-ray diffrac-tion by single crystals of mixtures of transfer ribonucleic acids. *Nature, Lond.*, **225**, 32–35. (*265*)

BLAKESLEE, A. F. (1922). Variations in *Datura* due to changes in chromosome number. *Am. Nat.*, **56**, 16–31. (*70, 453*)

BLATTNER, F. R. and DAHLBERG, J. E. (1972). RNA synthesis startpoints in bacterio-phage λ: are the promoter and operator transcribed? *Nature New Biol.*, **237**, 227–232. (*314*)

BLATTNER, F. R., DAHLBERG, J. E., BOETTIGER, J. K., FIANDT, M. and SZYBALSKI, W. (1972). Distance from a promoter mutation to an RNA synthesis startpoint on bacteriophage λ DNA. *Nature New Biol.*, **237**, 232–236. (*315*)

BLUMENFELD, M. and FORREST, H. S. (1971). Is *Drosophila* dAT on the Y chromosome? *Proc. natn. Acad. Sci. U.S.A.*, **68**, 3145–3149. (*406, 407*)

BODMER, W. F. and GANESAN, A. T. (1964). Biochemical and genetic studies of integra-tion and recombination in *Bacillus subtilis* transformation. *Genetics, Princeton*, **50**, 717–738. (*346*)

BOIVIN, A., VENDRELY, R. and VENDRELY, C. (1948). L'acide désoxyribonucléique du noyau cellulaire, dépositaire des caractères héréditaires; arguments d'ordre analytique. *C.r. hebd. Séanc. Acad. Sci., Paris*, **226**, 1061–1063. (*176*)

BOLE-GOWDA, B. N., PERKINS, D. D. and STRICKLAND, W. N. (1962). Crossing-over and interference in the centromere region of linkage group I of *Neurospora*. *Genetics, Princeton*, **47**, 1243–1252. (*114*)

BONNER, D. M., SUYAMA, Y. and DE MOSS, J. A. (1960). Genetic fine structure and enzyme formation. *Fedn Proc. Fedn Am. Socs exp. Biol.*, **19**, 926–930.          (*227*)

BONNER, J., HUANG, R. C. and GILDEN, R. V. (1963). Chromosomally directed protein synthesis. *Proc. natn. Acad. Sci. U.S.A.* **50**, 893–900.          (*422*)

BOON, T. and ZINDER, N. D. (1971). Genotypes produced by individual recombination events involving bacteriophage f1. *J. molec. Biol.*, **58**, 133–151.          (*352, 369*)

BOULTER, D., ELLIS, R. J. and YARWOOD, A. (1972). Biochemistry of protein synthesis in plants. *Biol. Rev.*, **47**, 113–175.          (*427, 429*)

BOURGAUX, P. and BOURGAUX-RAMOISY, D. (1971). A symmetrical model for polyoma virus DNA replication. *J. molec. Biol.*, **62**, 513–524.          (*208*)

BOVERI, T. (1888). Zellen-Studien. *Jena. Z. Naturw.*, **22**, 685–882.          (*59*)

BOVERI, T. (1889). Ein geschlechtlich erzeugter Organismus ohne mütterliche Eigenschaften. *Gesell. für Morph. Physiol. Münch.* (English translation entitled 'An organism produced sexually without characteristics of the mother.' *Am. Nat.*, 1893, **27**, 222–232.)          (*41–42, 44*)

BOVERI, T. (1902). Uber mehrpolige Mitosen als Mittel zur Analyse des Zellkerns. *Verh. phys.-med. Ges. Würzb. N.F.*, **35**, 67–90.          (*70, 71*)

BOYCE, R. P. and HOWARD-FLANDERS, P. (1964a). Release of ultraviolet light-induced thymine dimers from DNA in *E. coli* K12. *Proc. natn. Acad. Sci. U.S.A.*, **51**, 293–300.          (*354*)

BOYCE, R. P. and HOWARD-FLANDERS, P. (1964b). Genetic control of DNA breakdown and repair in *E. coli* K-12 treated with mitomycin C or ultraviolet light. *Z. VererbLehre*, **95**, 345–350.          (*354*)

BOYCOTT, A. E. and DIVER, C. (1923). On the inheritance of sinistrality in *Limnaea peregra*. *Proc. R. Soc. B*, **95**, 207–213.          (*47–48*)

BOYCOTT, A. E., DIVER, C., GARSTANG, S. L. and TURNER, F. M. (1930). The inheritance of sinistrality in *Limnaea peregra* (Mollusca, Pulmonata). *Phil. Trans. R. Soc.*, **219**, 51–131.          (*48*)

BOYER, S. H., NOYES, A. N., VRABLIK, G. R., DONALDSON, L. J., SCHAEFER, E. W., GRAY, C. W. and THURMON, T. F. (1971). Silent hemoglobin alpha genes in apes: potential source of thalassemia. *Science, N.Y.*, **171**, 182–185.          (*325*)

BOYLE, J. M., PATERSON, M. C. and SETLOW, R. B. (1970). Excision-repair properties of an *Escherichia coli* mutant deficient in DNA polymerase. *Nature, Lond.*, **226**, 708–710.          (*356*)

BRACHET, J. (1957). *Biochemical Cytology*. New York, Academic Press, 516 pp.    (*242*)

BRAM, S. and RIS, H. (1971). On the structure of nucleohistone. *J. molec. Biol.*, **55**, 325–336.          (*386*)

BRAMMAR, W. J., BERGER, H. and YANOFSKY, C. (1967). Altered amino acid sequences produced by reversion of frameshift mutants of tryptophan synthetase A gene of *E. coli*. *Proc. natn. Acad. Sci. U.S.A.*, **58**, 1499–1506.          (*276–77*)

BREGGER, T. (1918). Linkage in maize: the C aleurone factor and waxy endosperm. *Am. Nat.*, **52**, 57–61.          (*87–89*)

BREMER, H., KONRAD, M. W., GAINES, K. and STENT, G. S. (1965). Direction of chain growth in enzymic RNA synthesis. *J. molec. Biol.*, **13**, 540–553.          (*254, 255*)

BRENNER, S. (1957). On the impossibility of all overlapping triplet codes in information transfer from nucleic acid to proteins. *Proc. natn. Acad. Sci. U.S.A.*, **43**, 687–694.          (*234–35*)

BRENNER, S., BARNETT, L., CRICK, F. H. C. and ORGEL, A. (1961). The theory of mutagenesis. *J. molec. Biol.*, **3**, 121–124. (*239*)

BRENNER, S., BARNETT, L., KATZ, E. R. and CRICK, F. H. C. (1967). UGA: a third
nonsense triplet in the genetic code. *Nature, Lond.*, **213**, 449–450. (*296*)

BRENNER, S. and BECKWITH, J. R. (1965). *Ochre* mutants, a new class of suppressible
nonsense mutants. *J. molec. Biol.*, **13**, 629–637. (*295*)

BRENNER, S., BENZER, S. and BARNETT, L. (1958). Distribution of proflavin-induced
mutations in the genetic fine structure. *Nature, Lond.*, **182**, 983–985. (*236*)

BRENNER, S., JACOB, F. and MESELSON, M. (1961). An unstable intermediate carrying
information from genes to ribosomes for protein synthesis. *Nature, Lond.*, **190**,
576–581. (*250*)

BRENNER, S., STREISINGER, G., HORNE, R. W., CHAMPE, S. P., BARNETT, L., BENZER, S.
and REES, M. W. (1959). Structural components of bacteriophage. *J. molec. Biol.*,
**1**, 281–292. (*174–75*)

BRENNER, S. and STRETTON, A. O. W. (1965). Phase shifting of *amber* and *ochre* mutants.
*J. molec. Biol.*, **13**, 944–946. (*295*)

BRENNER, S., STRETTON, A. O. W. and KAPLAN, S. (1965). Genetic code: 'nonsense' triplets for chain termination and their suppression. *Nature, Lond.*, **206**, 994–8. (*295*)

BRESLER, S. E., KRENEVA, R. A. and KUSHEV, V. V. (1968). Correction of molecular
heterozygotes in the course of transformation. *Molec. gen. Genet.*, **102**, 257–268.
(*366*)

BRETSCHER, M. S. (1968a). Polypeptide chain termination: an active process. *J.
molec. Biol.*, **34**, 131–136. (*299*)

BRETSCHER, M. S. (1968b). Translocation in protein synthesis: a hybrid structure
model. *Nature, Lond.*, **218**, 675–677. (*289*)

BRETSCHER, M. S. and MARCKER, K. A. (1966). Polypeptidyl-s-ribonucleic acid and
amino-acyl-s-ribonucleic acid binding sites on ribosomes. *Nature, Lond.*, **211**,
380–384. (*270, 288, 290*)

BREWEN, J. G. and PEACOCK, W. J. (1969). Restricted rejoining of chromosomal sub-
units in aberration formation: a test for subunit dissimilarity. *Proc. natn. Acad. Sci.
U.S.A.*, **62**, 389–394. (*374–75, 384*)

BREWIN, N. (1972). Catalytic role for RNA in DNA replication. *Nature New Biol.*, **236**,
101. (*202*)

BRIDGES, C. B. (1914). Direct proof through non-disjunction that the sex-linked genes
of *Drosophila* are borne by the X-chromosome. *Science, N.Y.*, **40**, 107–109. (*76*)

BRIDGES, C. B. (1916). Non-disjunction as proof of the chromosome theory of heredity.
*Genetics, Princeton*, **1**, 1–52 and 107–163. (*76–79, 99–103*)

BRIDGES, C. B. (1922). The origin of variations in sexual and sex-limited characters.
*Am. Nat.*, **56**, 51–63. (*78*)

BRIDGES, C. B. (1936). The bar "gene" a duplication. *Science, N.Y.*, **83**, 210–211. (*160*)

BRIDGES, C. B. and ANDERSON, E. G. (1925). Crossing over in the X chromosomes of
triploid females of *Drosophila melanogaster*. *Genetics, Princeton*, **10**, 418–441.
(*100–103, 109, 118*)

BRINK, R. A. and MACGILLIVRAY, J. H. (1924). Segregation for the waxy character in
maize pollen and differential development of the male gametophyte. *Am. J. Bot.*,
**11**, 465–469. (*46*)

BRINK, R. A., STYLES, E. D. and AXTELL, J. D. (1968). Paramutation: directed genetic
change. *Science, N.Y.*, **159**, 161–170. (*394*)

BRODY, s. and YANOFSKY, c. (1963). Suppressor gene alteration of protein primary structure. *Proc. natn. Acad. Sci. U.S.A.* **50,** 9–16.                    (*283*)

BROKER, T. R. and LEHMAN, I. R. (1971). Branched DNA molecules: intermediates in T4 recombination. *J. molec. Biol.,* **60,** 131–149.                    (*351–53*)

BROWN, D. D. and GURDON, J. B. (1964). Absence of ribosomal RNA synthesis in the anucleolate mutant of *Xenopus laevis. Proc. natn. Acad. Sci. U.S.A.,* **51,** 139–146.                    (*403*)

BROWN, D. D. and WEBER, C. S. (1968). Gene linkage by RNA-DNA hybridization. *J. molec. Biol.,* **34,** 661–697.                    (*403, 406*)

BROWN, D. D., WENSINK, P. C. and JORDAN, E. (1971). Purification and some characteristics of 5S DNA from *Xenopus laevis. Proc. natn. Acad. Sci. U.S.A.,* **68,** 3175–3179.                    (*406*)

BROWN, D. D., WENSINK, P. C. and JORDAN, E. (1972). A comparison of the ribosomal DNA's of *Xenopus laevis* and *Xenopus mulleri*: the evolution of tandem genes. *J. molec. Biol.,* **63,** 57–73.                    (*403*)

BROWN, J. C. and SMITH, A. E. (1970). Initiator codons in eukaryotes. *Nature, Lond.,* **226,** 610–612.                    (*293*)

BROWN, R. (1833). On the organs and mode of fecundation in Orchideae and Asclepiadeae. *Trans. Linn. Soc. Lond.,* **16,** 685–745.                    (*40, 451*)

BROWN, S. W. (1966). Heterochromatin. *Science, N.Y.,* **151,** 417–425.                    (*421*)

BROWN, S. W., and CHANDRA, H. S. (1973). Inactivation system of the mammalian X chromosome. *Proc. natn. Acad. Sci. U.S.A.,* **70,** 195–199.                    (*421*)

BROWN, S. W. and ZOHARY, D. (1955). The relationship of chiasmata and crossing over in *Lilium formosanum. Genetics, Princeton,* **40,** 850–873.                    (*114, 122, 127–29*)

BROWNLEE, G. G., SANGER, F. and BARRELL, B. G. (1967). Nucleotide sequence of 5S-ribosomal RNA from *Escherichia coli. Nature, Lond.,* **215,** 735–736.                    (*247*)

BUNN, C. L., MITCHELL, C. H., LUKINS, H. B. and LINNANE, A. W. (1970). Biogenesis of mitochondria, XVIII. A new class of cytoplasmically determined antibiotic resistant mutants in *Saccharomyces cerevisiae, Proc. natn. Acad. Sci. U.S.A.,* **67,** 1233–1240.                    (*430*)

BURGESS, R. R., TRAVERS, A. A., DUNN, J. J. and BAUTZ, E. K. F. (1969). Factor stimulating transcription by RNA polymerase. *Nature, Lond.,* **221,** 43–46.                    (*256*)

CAIRNS, J. (1961). An estimate of the length of the DNA molecule of T2 bacteriophage by autoradiography. *J. molec. Biol.,* **3,** 756–761.                    (*203*)

CAIRNS, J. (1963). The bacterial chromosome and its manner of replication as seen by autoradiography. *J. molec. Biol.,* **6,** 208–213.                    (*200–1, 204*)

CAIRNS, J. (1964). The chromsome of *Escherichia coli. Cold Spring Harb. Symp. quant. Biol.,* **28,** 43–46.                    (*204, 205*)

CAIRNS, J. (1966). Autoradiography of HeLa cell DNA. *J. molec. Biol.,* **15,** 372–3.    (*206*)

CALEF, E. (1957). Effect on linkage maps of selection of crossovers between closely linked markers. *Heredity,* **11,** 265–279.                    (*334*)

CALLAHAN, R., BLUME, A. J. and BALBINDER, E. (1970). Evidence for the order promoter–operator–first structural gene in the tryptophan operon of *Salmonella. J. molec. Biol.,* **51,** 709–715.                    (*315*)

CALLAN, H. G. (1963). The nature of lampbrush chromosomes. *Int. Rev. Cytol.,* **15,** 1–34.                    (*377–80, 384, 385, 401, 408*)

CALLAN, H. G. (1966). Chromosomes and nucleoli of the axolotl, *Ambystoma mexicanum. J. Cell Sci.,* **1,** 85–108.                    (*405, 411*)

CALLAN, H. G. (1967). On the organization of genetic units in chromosomes. *J. Cell Sci.*, **2**, 1–7.                                        *383, 399–403, 408, 410, 411*)

CALLAN, H. G. (1972) Replication of DNA in the chromosomes of eukaryotes. *Proc. R. Soc.* B, **181**, 19–41.                                        *(207, 415)*

CALLAN, H. G. and LLOYD, L. (1956). Visual demonstration of allelic differences within cell nuclei. *Nature, Lond.*, **178**, 355–357.                        *(121)*

CALLAN, H. G. and LLOYD, L. (1960). Lampbrush chromosomes of crested newts *Triturus cristatus* (Laurenti). *Phil. Trans. R. Soc. B*, **243**, 135–219. *(381,399,400)*

CALLAN, H. G. and MACGREGOR, H. C. (1958). Action of deoxyribonuclease on lampbrush chromosomes. *Nature, Lond.*, **181**, 1479–1480.                *(380)*

CALVO, J. M. and FINK, G. R. (1971). Regulation of biosynthetic pathways in bacteria and fungi. *A. Rev. Biochem.*, **40**, 943–968.                        *(398)*

CAMENZIND, R. and NICKLAS, R. B. (1968). The non-random chromosome segregation in spermatocytes of *Gryllotalpa hexadactyla*. A micromanipulation analysis. *Chromosoma*, **24**, 324–335.                                        *(392)*

CAMPBELL, A. M. (1962). Episomes. *Adv. Genet.*, **11**, 101–145.                *(336)*

CAPECCHI, M. R. (1967). Polypeptide chain termination *in vitro*: isolation of a release factor. *Proc. natn. Acad. Sci. U.S.A.*, **58**, 1144–1151.            *(298–99)*

CAPECCHI, M. R. and GUSSIN, G. N. (1965). Suppression in vitro: identification of a serine-sRNA as a 'nonsense' suppressor. *Science, N.Y.*, **149**, 417–422.    *(296)*

CARLSON, E. A. (1961). Limitations of geometrical models for complementation mapping of allelic series. *Nature, Lond.*, **191**, 788–790.            *(228–29)*

CARLSON, P. S. (1971). A genetic analysis of the *rudimentary* locus of *Drosophila melanogaster*. *Genet. Res.*, **17**, 53–81.                                *(370)*

CAROTHERS, E. E. (1913). The Mendelian ratio in relation to certain Orthopteran chromosomes. *J. Morph.*, **24**, 487–511.                        *(72, 110, 118)*

CARTER, T. C. (1955). The estimation of total genetical map lengths from linkage test data. *J. Genet.*, **53**, 21–28.                                *(117)*

CARTER, T. C. and ROBERTSON, A. (1952). A mathematical treatment of genetical recombination using a four-strand model. *Proc. R. Soc.* B, **139**, 410–426.    *(113)*

CASE, M. E. and GILES, N. H. (1958). Evidence from tetrad analysis for both normal and aberrant recombination between allelic mutants in *Neurospora crassa*. *Proc. natn. Acad. Sci. U.S.A.*, **44**, 378–390.                            *(332)*

CASE, M. E. and GILES, N. H. (1960). Comparative complementation and genetic maps of the *pan*-2 locus in *Neurospora crassa*. *Proc. natn. Acad. Sci. U.S.A.*, **46**, 659–676.                                                    *(227–28)*

CASE, M. E. and GILES, N. H. (1964). Allelic recombination in *Neurospora*: tetrad analysis of a three-point cross within the *pan*-2 locus. *Genetics, Princeton*, **49**, 529–40. *(362)*

CASHMORE, T. (1971). Interaction between loops I and III in the tyrosine suppressor tRNA. *Nature New Biol.*, **230**, 236–239.                        *(268)*

CASKEY, C. T., REDFIELD, B. and WEISSBACH, H. (1967). Formylation of guinea pig liver methionyl-sRNA. *Archs Biochem. Biophys.*, **120**, 119–123.            *(291)*

CASPERSSON, T., FARBER, S., FOLEY, G. E., KUDYNOWSKI, J., MODEST, E. J., SIMONSSON, E., WAGH, U. and ZECH, L. (1968). Chemical differentiation along metaphase chromosomes. *Expl Cell Res.*, **49**, 219–222.                        *(407)*

CASPERSSON, T. and SCHULTZ, J. (1938). Nucleic acid metabolism of the chromosomes in relation to gene reproduction. *Nature, Lond.*, **142**, 294–295.            *(393)*

CASSUTO, E., LASH, T., SRIPRAKASH, K. S. and RADDING, C. M. (1971). Role of exonuclease and β protein of phage λ in genetic recombination, V. Recombination of λ DNA *in vitro. Proc. Natn. Acad. Sci. U.S.A.*, **68**, 1639–1643.                                (*348*)

CASSUTO, E. and RADDING, C. M. (1971). Mechanism for the action of λ exonuclease in genetic recombination. *Nature New Biol.*, **229**, 13–16.                    (*347, 349*)

CATCHESIDE, D. E. A. (1968). Regulation of the *am*-l locus in *Neurospora*: evidence of independent control of allelic recombination and gene expression. *Genetics, Princeton*, **59**, 443–452.                                                    (*369*)

CATCHESIDE, D. G. (1939). A position effect in *Oenothera. J. Genet.*, **38**, 345–352.
                                                                          (*390*)

CATCHESIDE, D. G. (1947). The *P*-locus position effect in *Oenothera. J. Genet.*, **48**, 31–42.
                                                                        (*389–91*)

CATCHESIDE, D. G. (1960). Complementation among histidine mutants of *Neurospora crassa. Proc. R. Soc.* B, **153**, 179–194.                                        (*224*)

CATCHESIDE, D. G. (1965). Multiple enzymic functions of a gene in *Neurospora crassa. Biochem. biophys. Res. Commun.*, **18**, 648–651.                            (*320*)

CATCHESIDE, D. G. (1966). A second gene controlling allelic recombination in *Neurospora crassa. Aust. J. Biol. Sci.*, **19**, 1039–1046.                          (*368*)

CATCHESIDE, D. G. and AUSTIN, B. (1971). Common regulation of recombination at the *amination-1* and *histidine-2* loci in *Neuropora crassa. Aust. J. biol. Sci.*, **24**, 107–115.
                                                                          (*368*)

CATCHESIDE, D. G. and OVERTON, A. (1959). Complementation between alleles in heterocaryons. *Cold Spring Harb. Symp. quant. Biol.*, **23**, 137–140.        (*222–23*)

CATTANACH, B. M. and ISAACSON, J. H. (1967). Controlling elements in the mouse X chromosome. *Genetics, Princeton*, **57**, 331–346.                            (*420*)

CAVALIER-SMITH, T. (1970). Electron microscopic evidence for chloroplast fusion in zygotes of *Chlamydomonas reinhardii. Nature, Lond.*, **228**, 333–335.        (*430*)

CHADWICK, P., PIRROTTA, V., STEINBERG, R., HOPKINS, N. and PTASHNE, M. (1971). The λ and 434 phage repressors. *Cold Spring Harb. Symp. quant. Biol.*, **35**, 283–294.
                                                                          (*314*)

CHAMBERLIN, M. and BERG, P. (1962). Deoxyribonucleic acid-directed synthesis of ribonucleic acid by an enzyme from *Escherichia coli. Proc. natn. Acad. Sci. U.S.A.*, **48**, 81–94.                                                              (*249*)

CHAMPE, S. P. and BENZER, S. (1962). Reversal of mutant phenotypes by 5-fluorouracil: an approach to nucleotide sequences in messenger-RNA. *Proc. natn. Acad. Sci. U.S.A.*, **48**, 532–546.                                                        (*251*)

CHAMPOUX, J. J. and DULBECCO, R. (1972). An activity from mammalian cells that untwists superhelical DNA—a possible swivel for DNA replication. *Proc. natn. Acad. Sci. U.S.A.*, **69**, 143–146.                                              (*204*)

CHAPEVILLE, F., LIPMANN, F., VON EHRENSTEIN, G., WEISBLUM, B., RAY, W. J. and BENZER, S. (1962). On the role of soluble ribonucleic acid in coding for amino acids. *Proc. natn. Acad. Sci. U.S.A.*, **48**, 1086–1092.                          (*259*)

CHARGAFF, E. (1950). Chemical specificity of nucleic acids and mechanism of their enzymatic degradation. *Experientia*, **6**, 201–209.                          (*178*)

CHARGAFF, E., BUCHOWICZ, J., TÜRLER, H. and SHAPIRO, H. S. (1965). A direct test of antiparallelism in complementary sequences of calf thymus deoxyribonucleic acid. *Nature, Lond.*, **206**, 145–147.                                        (*200*)

CHARGAFF, E., VISCHER, E., DONIGER, R, GREEN, C. and MISANI, F. (1949). The composition of the desoxypentose nucleic acids of thymus and spleen. *J. biol. Chem.*, **177**, 405–416. *(178)*

CHASE, M. and DOERMANN, A. H. (1958). High negative interference over short segments of the genetic structure of bacteriophage T4. *Genetics, Princeton,* **43**, 332–353. *(339)*

CHEN, B., DE CROMBRUGGHE, B., ANDERSON, W. B., GOTTESMAN, M. E., PASTAN, I. and PERLMAN, R. L. (1971). On the mechanism of action of *lac* repressor. *Nature New Biol.*, **233**, 67–70. *(314)*

CHOVNICK, A., BALLANTYNE, G. H. and HOLM, D. G. (1971). Studies on gene conversion and its relationship to linked exchange in *Drosophila melanogaster*. *Genetics, Princeton,* **69**, 179–209. *(370)*

CLARK, A. J., CHAMBERLIN, M., BOYCE, R. P. and HOWARD-FLANDERS, P. (1966). Abnormal metabolic response to ultraviolet light of a recombination deficient mutant of *Escherichia coli* K12. *J. molec. Biol.*, **19**, 442–454. *(355)*

CLARK, A. J. and MARGULIES, A. D. (1965). Isolation and characterization of recombination-deficient mutants of *Escherichia coli* K12. *Proc. natn. Acad. Sci. U.S.A.*, **53**, 451–459. *(355)*

CLARK, B. F. C., DOCTOR, B. P., HOLMES, K. C., KLUG, A., MARCKER, K. A., MORRIS, S. J. and PARADIES, H. H. (1968*b*). Crystallization of transfer RNA. *Nature, Lond.*, **219**, 1222–1224. *(265)*

CLARK, B. F. C., DUBE, S. K. and MARCKER, K. A. (1968*a*). Specific codon-anticodon interaction of an initiator-tRNA fragment. *Nature, Lond.*, **219**, 484–5. *(288)*

CLARK, B. F. C. and MARCKER, K. A. (1966). The role of N-formyl-methionyl-sRNA in protein biosynthesis. *J. molec. Biol.*, **17**, 394–406. *(287)*

CLARK, J. L. and STEINER, D. F. (1969). Insulin biosynthesis in the rat: demonstration of two proinsulins. *Proc. natn. Acad. Sci. U.S.A.*, **62**, 278–285. *(325)*

CLEAVER, J. E. (1968). Defective repair replination of DNA in xeroderma pigmentosum. *Nature, Lond.*, **218**, 652–656. *(358)*

CLEGG, J. B. (1970). Horse haemoglobin polymorphism: evidence for two linked non-allelic α-chain genes. *Proc. R. Soc. B,* **176**, 235–246. *(325)*

CLEGG, J. B., WEATHERALL, D. J., NA-NAKORN, S. and WASI, P. (1968). Haemoglobin synthesis in β-thalassaemia. *Nature Lond.*, **220**, 664–668. *(324)*

CLEVER, U. (1961). Genaktivitäten in den Riesenchromosomen von *Chironomus tentans* und ihre Beziehungen zur Entwicklung. 1. Genaktivierung durch Ecdyson. *Chromosoma*, **12**, 607–675. *(401)*

CODDINGTON, A. and FINCHAM, J. R. S. (1965). Proof of hybrid enzyme formation in a case of inter-allelic complementation in *Neurospora crassa*. *J. molec. Biol.*, **12**, 152–161. *(224)*

COEN, D., DEUTSCH, J., NETTER, P., PETROCHILO, E. and SLONIMSKI, P. P. (1970). Mitochondrial genetics. I. Methodology and phenomenology. *Symp. Soc. exp. Biol.*, **24**, 449–496. *(430)*

COLLINS, G. N. (1912). Gametic coupling as a cause of correlations. *Am. Nat.*, **46**, 569–590. *(87)*

COMINGS, D. E. (1967). Histones of genetically active and inactive chromatin. *J. Cell Biol.*, **35**, 699–708. *(421)*

COMINGS, D. E. (1971). Isolabelling and chromosome strandedness. *Nature New Biol.*, **229**, 24–25. *(373)*

COOPER, K. W. (1949). The cytogenetics of meiosis in *Drosophila*. Mitotic and meiotic autosomal chiasmata without crossing over in the male. *J. Morph.*, **84,** 81–122. *(134)*

COOPER, S. and HELMSTETTER, C. E. (1968). Chromosome replication and the division cycle of *Escherichia coli* B/r. *J. molec. Biol.*, **31,** 519–540.                    *(204, 205)*

CORRENS, C. (1900). G. Mendels Regel über das Verhalten der Nachkommenschaft der Rassenbartarde. *Ber. dt. bot. Ges.*, **18,** 158–168. (English translation entitled 'G. Mendel's law concerning the behavior of progeny of varietal hybrids' in *Genetics, Princeton* (1950), **35,** suppl. pp. 33–41.)                    *(9)*

CORRENS, C. (1909). Vererbungsversuche mit blass(gelb)grünen und buntblättrigen Sippen bei *Mirabilis Jalapa, Urtica pilulifera* und *Lunaria annua. Z. VererbLehre,* **1,** 291–329.                    *(14–15, 51, 53)*

CORY, S., MARCKER, K. A., DUBE, S. K. and CLARK, B. F. C. (1968). Primary structure of a methionine transfer RNA from *Escherichia coli. Nature, Lond.*, **220,** 1039–1040.
                    *(290)*

COVE, D. J. and PATEMAN, J. A. (1969). Autoregulation of the synthesis of nitrate reductase in *Aspergillus nidulans. J. Bact.*, **97,** 1374–1378.                    *(322)*

CRAWFORD, I. P. and YANOFSKY, C. (1958). On the separation of the tryptophan synthetase of *Escherichia coli* into two protein components. *Proc. natn. Acad. Sci. U.S.A.*, **44,** 1161–1170.                    *(227)*

CREIGHTON, H. B. and MCCLINTOCK, B. (1931). A correlation of cytological and genetical crossing-over in *Zea mays. Proc. natn. Acad. Sci. U.S.A.*, **17,** 492–7.*(89–91)*

CRICK, F. H. C. (1957). The structure of nucleic acids and their role in protein synthesis. *Biochem. Soc. Symp.*, **14,** 25–26.                    *(257–59)*

CRICK, F. H. C. (1963). The recent excitement in the coding problem. *Prog. Nucleic Acid. Res.*, **1,** 163–217.                    *(447)*

CRICK, F. H. C. (1966). Codon-anticodon pairing: the wobble hypothesis. *J. molec. Biol.*, **19,** 548–555.                    *(282, 283, 286)*

CRICK, F. H. C. (1967). The genetic code. *Proc. R. Soc. B*, **167,** 331–347.    *(275, 277)*

CRICK, F. H. C., BARNETT, L., BRENNER, S. and WATTS-TOBIN, R. J. (1961). General nature of the genetic code for proteins. *Nature, Lond.*, **192,** 1227–1232.
                    *(239–42, 276, 294)*

CRICK, F. H. C. and BRENNER, S. (1967). The absolute sign of certain phase-shift mutants in bacteriophage T4. *J. molec. Biol.*, **26,** 361–363.                    *(242)*

CRICK, F. H. C. and ORGEL, L. E. (1964). The theory of inter-allelic complementation. *J. molec. Biol.*, **8,** 161–165.                    *(229–30)*

CRICK, F. H. C. and WATSON, J. D. (1954). The complementary structure of deoxy-ribonucleic acid. *Proc. R. Soc. A*, **223,** 80–96.                    *(182)*

CUÉNOT, L. (1902). La loi de Mendel et l'hérédité de la pigmentation chez les souris. *Archs Zool. exp. gén.*, 3ᵉ Serie, **10,** Notes et Revue, pp. 27–30.                    *(11, 21)*

CUÉNOT, L. (1904). L'Hérédité de la pigmentation chez les souris (3ᵐᵉ Note). *Archs Zool. exp. gén.*, 4ᵉ Serie, **2,** Notes et Revue, pp. 45–56.                    *(137)*

CYBIS, J. and WEGLENSKI, P. (1969). Effects of lysine on arginine uptake and metabolism in *Aspergillus nidulans. Molec. gen. Genet.*, **104,** 282–287.                    *(171)*

DARLINGTON, C. D. (1929*a*). Ring-formation in *Oenothera* and other genera. *J. Genet.*, **20,** 345–363.                    *(133)*

DARLINGTON, C. D. (1929*b*). Meiosis in polyploids. Part II. Aneuploid hyacinths. *J. Genet.*, **21,** 17–56.                    *(63, 124, 133)*

DARLINGTON, C. D. (1930). A cytological demonstration of "genetic" crossing-over. *Proc. R. Soc. B*, **107**, 50–59. *(124–25)*

DARLINGTON, C. D. (1931*a*). Meiosis in diploid and tetraploid *Primula sinensis*. *J. Genet.*, **24**, 65–96. *(131–33)*

DARLINGTON, C. D. (1931*b*). Cytological theory in relation to heredity. *Nature, Lond.*, **127**, 709–712. *(133, 178)*

DARLINGTON, C. D. (1931*c*). The cytological theory of inheritance in *Oenothera*. *J. Genet.*, **24**, 405–474. *(129, 133)*

DARLINGTON, C. D. (1932). *Recent Advances in Cytology*, London, J. & A. Churchill, 559 pp. *(118, 120)*

DARLINGTON, C. D. (1934*a*). Anomalous chromosome pairing in the male *Drosophila pseudo-obscura*. *Genetics, Princeton*, **19**, 95–118. *(134)*

DARLINGTON, C. D. (1934*b*). The origin and behaviour of chiasmata. VII. *Zea mays*. *Z. VererbLehre*, **67**, 96–114. *(116)*

DARLINGTON, C. D. (1935). The time, place and action of crossing-over. *J. Genet.*, **31**, 185–212. *(129)*

DARLINGTON, C. D. (1936*a*). The external mechanics of the chromosomes. *Proc. R. Soc. B*, **121**, 264–319. *(445)*

DARLINGTON, C. D. (1936*b*). Crossing-over and its mechanical relationships in *Chorthippus* and *Stauroderus*. *J. Genet.*, **33**, 465–500. *(127)*

DARWIN, C. (1868). *The variation of animals and plants under domestication*. London, John Murray. *(2, 11–12, 433, 448)*

DARWIN, C. (1875). *The variation of animals and plants under domestication*. 2nd edn. London, John Murray. *(2)*

DAVIDSON, R. G., NITOWSKY, H. M. and CHILDS, B. (1963). Demonstration of two populations of cells in the human female heterozygous for glucose-6-phosphate dehydrogenase variants. *Proc. natn. Acad. Sci. U.S.A.*, **50**, 481–485. *(420)*

DAVIS, R. W. and HYMAN, R. W. (1971). A study in evolution: the DNA base sequence homology between coliphages T7 and T3. *J. molec. Biol.*, **62**, 287–301. *(351)*

DAVIS, R. W. and PARKINSON, J. S. (1971). Deletion mutants of bacteriophage lambda. III. Physical structure of *att*$^\Phi$. *J. molec. Biol.*, **56**, 403–423. *(347)*

DAWID, I. B., BROWN, D. D. and REEDER, R. H. (1970). Composition and structure of chromosomal and amplified ribosomal DNA's of *Xenopus laevis*. *J. molec. Biol.*, **51**, 341–360. *(411)*

DAWSON, G. W. P. and SMITH-KEARY, P. F. (1963). Episomic control of mutation in *Salmonella typhimurium*. *Heredity*, **18**, 1–20. *(394–95, 397)*

DE CROMBRUGGHE, B., CHEN, B., ANDERSON, W., NISSLEY, P., GOTTESMAN, M., PASTAN, I. and PERLMAN, R. (1971). *Lac* DNA, RNA polymerase and cyclic AMP receptor protein, cyclic AMP, *lac* repressor and inducer are the essential elements for controlled *lac* transcription. *Nature New Biol.*, **231**, 139–142. *(316)*

DE LANGE, R. J., FAMBROUGH, D. M., SMITH, E. L. and BONNER, J. (1969). Calf and pea histone IV. III. Complete amino acid sequence of pea seedling histone IV; comparison with the homologous calf thymus histone. *J. biol. Chem.*, **244**, 5669–5679. *(423)*

DE LANGE, R. J., HOOPER, J. A. and SMITH, E. L. (1972). Complete amino-acid sequence of calf-thymus histone III. *Proc. natn. Acad. Sci. U.S.A.*, **69**, 882–884. *(423)*

DE LANGE, R. J. and SMITH, E. L. (1971). Histones: structure and function. *A. Rev. Biochem.*, **40**, 279–314.    (*423*)

DELBRÜCK, M. and BAILEY, W. T. (1947). Induced mutations in bacterial viruses. *Cold Spring Harb. Symp. quant. Biol.*, **11**, 33–37.    (*213, 338*)

DELBRÜCK, M. and STENT, G. S. (1957). On the mechanism of DNA replication. In *The Chemical Basis of Heredity* edited by W. D. MCELROY and B. GLASS, Baltimore, Johns Hopkins Press, pp. 699–736.    (*186, 452*)

DE LUCIA, P. and CAIRNS, J. (1969). Genetic analysis of an *E. coli* strain with a mutation affecting DNA polymerase. *Nature, Lond.*, **224**, 1164–1166.    (*202*)

DEMEREC, M. (1924). A case of pollen dimorphism in maize. *Am. J. Bot.*, **11**, 461–4. (*46*)

DEMEREC, M. (1962). "Selfers"—attributed to unequal crossovers in *Salmonella*. *Proc. natn. Acad. Sci. U.S.A.*, **48**, 1696–1704.    (*366*)

DEMEREC, M., BLOMSTRAND, I. and DEMEREC, Z. E. (1955). Evidence of complex loci in *Salmonella*. *Proc. natn. Acad. Sci. U.S.A.*, **41**, 359–364.    (*301*)

DEMEREC, M. and DEMEREC, Z. E. (1956). Analysis of linkage relationships in *Salmonella* by transduction techniques. *Brookhaven Symp. Biol.*, **8**, 75–87.    (*301*)

DEMOSS, J. A., JACKSON, R. W. and CHALMERS, J. H. (1967). Genetic control of the structure and activity of an enzyme aggregate in the tryptophan pathway of *Neurospora crassa*. *Genetics, Princeton*, **56**, 413–424.    (*321*)

DE SERRES, F. J. (1964). Mutagenesis and chromosome structure. *J. cell. comp. Physiol.*, **64**, suppl. 1, 33–42.    (*319*)

DE VRIES, H. (1900). Das Spaltungsgesetz der Bastarde. *Ber. dt. bot. Ges.*, **18**, 83–90. (English translation entitled 'The law of separation of characters in crosses' in *Jl R. hort. Soc.*, 1901, **25**, 243–248.)    (*9, 10, 18*)

DE VRIES, H. (1901). *Die Mutationstheorie.* (English translation entitled *The Mutation Theory*; 2 vols. London, Kegan Paul, Trench, Trübner & Co., 1910, 1911, 582 and 683 pp.)    (*140, 450*)

DE VRIES, H. (1903). *Befruchtung und Bastardierung*, 62 pp., Leipzig. (English translation by C. S. GAGER entitled *Fertilization and Hybridization* as pp. 217–263 of *Intracellular pangenesis including a paper on fertilization and hybridization*, Chicago, Open Court Publ. Co., 1910.)    (*80–82, 98*)

DIERS, L. (1970). Origin of plastids: cytological results and interpretations including some genetical aspects. *Symp. Soc. exp. Biol.*, **24**, 129–145.    (*431*)

DINTZIS, H. M. (1961). Assembly of the peptide chains of hemoglobin. *Proc. natn. Acad. Sci. U.S.A.*, **47**, 247–261.    (*259, 289*)

DJORDJEVIC, B. and SZYBALSKI, W. (1960). Genetics of human cell lines. III. Incorporation of 5-bromo- and 5-iododeoxyuridine into the deoxyribonucleic acid of human cells and its effect on radiation sensitivity. *J. exp. Med.*, **112**, 509–31. (*194*)

DOBZHANSKY, T. (1929). Genetical and cytological proof of translocations involving the third and fourth chromosome of *Drosophila melanogaster*. *Biol. Zbl.*, **49**, 408–419.    (*147*)

DODGE, B. O. (1929). The nature of giant spores and segregation of sex factors in *Neurospora*. *Mycologia*, **21**, 222–231.    (*104*)

DODGE, J. D. (1964). Chromosome structure in the Dinophyceae. II. Cytochemical studies. *Arch. Mikrobiol.*, **48**, 66–80.    (*386*)

DOERFLER, W. and HOGNESS, D. S. (1968). Gene orientation in bacteriophage lambda as determined from the genetic activities of heteroduplex DNA formed *in vitro*. *J. molec. Biol.*, **33**, 661–678.    (*365*)

DOERMANN, A. H. (1944). A lysineless mutant of *Neurospora* and its inhibition by arginine. *Archs Biochem.*, **5**, 373–384. *(171)*

DOLZMANN, P. (1968). Photosynthese Reaktionen einiger Plastom-Mutanten von *Oenothera*. III. Strukturelle Aspekte. *Z. PflPhysiol.*, **58**, 300–309. *(431)*

DONCASTER, L. (1908). On sex-inheritance in the moth *Abraxas grossulariata* and its var. *lacticolor*. *Rep. Evol. Comm. R. Soc.*, **4**, 53–57. *(21, 22)*

DONCASTER, L. and RAYNOR, G. H. (1906). On breeding experiments with Lepidoptera. *Proc. zool. Soc. Lond.*, **1**, 125–133. *(20–22)*

DOUGLAS, L. T. and KROES, H. W. (1969). Meiosis. VI: A reinterpretation of tritium thymidine labelling of meiotic chromosomes, supporting the chiasmatype hypothesis. *Genetica*, **40**, 503–507. *(341)*

DOUNCE, A. L. (1952). Duplicating mechanism for peptide chain and nucleic acid synthesis. *Enzymologia*, **15**, 251–258. *(210, 218, 228, 233, 234)*

DRAKE, J. W. (1966). Heteroduplex heterozygotes in bacteriophage T4 involving mutations of various dimensions. *Proc. natn. Acad. Sci. U.S.A.*, **55**, 506–12. *(361)*

DRAKE, J. W. (1970). *The molecular basis of mutation.* San Francisco, Holden-Day. 273 pp. *(239)*

DRAKE, J. W. and GREENING, E. O. (1970). Suppression of chemical mutagenesis in bacteriophage T4 by genetically modified DNA polymerases. *Proc. natn. Acad. Sci. U.S.A.*, **99**, 823–829. *(202)*

DRAPEAU, G. R., BRAMMAR, W. J. and YANOFSKY, C. (1968). Amino acid replacements of the glutamic acid residue at position 48 in the tryptophan synthetase A protein of *Escherichia coli*. *J. molec. Biol.*, **35**, 357–367. *(363–64)*

DUBE, S. K., MARCKER, K. A., CLARK, B. F. C. and CORY, S. (1968). Nucleotide sequence of N-formyl-methionyl-transfer RNA. *Nature, Lond.*, **218**, 232–3. *(263, 285, 290)*

DUBE, S. K. and RUDLAND, P. S. (1970). Control of translation by T4 phage: altered binding of disfavoured messengers. *Nature, Lond.*, **226**, 820–823. *(291)*

DUDOCK, B. S., DIPERI, C., SCILEPPI, K. and RESZELBACH, R. (1971). The yeast phenyl-alanyl-transfer RNA synthetase recognition site: the region adjacent to the dihydrouridine loop. *Proc. natn. Acad. Sci. U.S.A.*, **68**, 681–684. *(268)*

DUDOCK, B. S., KATZ, G., TAYLOR, E. K. and HOLLEY, R. W. (1969). Primary structure of wheat germ phenylalanine transfer RNA. *Proc. natn. Acad. Sci. U.S.A.*, **62**, 941–945. *(268)*

DUNN, D. B. and SMITH, J. D. (1954). Incorporation of halogenated pyrimidines into the deoxyribonucleic acids of *Bacterium coli* and its bacteriophages. *Nature, Lond.*, **174**, 305–306. *(196, 235)*

DUPRAW, E. J. (1965). The organization of nuclei and chromosomes in honeybee embryonic cells. *Proc. natn. Acad. Sci. U.S.A.*, **53**, 161–168. *(373, 386)*

DURRANT, A. (1962). The environmental induction of heritable change in *Linum*. *Heredity*, **17**, 27–61. *(2)*

DUTT, M. K. and KAUFMANN, B. P. (1959). Degradational action of deoxyribonuclease on unfixed chromosomes of *Drosophila* and grasshoppers. *Nucleus, Calcutta*, **2**, 85–98. *(380)*

DUTTA, S. K., RICHMAN, N., WOODWARD, V. W. and MANDEL, M. (1967). Relatedness among species of fungi and higher plants measured by DNA hybridization and base ratios. *Genetics, Princeton*, **57**, 719–729. *(278)*

EAST, E. M. (1910). A Mendelian interpretation of variation that is apparently continuous. *Am. Nat.*, **44**, 65–82. *(33–5, 37)*

ECHLIN, P. and MORRIS, I. (1965). The relationships between blue-green algae and bacteria. *Biol. Rev.*, **40**, 143–187. *(435)*

ECHOLS, H. (1971). Lysogeny: viral repression and site-specific recombination. *A. Rev. Biochem.*, **40**, 827–854. *(336, 347)*

EDGAR, R. S. and LIELAUSIS, I. (1968). Some steps in the assembly of bacteriophage T4. *J. molec. Biol.*, **32**, 263–276. *(317, 425–26)*

EDSTRÖM, J.-E. and BEERMANN, W. (1962). The base composition of nucleic acids in chromosomes, puffs, nucleoli, and cytoplasm of *Chironomus* salivary gland cells. *J. Cell Biol.*, **14**, 371–380. *(401)*

EHRENSTEIN, G. VON, and LIPMANN, F. (1961). Experiments on hemoglobin biosynthesis. *Proc. natn. Acad. Sci. U.S.A.*, **47**, 941–950. *(279)*

EHRENSTEIN, G. VON, WEISBLUM, B. and BENZER, S. (1963). The function of sRNA as amino acid adaptor in the synthesis of hemoglobin. *Proc. natn. Acad. Sci. U.S.A.*, **49**, 669–675. *(259)*

ELLISON, J. R. and BARR, H. J. (1972). Quinacrine fluorescence of specific chromosome regions. Late replication and high A:T content in *Samoaia leonensis*. *Chromosoma*, **36**, 375–390. *(407, 408)*

ELSDALE, T. R., FISCHBERG, M. and SMITH, S. (1958). A mutation that reduces nucleolar number in *Xenopus laevis*. *Expl Cell Res.*, **14**, 642–643. *(402)*

ELSTON, R. C. and GLASSMAN, E. (1967). An approach to the problem of whether clustering of functionally related genes occurs in higher organisms. *Genet. Res.*, **9**, 141–147. *(325)*

EMERSON, R. A. (1911). Genetic correlation and spurious allelomorphism in maize. *Rep. Neb. agric. Exp. Stn*, **24**, 58–90. *(137)*

EMERSON, R. A. (1914). The inheritance of a recurring somatic variation in variegated ears of maize. *Am. Nat.*, **48**, 87–115. *(388)*

EMERSON, R. A. and EAST, E. M. (1913). The inheritance of quantitative characters in maize. *Res. Bull. Neb. agric. Exp. Stn*, **2**, 1–120. *(35–37)*

EMERSON, S. (1966). Quantitative implications of the DNA-repair model of gene conversion. *Genetics, Princeton*, **53**, 475–485. *(364)*

EMERSON, S. and BEADLE, G. W. (1933). Crossing-over near the spindle fiber in attached-X chromosomes of *Drosophila melanogaster*. *Z. VererbLehre*, **45**, 129–140. *(114)*

EMMER, M., DE CROMBRUGGHE, B., PASTAN, I. and PERLMAN, R. (1970). Cyclic AMP receptor protein of *E. coli*: its role in the synthesis of inducible enzymes. *Proc. natn. Acad. Sci. U.S.A.*, **66**, 480–487. *(315)*

ENGLESBERG, E., IRR, J., POWER, J. and LEE, N. (1965). Positive control of enzyme synthesis by gene C in the L-arabinose system. *J. Bacteriol.*, **90**, 946–957. *(316, 322)*

ENGLESBERG, E., SQUIRES, C. and MERONK, F. (1969). The L-arabinose operon in *Escherichia coli* B/r: a genetic demonstration of two functional states of the product of a regulator gene. *Proc. natn. Acad. Sci. U.S.A.*, **62**, 1100–1107. *(316)*

EPHRUSSI, B. (1953). *Nucleo-cytoplasmic Relations in Micro-organisms*. Oxford, Clarendon Press, 127 pp. *(56–57)*

EPHRUSSI-TAYLOR, H. and GRAY, T. (1966). Genetic studies of recombining DNA in pneumococcal transformation. *J. gen. Physiol.*, **49**, no. 6, suppl., 211–231. *(364)*

EPSTEIN, R. H., BOLLE, A., STEINBERG, C. M., KELLENBERGER, E., BOY DE LA TOUR, E., CHEVALLEY, R., EDGAR, R. S., SUSMAN, M., DENHARDT, G. H. and LIELAUSIS, A. (1964). Physiological studies of conditional lethal mutants of bacteriophage T4D. *Cold Spring Harb. Symp. quant. Biol.*, **28**, 375–394. (*338, 425–26*)

EVANS, H. J. (1967). Repair and recovery from chromosome damage induced by fractionated X-ray exposures. In *Radiat. Res.* 1966, pp. 482–501. North-Holland Pub. Co., Amsterdam. (*376*)

EVANS, H. J. and SAVAGE, J. R. K. (1963). The relation between DNA synthesis and chromosome structure as resolved by X-ray damage. *J. Cell Biol.*, **18**, 525–540. (*376*)

EVANS, H. J. and SCOTT, D. (1969). The induction of chromosome aberrations by nitrogen mustard and its dependence on DNA synthesis. *Proc. R. Soc.* B, **173**, 491–512. (*157*)

EVANS, T. E. (1966). Synthesis of a cytoplasmic DNA during the G2 interphase of *Physarum polycephalum. Biochem. biophys. Res. Commun.*, **22**, 678–683. (*278*)

FAREED, G. C. and RICHARDSON, C. C. (1967). Enzymatic breakage and joining of deoxyribonucleic acid, II. The structural gene for polynucleotide ligase in bacteriophage T4. *Proc. natn. Acad. Sci. U.S.A.*, **58**, 665–672. (*346*)

FARGIE, B. and HOLLOWAY, B. W. (1965). Absence of clustering of functionally related genes in *Pseudomonas aeruginosa. Genet. Res.*, **6**, 284–299. (*301*)

FARMER, J. B. and MOORE, J. E. S. (1905). On the maiotic phase (reduction divisions) in animals and plants. *Q. Jl microsc. Sci.*, **48**, 489–557. (*63, 450*)

FEUGHELMAN, M., LANGRIDGE, R., SEEDS, W. E., STOKES, A. R., WILSON, H. R., HOOPER, C. W., WILKINS, M. H. F., BARCLAY, R. K. and HAMILTON, L. D, (1955). Molecular structure of deoxyribose nucleic acid and nucleoprotein. *Nature, Lond.*, **175**, 834–838. (*185, 386*)

FINCHAM, J. R. S. (1959*a*). On the nature of the glutamic dehydrogenase produced by inter-allele complementation at the *am* locus of *Neurospora crassa. J. gen. Microbiol.*, **21**, 600–611. (*222, 224, 447*)

FINCHAM, J. R. S. (1959*b*). The biochemistry of genetic factors. *A. Rev. Biochem.*, **28**, 343–364. (*168*)

FINCHAM, J. R. S. (1967). Mutable genes in the light of Callan's hypothesis of serially repeated gene copies. *Nature, Lond.*, **215**, 864–866. (*402*)

FINCHAM, J. R. S. (1970). The regulation of gene mutation in plants. *Proc. R. Soc.* B, **179**, 295–302. (*433*)

FINCHAM, J. R. S. and CODDINGTON, A. (1963). Complementation at the *am* locus of *Neurospora crassa*: a reaction between different mutant forms of glutamate dehydrogenase. *J. molec. Biol.*, **6**, 361–373. (*224*)

FINCHAM, J. R. S. and DAY, P. R. (1971). *Fungal Genetics.* 3rd edn. Oxford, Blackwell, 402 pp. (*168, 397*)

FINCHAM, J. R. S. and HOLLIDAY, R. (1970). An explanation of fine structure map expansion in terms of excision repair. *Molec. gen. Genet.*, **109**, 309–322. (*362, 363*)

FINCHAM, J. R. S. and PATEMAN, J. A. (1957). Formation of an enzyme through complementary action of mutant 'alleles' in separate nuclei in a heterocaryon. *Nature, Lond.*, **179**, 741–742. (*218, 222–23*)

FINK, G. R. (1966). A cluster of genes controlling three enzymes in histidine biosynthesis in *Saccharomyces cerevisiae. Genetics, Princeton*, **53**, 445–459. (*319–20*)

FISHER, R. A. (1918). The correlation between relatives on the supposition of Mendelian inheritance. *Trans. R. Soc. Edinb.*, **52**, 399–433. Reprinted with commentary by P. A. P. Moran and C. A. B. Smith (1966). Cambridge Univ. Press. 62 pp.                                                                      (*31, 36–37*)

FISHER, R. A. (1936). Has Mendel's work been rediscovered? *Ann. Sci.*, **1**, 115–137.                                                                              (*28–9*)

FISHER, R. A. (1963). *Statistical Methods for Research Workers*. 13th edn. Edinburgh, Oliver and Boyd.                                                                     (*23, 27*)

FITCH, W. M. (1967). Evidence suggesting a non-random character to nucleotide replacements in naturally occurring mutations. *J. molec. Biol.*, **26**, 499–507.                                                                                          (*280*)

FLEMMING, W. (1879), Beiträge zur Kenntniss der Zelle und ihre Lebenserscheinungen. *Arch. mikrosk. Anat.*, **16**, 302–436.                                               (*58*)

FLEMMING, W. (1882). *Zellsubstanz, Kern- und Zelltheilung*. Leipzig. 424 pp. (*58–59, 450*)

FLEMMING, W. (1887). Neue Beiträge zur Kenntniss der Zelle. *Arch. mikrosk. Anat.*, **29**, 389–463.                                                                       (*62*)

FOGEL, S. and MORTIMER, R. K. (1969). Informational transfer in meiotic gene conversion. *Proc. natn. Acad. Sci. U.S.A.*, **62**, 96–103.                       (*362, 363*)

FOGEL, S. and MORTIMER, R. K. (1970). Fidelity of meiotic gene conversion in yeast. *Molec. gen. Genet.*, **109**, 177–185.                                               (*366*)

FOLK, W. R. and YANIV, M. (1972). Coding properties and nucleotide sequences of *E. coli* glutamine tRNAs. *Nature New Biol.*, **237**, 165–166.                      (*285*)

FORD, C. E. and HAMERTON, J. L. (1956). The chromosomes of man. *Nature, Lond.*, **178**, 1020–1023.                                                                    (*71*)

FORGET, B. G. and WEISSMAN, S. M. (1969). The nucleotide sequence of ribosomal 5S ribonucleic acid from KB cells. *J. biol. Chem.*, **244**, 3148–3165.                 (*247*)

FORK, D. C. and HEBER, U. W. (1968). Studies on electron-transport reactions of photosynthesis in plastome mutants of *Oenothera. Pl. Physiol.*, Lancaster, **43**, 606–612. (*431*)

FORTUIN, J. J. H. (1971). Another two genes controlling mitotic intragenic recombination and recovery from UV damage in *Aspergillus nidulans*. II. Recombination behaviour and X-ray-sensitivity of *uvsD* and *uvsE* mutants. *Mutation Res.*, **11**, 265–277.                                                                            (*357*)

FOX, E. and MESELSON, M. (1963). Unequal photosensitivity of the two strands of DNA in bacteriophage λ. *J. molec. Biol.*, **7**, 583–589.                            (*252*)

FOX, M. S. and ALLEN, M. K. (1964). On the mechanism of deoxyribonucleate integration in pneumococcal transformation. *Proc. natn. Acad. Sci. U.S.A.*, **52**, 412–419.                                                                                 (*338, 346*)

FRAENKEL-CONRAT, H. (1956). The role of the nucleic acid in the reconstitution of active tobacco mosaic virus. *J. Am. chem. Soc.*, **78**, 882–883.                  (*244*)

FRAENKEL-CONRAT, H. and SINGER, B. (1957). Virus reconstitution. II. Combination of protein and nucleic acid from different strains. *Biochim. biophys. Acta.*, **24**, 540–548.                                                                              (*243–44*)

FRANKLIN, E. C. and FRANGIONE, B. (1971). The molecular defect in a protein (CRA) found in γ1 heavy chain disease, and its genetic implications. *Proc. natn. Acad. Sci. U.S.A.*, **68**, 187–191.                                                           (*419*)

FRANKLIN, N. C. and LURIA, S. E. (1961). Transduction by bacteriophage P1 and the properties of the *lac* genetic region in *E. coli* and *S. dysenteriae. Virology*, **15,** 299–311. *(308)*

FRANKLIN, R. E. and GOSLING, R. G. (1953). Molecular configuration in sodium thymonucleate. *Nature, Lond.*, **171,** 740–741. *(182, 184, 424)*

FREESE, E. (1957). The correlation effect for a histidine locus of *Neurospora crassa. Genetics, Princeton,* **42,** 671–684. *(340)*

FREESE, E. (1959). The difference between spontaneous and base-analogue induced mutations of phage T4. *Proc. natn. Acad. Sci. U.S.A.*, **45,** 622-633.
*(236, 239, 453)*

FREESE, E., BAUTZ, E. and BAUTZ-FREESE, E. (1961). The chemical and mutagenic specificity of hydroxylamine. *Proc. natn. Acad. Sci. U.S.A.*, **47,** 845–855. *(238)*

FULLER, W. and HODGSON, A. (1967). Conformation of the anticodon loop in tRNA. *Nature, Lond.*, **215,** 817–821. *(286)*

GALL, J. G. (1955). Problems of structure and function in the amphibian oocyte nucleus. *Symp. Soc. exp. Biol.*, **9,** 358–370. *(400)*

GALL, J. G. (1956). On the submicroscopic structure of chromosomes. *Brookhaven Symp. Biol.*, **8,** 17–32. *(377, 380, 385, 408)*

GALL, J. G. (1963). Kinetics of deoxyribonuclease action on chromosomes. *Nature, Lond.*, **198,** 36–38. *(380–81, 384, 385)*

GALL, J. G. (1966). Chromosome fibers studied by a spreading technique. *Chromosoma,* **20,** 221–233. *(373, 386)*

GALL, J. G. and CALLAN, H. G. (1962). $H^3$ uridine incorporation in lampbrush chromosomes. *Proc. natn. Acad. Sci. U.S.A.*, **48,** 562–570. *(399, 400)*

GALLY, J. A. and EDELMAN, G. M. (1970). Somatic translocation of antibody genes. *Nature, Lond.*, **227,** 341–8. *(415)*

GALTON, F. (1889). *Natural Inheritance.* London, Macmillan, 259 pp. *(30–31, 37)*

GAMOW, G. (1954). Possible relation between deoxyribonucleic acid and protein structures. *Nature, Lond.*, **173,** 318. *(210, 218, 228, 233, 234)*

GAMOW, G. and YCAS, M. (1955). Statistical correlation of protein and ribonucleic acid composition. *Proc. natn Acad. Sci. U.S.A.* **41,** 1011–1019. *(234)*

GANNER, E. and EVANS, H. J. (1971). The relationship between patterns of DNA replication and of quinacrine fluorescence in the human chromosome complement. *Chromosoma,* **35,** 326–341. *(407, 421)*

GANSCHOW, R. and PAIGEN, K. (1967). Separate genes determining the structure and intracellular location of hepatic glucuronidase. *Proc. natn. Acad. Sci. U.S.A.*, **58,** 938–945. *(425)*

GARDNER, R. S., WAHBA, A. J., BASILIO, C., MILLER, R. S., LENGYEL, P. and SPEYER, J. F. (1962). Synthetic polynucleotides and the amino acid code. VII. *Proc. natn. Acad. Sci. U.S.A.*, **48,** 2087–2094. *(272)*

GAREN, A. and OTSUJI, N. (1964). Isolation of a protein specified by a regulator gene. *J. molec. Biol.*, **8,** 841–852. *(306, 316–17)*

GAREN, A. and SIDDIQI, O. (1962). Suppression of mutations in the alkaline phosphatase structural cistron of *E. coli. Proc. natn. Acad. Sci. U.S.A.*, **48,** 1121–1127. *(294)*

GARROD, A. E. (1909). *Inborn Errors of Metabolism.* London, Oxford Univ. Press. *(164)*

GARTLER, S. M., CHEN, S. H., FIALKOW, P. J., GIBLETT, E. R. and SINGH, S. (1972). *X* chromosome inactivation in cells from an individual heterozygous for two X-linked genes. *Nature New Biol.*, **236**, 149–150.                                    (*420*)

GEFTER, M. L., HIROTA, Y., KORNBERG, T., WECHSLER, J. A. and BARNOUX, C. (1971). Analysis of DNA polymerases II and III in mutants of *Escherichia coli* thermosensitive for DNA synthesis. *Proc. natn. Acad. Sci. U.S.A.*, **68**, 3150–3153.   (*202*)

GELLERT, M. (1967). Formation of covalent circles of lambda DNA by *E. coli* extracts. *Proc. natn. Acad. Sci. U.S.A.*, **57**, 148–155.                                    (*412*)

GHOSH, H. P. and KHORANA, H. G. (1967). Studies on polynucleotides, LXXXIV. On the role of ribosomal subunits in protein synthesis. *Proc. natn. Acad. Sci. U.S.A.*, **58**, 2455–2461.                                    (*289*)

GIERER, A. (1963). Function of aggregated reticulocyte ribosomes in protein synthesis. *J. molec. Biol.* **6**, 148–157.                                    (*271*)

GIERER, A. (1966). Model for DNA and protein interactions and the function of the operator. *Nature, Lond.*, **212**, 1480–1481.                                    (*307, 352*)

GIERER, A. and SCHRAMM G. (1956). Infectivity of ribonucleic acid from tobacco mosaic virus. *Nature, Lond.*, **177**, 702–703.                                    (*244*)

GILBERT, W. (1963). Polypeptide synthesis in *Escherichia coli*. II. The polypeptide chain and S-RNA. *J. molec. Biol.*, **6**, 389–403.                                    (*268, 270*)

GILBERT, W. and MÜLLER-HILL, B. (1966). Isolation of the *lac* repressor. *Proc. natn. Acad. Sci. U.S.A.*, **56**, 1891–1898.                                    (*304–6*)

GILBERT, W. and MÜLLER-HILL, B. (1967). The *lac* operator is DNA. *Proc. natn. Acad. Sci. U.S.A.*, **58**, 2415–2421.                                    (*305–6*)

GILES, N. H. (1952). Studies on the mechanism of reversion in biochemical mutants of *Neurospora crassa*. *Cold Spring Harb. Symp. quant. Biol.*, **16**, 283–313.   (*333, 340*)

GILES, N. H., CASE, M. E., PARTRIDGE, C. W. H. and AHMED, S. I. (1967). A gene cluster in *Neurospora crassa* coding for an aggregate of five aromatic synthetic enzymes. *Proc. natn. Acad. Sci. U.S.A.*, **58**, 1453-1460.                                    (*320–21*)

GILES, N. H., PARTRIDGE, C. W. H., AHMED, S. I. and CASE, M. E. (1967). The occurrence of two dehydroquinases in *Neurospora crassa*, one constitutive and one inducible. *Proc. natn. Acad. Sci. U.S.A.*, **58**, 1930–1937.                                    (*321*)

GILES, N. H., PARTRIDGE, C. W. H. and NELSON, N. J. (1957). The genetic control of adenylosuccinase in *Neurospora crassa*. *Proc. natn. Acad. Sci. U.S.A.*, **43**, 305–317.                                    (*218, 222–23*)

GILLHAM, N. W., BOYNTON, J. E. and BURKHOLDER, B. (1970). Mutations altering chloroplast ribosome phenotype in *Chlamydomonas*, I. Non-Mendelian mutations. *Proc. natn. Acad. Sci. U.S.A.*, **67**, 1026–1033.                                    (*430*)

GILLIE, O. J. (1966). The interpretation of complementation data. *Genet. Res.*, **8**, 9–31.                                    (*229–30*)

GILLIES, C. B. (1972). Reconstruction of the *Neurospora crassa* pachytene karyotype from serial sections of synaptonemal complexes. *Chromosoma*, **36**, 119–130. (*136*)

GILMORE, R. A., STEWART, J. W. and SHERMAN, F. (1968). Amino acid replacements resulting from super-suppression of a nonsense mutant of yeast. *Biochem. biophys. Acta*, **161**, 270–272.                                    (*298, 406*)

GOLDBERG, I. H., RABINOWITZ, M. and REICH, E. (1962). Basis of actinomycin action, I. DNA binding and inhibition of RNA polymerase synthetic reactions by actinomycin. *Proc. natn. Acad. Sci. U.S.A.*, **48**, 2094–2101.                                    (*245*)

GOLDMARK, P. J. and LINN, S. (1970). An endonuclease activity from *Escherichia coli* absent from certain *rec⁻* strains. *Proc. natn. Acad. Sci. U.S.A.*, **67**, 434–441. *(348)*

GOLDMARK, P. J. and LINN, S. (1972). Purification and properties of the *recBC* DNase of *Escherichia coli* K-12. *J. biol. Chem.*, **247**, 1849–1860. *(348)*

GOLDSCHMIDT, R. B. (1938). *Physiological Genetics.* New York, McGraw-Hill, 375 pp. *(158)*

GOLDSTEIN, A., KIRSCHBAUM, J. B. and ROMAN, A. (1965). Direction of synthesis of messenger RNA in cells of *Escherichia coli. Proc. natn. Acad. Sci. U.S.A.*, **54**, 1669–1675. *(255)*

GOLDSTEIN, J. L., MARKS, J. F. and GARTLER, S. M. (1971). Expression of two X-linked genes in human hair follicles of double heterozygotes. *Proc. natn. Acad. Sci. U.S.A.*, **68**, 1425–1427. *(420)*

GOLLUB, E., ZALKIN, H. and SPRINSON, D. B. (1967). Correlation of genes and enzymes, and studies on regulation of the aromatic pathway in *Salmonella. J. biol. Chem.*, **242**, 5323–5328. *(320)*

GOODMAN, H. M., ABELSON, J., LANDY, A., BRENNER, S. and SMITH, J. D. (1968). Amber suppression: a nucleotide change in the anticodon of a tyrosine transfer RNA. *Nature, Lond.*, **217**, 1019–1024. *(263, 285, 297, 406)*

GOSS, J. (1824). On the variation in the colour of peas, occasioned by cross impregnation. *Trans. hort. Soc. Lond.*, **5**, 234–236. *(2–6, 9)*

GOSSOP, G. H., YUILL, E. and YUILL, J. L. (1940). Heterogeneous fructifications in species of *Aspergillus. Trans. Br. mycol. Soc.*, **24**, 337–344. *(46–47)*

GRAY, C. H. and TATUM, E. L. (1944). X-ray induced growth factor requirements in bacteria. *Proc. natn. Acad. Sci. U.S.A.*, **30**, 404–410. *(168)*

GREEN, M. M. (1969). Controlling element mediated transpositions of the *white* gene in *Drosophila melanogaster. Genetics, Princeton*, **61**, 429–441. *(395)*

GREEN, M. M. and GREEN, K. C. (1949). Crossing-over between alleles at the lozenge locus in *Drosophila melanogaster. Proc. natn. Acad. Sci. U.S.A.*, **35**, 586–591. *(213)*

GREENAWAY, P. J. and MURRAY, K. (1971). Heterogeneity and polymorphism in chicken erythrocyte histone fraction V. *Nature New Biol.*, **229**, 233–238. *(423)*

GREENBLATT, J. and SCHLEIF, R. (1971). Arabinose C protein: regulation of the arabinose operon *in vitro. Nature New Biol.*, **233**, 166–170. *(316)*

GRÉGOIRE, V. (1904). La réduction numérique des chromosomes et les cinèses de maturation. *Cellule*, **21**, 297–314. *(63, 450)*

GRÉGOIRE, V. (1907). La formation des gemini hétérotypiques dans les végétaux. *Cellule*, **24**, 369–420. *(63, 454)*

GRIFFITH, F. (1928). Significance of pneumococcal types. *J. Hyg., Camb.*, **27**, 113–159. *(173, 336, 453)*

GROSS, J. D., GRUNSTEIN, J. and WITKIN, E. M. (1971). Inviability of *recA⁻* derivatives of the DNA polymerase mutant of De Lucia and Cairns. *J. molec. Biol.*, **58**, 631–634. *(356)*

GROSS, S. R. (1962). On the mechanism of complementation at the *leu*-2 locus of *Neurospora. Proc. natn. Acad. Sci. U.S.A.*, **48**, 922–930. *(229)*

GRUN, P. (1958). Plant lampbrush chromosomes. *Expl Cell Res.*, **14**, 619–621. *(381)*

GRUNBERG-MANAGO, M. and OCHOA, S. (1955). Enzymatic synthesis and breakdown of polynucleotides; polynucleotide phosphorylase. *J. Am. chem. Soc.*, **77**, 3165–3166. *(245, 271)*

GUEST, J. R., DRAPEAU, G. R., CARLTON, B. C. and YANOFSKY, C. (1967). The amino acid sequence of the A protein (α subunit) of the tryptophan synthetase of *Escherichia coli*. V. Order of tryptic peptides and the complete amino acid sequence. *J. biol. Chem.*, **242**, 5442–5446.                                 (*230*)

GUEST, J. R. and YANOFSKY, C. (1965). Amino acid replacements associated with reversion and recombination within a coding unit. *J. molec. Biol.*, **12**, 793–804.
(*275, 338*)

GUEST, J. R. and YANOFSKY, C. (1966). Relative orientation of gene, messenger and polypeptide chain. *Nature, Lond.*, **210**, 799–802.                              (*281*)

GUILD, W. R. and ROBISON, M. (1963). Evidence for message reading from a unique strand of pneumococcal DNA. *Proc. natn. Acad. Sci. U.S.A.*, **50**, 106–112.    (*252*)

GURDON, J. B. (1970). Nuclear transplantation and the control of gene activity in animal development. *Proc. R. Soc.* B, **176**, 303–314.                       (*398*)

GURDON, J. B., LANE, C. D., WOODLAND, H. R. and MARBAIX, G. (1971). Use of frog eggs and oocytes for the study of messenger RNA and its translation in living cells. *Nature, Lond.*, **233**, 177–182.                                               (*251*)

GURNEY, T. and FOX, M. S. (1968). Physical and genetic hybrids formed in bacterial transformation. *J. molec. Biol.*, **32**, 83–100.                              (*346*)

GUSSIN, G. N. (1966). Three complementation groups in bacteriophage R17. *J. molec. Biol.*, **21**, 435–453.                                                       (*312*)

GUTZ, H. (1971). Site specific induction of gene conversion in *Schizosaccharomyces pombe*. *Genetics, Princeton*, **69**, 317–337.                          (*362, 363, 370*)

GYURASITS, E. B., and WAKE, R. G. (1973). Bidirectional chromosome replication in *Bacillus subtilis*. *J. molec. Biol.*, **73**, 55–63.                            (*209*)

HÄCKER, V. (1897). Ueber weitere Uebereinstimmungen zwischen den Fortpflanzungs-vorgängen der Tiere und Pflanzen. *Biol. Zbl.*, **17**, 689–705, 721–745. (*67, 447*)

HAGA, T. (1944). Meiosis in *Paris*. I. Mechanism of chiasma formation. *J. Fac. Sci. Hokkaido Univ.*, Ser. V. (Botany), **5**, 121–198.                        (*120–22*)

HALDANE, J. B. S. (1919). The probable errors of calculated linkage values, and the most accurate method of determining gametic from certain zygotic series. *J. Genet.*, **8**, 291–297.                                                        (*20*)

HALDANE, J. B. S. (1931). The cytological basis of genetical interference. *Cytologia*, 3 54–65.                                                            (*112, 115*)

HALDANE, J. B. S. (1941). *New Paths in Genetics*. London, Allen and Unwin, 206 pp.
(*213, 446, 453*)

HALL, B. D. and SPIEGELMAN, S. (1961). Sequence complementarity of T2-DNA and T2-specific RNA. *Proc. natn. Acad. Sci. U.S.A.*, **47**, 137–146.    (*248–49, 309*)

HALLIER, U. W. (1968). Photosynthese-Reaktionen einiger Plastom-Mutanten von *Oenothera*. II. Die Bildung von ATP und NADPH. *Z. PflPhysiol.*, **58**, 289–299.
(*431*)

HAMILTON, L. D. (1968). DNA: models and reality. *Nature, Lond.*, **218**, 633–637.  (*424*)

HÄMMERLING, J. (1953). Nucleo-cytoplasmic relationships in the development of *Acetabularia*. *Int. Rev. Cytol.*, **2**, 475–498.                          (*42–44, 45*)

HÄMMERLING, J. (1963). Nucleo-cytoplasmic interactions in *Acetabularia* and other cells. *A. Rev. Pl. Physiol.*, **14**, 65–92.                            (*42–44, 45*)

HARDY, G. H. (1908). Mendelian proportions in a mixed population. *Science, N.Y.*, **28**, 49–50.                                                     (*13–14, 139, 399*)

HARRIS, H. and SABATH, L. D. (1964). Induced enzyme synthesis in the absence of concomitant ribonucleic acid synthesis. *Nature, Lond.*, **202**, 1078–1080.    (*317*)

HARRISON, B. J. and FINCHAM, J. R. S. (1964). Instability at the *pal* locus in *Antirrhinum majus*. 1. Effects of environment on frequencies of somatic and germinal mutation. *Heredity*, **19**, 237–258. (*433*)

HARTMAN, P. E. (1956). Linked loci in the control of consecutive steps in the primary pathway of histidine synthesis in *Salmonella typhimurium*. *Publs Carnegie Instn*, **612**, 35–61. (*301*)

HARTMAN, P. E., HARTMAN, Z., STAHL, R. C. and AMES, B. N. (1971). Classification and mapping of spontaneous and induced mutations in the histidine operon of *Salmonella*. *Adv. Genet.*, **16**, 1–34. (*308*)

HASTINGS, P. J. and WHITEHOUSE, H. L. K. (1964). A polaron model of genetic recombination by the formation of hybrid deoxyribonucleic acid. *Nature, Lond.*, **201**, 1052–1054. (*343*)

HAUT, W. F. and TAYLOR, J. H. (1967). Studies of bromouracil deoxyriboside substitution in DNA of bean roots (*Vicia faba*). *J. molec. Biol.*, **26**, 389–401. (*414*)

HAWTHORNE, D. C. and MORTIMER, R. K. (1963). Super-suppressors in yeast. *Genetics, Princeton*, **48**, 617–620. (*298*)

HAYASHI, M., HAYASHI, M. N. and SPIEGELMAN, S. (1964). DNA circularity and the mechanism of strand selection in the generation of genetic messages. *Proc. natn. Acad. Sci. U.S.A.*, **51**, 351–359. (*253*)

HAYASHI, M., SPIEGELMAN, S., FRANKLIN, N. C. and LURIA, S. E. (1963). Separation of the RNA message transcribed in response to a specific inducer. *Proc. natn. Acad. Sci. U.S.A.*, **49**, 729–736. (*304*)

HAYES, W. (1966a). Sex factors and viruses. *Proc. R. Soc. B*, **164**, 230–245. (*335–36*)

HAYES, W. (1966b). Some controversial aspects of bacterial sexuality. *Proc. R. Soc. B*, **165**, 1–19. (*335–36*)

HAYES, W. (1968). *The Genetics of Bacteria and their Viruses*. 2nd edn. Oxford, Blackwell, 925 pp. (*335, 337*)

HEDDLE, J. A., WHISSELL, D. and BODYCOTE, D. J. (1969). Changes in chromosome structure induced by radiations: a test of the two chief hypotheses. *Nature, Lond.*, **221**, 1158–1160. (*157*)

HEIDENHAIN, M. (1894). Neue Untersuchungen über die Centralkörper und ihre Beziehungen zum Kern- und Zellenprotoplasma. *Arch. mikrosk. Anat.*, **43**, 423–758. (*59, 453*)

HEITZ, E. (1928). Das Heterochromatin der Moose. I. *Jb. wiss. Bot.*, **69**, 762–818. (*391, 449*)

HENDERSON, S. A. (1964). RNA synthesis during male meiosis and spermiogenesis. *Chromosoma*, **15**, 345–366. (*381, 422*)

HENDERSON, S. A. (1966). Time of chiasma formation in relation to the time of deoxyribonucleic acid synthesis. *Nature, Lond.*, **211**, 1043–1047. (*341*)

HENKING, H. VON. (1891). Untersuchungen über die ersten Entwicklungsvorgänge in den Eiern der Insekten. II. Über Spermatogenese und deren Beziehung zur Eientwicklung bei *Pyrrhocoris apterus L. Z. wiss Zool.*, **51**, 685–736. (*72, 73*)

HENNIG, W. and WALKER, P. M. B. (1970). Variations in the DNA from two rodent families (Cricetidae and Muridae). *Nature, Lond.*, **225**, 915–919. (*407*)

HENNING, U. and YANOFSKY, C. (1962). Amino acid replacements associated with reversion and recombination within the A gene. *Proc. natn. Acad. Sci. U.S.A.*, **48**, 1497–1504. (*275, 338*)

HERREROS, B. and GIANNELLI, F. (1967). Spatial distribution of old and new chromatid sub-units and frequency of chromatid exchanges in induced human lymphocyte endoreduplications. *Nature, Lond.*, **216**, 286–288.          (*192, 374, 384, 414*)

HERRMANN, F. and BAUER-STÄB, G. (1969). Lamellarproteine mutierter Plastiden der Plastommutante albomaculata-1 von *Antirrhinum majus* L. *Flora, Jena*, A, **160**, 391–393.          (*431*)

HERSHEY, A. D. (1947). Spontaneous mutations in bacterial viruses. *Cold Spring Harb. Symp. quant. Biol.*, **11**, 67–77.          (*213, 338*)

HERSHEY, A. D. and CHASE, M. (1952a). Genetic recombination and heterozygosis in bacteriophage. *Cold Spring Harb. Symp. quant. Biol.*, **16**, 471–479.          (*339*)

HERSHEY, A. D. and CHASE, M. (1952b). Independent functions of viral protein and nucleic acid in growth of bacteriophage. *J. gen. Physiol.*, **36**, 39–56. (*176, 185*)

HERSHEY, A. D. and ROTMAN, R. (1948). Linkage among genes controlling inhibition of lysis in a bacterial virus. *Proc. natn. Acad. Sci. U.S.A.*, **34**, 89–96.     (*213–14*)

HERSHEY, A. D. and ROTMAN, R. (1949). Genetic recombination between host-range and plaque-type mutants of bacteriophage in single bacterial cells. *Genetics, Princeton*, **34**, 44–71.          (*339*)

HERTWIG, O. (1875). Beiträge zur Kenntniss der Bildung, Befruchtung und Theilung des thierischen Eies. *Morph. Jb.*, **1**, 347–434.          (*40–41*)

HESS, O. (1970). Independence between modification of genetic position effects and formation of lampbrush loops by the Y chromosome of *Drosophila hydei*. *Molec. gen. Genet.*, **107**, 224–242.          (*398*)

HESS, O. and MEYER, G. F. (1968). Genetic activities of the Y chromosome in *Drosophila* during spermatogenesis. *Adv. Genet.*, **14**, 171–223.          (*381, 401*)

HINDLEY, J. and STAPLES, D. H. (1969). Sequence of a ribosome binding site in bacteriophage Q$\beta$-RNA. *Nature, Lond.*, **224**, 964–967.          (*291, 292*)

HIPPOCRATES (ca. 400 B.C.). *On airs, waters and places*, chap. 14. (English translation by F. ADAMS, London, Baillière, Tindall and Cox, 1939.)          (*1–3, 12, 40*)

HIRSCH, D. (1971). Tryptophan transfer RNA as the UGA suppressor. *J. molec. Biol.*, **58**, 439–458.          (*268, 297, 298*)

HOAGLAND, M. B., ZAMECNIK, P. C. and STEPHENSON, M. L. (1957). Intermediate reactions in protein biosynthesis. *Biochim. biophys. Acta*, **24**, 215–216. (*258, 259*)

HOGNESS, D. S., DOERFLER, W., EGAN, J. B. and BLACK, L. W. (1967). The position and orientation of genes in $\lambda$ and $\lambda$dg DNA. *Cold Spring Harb. Symp. quant. Biol.*, **31**, 129–138.          (*256, 365*)

HOLLEY, R. W., APGAR, J., EVERETT, G. A., MADISON, J. T., MARQUISEE, M., MERRILL. s. H., PENSWICK, J. R. and ZAMIR, A. (1965). Structure of a ribonucleic acid. *Science, N.Y.*, **147**, 1462–1465.          (*262, 263, 285*)

HOLLIDAY, R. (1962). Mutation and replication in *Ustilago maydis*. *Genet. Res.*, **3**, 472–486.          (*358–59*)

HOLLIDAY, R. (1964). A mechanism for gene conversion in fungi. *Genet. Res.*, **5**, 282–304.          (*343, 344, 361, 362, 364*)

HOLLIDAY, R. (1967). Altered recombination frequencies in radiation sensitive strains of *Ustilago*. *Mutation Res.*, **4**, 275–288.          (*356*)

HOLLIDAY, R. (1968). Genetic recombination in fungi. In *Replication and recombination of genetic material*, ed. by W. J. Peacock and R. D. Brock, Australian Acad. Sci., Canberra, pp. 157–174.          (*362*)

HOLLIDAY, R. (1971). Biochemical measure of the time and frequency of radiation-induced allelic recombination in *Ustilago. Nature New Biol.*, **232**, 233–236. *(357)*

HOOD, L. and TALMAGE, D. W. (1970). Mechanism of antibody diversity: germ line basis for variability. *Science, N.Y.*, **168**, 325–334. *(415)*

HOPWOOD, D. A. (1967). Genetic analysis and genome structure in *Streptomyces coelicolor. Bact. Rev.*, **31**, 373–403. *(335)*

HOPWOOD, D. A. and SERMONTI, G. (1962). The genetics of *Streptomyces coelicolor. Adv. Genet.* **11**, 273–342. *(301)*

HOROWITZ, N. H., FLING, M., MACLEOD, H. and SUEOKA, N. (1961). A genetic study of two new structural forms of tyrosinase in *Neurospora. Genetics, Princeton,* **46**, 1015–1024. *(168)*

HOTTA, Y., ITO, M. and STERN, H. (1966). Synthesis of DNA during meiosis. *Proc. natn. Acad. Sci. U.S.A.*, **56**, 1184–1191. *(344)*

HOTTA, Y., PARCHMAN, L. G. and STERN, H. (1968). Protein synthesis during meiosis. *Proc. natn. Acad. Sci. U.S.A.*, **60**, 575–582. *(344)*

HOTTA, Y. and STERN, H. (1971a). Analysis of DNA synthesis during meiotic prophase in *Lilium. J. molec. Biol.*, **55**, 337–355. *(410, 412)*

HOTTA, Y. and STERN, H. (1971b). A DNA-binding protein in meiotic cells of *Lilium. Devl Biol.*, **26**, 87–99. *(345–46)*

HOTTA, Y. and STERN, H. (1971c). Meiotic protein in spermatocytes of mammals. *Nature New Biol.*, **234**, 83–86. *(345)*

HOWARD, A. and PELC, S. R. (1951a). Nuclear incorporation of P$^{32}$ as demonstrated by autoradiographs. *Expl Cell Res.*, **2**, 178–187. *(177, 186)*

HOWARD, A. and PELC, S. R. (1951b). Synthesis of nucleoprotein in bean root cells, *Nature, Lond.*, **167**, 599–601. *(178)*

HOWARD-FLANDERS, P. and BOYCE, R. P. (1966). DNA repair and genetic recombination: studies on mutants of *Escherichia coli* defective in these processes. *Radiat. Res.*, Suppl. **6**, 156–184. *(361)*

HOWARD-FLANDERS, P., BOYCE, R. P. and THERIOT, L. (1966). Three loci in *Escherichia coli* K-12 that control the excision of pyrimidine dimers and certain other mutagen products from DNA. *Genetics, Princeton*, **53**, 1119–1136. *(355)*

HOWELL, S. H. and STERN, H. (1971). The appearance of DNA breakage and repair activities in the synchronous meiotic cycle of *Lilium. J. molec. Biol.*, **55**, 357–378. *(344, 410)*

HUBERMAN, J. A. and RIGGS, A. D. (1968). On the mechanism of DNA replication in mammalian chromosomes. *J. molec. Biol.*, **32**, 327–341. *(206, 414)*

HUNNABLE, E. G. and COX, B. S. (1971). The genetic control of dark recombination in yeast. *Mutation Res.*, **13**, 297–309. *(357)*

HUNT, J. A. and INGRAM, V. M. (1958). Allelomorphism and the chemical differences of the human haemoglobins A, S and C. *Nature, Lond.*, **181**, 1062–1063. *(226)*

HUNT, J. A. and INGRAM, V. M. (1959). A terminal peptide sequence of human haemoglobin? *Nature, Lond.*, **184**, 640–641. *(221)*

HURWITZ, J., FURTH, J. J., ANDERS, M., ORITZ, P. J. and AUGUST, J. T. (1962). The enzymatic incorporation of ribonucleotides into RNA and the role of DNA. *Cold Spring Harb. Symp. quant. Biol.*, **26**, 91–100. *(249)*

ICHIKAWA, T. and SUNDARALINGAM, M. (1972). X-ray diffraction study of a new crystal form of yeast phenylalanine tRNA. *Nature New Biol.*, **236,** 174–175.   (*268*)

IMADA, M., INOUYE, M., EDA, M. and TSUGITA, A. (1970). Frameshift mutation in the lysozyme gene of bacteriophage T4: demonstration of the insertion of four bases and the preferential occurrence of base addition in acridine mutagenesis. *J. molec. Biol.*, **54,** 199–217.   (*242, 276, 277*)

IMAMOTO, F., MORIKAWA, N. and SATO, K. (1965). On the transcription of the tryptophan operon in *Escherichia coli*. III. Multicistronic messenger RNA and polarity for transcription. *J. molec. Biol.*, **13,** 169–182.   (*309*)

IMAMOTO, F. and YANOFSKY, C. (1967). Transcription of the typtophan operon in polarity mutants of *Escherichia coli*. *J. molec. Biol.*, **28,** 1–35.   (*302, 311*)

INGRAM, V. M. (1956). A specific chemical difference between the globins of normal human and sickle-cell anaemia haemoglobin. *Nature, Lond.*, **178,** 792–794.   (*219, 221*)

INGRAM, V. M. (1957). Gene mutations in human haemoglobin: the chemical difference between normal and sickle cell haemoglobin. *Nature, Lond.*, **180,** 326–328.   (*221*)

INGRAM, V. M. and STRETTON, A. O. W. (1961). Human haemoglobin $A_2$: chemistry, genetics and evolution. *Nature, Lond.*, **190,** 1079–1084.   (*322*)

INMAN, R. B. (1966). A denaturation map of the $\lambda$ phage DNA molecule determined by electron microscopy. *J. molec. Biol.*, **18,** 464–476.   (*428*)

INMAN, R. B. and SCHNÖS, M. (1971). Structure of branch points in replicating DNA: presence of single-stranded connections in $\lambda$ DNA branch points. *J. molec. Biol.*, **56,** 319–325.   (*208*)

IPPEN, K., MILLER, J. H., SCAIFE, J. and BECKWITH, J. (1968). New controlling element in the *lac* operon of *E. coli*. *Nature, Lond.*, **217,** 825–827.   (*314*)

ITANO, H. A. and NEEL, J. V. (1950). A new inherited abnormality of human hemoglobin. *Proc. natn. Acad. Sci. U.S.A.*, **36,** 613–617.   (*225*)

ITANO, H. A. and ROBINSON, E. A. (1959). Formation of normal and doubly abnormal haemoglobins by recombination of haemoglobin I with S and C. *Nature, Lond.*, **183,** 1799–1800.   (*226*)

ITANO, H. A. and ROBINSON, E. A. (1960). Genetic control of the α- and β-chains of hemoglobin. *Proc. natn. Acad. Sci. U.S.A.*, **46,** 1492–1501.   (*226*)

ITO, J. and CRAWFORD, I. P. (1965). Regulation of the enzymes of the tryptophan pathway in *Escherichia coli*. *Genetics, Princeton*, **52,** 1303–1316.   (*310*)

ITO, M., HOTTA, Y. and STERN, H. (1967). Studies of meiosis *in vitro*. II. Effect of inhibiting DNA synthesis during meiotic prophase on chromosome structure and behavior. *Devl Biol.*, **16,** 54–77.   (*410, 412*)

IZAWA, M., ALLFREY, B. G. and MIRSKY, A. E. (1963). The relationship between RNA synthesis and loop structure in lampbrush chromosomes. *Proc. natn. Acad. Sci. U.S.A.*, **49,** 544–551.   (*422*)

JACKSON, E. N. and YANOFSKY, C. (1972). Internal promoter of the tryptophan operon of *Escherichia coli* is located in a structural gene. *J. molec. Biol.*, **69,** 307–313.   (*315*)

JACKSON, R. and HUNTER, T. (1970). Role of methionine in the initiation of haemoglobin synthesis. *Nature, Lond.*, **227,** 672–676.   (*293*)

JACKSON, R. C. (1957). New low chromosome number for plants. *Science, N.Y.*, **126,** 1115–1116. (*72*)

JACOB, F. and BRENNER, S. (1963). Sur la régulation de la synthèse du DNA chez les bactéries: l'hypothèse du réplicon. *C.r. hebd. Séanc. Acad. Sci., Paris*, **256,** 298–300. (*204, 206, 318, 452*)

JACOB F., BRENNER S. and CUZIN F. (1964) On the regulation of DNA replication in bacteria. *Cold Spring Harb. Symp. quant. Biol.* **28,** 329–348. (*318*)

JACOB, F. and MONOD, J. (1961). Genetic regulatory mechanisms in the synthesis of proteins. *J. molec. Biol.*, **3,** 318–356. (*302–8, 318, 395*)

JACOB, F. and MONOD, J. (1962). On the regulation of gene activity. *Cold Spring Harb. Symp. quant. Biol.*, **26,** 193–211. (*302–8, 318, 395*)

JACOB, F., PERRIN, D., SANCHEZ, C., and MONOD, J. (1960). L'opéron: groupe de gènes à expression coordonée pare un opérateur. *C.r. hebd. Séanc. Acad. Sci., Paris*, **250,** 1727–1729. (*302, 451*)

JACOB, F., ULLMAN, A. and MONOD, J. (1964). Le promoteur, élément génétique nécessaire à l'expression d'un opéron. *C.r. hebd. Séanc. Acad. Sci., Paris*, **258,** 3125–3128. (*313, 451*)

JACOB, F. and WOLLMAN, E. L. (1958a). Genetic and physical determinations of chromosomal segments in *E. coli. Symp. Soc. exp. Biol.*, **12,** 75–92. (*335, 338*)

JACOB, F. and WOLLMAN, E. L. (1958b). Les épisomes, éléments génétique ajoutés. *C.r. hebd. Séanc. Acad. Sci., Paris*, **247,** 154–156. (*335, 394, 448*)

JACOBS, P. A. and STRONG, J. A. (1959). A case of human intersexuality having a possible XXY sex-determining mechanism. *Nature, Lond.*, **183,** 302–303. (*79*)

JACOBSON, K. B. (1971). Role of an isoacceptor transfer ribonucleic acid as an enzyme inhibitor: effect on tryptophan pyrrolase of *Drosophila. Nature New Biol.*, **231,** 17–19. (*286*)

JACOBY, G. A. and GORINI, L. (1969). A unitary account of the repression mechanism of arginine biosynthesis in *Escherichia coli*. I. The genetic evidence. *J. molec. Biol.*, **39,** 73–87. (*308*)

JAENISCH, R., MAYER, A. and LEVINE, A. (1971). Replicating SV40 molecules containing closed circular template DNA strands. *Nature New Biol.*, **233,** 72–75. (*204, 207–8*)

JANSEN, G. J. O. (1970). Abnormal frequencies of spontaneous mitotic recombination in *uvsB* and *uvsC* mutants of *Aspergillus nidulans. Mutation Res.*, **10,** 33–41. (*357*)

JANSSENS, F. A. (1909). Spermatogénèse dans les Batraciens. V. La théorie de la chiasmatypie. Nouvelles interprétation des cinèses de maturation. *Cellule*, **25,** 387–411. (*94–99, 110, 117, 118, 131, 136, 445*)

JENKINS, M. T. (1924). Heritable characters of maize. XX. Iojap-striping, a chlorophyll defect. *J. Hered.*, **15,** 467–472. (*50*)

JESSOP, A. P. and CATCHESIDE, D. G. (1965). Interallelic recombination at the *his*-1 locus in *Neurospora crassa* and its genetic control. *Heredity*, **20,** 237–256. (*368*)

JINKS, J. L. (1954). Somatic selection in fungi. *Nature, Lond.*, **174,** 409–410. (*55*)

JINKS, J. L. (1964). *Extrachromosomal inheritance.* New Jersey, Prentice-Hall, 177 pp. (*55, 57*)

JOHANNSEN, W. (1903). *Über Erblichkeit in Populationen und in reinen Linien.* Jena, Gustav Fischer. (*32–33*)

JOHANNSEN, W. (1909). *Elemente der exakten Erblichkeitslehre.* Jena, Gustav Fischer. (*23–24, 31–33, 448, 449, 451*)

JOHN, H. A., BIRNSTIEL, M. L. and JONES, K. W. (1969). RNA–DNA hybrids at the cytological level. *Nature, Lond.*, **223**, 582–587.                                    (*407*)

JOHNS, E. W. (1972). Histones, chromatin structure and RNA synthesis. *Nature New Biol.*, **237**, 87–88.                                                          (*423*)

JOHNSON, L. and SÖLL, D. (1970). *In vitro* biosynthesis of pseudouridine at the poly-nucleotide level by an enzyme extract from *Escherichia coli. Proc. natn. Acad. Sci. U.S.A.*, **67**, 943–950.                                                     (*264*)

JOHNSON, R. T. and HARRIS, H. (1969). DNA synthesis and mitosis in fused cells, II. HeLa–chick erythrocyte heterokaryons. *J. Cell Sci.*, **5**, 625–643.          (*399*)

JONES, A. S. and THOMPSON, T. W., (1963). The deoxyribonucleic acids of some Protozoa and a mould. *J. Protozool.*, **10**, 91–93.                                 (*278*)

JONES, G. H. (1971). The analysis of exchanges in tritium-labelled meiotic chromo-somes. II. *Stethophyma grossum. Chromosoma*, **34**, 367–382.          (*136, 342*)

JONES, K. W. (1970). Chromosomal and nuclear location of mouse satellite DNA in individual cells. *Nature, Lond.*, **225**, 912–915.                              (*407*)

JONES, K. W. and ROBERTSON, F. W. (1970). Localisation of reiterated nucleotide sequences in *Drosophila* and mouse by *in situ* hybridisation of complementary RNA. *Chromosoma*, **31**, 331–345.                                              (*408*)

JOSSE, J., KAISER, A. D. and KORNBERG, A. (1961). Enzymatic synthesis of deoxy-ribonucleic acid. VIII. Frequencies of nearest neighbor base sequences in deoxy-ribonucleic acid. *J. biol. Chem.*, **236**, 864–875.                       (*197–200*)

JUDD, B. H., SHEN, N. W. and KAUFMAN, T. C. (1972). The anatomy and function of a segment of the *X* chromosome of *Drosophila melanogaster. Genetics, Princeton* **71**, 139–156.                                                                     (*401*)

KAEMPFER, R. O. R., MESELSON, M. and RASKAS, H. J. (1968). Cyclic dissociation into stable subunits and reformation of ribosomes during bacterial growth. *J. molec. Biol.*, **31**, 277–289.                                                  (*289*)

KAISER, D., TABOR, H. and TABOR, C. W. (1963). Spermine protection of coliphage λ DNA against breakage by hydrodynamic shear. *J. molec. Biol.*, **6**, 141–147.  (*387*)

KAPLAN, S., STRETTON, A. O. W. and BRENNER, S. (1965). *Amber* suppressors: efficiency of chain propagation and suppressor specific amino acids. *J. molec. Biol.*, **14**, 528–533.                                                                         (*297*)

KAPULER. A. M. and BERNSTEIN, H, (1963). A molecular model for an enzyme based on a correlation between the genetic and complementation maps of the locus specifying the enzyme. *J. molec. Biol.*, **6**, 443–451.                          (*229*)

KAUFMANN, B. P. and MCDONALD, M. R. (1957). Organization of the chromosome. *Cold Spring Harb. Symp. quant. Biol.*, **21**, 233–246.                          (*372–73*)

KAYANO, H. (1960a). Chiasma studies in structural hybrids. III. Reductional and equational separation in *Disporum sessile. Cytologia*, **25**, 461–467.     (*121–23*)

KAYANO, H. (1960b). Chiasma studies in structural hybrids. IV. Crossing-over in *Disporum sessile. Cytologia*, **25**, 468–475.                              (*129–30*)

KAYANO, H. and NAKAMURA, K. (1960). Chiasma studies in structural hybrids. V. Heterozygotes for a centric fusion and for a translocation in *Acrida lata. Cytologia*, **25**, 476–480.                                                               (*123*)

KEDES, L. H. and BIRNSTIEL, M. L. (1971). Reiteration and clustering of DNA sequences complementary to histone messenger RNA. *Nature New Biol.*, **230**, 165–9.   (*402*)

KELLENBERGER, E. (1961). Vegetative bacteriophage and the maturation of the virus particles. *Adv. Virus Res.*, **8**, 1–61. (*387*)

KELLENBERGER, G., ZICHICHI, M. L. and WEIGLE, J. (1961). Exchange of DNA in the recombination of bacteriophage λ. *Proc. natn. Acad. Sci. U.S.A.*, **47**, 869–78. (*339*)

KELLEY, W. S. and WHITFIELD, H. J. (1971). Purification of an altered DNA polymerase from an *E. coli* strain with a *pol* mutation. *Nature, Lond.*, **230**, 33–36. (*202*)

KELLY, R. B., ATKINSON, M. R., HUBERMAN, J. A. and KORNBERG, A. (1969). Excision of thymine dimers and other mismatched sequences by DNA polymerase of *Escherichia coli*. *Nature, Lond.*, **224**, 495–501. (*202, 355*)

KELLY, T. J. and SMITH, H. O. (1970). A restriction enzyme from *Hemophilus influenzae*. II. Base sequence of the recognition site. *J. molec. Biol.*, **51**, 393–409. (*368*)

KELNER, A. (1949). Effect of visible light on the recovery of *Streptomyces griseus* conidia from ultraviolet irradiation injury. *Proc. natn. Acad. Sci. U.S.A.*, **35**, 73–79. (*354*)

KEYL, H.-G. (1965a). A demonstrable local and geometric increase in the chromosomal DNA of *Chironomus*. *Experientia*, **21**, 191–193.
(*206, 383–84, 400, 408, 410, 411*)

KEYL, H.-G. (1965b). Duplikationen von Untereinheiten der Chromosomalen DNS während der Evolution von *Chironomus thummi*. *Chromosoma*, **17**, 139–180.
(*206, 383–84, 400, 408, 410, 411*)

KEYL, H.-G. and PELLING, C. (1963). Differentielle DNS-Replikation in den Speicheldrüsen-Chromosomen von *Chironomus thummi*. *Chromosoma*, **14**, 347–59. (*384, 411*)

KHORANA, H. G., BÜCHI, H., GHOSH, H., GUPTA, N., JACOB, T. M., KÖSSEL, H., MORGAN, R., NARANG, S. A., OHTSUKA, E., and WELLS, R. D. (1967). Polynucleotide synthesis and the genetic code. *Cold Spring Harb. Symp. quant. Biol.*, **31**, 39–49. (*273–74*)

KIHLMAN, B. A. (1961). Biochemical aspects of chromosome breakage. *Adv. Genet.*, **10**, 1–59. (*150*)

KIHLMAN, B. A. and HARTLEY, B. (1967). 'Sub-chromatid' exchanges and the 'folded fibre' model of chromosome structure. *Hereditas*, **57**, 289–294. (*382*)

KIHO, Y. and RICH, A. (1965). A polycistronic messenger RNA associated with β-galactosidase induction. *Proc. natn. Acad. Sci. U.S.A.*, **54**, 1751–1758. (*309*)

KING, J. (1968). Assembly of the tail of bacteriophage T4. *J. molec. Biol.*, **32**, 231–262.
(*317, 425–26*)

KING, J. (1971). Bacteriophage T4 tail assembly: four steps in core formation. *J. molec. Biol.*, **58**, 693–709. (*427*)

KING, J. and LAEMMLI, U. K. (1971). Polypeptides of the tail fibres of bacteriophage T4. *J. molec. Biol.*, **62**, 465–477. (*427*)

KING, T. J. and BRIGGS, R. (1955). Changes in the nuclei of differentiating gastrula cells, as demonstrated by nuclear transplantation. *Proc. natn. Acad. Sci. U.S.A.*, **41**, 321–325. (*398*)

KIRK, J. T. O. and TILNEY-BASSETT, R. A. E. (1967). *The plastids: their chemistry, structure, growth and inheritance.* London and San Francisco, W. H. Freeman, 608 pp.
(*430*)

KIT, S. (1961). Equilibrium sedimentations in density gradients of DNA preparations from animal tissues. *J. molec. Biol.*, **3**, 711–716. (*406*)

KITANI, Y. and OLIVE, L. S. (1967). Genetics of *Sordaria fimicola*. VI. Gene conversion at the *g* locus in mutant × wild type crosses. *Genetics, Princeton*, **57**, 767–782. (*370*)

KITANI, Y., OLIVE, L. S. and EL-ANI, A. S. (1962). Genetics of *Sordaria fimicola*. V. Aberrant segregation at the *g* locus. *Am. J. Bot.*, **49**, 697–706.
(*329–30, 341, 359, 364*)

KLEIN, H. A. and CAPECCHI, M. R. (1971). Polypeptide chain termination. Purification of the release factors, $R_1$ and $R_2$, from *Escherichia coli. J. biol. Chem.*, **246**, 1055–1061.   *(299)*

KLEINSCHMIDT, A. K., LANG, D., JACHERTS, D. and ZAHN, R. K. (1962). Darstellung und Langenmessungen des gesamten Desoxyribonucleinsäure-inhaltes von T2-Bakteriophagen. *Biochim. biophys. Acta*, **61**, 857–864.   *(175, 203, 373)*

KNAPP, E. and SCHREIBER, H. (1939). Quantitative Analyse der mutationsauslösenden Wirkung monochromatischen UV-Lichtes in Spermatozoiden von *Sphaerocarpus. Naturwissenschaften*, **27**, 304; *Proc. 7th int. genet. Congr.*, *Edinburgh*, pp. 175–176.   *(172, 185, 385)*

KNIGHT, T. A. (1799). An account of some experiments on the fecundation of vegetables. *Phil. Trans. R. Soc.*, **1**, 195–204.   *(2–6, 9)*

KNIPPERS, R. (1970). DNA polymerase II. *Nature, Lond.*, **228**, 1050–1053.   *(202)*

KÖHLER, H., SHIMIZU, A., PAUL, C., MOORE, V. and PUTNAM, F. W. (1970). Three variable-gene pools common to IgM, IgG and IgA immunoglobulins. *Nature, Lond.*, **227**, 1318–1320.   *(418, 419)*

KOLLER, P. C. (1935). The internal mechanics of the chromosomes, IV. Pairing and coiling in salivary gland nuclei of *Drosophila. Proc. R. Soc. B.*, **118**, 371–397.   *(451)*

KOLLER, P. C. (1938). The genetical and mechanical properties of the sex-chromosomes. IV. The golden hamster. *J. Genet.*, **36**, 177–195.   *(122)*

KÖLREUTER, J. G. (1763). Vorläufige Nachricht von einigen das Geschlecht der Pflanzen betreffenden Versuchen und Beobachten, Forsetzungen 1. (Ostwald's Klassiker der exacten Wissenschaften. 41. Leipzig, 1893.)   *(3, 6, 35)*

KORNBERG, T. and GEFTER, M. L. (1971). Purification and DNA synthesis in cell-free extracts: properties of DNA polymerase II. *Proc. natn. Acad. Sci. U.S.A.*, **68**, 761–764.   *(202)*

KOZINSKI, A. W., KOZINSKI, P. B. and JAMES, R. (1967). Molecular recombination in T4 bacteriophage deoxyribonucleic acid. I. Tertiary structure of early replicative and recombining deoxyribonucleic acid. *J. Virol.*, **1**, 758–770.   *(347)*

KUBINSKI, H., OPARA-KUBINSKA, Z. and SZYBALSKI, W. (1966). Patterns of interaction between polyribonucleotides and individual DNA strands derived from several vertebrates, bacteria and bacteriophages. *J. molec. Biol.*, **20**, 313–329.   *(255)*

KÜHN, A. (1937). Entwicklungsphysiologisch-genetische Ergebnisse an *Ephestia kuhniella. Z. VererbLehre*, **73**, 419–455.   *(48–49)*

KUHN, W. (1957). Zeitbedarf der Längsteilung von miteinander verzwirnten Fadenmolekülen. *Experientia*, **13**, 301–307.   *(203)*

KUNKEL, H. G., NATVIG, J. B. and JOSLIN, F. G. (1969). A 'Lepore' type of hybrid $\gamma$ globulin. *Proc. natn. Acad. Sci. U.S.A.*, **62**, 144–149.   *(325, 416)*

KÜNTZEL, H. and SCHÄFER, K. P. (1971). Mitochondrial RNA polymerase from *Neurospora crassa. Nature New Biol.*, **231**, 265–269.   *(429)*

KUNZ, W. (1967). Lampenbürstenchromosomen und multiple nukleolen bei Orthopteren. *Chromosoma*, **21**, 446–462.   *(411)*

KUSHNER, S. R., NAGAISHI, H. and CLARK, A. J. (1972). Indirect suppression of *recB* and *recC* mutations by exonuclease I deficiency. *Proc. natn. Acad. Sci. U.S.A.*, **69**, 1366–1370.   *(351)*

KUSHNER, S. R., NAGAISHI, H., TEMPLIN, A. and CLARK, A. J. (1971). Genetic recombination in *Escherichia coli*: the role of exonuclease I. *Proc. natn. Acad. Sci. U.S.A.*, **68**, 824–827.   *(351)*

KUWANO, M., SCHLESSINGER, D. and APIRION, D. (1970). Ribonuclease V of *Escherichia coli* IV. Exonucleolytic cleavage in the 5′ to 3′ direction with production of 5′-nucleotide monophosphates. *J. molec. Biol.*, **51**, 75–82. (*312*)

KUWANO, M., SCHLESSINGER, D. and MORSE, D. E. (1971). Loss of dispensable endonuclease activity in relief of polarity by *suA*. *Nature New Biol.*, **231**, 214–217. (*312*)

LABIE, D., SCHROEDER, W. A. and HUISMAN, T. H. J. (1966). The amino acid sequence of the $\delta$-$\beta$ chains of hemoglobin Lepore Augusta = Lepore Washington. *Biochim. biophys. Acta*, **127**, 428–437. (*323*)

LA COUR, L. F. and PELC, S. R. (1958). Effect of colchicine on the utilization of labelled thymidine during chromosomal reproduction. *Nature, Lond.*, **182**, 506–508. (*373*)

LA COUR, L. F. and WELLS, B. (1967). The loops and ultrastructure of the nucleolus of *Ipheion uniflorum. Z. Zellforsch. mikrosk. Anat.*, **82**, 25–45. (*405*)

LAMARCK, J. B. (1809). *Philosophie Zoologique*. Paris. ('Translation, entitled 'Zoological Philosophy,' by H. ELLIOT, London, Macmillan, 1914, 410 pp.) (*40*)

LANDSTEINER, K. and WIENER, A. S. (1941). Studies on an agglutinogen (Rh) in human blood reacting with anti-rhesus sera and with human isoantibodies. *J. exp. Med.*, **74**, 309–320. (*163*)

LARK, K. G. (1972). Evidence for the direct involvement of RNA in the initiation of DNA replication in *Escherichia coli* 15T⁻. *J. molec. Biol.*, **64**, 47–60. (*202*)

LATARJET, R., MUEL, B., HAIG, D. A., CLARKE, M. C. and ALPER, T. (1970). Inactivation of the scrapie agent by near monochromatic ultraviolet light. *Nature, Lond.*, **227**, 1341–1343. (*431*)

LAWRENCE, W. J. C. (1950). Genetic control of biochemical synthesis as exemplified by plant genetics—flower colours. *Biochem. Soc. Symp.*, **4**, 3–9. (*165*)

LEBLON, G. (1972*a*). Mechanism of gene conversion in *Ascobolus immersus*. I. Existence of a correlation between the origin of mutants induced by different mutagens and their conversion spectrum. *Molec. gen. Genet.*, **115**, 36–48. (*362*)

LEBLON, G. (1972*b*). Mechanism of gene conversion in *Ascobolus immersus*. II. The relationships between the genetic alterations in $b_1$ or $b_2$ mutants and their conversion spectrum. *Molec. gen. Genet.*, **116**, 322–335. (*363*)

LEDERBERG, J. (1955). Recombination mechanisms in bacteria. *J. cell. comp. Physiol.*, **45**, suppl. 2, 75–107. (*340*)

LEDERBERG, J., LEDERBERG, E. M., ZINDER, N. D. and LIVELY, E. R. (1952). Recombination analysis of bacterial heredity. *Cold Spring Harb. Symp. quant. Biol.*, **16**, 413–443. (*336, 453*)

LEDERBERG, J. and TATUM, E. L. (1946). Gene recombination in *Escherichia coli. Nature, Lond.*, **158**, 558. (*168, 334*)

LEHMAN, I. R., BESSMAN, M. J., SIMMS, E. S. and KORNBERG, A. (1958). Enzymatic synthesis of deoxyribonucleic acid. I. Preparation of substrates and partial purification of an enzyme from *Escherichia coli. J. biol. Chem.*, **233**, 163–170. (*194*)

LEHMAN, I. R., ZIMMERMAN, S. B., ADLER, J., BESSMAN, M. J., SIMMS, E. S. and KORNBERG, A. (1958). Enzymatic synthesis of deoxyribonucleic acid. V. Chemical composition of enzymatically synthesized deoxyribonucleic acid. *Proc. natn. Acad. Sci. U.S.A.*, **44**, 1191–1196. (*194*)

LEHMANN, A. R. (1972). Postreplication repair of DNA in ultraviolet-irradiated mammalian cells. *J. molec. Biol.*, **66**, 319–337. (*358*)

LEIGHTON, T. J., DILL, B. C., STOCK, J. J. and PHILLIPS, P. (1971). Absence of histones from the chromosomal proteins of fungi. *Proc. natn. Acad. Sci. U.S.A.*, **68,** 677–680.
(*386*)

LEJEUNE, J., GAUTIER, M. and TURPIN, R. (1959). Étude des chromosomes somatiques de neuf enfants mongoliens. *C.r. hebd. Séanc. Acad. Sci., Paris*, **248,** 1721–1722.
(*71*)

LENGYEL, P., SPEYER, J. F. and OCHOA, S. (1961). Synthetic polynucleotides and the amino acid code. *Proc. natn. Acad. Sci. U.S.A.*, **47,** 1936–1942.      (*272*)

LEVINTHAL, C. (1954). Recombination in phage T2: its relationship to heterozygosis and growth. *Genetics, Princeton*, **39,** 169–184.      (*339, 346*)

LEVINTHAL, C. and CRANE, H. R. (1956). On the unwinding of DNA. *Proc. natn. Acad. Sci. U.S.A.*, **42,** 436–438.      (*203*)

LEVITT, M. (1969). Detailed molecular model for transfer ribonucleic acid. *Nature, Lond.*, **224,** 759–763.      (*268, 269*)

LEWIS, E. B. (1941). Another case of unequal crossing-over in *Drosophila melanogaster*. *Proc. natn. Acad. Sci. U.S.A.*, **27,** 31–35.      (*161–63*)

LEWIS, E. B. (1945). The relation of repeats to position effect in *Drosophila melanogaster*. *Genetics, Princeton*, **30,** 137–166.      (*162, 213*)

LEWIS, E. B. (1952). The pseudoallelism of white and apricot in *Drosophila melanogaster*. *Proc. natn. Acad. Sci. U.S.A.*, **38,** 953–961.      (*213*)

LIMA-DE-FARIA, A. (1959). Differential uptake of tritiated thymidine into hetero- and euchromatin in *Melanoplus* and *Secale*. *J. biophys. biochem. Cytol.*, **6,** 457–466.
(*392*)

LIMA-DE-FARIA, A. and JAWORSKA, H. (1968). Late DNA synthesis in heterochromatin. *Nature, Lond.*, **217,** 138–142.      (*392*)

LINDEGREN, C. C. (1932). The genetics of *Neurospora*—II. Segregation of the sex factors in asci of *N. crassa, N. sitophila*, and *N. tetrasperma*. *Bull. Torrey bot. Club*, **59,** 119–138.      (*104–6*)

LINDEGREN, C. C. (1933). The genetics of *Neurospora*—III. Pure bred stocks and crossing-over in *N. crassa*. *Bull. Torrey bot. Club*, **60,** 133–154.      (*103–9, 328*)

LINDEGREN, C. C. (1953). Gene conversion in *Saccharomyces*. *J. Genet.*, **51,** 625–637.
(*328, 447*)

LING, V., JERGIL, B. and DIXON, G. H. (1971). The biosynthesis of protamine in trout testis. III. Characterization of protamine components and their synthesis during testis development. *J. biol. Chem.*, **246,** 1168–1176.      (*387*)

LISSOUBA, P., MOUSSEAU, J., RIZET, G. and ROSSIGNOL, J. L. (1962). Fine structure of genes in the ascomycete *Ascobolus immersus*. *Adv. Genet.*, **11,** 343–380.      (*332*)

LISSOUBA, P. and RIZET, G. (1960). Sur l'existence d'une génétique polarisée ne subissant que des échanges non réciproques. *C.r. hebd. Séanc. Acad. Sci., Paris*, **250,** 3408–3410.      (*331–33, 361*)

LITMAN, R. M. and PARDEE, A. B. (1956). Production of bacteriophage mutants by a disturbance of deoxyribonucleic acid metabolism. *Nature, Lond.*, **178,** 529–531.
(*235*)

LIVINGSTONE, F. B. (1971). Malaria and human polymorphisms. *A. Rev. Genet.*, **5,** 33–64.      (*14*)

LODISH, H. F. (1970). Secondary structure of bacteriophage f2 ribonucleic acid and the initiation of *in vitro* protein biosynthesis. *J. molec. Biol.*, **50,** 689–702.   (*291*)

LODISH, H. F. and ZINDER, N. D. (1966). Mutants of the bacteriophage f2. VIII. Control mechanisms for phage-specific syntheses. *J. molec. Biol.*, **19**, 333–348. *(312)*

LONGLEY, A. E. (1945). Abnormal segregation during megasporogenesis in maize. *Genetics, Princeton*, **30**, 100–113. *(392)*

LONGUET-HIGGINS, H. C. and ZIMM, B. H. (1960). Calculation of the rate of uncoiling of the DNA molecule. *J. molec. Biol.*, **2**, 1–4. *(203)*

LOPER, J. C. (1961). Enzyme complementation in mixed extracts of mutants from the *Salmonella* histidine B locus. *Proc. natn. Acad. Sci. U.S.A.*, **47**, 1440–1450. *(227)*

LU, P. and RICH, A. (1971). The nature of the polypeptide chain termination signal. *J. molec. Biol.*, **58**, 513–531. *(298)*

LWOFF, A. and TOURNIER, P. (1966). The classification of viruses. *A. Rev. Microbiol.*, **20**, 45–74. *(444)*

LYON, M. F. (1961). Gene action in the X-chromosome of the mouse (*Mus musculus* L.) *Nature, Lond.*, **190**, 372–373. *(419–20)*

LYON, M. F. (1962). Sex chromatin and gene action in the mammalian X-chromosome. *Am. J. hum. Genet.*, **14**, 135–148. *(419–20)*

McCLINTOCK, B. (1931). Cytological observations of deficiencies involving known genes, translocations and an inversion in *Zea mays*. *Res. Bull. Mo. agric. Exp. Stn*, **163**, 1–30. *(125)*

McCLINTOCK, B. (1933). The association of non-homologous parts of chromosomes in the mid-prophase of meiosis in *Zea mays*. *Z. Zellforsch. mikrosk. Anat.*, **19**, 191–237. *(125)*

McCLINTOCK, B. (1934). The relation of a particular chromosomal element to the development of the nucleoli in *Zea mays*. *Z. Zellforsch. mikrosk. Anat.*, **21**, 294–328. *(405)*

McCLINTOCK, B. (1944). The relation of homozygous deficiencies to mutations and allelic series in maize. *Genetics, Princeton*, **29**, 478–502. *(162–63)*

McCLINTOCK, B. (1950). The origin and behavior of mutable loci in maize. *Proc. natn. Acad. Sci. U.S.A.*, **36**, 344–355. *(393–94, 421)*

McCLINTOCK, B. (1952). Chromosome organization and genic expression. *Cold Spring Harb. Symp. quant. Biol.*, **16**, 13–47. *(394)*

McCLINTOCK, B. (1957). Controlling elements and the gene. *Cold Spring Harb. Symp. quant. Biol.*, **21**, 197–216. *(394)*

McCLINTOCK, B. (1965). The control of gene action in maize. *Brookhaven Symp. Biol.*, **18**, 162–184. *(394)*

McCLINTOCK, B. (1968). Genetic systems regulating gene expression during development. *Symp. Soc. dev. Biol.*, **26**, 84–112. *(394)*

McCLUNG, C. E. (1900). The spermatocyte divisions of the Acrididae. *Kans. Univ. Q.*, **9**, 73–100. *(446)*

McCLUNG, C. E. (1902). The accessory chromosome—sex determinant? *Biol. Bull. mar. biol. Lab., Woods Hole*, **3**, 43–84. *(74)*

MACGREGOR, H. C. (1968). Nucleolar DNA in oocytes of *Xenopus laevis*. *J. Cell Sci.*, **3**, 437–444. *(411)*

MACGREGOR, H. C. and CALLAN, H. G. (1962). The action of enzymes on lampbrush chromosomes. *Q. Jl microsc. Sci.*, **103**, 173–203. *(380)*

MACKENDRICK, M. E. and PONTECORVO, G. (1952). Crossing-over between alleles at the *w* locus in *Drosophila melanogaster*. *Experientia*, **8**, 390–391.                    (*213*)

McLAUGHLIN, C., MAGEE, P. T. and HARTWELL, L. H. (1969). Role of isoleucyl-transfer ribonucleic acid synthetase in ribonucleic acid synthesis and enzyme repression in yeast. *J. Bact.*, **100**, 579–584.                    (*398*)

McLEISH, J. and SUNDERLAND, N. (1961). Measurements of deoxyribosenucleic acid (DNA) in higher plants by Feulgen photometry and chemical methods. *Expl Cell Res.*, **24**, 527–540.                    (*382*)

MADISON, J. T., EVERETT, G. A. and KUNG, H. (1966). Nucleotide sequence of a yeast tyrosine transfer RNA. *Science, N.Y.*, **153**, 531–534.                    (*263*)

MAEDA, T. (1930). On the configurations of gemini in the pollen mother cells of *Vicia faba* L. *Mem. Coll. Sci. Kyoto Univ. B.*, **5**, 125–137.                    (*115*)

MAGNI, G. E. (1963). The origin of spontaneous mutations during meiosis. *Proc. natn. Acad. Sci. U.S.A.*, **50**, 975–980.                    (*366*)

MAGNI, G. E. (1964). Origin and nature of spontaneous mutations in meiotic organisms. *J. cell. comp. Physiol.*, **64**, suppl. 1, 165–172.                    (*366*)

MAGNI, G. E. and BORSTEL, R. C. VON (1962). Different rates of spontaneous mutation during mitosis and meiosis in yeast. *Genetics, Princeton*, **47**, 1097–1108.        (*366*)

MAGNI, G. E. and PUGLISI, P. P. (1967). Mutagenesis of super-suppressors in yeast. *Cold Spring Harb. Symp. quant. Biol.*, **31**, 699–704.                    (*298*)

MAGUIRE, M. (1966). Double-strandedness of meiotic prophase chromatids to light microscope optics and its relationship to genetic recombination. *Proc. natn. Acad. Sci. U.S.A.*, **55**, 44–50.                    (*372*)

MAITRA, U. and HURWITZ, J. (1965). The role of DNA in RNA synthesis, IX. Nucleoside triphosphate termini in RNA polymerase products. *Proc. natn. Acad. Sci. U.S.A.*, **54**, 815–822.                    (*254, 255*)

MANGIAROTTI, G. and SCHLESSINGER, D. (1966). Polyribosome metabolism in *Escherichia coli*. I. Extraction of polyribosomes and ribosomal subunits from fragile, growing *Escherichia coli*. *J. molec. Biol.*, **20**, 123–143.                    (*289*)

MANTON, I. and SLEDGE, W. A. (1955). Observations on the cytology and taxonomy of the pteridophyte flora of Ceylon. *Phil. Trans. R. Soc. B.*, **238**, 127–185.        (*72*)

MARCKER, K. A. (1965). The formation of N-formyl-methionyl-sRNA. *J. molec. Biol.*, **14**, 63–70.                    (*287*)

MARCKER, K. A. and SANGER, F. (1964). *N*-Formyl-methionyl-S-RNA. *J. molec. Biol.*, **8**, 835–840.                    (*287*)

MARMUR, J. and LANE, D. (1960). Strand separation and specific recombination in deoxyribonucleic acids: biological studies. *Proc. natn. Acad. Sci. U.S.A.*, **46**, 453–461.                    (*248, 343*)

MARTIN, P. G. and SHANKS, R. (1966). Does *Vicia faba* have multistranded chromosomes? *Nature, Lond.*, **211**, 650–651.                    (*382*)

MARTIN, R. G. (1964). The one operon—one messenger theory of transcription. *Cold Spring Harb. Symp. quant. Biol.*, **28**, 357–361.                    (*309*)

MARTIN, R. G. (1967). Frameshift mutants in the histidine operon of *Salmonella typhimurium*. *J. molec. Biol.*, **26**, 311–328.                    (*310*)

MARTIN, R. G., SILBERT, D. F., SMITH, D. W. E. and WHITFIELD, H. J. (1966). Polarity in the histidine operon. *J. molec. Biol.*, **21**, 357–369.                    (*309–10*)

MASTERS, M. and BRODA, P. (1971). Evidence for the bidirectional replication of the *Escherichia coli* chromosome. *Nature New Biol.*, **232**, 137–140.                    (*207*)

MATHER, K. (1933a). The relation between chiasmata and crossing-over in diploid and triploid *Drosophila melanogaster*. *J. Genet.*, **27**, 243–259.     *(113–15, 449)*

MATHER, K. (1933b). Interlocking as a demonstration of the occurrence of genetical crossing-over during chiasma formation. *Am. Nat.*, **67**, 476–479.     *(129)*

MATHER, K. (1935). Meiosis in *Lilium*. *Cytologia*, **6**, 354–380.     *(129)*

MATHER, K. (1944). The genetical activity of heterochromatin. *Proc. R. Soc. B*, **132**, 308–332.     *(392)*

MATHER, K. (1949). Chapter 4 in DARLINGTON, C. D. and MATHER, K. *The Elements of Genetics*. London, Allen and Unwin.     *(39)*

MATHER, K. and JINKS, J. L. (1971). *Biometrical Genetics*. 2nd edn. London, Chapman and Hall, 382 pp.     *(37)*

MAZIA, D. (1955). The organization of the mitotic apparatus. *Symp. Soc. exp. Biol.*, **9**, 335–357.     *(62)*

MEEZAN, E. and WOOD, W. B. (1971). The sequence of gene product interaction in bacteriophage T4 tail core assembly. *J. molec. Biol.*, **58**, 685–692.     *(427)*

MENDEL, G. (1866). Versuche über Pflanzenhybriden. *Verh. naturf. Ver. Brünn*, **4**, 3–44. (Reprinted in *Flora, Jena*, 1901, **89**, 364–403. English translations entitled 'Experiments in plant-hybridisation' in *Jl R. hort. Soc.*, 1901, **26**, 1–32 and in Bateson (1909), pp. 317–361).     *(2–23, 27–29, 31, 44, 80, 84, 448, 452)*

MESELSON, M. (1964). On the mechanism of genetic recombination between DNA molecules. *J. molec. Biol.*, **9**, 734–745.     *(339, 346)*

MESELSON, M. (1965). The duplication and recombination of genes. *Proc. 16th int. Congr. Zool. 1963*. In *Ideas in Modern Biology*, pp 3–16, ed. J. A. Moore, The Natural History Press, New York.     *(343)*

MESELSON, M. and STAHL, F. W. (1958). The replication of DNA in *Escherichia coli*. *Proc. natn. Acad. Sci. U.S.A.*, **44**, 671–682.     *(192–94)*

MESELSON, M. and WEIGLE, J. J. (1961). Chromosome breakage accompanying genetic recombination in bacteriophage. *Proc. natn. Acad. Sci. U.S.A.*, **47**, 857–868.     *(339)*

MEYER, G. F. (1961). Fine structure of spermatocyte nuclei of *Drosophila melanogaster*. *Proc. 2nd Eur. Reg. Conf. Electron Microsc.*, pp. 951–954. Delft.     *(136)*

MEYER, G. F. (1964). A possible correlation between the submicroscopic structure of meiotic chromosomes and crossing over. *Proc. 3rd Eur. Reg. Conf. Electron Microsc.*, pp. 461–462. Prague.     *(136)*

MICHAELIS, P. (1954). Cytoplasmic inheritance in *Epilobium* and its theoretical significance. *Adv. Genet.*, **6**, 287–401.     *(54–55)*

MICHAELIS, P. (1967). The segregation of plastids as an example of plasmon analysis. *Nucleus, Calcutta*, **10**, 111–127.     *(431)*

MICHIE, D. and WALLACE, M. E. (1953). Affinity: a new genetic phenomenon in the house mouse. *Nature, Lond.*, **171**, 26–28.     *(392)*

MIESCHER, F. (1871). Ueber die chemische Zusammensetzung der Eiterzellen. In HOPPE-SEYLER, F., *Medicinisch-chemische Untersuchungen*, pp. 441–460, Berlin, August Hirschwald.     *(172)*

MIESCHER, F. (1897). In VOGEL F. C. W., *Die histochemischen und physiologischen Arbeiten von Friedrich Miescher*. Leipzig.     *(172)*

MILLER, O. L. (1965). Fine structure of lampbrush chromosomes. *Natn. Cancer Inst. Monogr.*, **18**, 79–99.     *(373, 384)*

MILLER, O. L. (1966). Structure and composition of peripheral nucleoli of salamander oocytes. *Natn. Cancer Inst. Monogr.*, **23**, 53–66.     *(411)*

MILLER, O. L. and BEATTY, B. R. (1969*a*). Visualization of nucleolar genes. *Science, N.Y.*, **164**, 955–957.                                                    (*403, 404, 411*)

MILLER, O. L. and BEATTY, B. R. (1969*b*). Portrait of a gene. *J. cell. Physiol.*, **74**, suppl. 1, 225–234.                                                              (*402, 403, 411*)

MILLER, O. L., HAMKALO, B. A. and THOMAS, C. A. (1970). Visualization of bacterial genes in action. *Science, N.Y.*, **169**, 392–395.                        (*312*)

MILSTEIN, C. and MUNRO, A. J. (1970). The genetic basis of antibody specificity. *A. Rev. Microbiol.*, **24**, 335–358.                                               (*415*)

MILSTEIN, C. and PINK, J. R. L. (1970). Structure and evolution of immunoglobulins. *Progr. Biophys. molec. Biol.*, **21**, 209–263.                  (*415, 417*)

MIN JOU, W., HAEGEMAN, G., YSEBAERT, M. and FIERS, W. (1972). Nucleotide sequence of the gene coding for the bacteriophage MS2 coat protein. *Nature, Lond.*, **237**, 82–88.                                                                                    (*279, 312–13*)

MINSON, A. C. and CREASER, E. H. (1969). Purification of a trifunctional enzyme, catalysing three steps of the histidine pathway, from *Neurospora crassa*. *Biochem. J.*, **114**, 49–56.                                                                         (*320*)

MIRSKY, A. E. and RIS, H. (1949). Variable and constant components of chromosomes. *Nature, Lond.*, **163**, 666–667.                                               (*177*)

MIRSKY, A. E. and RIS, H. (1951). The desoxyribonucleic acid content of animal cells and its evolutionary significance. *J. gen. Physiol.*, **34**, 451–462.    (*382*)

MITCHELL, H. K. and LEIN, J. (1948). A *Neurospora* mutant deficient in the enzymatic synthesis of tryptophan. *J. biol. Chem.*, **175**, 481–482.         (*168*)

MITCHELL, M. B. (1955). Aberrant recombination of pyridoxine mutants of *Neurospora*. *Proc. natn. Acad. Sci. U.S.A.*, **41**, 215–220.                     (*328*)

MITCHELL, M. B. (1956). A consideration of aberrant recombination in *Neurospora*. *C.r. Trav. Lab. Carlsberg, Ser. Physiol.*, **26**, 285–298.           (*333*)

MIURA, A. and TOMIZAWA, J. (1968). Studies on radiation-sensitive mutants of *E. coli*, III. Participation of the Rec system in induction of mutation by ultraviolet irradiation. *Molec. gen. Genet.*, **103**, 1–10.                                     (*356*)

MIZUTANI, S., BOETTIGER, D. and TEMIN, H. M. (1970). A DNA-dependent DNA polymerase and a DNA endonuclease in virions of Rous sarcoma virus. *Nature, Lond.*, **228**, 424–427.                                                                  (*246*)

MONESI, V. (1962). Autoradiographic study of DNA synthesis and the cell cycle in spermatogonia and spermatocytes of mouse testis using tritiated thymidine. *J. Cell Biol.*, **14**, 1–18.                                                                     (*178*)

MONOD, J. and JACOB, F. (1962). General conclusions: teleonomic mechanisms in cellular metabolism, growth, and differentiation. *Cold Spring Harb. Symp. quant. Biol.*, **26**, 389–401.                                                            (*396–97, 431*)

MONRO, R. E. (1967). Catalysis of peptide bond formation by 50 S ribosomal subunits from *Escherichia coli*. *J. molec. Biol.*, **26**, 147–151.               (*259*)

MONTGOMERY, T. H. (1901). A study of the chromosomes of the germ cells of Metazoa. *Trans. Am. phil. Soc.*, **20**, 154–236.                                     (*69–71*)

MONTGOMERY, T. H. (1906). The terminology of aberrant chromosomes and their behavior in certain Hemiptera. *Science, N.Y.*, **23**, 36–38.              (*445*)

MORA, G., DONNER, D., THAMMANA, P., LUTTER, L., KURLAND, C. G. and CRAVEN, G. R. (1971). Purification and characterization of 50 S ribosomal proteins of *Escherichia coli*. *Molec. gen. Genet.*, **112**, 229–242.                              (*247*)

MORGAN, A. R., WELLS, R. D. and KHORANA, H. G. (1966). Studies on polynucleotides, LIX. Further codon assignments from amino acid incorporations directed by ribopolynucleotides containing repeating trinucleotide sequences. *Proc. natn. Acad. Sci. U.S.A.*, **56,** 1899–1906. *(273, 296)*

MORGAN, T. H. (1910). Sex limited inheritance in *Drosophila. Science, N.Y.*, **32,** 120–122. *(22, 75, 76, 140)*

MORGAN, T. H. (1911a). The application of the conception of pure lines to sex-limited inheritance and to sexual dimorphism. *Am. Nat.*, **45,** 65–78. *(74–76, 81, 93, 98)*

MORGAN, T. H. (1911b). An attempt to analyze the constitution of the chromosomes on the basis of sex-limited inheritance in *Drosophila. J. exp. Zool*, **11,** 365–414. *(81–86, 111, 138)*

MORGAN, T. H. (1912). Further experiments with mutations in eye-color of *Drosophila:* the loss of the orange factor. *J. Acad. nat. Sci. Philad.*, **15,** 321–346. *(138, 140)*

MORGAN, T. H. (1919). *The Physical Basis of Heredity.* Philadelphia, J. B. Lippincott, 305 pp. *(22, 84, 85, 111, 116)*

MORGAN, T. H. (1926). *The Theory of the Gene.* New Haven, Yale Univ. Press, 343 pp. *(109)*

MORGAN, T. H. and CATTELL, E. (1912). Data for the study of sex-linked inheritance in *Drosophila. J. exp. Zool.*, **13,** 79–101. *(84, 447)*

MORGAN, T. H., STURTEVANT, A. H., MULLER, H. J. and BRIDGES, C. B. (1915). *The Mechanism of Mendelian Heredity.* New York, Henry Holt, 262 pp.
*(115–16, 134, 139)*

MORIKAWA, N. and IMAMOTO, F. (1969). On the degradation of messenger RNA for the tryptophan operon in *Escherichia coli. Nature, Lond.*, **223,** 37–40. *(311)*

MORSE, D. E. and PRIMAKOFF, P. (1970). Relief of polarity in *E. coli* by 'suA'. *Nature, Lond.*, **226,** 28–31. *(311, 312)*

MORSE, D. E. and YANOFSKY, C. (1969). Polarity and the degradation of mRNA. *Nature, Lond.*, **224,** 329–331. *(311)*

MOSES, M. J. (1956). Chromosomal structures in crayfish spermatocytes. *J. biophys. biochem. Cytol.*, **2,** 215–217. *(134)*

MOSES, M. J. (1958). The relation between the axial complex of meiotic prophase chromosomes and chromosome pairing in a salamander (*Plethodon cinereus*). *J. biophys. biochem. Cytol.*, **4,** 633–638. *(134, 452)*

MOSES, M. J. and COLEMAN, J. R. (1964). Structural patterns and the functional organization of chromosomes. In *The Role of Chromosomes in Development* edited by M. LOCKE, New York, Academic Press, pp. 11–49. *(134–35)*

MOTULSKY, A. G. (1965). Current concepts of the genetics of the thalassemias. *Cold Spring Harb. Symp. quant. Biol.*, **29,** 399–413. *(324)*

MOUNOLOU, J. C., JAKOB, H. and SLONIMSKI, P. P. (1966). Mitochondrial DNA from yeast 'petite' mutants: specific changes of buoyant density corresponding to different cytoplasmic mutations. *Biochem. biophys. Res. Commun.*, **24,** 218–224.
*(430)*

MOUSSET, S. and THOMAS, R. (1969). Ter, a function which generates the ends of the mature λ chromosome. *Nature, Lond.*, **221,** 242–244. *(412)*

MUELLER, G. C. and KAJIWARA, K. (1966). Early- and late-replicating deoxyribonucleic acid complexes in HeLa nuclei. *Biochim. biophys. Acta*, **114,** 108–15. *(206)*

MULLER, H. J. (1916). The mechanism of crossing-over. *Am. Nat.*, **50,** 193–221, 284–305, 350–366, 421–434. *(88, 98, 111–12, 131, 334, 449)*

MULLER, H. J. (1927). Artificial transmutation of the gene. *Science, N.Y.*, **66**, 84–87.
(*143–45, 147*)

MULLER, H. J. (1928a). The measurement of gene mutation rate in *Drosophila*, its high variability, and its dependence upon temperature. *Genetics, Princeton*, **13**, 279–357.
(*141–42, 145–46*)

MULLER, H. J. (1928b). The problem of genic modification. *Z. VererbLehre*, Suppl. 1, 234–260.
(*143–45*)

MULLER, H. J. (1930a). Types of visible variations induced by X-rays in *Drosophila*. *J. Genet.*, **22**, 299–334.
(*390*)

MULLER, H. J. (1930b). Radiation and genetics. *Am. Nat.*, **64**, 220–251.
(*145*)

MULLER, H. J. and ALTENBURG, E. (1930). The frequency of translocations produced by X-rays in *Drosophila*. *Genetics, Princeton*, **15**, 283–311.
(*147*)

MULLER, H. J. and PAINTER, T. S. (1929). The cytological expression of changes in gene alignment produced by X-rays in *Drosophila*. *Am. Nat.*, **63**, 193–200.
(*147*)

MULLER, H. J., PROKOFYEVA, A. A. and KOSSIKOV, K. V. (1936). Unequal crossing-over in the bar mutant as a result of duplication of a minute chromosome section. *C.r. Acad. Sci. U.R.S.S.*, **2**, 78.
(*160*)

MURRAY, N. E. (1963). Polarized recombination and fine structure within the *me*-2 gene of *Neurospora crassa*. *Genetics, Princeton*, **48**, 1163–1183.
(*332*)

MURRAY, N. E. (1968). Polarized intragenic recombination in chromosome rearrangements of *Neurospora*. *Genetics, Princeton*, **58**, 181–191.
(*332*)

MURRAY, N. E. (1970). Recombination events that span sites within neighbouring gene loci of *Neurospora*. *Genet. Res.*, **15**, 109–121.
(*369*)

MYERS, G. L. and SADLER, J. R. (1971). Mutational inversion of control of the lactose operon of *Escherichia coli*. *J. molec. Biol.*, **58**, 1–28.
(*307*)

NATHAN, D. G., LODISH, H. F., KAN, Y. W. and HOUSMAN, D. (1971). Beta thalassemia and translation of globin messenger RNA. *Proc. natn. Acad. Sci. U.S.A.*, **68**, 2514–2518.
(*324*)

NATHANS, D. and DANNA, K. J. (1972). Specific origin in SV40 DNA replication. *Nature New Biol.*, **236**, 200–202.
(*209*)

NAZARIO, M., KINSEY, J. A. and AHMAD, M. (1971). *Neurospora* mutant deficient in tryptophanyl-transfer ribonucleic acid synthetase activity. *J. Bact.*, **105**, 121–126.
(*398*)

NEEL, J. V. (1949). Inheritance of sickle cell anemia. *Science, N.Y.*, **110**, 64–66.
(*218–19, 225*)

NEWTON, W. A., BECKWITH, J. R., ZIPSER, D. and BRENNER, S. (1965). Nonsense mutants and polarity in the *lac* operon of *Escherichia coli*. *J. molec. Biol.*, **14**, 290–296.
(*309, 311*)

NEWTON, W. C. F. and DARLINGTON, C. D. (1929). Meiosis in polyploids. Part I. Triploid and pentaploid tulips. *J. Genet.*, **21**, 1–15.
(*124, 446*)

NICHOLS, J. L. (1970). Nucleotide sequence from the polypeptide chain termination region of the coat protein cistron in bacteriophage R17 RNA. *Nature, Lond.*, **225**, 147–151.
(*292, 298*)

NILSSON-EHLE, H. (1909). Kreuzungsuntersuchungen an Hafer und Weizen. *Acta Univ. lund.*, ser 2, **5**, no. 2, 1–122.
(*33–35*)

NIRENBERG, M. W. and LEDER, P. (1964). RNA codewords and protein synthesis. *Science, N.Y.*, **145**, 1399–1407.
(*273, 274*)

NIRENBERG, M. W. and MATTHAEI, J. H. (1961). The dependence of cell-free protein synthesis in *E. coli* upon naturally occurring or synthetic polyribonucleotides. *Proc. natn. Acad. Sci. U.S.A.*, **47**, 1588–1602. *(250, 271, 274, 422)*

NISHIMURA, S., JONES, D. S. and KHORANA, H. G. (1965). Studies on polynucleotides, XLVIII. The *in vitro* synthesis of a co-polypeptide containing two amino acids in alternating sequence dependent upon a DNA-like polymer containing two nucleotides in alternating sequence. *J. molec. Biol.*, **13**, 302–324. *(274)*

NISHIMURA, S., JONES, D. S., OHTSUKA, E., HAYATSU, H., JACOB, T. M. and KHORANA, H. G. (1965). Studies on polynucleotides, XLVII. The *in vitro* synthesis of homopeptides as directed by a ribopolynucleotide containing a repeating trinucleotide sequence. New codon sequences for lysine, glutamic acid and arginine. *J. molec. Biol.*, **13**, 283–301. *(274)*

NODA, S. (1960). Chiasma studies in structural hybrids. II. Reciprocal translocation in *Lilium maximowiczii. Cytologia*, **25**, 456–460. *(123)*

NODA, S. (1961). Chiasma studies in structural hybrids. VII. Reciprocal translocation in *Scilla scilloides. Cytologia*, **26**, 74–77. *(121–23)*

NOMURA, M. and LOWRY, C. V. (1967). Phage f2 RNA-directed binding of formylmethionyl-tRNA to ribosomes and the role of 30 S ribosomal subunits in initiation of protein synthesis. *Proc. natn. Acad. Sci. U.S.A.*, **58**, 946–953. *(288, 289)*

NUFFER, M. G. (1955). Dosage effect of multiple *Dt* loci on mutation of *a* in the maize endosperm. *Science, N.Y.*, **121**, 399–400. *(393)*

NUR, U. (1961). Meiotic behavior of an unequal bivalent in the grasshopper *Calliptamus palaestinensis* Bdhr. *Chromosoma*, **12**, 272–279. *(119–20)*

OEHLKERS, F. (1943). Die Auslösung von Chromosomenmutationen in der Meiosis durch Einwirkung von Chemikalien. *Z. VererbLehre*, **81**, 313–341. *(149–50)*

OGAWA, T., TOMIZAWA, J. and FUKE, M. (1968). Replication of bacteriophage DNA, II. Structure of replicating DNA of phage lambda. *Proc. natn. Acad. Sci. U.S.A.*, **60**, 861–865. *(204)*

OHASHI, Z., SANEYOSHI, M., HARADA, F., HARA, H. and NISHIMURA, S. (1970). Presumed anticodon structure of glutamic acid tRNA from *E. coli*: a possible location of a 2-thiouridine derivative in the first position of the anticodon. *Biochem. biophys. Res. Commun.*, **40**, 866–872. *(285)*

OHNO, S. (1971). Simplicity of mammalian regulatory systems inferred by single gene determination of sex phenotypes. *Nature, Lond.*, **234**, 134–137. *(397)*

OKADA, Y., TERZAGHI, E., STREISINGER, G., EMRICH, J., INOUYE, M. and TSUGITA, A. (1966). A frame-shift mutation involving the addition of two base pairs in the lysozyme gene of phage T4. *Proc. natn. Acad. Sci. U.S.A.*, **56**, 1692–1698. *(276–7)*

OKAZAKI, R., OKAZAKI, T., SAKABE, K., SUGIMOTO, K. and SUGINO, A. (1968). Mechanism of DNA chain growth, I. Possible discontinuity and unusual secondary structure of newly synthesized chains. *Proc. natn. Acad. Sci. U.S.A.*, **59**, 598–605. *(202)*

OKAZAKI, R., SUGIMOTO, K., OKAZAKI, T., IMAE, Y. and SUGINO, A. (1970). DNA chain growth: *in vivo* and *in vitro* synthesis in a DNA polymerase-negative mutant of *E. coli. Nature, Lond.*, **228**, 223–226. *(202)*

OLIVE, L. S. (1959). Aberrant tetrads in *Sordaria fimicola. Proc. natn. Acad. Sci. U.S.A.*, **45**, 727–732. *(329)*

OLIVER, C. P. (1940). A reversion to wild-type associated with crossing-over in *Drosophila melanogaster. Proc. natn. Acad. Sci. U.S.A.*, **26**, 452–4. *(161–63, 213)*

OPPENHEIM, A. B. and RILEY, M. (1967). Covalent union of parental DNA's following conjugation in *Escherichia coli*. *J. molec. Biol.*, **28**, 503–511.                    (*346*)

ÖSTERGREN, G. and WAKONIG, T. (1954). True or apparent sub-chromatid breakage and the induction of labile states in cytological chromosome loci. *Botan. Notiser*, **107**, 357–375.                    (*382*)

OSTERTAG, W., EHRENSTEIN, G. VON and CHARACHE, S. (1972). Duplicated α-chain genes in Hopkins-2 haemoglobin of man and evidence for unequal crossing over between them. *Nature New Biol.*, **237**, 90–94.                    (*325*)

OSTERTAG, W. and SMITH, E. W. (1969). Hemoglobin-Lepore Baltimore, a third type of a $\delta\beta$ crossover ($\delta^{50}$, $\beta^{86}$). *Eur. J. Biochem.*, **10**, 371–376.                    (*323*)

OZEKI, H. and IKEDA, H. (1968). Transduction mechanisms. *A. Rev. Genet.*, **2**, 245–278.                    (*336*)

PADMANABHAN, R. and WU, R. (1972). Nucleotide sequence analysis of DNA. IV. Complete nucleotide sequence of the left-hand cohesive end of coliphage 186 DNA. *J. molec. Biol.*, **65**, 447–467.                    (*412*)

PAINTER, T. S. (1934). A new method for the study of chromosome aberrations and the plotting of chromosome maps in *Drosophila melanogaster*. *Genetics, Princeton*, **19**, 175–188.                    (*147–8*)

PALADE, G. E. (1955). A small particulate component of the cytoplasm. *J. biophys. biochem. Cytol.*, **1**, 59–68.                    (*246*)

PARDEE, A. B., JACOB, F. and MONOD, J. (1959). The genetic control and cytoplasmic expression of inducibility in the synthesis of β-galactosidase by *E. coli*. *J. molec. Biol.*, **1**, 165–178.                    (*303*)

PARDON, J. F., and WILKINS, M. H. F. (1972). A super-coil model for nucleohistone. *J. molec. Biol.*, **68**, 115–124.                    (*387*)

PARDON, J. F., WILKINS, M. H. F. and RICHARDS, B. M. (1967). Super-helical model for nucleohistone. *Nature, Lond.*, **215**, 508–509.                    (*386*)

PARDUE, M. L. and GALL, J. G. (1969). Molecular hybridization of radioactive DNA to the DNA of cytological preparations. *Proc. natn. Acad. Sci. U.S.A.*, **64**, 600–604.                    (*226, 407*)

PARDUE, M. L. and GALL, J. G. (1970). Chromosomal localization of mouse satellite DNA. *Science, N.Y.*, **168**, 1356–1358.                    (*408*)

PARKS, J. S., GOTTESMAN, M., SHIMADA, K., WEISBERG, R. A., PERLMAN, R. L. and PASTAN, I. (1971). Isolation of the *gal* repressor. *Proc. natn. Acad. Sci. U.S.A.*, **68**, 1891–1895.                    (*306*)

PARNELL, F. R. (1921). Note on the detection of segregation by examination of the pollen of rice. *J. Genet.*, **11**, 209–212.                    (*45, 46*)

PASCHER, A. (1918). Ueber die Beziehung der Reduktionsteilung zur Mendelschen Spaltung. *Ber. dt. bot. Ges.*, **36**, 163–168.                    (*103*)

PASTAN, I. and PERLMAN, R. L. (1968). The role of the lac promotor locus in the regulation of β-galactosidase synthesis by cyclic 3',5'-adenosine monophosphate. *Proc. natn. Acad. Sci. U.S.A.*, **61**, 1336–1342.                    (*315*)

PASZEWSKI, A. (1967). A study on simultaneous conversions in linked genes in *Ascobolus immersus*. *Genet. Res.*, **10**, 121–126.                    (*362*)

PASZEWSKI, A. and SURZYCKI, S. (1964). 'Selfers' and high mutation rate during meiosis in *Ascobolus immersus*. *Nature, Lond.*, **204**, 809.                    (*367*)

PATEMAN, J. A. and FINCHAM, J. R. S. (1958). Gene-enzyme relationships at the *am* locus in *Neurospora crassa*. *Heredity*, **12**, 317–332.                    (*222–23*)

PATTISON, I. H. and JONES, K. M. (1968). Detection of the scrapie agent in tissues of normal mice and in tumours of tumour-bearing but otherwise normal mice. *Nature, Lond.*, **218**, 102–104. (*432*)

PAUL, J. and GILMOUR, R. S. (1968). Organ-specific restriction of transcription in mammalian chromatin. *J. molec. Biol.*, **34**, 305–316. (*422–23*)

PAULING, L., ITANO, H. A., SINGER, S. J. and WELLS, I. C. (1949). Sickle cell anemia, a molecular disease. *Science, N.Y.*, **110**, 543–548. (*219*)

PEACOCK, W. J. (1963). Chromosome duplication and structure as determined by autoradiography. *Proc. natn. Acad. Sci. U.S.A.*, **49**, 793–801. (*373*)

PEACOCK, W. J. (1970). Replication, recombination and chiasmata in *Goniaea australasiae* (Orthoptera: Acrididae). *Genetics, Princeton*, **65**, 593–617. (*136, 342*)

PEARSON, K. (1897). Mathematical contributions to the theory of evolution. III. Regression, heredity and panmixia. *Phil. Trans. R. Soc. A*, **187**, 253–318. (*31*)

PEARSON, K. (1900). On the criterion that a given system of deviations from the probable in the case of a correlated system of variables is such that it can be reasonably supposed to have arisen from random sampling. *Phil. Mag.*, Ser. V, **50**, 157–175. (*26*)

PEARSON, K. (1904). Mathematical contributions to the theory of evolution. XII. On a generalised theory of alternative inheritance, with special reference to Mendel's laws. *Phil. Trans. R. Soc. A*, **203**, 53–86. (*31, 36*)

PEARSON, K. and LEE A. (1903) On the laws of inheritance in man. I. Inheritance of physical characters. *Biometrika*, **2**, 357–462. (*31, 36–37*)

PEARSON, P. L., ELLIS, J. D. and EVANS, H. J. (1970). A gross reduction in chiasma formation during meiotic prophase and a defective DNA repair mechanism associated with a case of human male infertility. *Cytogenetics*, **9**, 460–467. (*358*)

PELLING, C. (1966). A replicative and synthetic chromosomal unit—the modern concept of the chromomere. *Proc. R. Soc. B*, **164**, 279–289. (*206, 383, 411*)

PENROSE, L. S. (1963). *The Biology of Mental Defect*, 3rd edn. London, Sidgwick and Jackson. (*31*)

PETERSON, P. A. (1966). Phase variation of regulatory elements in maize. *Genetics, Princeton*, **54**, 249–266. (*397*)

PETES, T. D. and FANGMAN, W. L. (1972). Sedimentation properties of yeast chromosomal DNA. *Proc. natn. Acad. Sci. U.S.A.*, **69**, 1188–1191. (*385*)

PETTIJOHN, D. and HANAWALT, P. C. (1964). Evidence for repair-replication of ultra-violet damaged DNA in bacteria. *J. molec. Biol.*, **9**, 395–410. (*355*)

PICARD, M. (1971). Genetic evidence for a polycistronic unit of transcription in the complex locus '14' in *Podospora anserina*. I. Genetic and complementation maps. *Molec. gen. Genet.*, **111**, 35–50. (*322*)

PIPERNO, G., FONTY, G. and BERNARDI, G. (1972). The mitochondrial genome of wild-type yeast cells. II. Investigations on the compositional heterogeneity of mitochondrial DNA. *J. molec. Biol.*, **65**, 191–205. (*428*)

PIRROTTA, V. and PTASHNE, M. (1969). Isolation of the 434 phage repressor. *Nature, Lond.*, **222**, 541–544. (*305–6*)

POLANI, P. E. (1969). Abnormal sex chromosomes and mental disorder. *Nature, Lond.*, **223**, 680–686. (*421*)

PONTECORVO, G. (1947). Genetic systems based on heterocaryosis. *Cold Spring Harb. Symp. quant. Biol.*, **11**, 193–201. (*46*)

PONTECORVO, G. (1950). New fields in the biochemical genetics of microorganisms. *Biochem. Soc. Symp.*, **4**, 40–50.                                    (*213, 215, 446, 453*)

PONTECORVO, G. (1952). The genetic formulation of gene structure and action. *Adv. Enzymol.*, **13**, 121–149.                                                    (*211*)

PONTECORVO, G. (1953). The genetics of *Aspergillus nidulans*. *Adv. Genet.*, **5**, 141–238.                                                                                    (*171*)

PONTECORVO, G. (1958). *Trends in Genetic Analysis*. New York, Columbia Univ. Press.                                                                            (*360*)

PONTECORVO, G. and GEMMELL, A. R. (1944). Genetic proof of heterokaryosis in *Penicillium notatum*. *Nature, Lond.*, **154**, 514–515.                    (*47*)

PREER, J. R. (1971). Extrachromosomal inheritance: hereditary symbionts, mitochondria, chloroplasts. *A. Rev. Genet.*, **5**, 361–406.            (*430, 431*)

PRICE, P. M., CONOVER, J. H. and HIRSCHHORN, K. (1972). Chromosomal localization of human haemoglobin structural genes. *Nature, Lond.*, **237**, 340–342.    (*226*)

PRINGSHEIM, E. G. and PRINGSHEIM, O. (1952). Experimental elimination of chromatophores and eye-spot in *Euglena gracilis*. *New Phytol.*, **51**, 65–76.    (*56*)

PRITCHARD, R. H. (1955). The linear arrangement of a series of alleles of *Aspergillus nidulans*. *Heredity*, **9**, 343–371.            (*211, 327, 333, 334, 340*)

PROVASOLI, L., HUTNER, S. H. and SCHATZ, A. (1948). Streptomycin-induced chlorophyll-less races of *Euglena*. *Proc. Soc. exp. Biol. Med.*, **69**, 279–282.    (*56*)

PTASHNE, M. (1967a). Isolation of the λ phage repressor. *Proc. natn. Acad. Sci. U.S.A.*, **57**, 306–313.                                                (*304, 306*)

PTASHNE, M. (1967b). Specific binding of the λ phage repressor to λ DNA. *Nature, Lond.*, **214**, 232–234.                                            (*305–6*)

PTASHNE, M. and HOPKINS, N. (1968). *Proc. natn. Acad. Sci. U.S.A.*, **60**, 1282–7.    (*304*)

PUNNETT, R. C. (1917). Reduplication series in sweet peas. II. *J. Genet.*, **6**, 185–93. (*19*)

PUNNETT, R. C. and BATESON, W. (1908). The heredity of sex. *Science, N.Y.*, **27**, 785–787.                                                            (*20–2, 74, 452*)

QUAGLIAROTTI, G. and RITOSSA, F. M. (1968). On the arrangement of genes for 28S and 18S ribosomal RNA's in *Drosophila melanogaster*. *J. molec. Biol.*, **36**, 57–69.    (*403*)

RACE, R. R. (1944). An 'incomplete' antibody in human serum. *Nature, Lond.*, **153**, 771–772.                                                                    (*163*)

RACE, R. R. and SANGER, R. (1962). *Blood Groups in Man*, 4th edn. Oxford, Blackwell.                                                                        (*140, 163*)

RADFORD, A. (1969a). Polarised complementation at the pyrimidine-3 locus of *Neurospora*. *Molec. gen. Genet.*, **104**, 288–294.                (*319*)

RADFORD, A. (1969b). Information from ICR-170-induced mutations on the structure of the pyrimidine-3 locus in Neurospora. *Mutation Res.*, **8**, 537–544.    (*319*)

RAFFEL, D. and MULLER, H. J. (1940). Position effect and gene divisibility considered in connection with three strikingly similar scute mutations. *Genetics, Princeton*, **25**, 541–583.                                                            (*158, 161*)

RAJBHANDARY, U. L., CHANG, S. H., STUART, A., FAULKNER, R. D., HOSKINSON, R. M. and KHORANA, H. G. (1967). Studies on polynucleotides, LXVIII. The primary structure of yeast phenylalanine transfer RNA. *Proc. natn. Acad. Sci. U.S.A.*, **57**, 751–758.                                                            (*263*)

RANDALL, J. (1969). The flagellar apparatus as a model organelle for the study of growth and morphopoiesis. *Proc. R. Soc. B*, **173**, 31–62.            (*427*)

RANDOLPH, L. F. (1928). Types of supernumerary chromosomes in maize. *Anat. Rec.*, **41**, 102. *(446)*

RAPOPORT, J. A. (1946). Carbonyl compounds and the chemical mechanism of mutation. *C.r. Acad. Sci. U.R.S.S.*, **54**, 65. *(146)*

RASMUSSEN, R. E. and PAINTER, R. B. (1964). Evidence for repair of ultra-violet damaged deoxyribonucleic acid in cultured mammalian cells. *Nature, Lond.*, **203**, 1360–1362. *(357)*

REES, H., CAMERON, F. M., HAZARIKA, M. H. and JONES, G. H. (1966). Nuclear variation between diploid Angiosperms. *Nature, Lond.*, **211**, 828–830. *(382)*

REICH, E., FRANKLIN, R. M., SHATKIN, A. J. and TATUM, E. L. (1962). Action of actinomycin D on animal cells and viruses. *Proc. natn. Acad. Sci. U.S.A.*, **48**, 1238–45. *(424)*

RENNER, O. (1924). Die Scheckung der Oenotherenbastarde. *Biol. Zbl.*, **44**, 309–336. *(52, 54)*

REVELL, S. H. (1955). A new hypothesis for 'chromatid' changes. In *Radiobiology Symposium* 1954 (eds. BACQ, Z. M. and ALEXANDER, P.) pp. 243–253. London, Butterworth. *(151–57)*

REVELL, S. H. (1959). The accurate estimation of chromatid breakage, and its relevance to a new interpretation of chromatid aberrations induced by ionizing radiations. *Proc. R. Soc. B.*, **150**, 563–589. *(155–56)*

RHOADES, M. M. (1936). The effect of varying gene dosage on aleurone colour in maize. *J. Genet.*, **33**, 347–354. *(388, 398)*

RHOADES, M. M. (1938). Effect of the *Dt* gene on the mutability of the $a_1$ allele in maize. *Genetics, Princeton*, **23**, 377–397. *(388–90)*

RHOADES, M. M. (1942). Preferential segregation in maize. *Genetics, Princeton*, **27**, 395–407. *(392)*

RHOADES, M. M. (1943). Genic induction of an inherited cytoplasmic difference. *Proc. natn. Acad. Sci. U.S.A.*, **29**, 327–329. *(50–1)*

RHOADES, M. M. (1945). On the genetic control of mutability in maize. *Proc. natn. Acad. Sci. U.S.A.*, **31**, 91–95. *(390)*

RHOADES, M. M. (1947). Plastid mutations. *Cold Spring Harb. Symp. quant. Biol.*, **11**, 202–207. *(53)*

RHOADES, M. M. (1950). Meiosis in maize. *J. Hered.*, **41**, 58–67. *(116–17)*

RICHARDSON, B. J., CZUPPON, A. B. and SHARMAN, G. B. (1971). Inheritance of glucose-6-phosphate dehydrogenase variation in kangaroos. *Nature New Biol.*, **230**, 154–155. *(421)*

RICHARDSON, C. C., SCHILDKRAUT, C. L. and KORNBERG, A. (1964). Studies on the replication of DNA by DNA polymerases. *Cold Spring Harb. Symp. quant. Biol.*, **28**, 9–19. *(197)*

RIFKIN, M. R. and LUCK, D. J. L. (1971). Defective production of mitochondrial ribosomes in the *poky* mutant of *Neurospora crassa*. *Proc. natn. Acad. Sci. U.S.A.*, **68**, 287–290. *(430)*

RIGGS, A. D. and BOURGEOIS, S. (1968). On the assay, isolation and characterization of the *lac* repressor. *J. molec. Biol.*, **34**, 361–364. *(307)*

RILEY, R. and CHAPMAN, V. (1958). Genetic control of the cytologically diploid behaviour of hexaploid wheat. *Nature, Lond.*, **182**, 713–715. *(63)*

RILEY, R. and MILLER, T. E. (1966). The differential sensitivity of desynaptic and normal genotypes of barley to X-rays. *Mutation Res.*, **3**, 355–359. *(358)*

RINES, H. W., CASE, M. E. and GILES, N. H. (1969). Mutants in the *arom* gene cluster of *Neurospora crassa* specific for biosynthetic dehydroquinase. *Genetics, Princeton,* **61,** 789–800.                                                                                                (*321*)

RIS, H. (1949). The anaphase movement of chromosomes in the spermatocytes of the grasshopper. *Biol. Bull. mar. biol. Lab., Woods Hole,* **96,** 90–106.                (*62*)

RIS, H. (1957). Chromosome structure. In *The Chemical Basis of Heredity* edited by W. D. MCELROY and B. GLASS, Baltimore, Johns Hopkins Press, pp. 23–69.    (*372, 393*)

RIS, H. (1962). Interpretation of ultrastructure in the cell nucleus. *The interpretation of ultrastructure* (ed. R. J. C. Harris), pp. 69–88. Academic Press, New York and London.                                                                                        (*386*)

RIS, H. and MIRSKY, A. E. (1949). Quantitative cytochemical determination of desoxyribonucleic acid with the Feulgen nucleal reaction. *J. gen. Physiol.,* **33,** 125–146.                                                                                                        (*177*)

RITOSSA, F. M., ATWOOD, K. C. and SPIEGELMAN, S. (1966a). On the redundancy of DNA complementary to amino acid transfer RNA and its absence from the nucleolar organizer region of *Drosophila melanogaster. Genetics, Princeton,* **54,** 663–676.                                                                                        (*406*)

RITOSSA, F. M., ATWOOD, K. C. and SPIEGELMAN, S. (1966b). A molecular explanation of the bobbed mutants of Drosophila as partial deficiencies of 'ribosomal' DNA. *Genetics, Princeton,* **54,** 819–834.                                                        (*405*)

RITOSSA, F. M. and SPIEGELMAN, S. (1965). Localization of DNA complementary to ribosomal RNA in the nucleolus organizer region of *Drosophila melanogaster. Proc. natn. Acad. Sci. U.S.A.,* **53,** 737–745.                                            (*403*)

RIZET, G. and ROSSIGNOL, J. L. (1966). Sur la dimension probable des échanges réciproques au sein d'un locus complexe d'*Ascobolus immersus. C.r. hebd. Séanc. Acad. Sci., Paris,* **262,** 1250–1253.                                                (*359*)

ROBERTS, K. and NORTHCOTE, D. H. (1970). Structure of the nuclear pore in higher plants. *Nature, Lond.,* **228,** 385–386.                                                (*59*)

ROBERTS, R. B. (1958). In introduction to *Microsomal Particles and Protein Synthesis,* edited by R. B. ROBERTS, Washington, Pergamon Press, 168 pp.        (*246, 452*)

RODARTE-RAMÓN, U. S. (1972). Radiation-induced recombination in *Saccharomyces*: the genetic control of recombination in mitosis and meiosis. *Radiation Res.,* **49,** 148–154.                                                                                        (*371*)

RODARTE-RAMÓN, U. S. and MORTIMER, R. K. (1972). Radiation-induced recombination in *Saccharomyces*: isolation and genetic study of recombination-deficient mutants. *Radiation Res.,* **49,** 133–147.                                            (*357, 371*)

ROESS, W. B. and DEBUSK, A. G. (1968). Properties of a basic amino acid permease in *Neurospora crassa. J. gen. Microbiol.,* **52,** 421–432.                        (*171*)

ROGER, M. (1972). Evidence for conversion of heteroduplex transforming DNAs to homoduplexes by recipient pneumococcal cells. *Proc. natn. Acad. Sci. U.S.A.,* **69,** 466–470.                                                                                        (*364*)

ROPER, J. A. (1950). Search for linkage between genes determining a vitamin requirement. *Nature, Lond.,* **166,** 956.                                                    (*211, 214*)

ROPER, J. A. (1953). Pseudo-allelism. In PONTECORVO, G., The genetics of *Aspergillus nidulans. Adv. Genet.,* **5,** 208–218.                                                (*211*)

ROPER, J. A. and PRITCHARD, R. H. (1955). Recovery of the complementary products of mitotic crossing-over. *Nature, Lond.,* **175,** 639.                            (*212*)

ROSSEN, J. M. and WESTERGAARD, M. (1966). Studies on the mechanism of crossing over. II. Meiosis and the time of meiotic chromosome replication in the ascomycete *Neottiella rutilans* (Fr.) Dennis. *C.r. Trav. Lab. Carlsberg*, **35**, 233–260. *(341)*

ROSSIGNOL, J. L. (1969). Existence of homogeneous categories of mutants exhibiting various conversion patterns in gene 75 of *Ascobolus immersus*. *Genetics, Princeton*, **63**, 795–805. *(362)*

ROTH, R. and FOGEL, S. (1971). A system selective for yeast mutants deficient in meiotic recombination. *Molec. gen. Genet.*, **112**, 295–305. *(371)*

ROTHFELS, K., SEXSMITH, E., HEIMBURGER, M. and KRAUSE, M. O. (1966). Chromosome size and DNA content of species of *Anemone* L. and related genera (Ranunculaceae). *Chromosoma*, **20**, 54–74. *(382)*

RUPERT, C. S. (1960). Photoreactivation of transforming DNA by an enzyme from bakers' yeast. *J. gen. Physiol.*, **43**, 573–595. *(354)*

RUPERT, C. S., GOODGAL, S. H. and HERRIOTT, R. M. (1958). Photoreactivation in vitro of ultraviolet inactivated *Hemophilus influenzae* transforming factor. *J. gen. Physiol.*, **41**, 451–471. *(354)*

RUPP, W. D. and HOWARD-FLANDERS, P. (1968). Discontinuities in the DNA synthesized in the excision-defective strain of *Escherichia coli* following ultraviolet irradiation. *J. molec. Biol.*, **31**, 291–304. *(355)*

RUPP, W. D., WILDE, C. E., RENO, D. L. and HOWARD-FLANDERS, P. (1971). Exchanges between DNA strands in ultraviolet-irradiated *Escherichia coli*. *J. molec. Biol.*, **61**, 25–44. *(355–6)*

RUSSELL, L. B. and CHU, E. H. Y. (1961). An XXY male in the mouse. *Proc. natn. Acad. Sci. U.S.A.*, **47**, 571–575. *(79)*

SADLER, J. R. and SMITH, T. F. (1971). Mapping of the lactose operator. *J. molec. Biol.*, **62**, 139–169. *(306–7)*

SAGER, R. (1962). Streptomycin as a mutagen for nonchromosomal genes. *Proc. natn. Acad. Sci. U.S.A.*, **48**, 2018–2026. *(56)*

SAGER, R. and LANE, D. (1972). Molecular basis of maternal inheritance. *Proc. natn. Acad. Sci. U.S.A.*, **69**, 2410–2413. *(431)*

SAGER, R. and RAMANIS, Z. (1963). The particulate nature of nonchromosomal genes in *Chlamydomonas*. *Proc. natn. Acad. Sci. U.S.A.*, **50**, 260–268. *(56)*

SAGER, R. and RAMANIS, Z. (1965). Recombination of nonchromosomal genes in *Chlamydomonas*. *Proc. natn. Acad. Sci. U.S.A.*, **53**, 1053–1061. *(430)*

SAGER, R. and RAMANIS, Z. (1967). Biparental inheritance of nonchromosomal genes induced by ultraviolet irradiation. *Proc. natn. Acad. Sci. U.S.A.*, **58**, 931–7. *(430)*

SAGER, R. and RAMANIS, Z. (1970). Genetic studies of chloroplast DNA in *Chlamydomonas*. *Symp. Soc. exp. Biol.*, **24**, 401–417. *(430)*

ST. LAWRENCE, P. (1956). The *q* locus of *Neurospora crassa*. *Proc. natn. Acad. Sci. U.S.A.*, **42**, 189–194. *(333–34)*

SALAS, M., SMITH, M. A., STANLEY, W. M., WAHBA, A. J. and OCHOA, S. (1965). Direction of reading of the genetic message. *J. biol. Chem.*, **240**, 3988–3995. *(281)*

SAMBROOK, J. F., FAN, D. P. and BRENNER, S. (1967). A strong suppressor specific for UGA. *Nature, Lond.*, **214**, 452–453. *(296, 297, 309)*

SANGER, F., BROWNLEE, G. G. and BARRELL, B. G. (1965). A two-dimensional fractionation procedure for radioactive nucleotides. *J. molec. Biol.*, **13**, 373–398. *(262, 290)*

SANSOME, E. (1961). Meiosis in the oogonium and antheridium of *Pythium debaryanum* Hesse and its significance in the life history of the biflagellatae. *Trans. Br. mycol. Soc.*, **46,** 63–72.                                                                                          *(436)*

SARABHAI, A. S., STRETTON, A. O. W., BRENNER, S. and BOLLE, A. (1964). Co-linearity of the gene with the polypeptide chain. *Nature, Lond.*, **201,** 13–17.           *(230–31)*

SAX, K. (1930). Chromosome structure and the mechanism of crossing-over. *J. Arnold Arbor.*, **11,** 193–220.                                                                              *(133)*

SAX, K. (1936). Chromosome coiling in relation to meiosis and crossing-over. *Genetics, Princeton*, **21,** 324–338.                                                                        *(133)*

SCAIFE, J. and BECKWITH, J. R. (1967). Mutational alteration of the maximal level of *lac* operon expression. *Cold Spring Harb. Symp. quant. Biol.*, **31,** 403–408.      *(311)*

SCHEKMAN, R., WICKNER, W., WESTERGAARD, O., BRUTLAG, D., GEIDER, K., BERTSCH, L. and KORNBERG, A. (1972). Initiation of DNA synthesis: synthesis of $\phi$X174 replicative form requires RNA synthesis resistant to rifampicin. *Proc. natn. Acad. Sci. U.S.A.*, **69,** 2691–2695.                                                       *(209)*

SCHINDLER, A. M. and MIKAMO, K. (1970). Triploidy in man. Report of a case and a discussion on etiology. *Cytogenetics*, **9,** 116–130.                                      *(421)*

SCHLEIF, R., GREENBLATT, J. and DAVIS, R. W. (1971). Dual control of arabinose genes on transducing phage $\lambda$d *ara*. *J. molec. Biol.*, **59,** 127–150.              *(348)*

SCHLESINGER, M. J. and LEVINTHAL, C. (1963). Hybrid protein formation of *E. coli* alkaline phosphatase leading to *in vitro* complementation. *J. molec. Biol.*, **7,** 1–12.                                                                                                                        *(224)*

SCHMID, W. (1963). DNA replication patterns of human chromosomes. *Cytogenetics*, **2,** 175–193.                                                                                    *(206, 392)*

SCHNÖS, M. and INMAN, R. B. (1970). Position of branch points in replicating $\lambda$ DNA. *J. molec. Biol.*, **51,** 61–73.                                                        *(207)*

SCHÖTZ, F. (1970). Effects of the disharmony between genome and plastome on the differentiation of the thylakoid system in *Oenothera*. *Symp. Soc. exp. Biol.*, **24,** 39–54.                                                                                                                        *(431)*

SCHRADER, F. and HUGHES-SCHRADER, S. (1956). Polyploidy and fragmentation in the chromosomal evolution of various species of *Thyanta* (Hemiptera). *Chromosoma*, **7,** 469–496.                                                                                                    *(382)*

SCHROEDER, W. A., HUISMAN, T. H. J., SHELTON, J. R., SHELTON, J. B., KLEIHAUER, E. F., DOZY, A. M. and ROBBERSON, B. (1968). Evidence for multiple structural genes for the $\gamma$ chain of human fetal hemoglobin. *Proc. natn. Acad. Sci. U.S.A.*, **60,** 537–544.                                                                                                                        *(325)*

SCHULTZ, J. (1936). Variegation in *Drosophila* and the inert chromosome regions. *Proc. natn. Acad. Sci. U.S.A.*, **22,** 27–33.                                                  *(393)*

SCHULTZ, J. (1948). The nature of heterochromatin. *Cold Spring Harb. Symp. quant. Biol.*, **12,** 179–191.                                                                          *(393, 398)*

SCHWARZACHER, H. G. and SCHNEDL, W. (1966). Position of labelled chromatids in diplochromosomes of endo-reduplicated cells after uptake of tritiated thymidine. *Nature, Lond.*, **209,** 107–108.                                                                  *(413)*

SCOLNICK, E., TOMPKINS, R., CASKEY, T. and NIRENBERG, M. (1968). Release factors differing in specificity for terminator codons. *Proc. natn. Acad. Sci. U.S.A.*, **61,** 768–774.                                                                                                            *(299)*

SCOTT-MONCRIEFF, R. (1936). A biochemical survey of some mendelian factors for flower colour. *J. Genet.*, **32,** 117–170.                                                      *(165)*

SEARS, E. R. and OKAMOTO, M. (1958). Intergenomic chromosome relationships in hexaploid wheat. *Proc. 10th int. Congr. Genet.*, **2**, 258–259. Univ. Toronto Press.
*(63, 66)*

SETLOW, J. K., BOLING, M. E. and BOLLUM, F. J. (1965). The chemical nature of photo-reactivable lesions in DNA. *Proc. natn. Acad. Sci. U.S.A.*, **53**, 1430–1436. *(354)*

SETLOW, R. B. and CARRIER, W. L. (1964). The disappearance of thymine dimers from DNA: an error-correcting mechanism. *Proc. natn. Acad. Sci. U.S.A.*, **51**, 226–231.
*(354)*

SETLOW, R. B., REGAN, J. D., GERMAN, J. and CARRIER, W. L. (1969). Evidence that xeroderma pigmentosum cells do not perform the first step in the repair of ultraviolet damage to their DNA. *Proc. natn. Acad. Sci. U.S.A.*, **64**, 1035–1041.
*(358)*

SHANFIELD, B. and KÄFER, A. (1969). UV-sensitive mutants increasing mitotic crossing-over in *Aspergillus nidulans*. *Mutation Res.*, **7**, 485–487. *(357)*

SHARMAN, G. B. (1971). Late DNA replication in the paternally derived *X* chromosome of female kangaroos. *Nature, Lond.*, **230**, 231–232. *(421)*

SHERMAN, F., STEWART, J. W., PARKER, J. H., PUTTERMAN, G. J., AGRAWAL, B. B. L. and MARGOLIASH, E. (1970). The relationship of gene structure and protein structure of iso-1-cytochrome *c* from yeast. *Symp. Soc. exp. Biol.*, **24**, 85–107. *(233, 293)*

SIDDIQI, O. H. (1962). The fine genetic structure of the *paba*-1 region of *Aspergillus nidulans*. *Genet. Res.*, **3**, 69–89. *(332)*

SIDDIQI, O. H. (1963). Incorporation of parental DNA into genetic recombinants of *E. coli*. *Proc. natn. Acad. Sci. U.S.A.*, **49**, 589–592. *(338)*

SIGNER, E. R. (1968). Lysogeny: the integration problem. *A. Rev. Microbiol.*, **22**, 451–488. *(336, 347)*

SILVERSTONE, A. E., MAGASANIK, B., REZNIKOFF, W. S., MILLER, J. H. and BECKWITH, J. R. (1969). Catabolite sensitive site of the *lac* operon. *Nature, Lond.*, **221**, 1012–1014.
*(315)*

SIMCHEN, G. and STAMBERG, J. (1969). Genetic control of recombination in *Schizophyllum commune*: specific and independent regulation of adjacent and non-adjacent chromosomal regions. *Heredity*, **24**, 369–381. *(369)*

SINSHEIMER, R. L. (1959). A single-stranded deoxyribonucleic acid from bacteriophage φX174. *J. molec. Biol.*, **1**, 43–53. *(250)*

SLAYTER, H. S., WARNER, J. R., RICH, A. and HALL, C. E. (1963). The visualization of polyribosomal structure. *J. molec. Biol.*, **7**, 652–657. *(270–71)*

SLIZYNSKA, H. and SLIZYNSKI, B. M. (1947). Genetical and cytological studies of lethals induced by chemical treatment in *Drosophila melanogaster*. *Proc. R. Soc. Edinb. B*, **62**, 234–242. *(149)*

SLIZYNSKI, B. M. (1964). Chiasmata in spermatocytes of *Drosophila melanogaster*. *Genet. Res.*, **5**, 80–84. *(134)*

SMITH, A. E. and MARCKER, K. A. (1970). Cytoplasmic methionine transfer RNAs from eukaryotes. *Nature, Lond.*, **226**, 607–610. *(291)*

SMITH, B. R. (1966). Genetic controls of recombination. I. The recombination-2 gene of *Neurospora crassa*. *Heredity*, **21**, 481–498. *(368)*

SMITH, G. P., HOOD, L. and FITCH, W. M. (1971). Antibody diversity. *A. Rev. Biochem.*, **40**, 969–1012. *(415)*

SMITH, P. A. and KING, R. C. (1968). Genetic control of synaptonemal complexes in *Drosophila melanogaster*. *Genetics, Princeton*, **60**, 335–351.                    *(136)*

SMITH, S. G. (1935). Chromosome fragmentation produced by crossing-over in *Trillium erectum* L. *J. Genet.*, **30**, 227–232.                    *(125, 127)*

SMITH, T. F. and SADLER, J. R. (1971). The nature of lactose operator constitutive mutations. *J. molec. Biol.*, **59**, 273–305.                    *(306–7)*

SMYTH, D. R. (1971). Effect of *rec-3* on polarity of recombination in the *amination-1* locus of *Neurospora crassa*. *Aust. J. biol. Sci.*, **24**, 97–106.                    *(368)*

SNEDECOR, G. W. (1956). *Statistical Methods*, 5th edn. Ames, Iowa State Univ. Press.                    *(23)*

SNYDER, L. H. (1931). Inherited taste deficiency. *Science, N.Y.*, **74**, 151–152.    *(13–14)*

SOBELL, H. M. (1972). Molecular mechanism for genetic recombination. *Proc. natn. Acad. Sci., U.S.A.*, **69**, 2483–2487.                    *(352, 369, 370)*

SOBELL, H. M., JAIN, S. C., SAKORE, T. D. and NORDMAN, C. E. (1971). Stereochemistry of actinomycin–DNA binding. *Nature New Biol.*, **231**, 200–205.                    *(307)*

SOKOLOFF, L., FRANCIS, C. M. and CAMPBELL, P. L. (1964). Thyroxine stimulation of amino acid incorporation into protein independent of any action on messenger RNA synthesis. *Proc. natn. Acad. Sci. U.S.A.*, **52**, 728–736.                    *(424)*

SÖLL, D., CHERAYIL, J. D. and BOCK, R. M. (1967). Studies on polynucleotides. LXXV. Specificity of tRNA for codon recognition as studied by the ribosomal binding technique. *J. molec. Biol.*, **29**, 97–112.                    *(283)*

SÖLL, D. and RAJBHANDARY, U. L. (1967). Studies on polynucleotides. LXXVI. Specificity of transfer RNA for codon recognition as studied by amino acid incorporation. *J. molec. Biol.*, **29**, 113–124.                    *(283)*

SOUTHERN, E. M. (1970). Base sequence and evolution of guinea-pig α-satellite DNA. *Nature, Lond.*, **227**, 794–798.                    *(407)*

SPARROW, F. K. (1959). Interrelationships and phylogeny of the aquatic Phycomycetes. *Mycologia*, **50**, 797–813.                    *(436)*

SPATZ, H. C. and CROTHERS, D. M. (1969). The rate of DNA unwinding. *J. molec. Biol.*, **42**, 191–219.                    *(204)*

SPATZ, H. C. and TRAUTNER, T. A. (1970). One way to do experiments on gene conversion? Transfection with heteroduplex *SPP*1 DNA. *Molec. gen. Genet.*, **109**, 84–106.                    *(365)*

SPENCER, T. and HARRIS, H. (1964). Regulation of enzyme synthesis in an enucleate cell. *Biochem. J.*, **91**, 282–286.                    *(424, 427)*

SPIEGELMAN, S., BURNY, A., DAS, M. R., KEYDAR, J., SCHLOM, J., TRAVNICEK, M. and WATSON, K. (1970*a*). Characterization of the products of RNA-directed DNA polymerases in oncogenic RNA viruses. *Nature, Lond.*, **227**, 563–567.                    *(245)*

SPIEGELMAN, J., BURNY, A., DAS, M. R., KEYDAR, J., SCHLOM, J., TRAVNICEK, M. and WATSON, K. (1970*b*). DNA-directed DNA polymerase activity in oncogenic RNA viruses. *Nature, Lond.*, **227**, 1029–1031.                    *(246)*

SPURWAY, H. (1948). Genetics and cytology of *Drosophila subobscura*. IV. An extreme example of delay in gene action, causing sterility. *J. Genet.*, **49**, 126–140.    *(49)*

SRB, A. M. and HOROWITZ, N. H. (1944). The ornithine cycle in *Neurospora* and its genetic control. *J. biol. Chem.*, **154**, 129–139.                    *(166–67)*

STADLER, D. R. and TOWE, A. M. (1963). Recombination of allelic cysteine mutants in *Neurospora*. *Genetics, Princeton*, **48**, 1323–1344. *(331, 332)*

STADLER, D. R. and TOWE, A. M. (1971). Evidence for meiotic recombination in *Ascobolus* involving only one member of a tetrad. *Genetics, Princeton*, **68**, 401–413. *(370)*

STADLER, L. J. (1926). The variability of crossing-over in maize. *Genetics, Princeton*, **11**, 1–37. *(88–89, 111)*

STADLER, L. J. (1928). Mutations in barley induced by X-rays and radium. *Science, N.Y.*. **68**, 186–187. *(145)*

STADLER, L. J. (1930). Some genetic effects of X-rays in plants. *J. Hered.*, **21**, 3–19. *(145)*

STAEHELIN, M. (1971). The primary structure of transfer ribonucleic acid. *Experientia*, **27**, 1–11. *(284)*

STAEHELIN, T., WETTSTEIN, F. O., OURA, H. and NOLL, H. (1964). Determination of the coding ratio based on molecular weight of messenger ribonucleic acid associated with ergosomes of different aggregate size. *Nature, Lond.*, **201**, 264–270. *(271)*

STAHL, F. W. and MURRAY, N. E. (1966). The evolution of gene clusters and genetic circularity in microorganisms. *Genetics, Princeton*, **53**, 569–576. *(318)*

STAHL, F. W., MURRAY, N. E., NAKATA, A. and CRASEMAN, J. M. (1966). Intergenic *cis-trans* position effects in bacteriophage T4. *Genetics, Princeton*, **54**, 223–32. *(317)*

STANIER, R. Y. (1964). Towards a definition of the bacteria. In *The Bacteria*, edited by I. C. GUNSALUS and R. Y. STANIER. New York, Academic Press, vol. 5, pp. 445–464. *(435)*

STANLEY, W. M., SALAS, M., WAHBA, A. J. and OCHOA, S. (1966). Translation of the genetic message: factors involved in the initiation of protein synthesis. *Proc. natn. Acad. Sci. U.S.A.*, **56**, 290–295. *(288)*

STAPLES, D. H. and HINDLEY, J. (1971). Ribosome binding site of Qβ RNA polymerase cistron. *Nature New Biol.*, **234**, 211–212. *(291, 292)*

STAPLES, D. H., HINDLEY, J., BILLETER, M. A. and WEISSMANN, C. (1971). Localization of Qβ maturation cistron ribosome binding site. *Nature New Biol.*, **234**, 202–204. *(291, 292)*

STEDMAN, E. and STEDMAN, E. (1951). The basic proteins of cell nuclei. *Phil. Trans. R. Soc. B*, **235**, 565–595. *(422)*

STEITZ, J. A. (1969). Polypeptide chain initiation: nucleotide sequences of the three ribosomal binding sites in bacteriophage R17 RNA. *Nature, Lond.*, **224**, 957–964. *(290, 292)*

STEITZ, J. A., DUBE, S. K. and RUDLAND, P. S. (1970). Control of translation by T4 phage: altered ribosome binding at R17 initiation sites. *Nature, Lond.*, **226**, 824–827. *(291)*

STERN, C. (1931). Zytologisch-genetische Untersuchungen als Beweise für die Morgansche Theorie des Faktorenaustauschs. *Biol. Zbl.*, **51**, 547–587. *(91–93)*

STERN, C. (1936). Somatic crossing over and segregation in *Drosophila melanogaster*. *Genetics, Princeton*, **21**, 625–730. *(212)*

STERN, H. and HOTTA, Y. (1967). Chromosome behavior during development of meiotic tissue. In *The Control of Nuclear Activity* (ed. L. Goldstein) pp. 47–76. Prentice-Hall, Englewood Cliffs, N.J. *(412)*

STEVENS, N. M. (1905). Studies in spermatogenesis with especial reference to the 'accessory chromosome.' *Publs Carnegie Instn*, **36**, 1–32. *(73, 74)*

STEVENS, N. M. (1908). A study of the germ-cells of certain Diptera, with reference to the heterochromosomes and the phenomena of synapsis. *J. exp. Zool.*, **5**, 359–374.
(*73, 74*)

STEVENS, W. L. (1936). The analysis of interference. *J. Genet.*, **32**, 51–64.   (*112*)

STORCK, R. (1966). Nucleotide composition of nucleic acids of fungi. II. Deoxyribonucleic acids. *J. Bact.*, **91**, 227–230.   (*278*)

STRASBURGER, E. (1879). *Die Angiopermen und die Gymnospermen.* Jena, Gustav Fischer, 173 pp.   (*40–41*)

STRASBURGER, E. (1882). Ueber den Theilungsvorgang der Zellkerne und das Verhältniss der Kerntheilung zur Zelltheilung. *Arch. mikrosk. Anat.*, **21**, 476–590.
(*58–59*)

STRASBURGER, E. (1884). Die Controversen der indirecten Kerntheilung. *Arch. mikrosk. Anat.*, **23**, 246–304.   (*58–59, 445, 450, 451*)

STRASBURGER, E. (1888). *Ueber Kern- und Zelltheilung im Pfanzenreiche, nebst einem Anhang über Befruchtung.* Histologische Beiträge, Heft I, Jena, Gustav Fischer, 258 pp.   (*62*)

STRASBURGER, E. (1905). Typische und allotypische Kernteilung. Ergebnisse und Erörterungen. *Jb. wiss. Bot.*, **42**, 1–71.   (*448, 449*)

STRASBURGER, E. (1910). Chromosomenzahl. *Flora, Jena*, **100**, 398–446.   (*451*)

STREISINGER, G., EMRICH, J. and STAHL, M. M. (1967). Chromosome structure in phage T4, III. Terminal redundancy and length determination. *Proc. natn. Acad. Sci. U.S.A.*, **57**, 292–295.   (*338*)

STREISINGER, G., OKADA, Y., EMRICH, J., NEWTON, J., TSUGITA, A., TERZAGHI, E. and INOUYE, M. (1967). Frameshift mutations and the genetic code. *Cold Spring Harb. Symp. quant. Biol.*, **31**, 77–84.   (*276, 367*)

STURTEVANT, A. H. (1913). The linear arrangement of six sex-linked factors in *Drosophila*, as shown by their mode of association. *J. exp. Zool.*, **14**, 43–59.
(*84–86, 108, 210*)

STURTEVANT, A. H. (1923). Inheritance of direction of coiling in *Limnaea. Science, N.Y.*, **58**, 269–270.   (*48*)

STURTEVANT, A. H. (1925). The effects of unequal crossing over at the Bar locus in *Drosophila. Genetics, Princeton*, **10**, 117–147.   (*158–60, 213*)

STURTEVANT, A. H. (1926). A cross-over reducer in *Drosophila melanogaster* due to inversion of a section of the third chromosome. *Biol. Zbl.*, **46**, 697–702.   (*449*)

STURTEVANT, A. H. and MORGAN, T. H. (1923). Reverse mutation of the bar gene correlated with crossing over. *Science, N.Y.*, **57**, 746–747.   (*159–61*)

SUBAK-SHARPE, H., BÜRK, R. R., CRAWFORD, L. V., MORRISON, J. M., HAY, J. and KEIR, H. M. (1967). An approach to evolutionary relationships of mammalian DNA viruses through analysis of the pattern of nearest neighbor base sequences. *Cold Spring Harb. Symp. quant. Biol.*, **31**, 737–748.   (*280–81*)

SUEOKA, N. (1960). Mitotic replication of deoxyribonucleic acid in *Chlamydomonas reinhardi. Proc. natn. Acad. Sci. U.S.A.*, **46**, 83–91.   (*193*)

SUEOKA, N. (1961a). Variation and heterogeneity of base composition of deoxyribonucleic acids: a compilation of old and new data. *J. molec. Biol.*, **3**, 31–40.(*406*)

SUEOKA, N. (1961b). Correlation between base composition of deoxyribonucleic acid and amino acid composition of protein. *Proc. natn. Acad. Sci. U.S.A.*, **47**, 1141–1149.   (*280*)

SUEOKA, N. (1964). Compositional variation and heterogeneity of nucleic acids and protein in Bacteria. In *The Bacteria*, edited by I. C. Gunsalus and R. Y. Stainer. New York, Academic Press, vol. **5**, pp. 419–443. *(278)*

SUGINO, A., HIROSE, S. and OKAZAKI, R. (1972). RNA-linked nascent DNA fragments in *Escherichia coli*. *Proc. natn. Acad. Sci. U.S.A.*, **69**, 1863–1867. *(209)*

SUNDARARAJAN, T. A. and THACH, R. E. (1966). Role of the formylmethionine codon AUG in phasing translation of synthetic messenger RNA. *J. molec. Biol.*, **19**, 74–90. *(290)*

SURZYCKI, S. J. and GILLHAM, N. W. (1971). Organelle mutations and their expression in *Chlamydomonas reinhardi*. *Proc. natn. Acad. Sci. U.S.A.*, **68**, 1301–1306. *(431)*

SUTTON, W. S. (1902). On the morphology of the chromosome group in *Brachystola magna*. *Biol. Bull. mar. biol. Lab.*, *Woods Hole*, **4**, 24–39. *(69)*

SUTTON, W. S. (1903). The chromosomes in heredity. *Biol. Bull. mar. biol. Lab.*, *Woods Hole*, **4**, 231–248. *(71, 80)*

SWANSON, C. P. (1947). X-ray and ultraviolet studies on pollen tube chromosomes. II. The quadripartite structure of the prophase chromosomes of *Tradescantia*. *Proc. natn. Acad. Sci. U.S.A.*, **33**, 229–232. *(381)*

SWARTZ, M. N., TRAUTNER, T. A. and KORNBERG, A. (1962). Enzymatic synthesis of deoxyribonucleic acid. XI. Further studies on nearest neighbor base sequences in deoxyribonucleic acids. *J. biol. Chem.*, **237**, 1961–1967. *(198–200, 406)*

SWIFT, H. H. (1950). The desoxyribose nucleic acid content of animal nuclei. *Physiol. Zoöl.*, **23**, 169–198. *(177)*

SZYBALSKI, W. (1964). Structural modifications of DNA: crosslinking, circularization and single-strand interruptions. *Abh. dt. Akad. Wiss. Berl.*, *Kl. Med.*, 1964, (4), 1–19. *(346)*

TAKANAMI, M. and OKAMOTO, T. (1963). Interaction of ribosomes and synthetic polyribonucleotides. *J. molec. Biol.*, **7**, 323–333. *(258)*

TAKEISHI, K., UKITA, T. and NISHIMURA, S. (1968). Characterization of two species of methionine transfer ribonucleic acid from bakers' yeast. *J. biol. Chem.*, **243**, 5761–5769. *(291)*

TATUM, E. L. (1964). Genetic determinants. *Proc. natn. Acad. Sci. U.S.A.*, **51**, 908–15. *(272)*

TATUM, E. L., BONNER, D. and BEADLE, G. W. (1944). Anthranilic acid and the biosynthesis of indole and tryptophan by *Neurospora*. *Archs Biochem.*, **3**, 477–478. *(168)*

TAYLOR, J. H. (1953). Autoradiographic detection of incorporation of P$^{32}$ into chromosomes during meiosis and mitosis. *Expl Cell Res.*, **4**, 164–173. *(177)*

TAYLOR, J. H. (1958). Sister chromatid exchanges in tritium-labeled chromosomes. *Genetics, Princeton*, **43**, 515–529. *(189, 382)*

TAYLOR, J. H. (1959a). The organization and duplication of genetic material. *Proc. 10th int. Congr. Genet.* **1**, 63–78. Univ. Toronto Press. *(189, 191, 384, 385)*

TAYLOR, J. H. (1959b). Autoradiographic studies of nucleic acids and proteins during meiosis in *Lilium longiflorum*. *Am. J. Bot.*, **46**, 477–484. *(178)*

TAYLOR, J. H. (1960). Asynchronous duplication of chromosomes in cultured cells of Chinese hamster. *J. biophys. biochem. Cytol.*, **7**, 455–463. *(206)*

TAYLOR, J. H. (1965). Distribution of tritium-labeled DNA among chromosomes during meiosis. I. Spermatogenesis in the grasshopper. *J. Cell Biol.* **25**, (2), 57–67. *(136, 341–42)*

TAYLOR, J. H. and TAYLOR, S. H. (1953). The autoradiograph—a tool for cytogeneticists. *J. Hered.*, **44**, 128–132.                    (*178*)

TAYLOR, J. H., WOODS, P. S. and HUGHES, W. L. (1957). The organization and duplication of chromosomes as revealed by autoradiographic studies using tritium-labeled thymidine. *Proc. natn. Acad. Sci. U.S.A.*, **43**, 122–128. (*186–91, 206, 385*)

TAYLOR, K., HRADECNA, Z. and SZYBALSKI, W. (1967). Asymmetric distribution of the transcribing regions on the complementary strands of coliphage λ DNA. *Proc. natn. Acad. Sci. U.S.A.*, **57**, 1618–1625.                    (*256*)

TEAS, H. J., HOROWITZ, N. H. and FLING, M. (1948). Homoserine as a precursor of threonine and methionine in *Neurospora*. *J. biol. Chem.*, **172**, 651–658.     (*169*)

TEMIN, H. M. (1963). The effects of actinomycin D on growth of Rous sarcoma virus. *Virology*, **20**, 577–582.                    (*245*)

TEMIN, H. M. and MIZUTANI, S. (1970). RNA-dependent DNA polymerase in virions of Rous sarcoma virus. *Nature, Lond.*, **226**, 1211–1213.          (*245*)

TERZAGHI, E. (1971). Alternative pathways of tail fiber assembly in bacteriophage T4? *J. molec. Biol.*, **59**, 319–327.                (*317, 425*)

TERZAGHI, E., OKADA, Y., STREISINGER, G., EMRICH, J., INOUYE, M. and TSUGITA, A. (1966). Change of a sequence of amino acids in phage T4 lysozyme by acridine-induced mutations. *Proc. natn. Acad. Sci. U.S.A.*, **56**, 500–507.     (*276–77, 281*)

THACH, R. E., CECERE, M. A., SUNDARARAJAN, T. A. and DOTY, P. (1965). The polarity of messenger translation in protein synthesis. *Proc. natn. Acad. Sci. U.S.A.*, **54**, 1167–1173.                    (*281*)

THODAY, J. M. and READ, J. (1947). Effect of oxygen on the frequency of chromosome aberrations produced by X-rays. *Nature, Lond.*, **160**, 608.          (*150*)

THOMAS, C. A. (1956). The enzymatic degradation of desoxyribose nucleic acid. *J. Am. chem. Soc.*, **78**, 1861–1868.                    (*381*)

THOMAS, C. A. (1966). Recombination of DNA molecules. *Progr. nucleic Acid Res. molec. Biol.*, **5**, 315–337.                    (*346*)

THOMAS, C. A. (1971). The genetic organization of chromosomes. *A. Rev. Genet.*, **5**, 237–256.                    (*401*)

THOMAS, C. A., HAMKALO, B. A., MISRA, D. N. and LEE, C. S. (1970). Cyclization of eucaryotic deoxyribonucleic acid fragments. *J. molec. Biol.*, **51**, 621–632.   (*402*)

THOMAS, P. L. and CATCHESIDE, D. G. (1969). Genetic controls of flanking marker behaviour in an allelic cross of *Neurospora crassa*. *Can. J. Genet. Cytol.*, **11**, 558–566.
                    (*368*)

TIMOFÉËFF-RESSOVSKY, N. W. (1934). The experimental production of mutations. *Biol. Rev.*, **9**, 411–457.                    (*144*)

TJIO, J. H. and LEVAN, A. (1956). The chromosome number of man. *Hereditas*, **42**, 1–6.
                    (*70*)

TONOMURA, B., MALKIN, R. and RABINOWITZ, J. C. (1965). Deoxyribonucleic acid base composition of clostridial species. *J. Bact.*, **89**, 1438–1439.          (*278*)

TRAUTNER, T. A., SPATZ, H. C., BEHRENS, B., PAWLEK, B. and BEHNCKE, M. (1973). Exchange between complementary strands of DNA? *Adv. Biosci.*, **8**, 79–87.
                    (*365–66*)

TRAVERS, A. A. (1969). Bacteriophage sigma factor for RNA polymerase. *Nature, Lond.*, **223**, 1107–1110.                    (*257*)

TRAVERS, A. A. (1971). RNA polymerase and T4 development. *Cold Spring Harb. Symp. quant. Biol.*, **35**, 241–251.                    (*257*)

TRAVERS, A. A. and BURGESS, R. R. (1969). Cyclic re-use of the RNA polymerase sigma factor. *Nature, Lond.*, **222**, 537–540. *(256–57)*

TROSKO, J. E. and WOLFF, S. (1965). Strandedness of *Vicia faba* chromosomes as revealed by enzyme digestion studies. *J. Cell Biol.*, **26**, 125–135. *(372)*

TSCHERMAK, E. VON. (1900). Über künstliche Kreuzung bei *Pisum sativum*. *Ber. dt. bot. Ges.*, **18**, 232–239. (English translation, entitled 'Concerning artificial crossing in *Pisum sativum*' in *Genetics, Princeton*, 1950, **35**, suppl. pp. 42–47.) *(9)*

TSUGITA, A. (1962). The proteins of mutants of TMV: classification of spontaneous and chemically evoked strains. *J. molec. Biol.*, **5**, 293–300. *(243)*

TSUGITA, A. and FRAENKEL-CONRAT, H. (1962). The composition of proteins of chemically evoked mutants of TMV RNA. *J. molec. Biol.*, **4**, 73–82. *(274)*

TWARDZIK, D. R., GRELL, E. H. and JACOBSON, K. B. (1971). Mechanism of suppression in *Drosophila*: a change in tyrosine transfer RNA. *J. molec. Biol.*, **57**, 231–245. *(286)*

ULLERICH, F.-H. (1966). Karyotyp und DNS-Gehalt von *Bufo bufo, B. viridis, B. bufo* × *B. viridis* und *B. calamita* (*Amphibia, Anura*). *Chromosoma*, **18**, 316–342. *(383)*

UMBARGER, H. E. (1962). Feedback control by endproduct inhibition. *Cold Spring Harb. Symp. quant. Biol.*, **26**, 301–312. *(396)*

VENDRELY, R. and VENDRELY, C. (1948). La teneur du noyau cellulaire en acide désoxyribonucleique à travers les organes, les individus et les espèces animales. *Experientia*, **4**, 434–436. *(176)*

VISCHER, E., ZAMENHOF, F. and CHARGAFF, E. (1949). Microbial nucleic acids: the desoxypentose nucleic acids of avian tubercle bacilli and yeast. *J. biol. Chem.*, **177**, 429–438. *(178)*

VOGEL, H. J. (1965). Lysine biosynthesis and evolution. In *Evolving Genes and Proteins* (eds. Bryson, V. and Vogel, H. J.) Academic Press, New York, pp. 25–40. *(435, 436)*

VOLKIN, E. and ASTRACHAN, L. (1957). RNA metabolism in T2-infected *Escherichia coli*, In *The Chemical Basis of Heredity* edited by W. D. MCELROY and B. GLASS, Baltimore. Johns Hopkins Press, pp. 686–695. *(247–48, 251)*

VON WETTSTEIN, D. (1971). The synaptinemal complex and four-strand crossing over. *Proc. natn. Acad. Sci. U.S.A.*, **68**, 851–855. *(135)*

WACKER, A., DELLWEG, H. and WEINBLUM, D. (1960). Stahlenchemische Veränderung der Bakterien-Desoxyribonucleinsäure in vivo. *Naturwissenschaften*, **47**, 477. *(354)*

WAGNER, R. P., RADHAKRISHNAN, A. N. and SNELL, E. E. (1958). The biosynthesis of isoleucine and valine in *Neurospora crassa*. *Proc. natn. Acad. Sci. U.S.A.*, **44**, 1047–1053. *(170)*

WALDEYER, W. (1888). Ueber Karyokinese und ihre Beziehung zu den Befruchtungs-vorgängen. *Arch. mikrosk. Anat.*, **32**, 1–122. (English translation entitled 'Karyokinesis and its relation to the process of fertilization' in *Q. Jl microsc. Sci.*, 1889, **30**, 159–281.) *(58, 446)*

WALEN, K. H. (1965). Spatial relationships in the replication of chromosomal DNA. *Genetics, Princeton*, **51**, 915–929. *(413)*

WALKER, P. M. B. (1971). Origin of satellite DNA. *Nature, Lond.*, **229**, 306–308. *(408)*

WALKER, P. M. B. and YATES, H. B. (1952). Nuclear components of dividing cells. *Proc. R. Soc. B.*, **140**, 274–299. *(177)*

WALL, A. M., RILEY, R. and GALE, M. D. (1971). The position of a locus on chromosome 5B of *Triticum aestivum* affecting homoeologous meiotic pairing. *Genet. Res.*, **18**, 329–339.                                                                                    (*66*)

WALLACE, H. and BIRNSTIEL, M. L. (1966). Ribosomal cistrons and the nucleolar organizer. *Biochim. biophys. Acta*, **114**, 296–310.                                    (*403*)

WANG, A. C., WILSON, S. K., HOPPER, J. E., FUDENBERG, H. H. and NISONOFF, A. (1970). Evidence for control of synthesis of the variable regions of the heavy chains of immunoglobulins G and M by the same gene. *Proc. natn. Acad. Sci. U.S.A.*, **66**, 337–343.                                                                                    (*419*)

WARING, M. and BRITTEN, R. J. (1966). Nucleotide sequence repetition: a rapidly reassociating fraction of mouse DNA. *Science, N.Y.*, **154**, 791–794.          (*407*)

WARNER, J. R. and RICH, A. (1964). The number of soluble RNA molecules on reticulocyte polyribosomes. *Proc. natn. Acad. Sci. U.S.A.*, **51**, 1134–1141.          (*270*)

WATSON, J. D. and CRICK, F. H. C. (1953a). A structure for deoxyribose nucleic acid. *Nature, Lond.*, **171**, 737–738.                        (*179–85, 196, 198, 201, 217*)

WATSON, J. D. and CRICK, F. H. C. (1953b). Genetical implications of the structure of deoxyribonucleic acid. *Nature, Lond.*, **171**, 964–967.

(*185–210, 235–6, 346, 386, 452*)

WATSON, J. D. and CRICK, F. H. C. (1954). The structure of DNA. *Cold Spring Harb. Symp. quant. Biol.*, **18**, 123–131.                                          (*185–6*)

WEBER, H., BILLETER, M. A., KAHANE, S., WEISSMANN, C., HINDLEY, J. and PORTER, A. (1972). Molecular basis for repressor activity of Q$\beta$ replicase. *Nature New Biol.*, **237**, 166–170.                                                                      (*396*)

WEBSTER, R. E., ENGELHARDT, D. L. and ZINDER, N. D. (1966). *In vitro* protein synthesis: chain initiation. *Proc. natn. Acad. Sci. U.S.A.*, **55**, 155–161.          (*287*)

WEIGERT, M. G. and GAREN, A. (1965). Base composition of nonsense codons in *E. coli*. Evidence from amino-acid substitutions at a tryptophan site in alkaline phosphatase. *Nature, Lond.*, **206**, 992–994.                                          (*295*)

WEIGERT, M. G., LANKA, E. and GAREN, A. (1965). Amino acid substitutions resulting from suppression of nonsense mutations. II. Glutamine insertion by the Su-2 gene; tyrosine insertion by the Su-3 gene. *J. molec. Biol.*, **14**, 522–527.    (*297*)

WEINBERG, W. (1908). Über den Nachweis der Vererbung beim Menschen. *Jh. Ver. vaterl. Naturk. Württ.*, **64**, 369–382.                        (*13–14, 139, 399*)

WEINBERG, W. (1909). Über Vererbungsgesetze beim Menschen. *Z. VererbLehre* **1**, 377–392 and 440–460.                                                          (*139*)

WEINER, A. M. and WEBER, K. (1971). Natural read-through at the UGA termination signal of Q$\beta$ coat protein cistron. *Nature New Biol.*, **234**, 206–209.    (*292, 298*)

WEINTRAUB, H. (1972). Bi-directional initiation of DNA synthesis in developing chick erythroblasts. *Nature New Biol.*, **236**, 195–197.                          (*207*)

WEISBLUM, B., BENZER, S. and HOLLEY, R. W. (1962). A physical basis for degeneracy in the amino acid code. *Proc. natn. Acad. Sci. U.S.A.*, **48**, 1449–1454.    (*272*)

WEISBLUM, B. and DE HASETH, P. L. (1972). Quinacrine, a chromosome stain specific for deoxyadenylate-deoxythymidylate-rich regions in DNA. *Proc. natn. Acad. Sci. U.S.A.*, **69**, 629–632.                                                          (*407*)

WEISMANN, A. (1883). *Über die Vererbung*. Jena. (English translation entitled 'On Heredity,' in 'Essays upon Heredity and kindred biological Problems,' 1889, Oxford, Clarendon Press, pp. 67–105.)                                          (*40*)

WEISMANN, A. (1885). *Die Kontinuität des Keimplasmas* . . . Jena. (English translation entitled 'The continuity of the germ-plasm as the foundation of a theory of heredity' in 'Essays upon Heredity and kindred biological Problems,' 1889, Oxford. Clarendon Press, pp. 161–248.) *(40–41)*

WEISMANN, A. (1887). *Ueber die Zahl der Richtungskörper und über ihre Bedeutung für die Vererbung*. Jena. (English translation entitled 'On the number of polar bodies and their significance in heredity' in 'Essays upon Heredity and kindred biological Problems,' 1889, Oxford, Clarendon Press, pp. 333–384.) *(62, 69)*

WEISS, B. and RICHARDSON, C. C. (1967). Enzymatic breakage and joining of deoxyribonucleic acid, I. Repair of single-strand breaks in DNA by an enzyme system from *Escherichia coli* infected with T4 bacteriophage. *Proc. natn. Acad. Sci. U.S.A.*, **57,** 1021–1028. *(346)*

WEISS, S. B. and NAKAMOTO, T. (1961). The enzymatic synthesis of RNA: nearest-neighbor base frequencies. *Proc. natn. Acad. Sci. U.S.A.*, **47,** 1400–1405. *(249)*

WEISSMANN, C., SIMON, L. and OCHOA, S. (1962). Induction by an RNA phage of an enzyme catalyzing incorporation of ribonucleotides into ribonucleic acid. *Proc. natn. Acad. Sci., U.S.A.*, **49,** 407–414. *(245)*

WELSHONS, W. J. and RUSSELL, L. B. (1959). The Y-chromosome as the bearer of male determining factors in the mouse. *Proc. natn. Acad. Sci. U.S.A.*, **45,** 560–566. *(79)*

WENRICH, D. H. (1916). The spermatogenesis of *Phrynotettix magnus*, with special reference to synapsis and the individuality of the chromosomes. *Bull. Mus. comp. Zool. Harv.*, **60,** 55–136. *(63, 118–20, 130–31)*

WERNER, R. (1971a). Mechanism of DNA replication. *Nature, Lond.*, **230,** 570–572. *(202)*

WERNER, R. (1971b). Nature of DNA precursors. *Nature New Biol.*, **233,** 99–103. *(202)*

WESTERGAARD, M. (1958). The mechanism of sex determination in dioecious flowering plants. *Adv. Genet.* **9,** 217–281. *(79)*

WESTERGAARD, O., BRUTLAG, D. and KORNBERG, A. (1973). Initiation of deoxyribonucleic acid synthesis. IV. Incorporation of the ribonucleic acid primer into the phage replicative form. *J. biol. Chem.*, **248,** 1361–1364. *(209)*

WETTSTEIN, R. and SOTELO, J. R. (1967). Electron microscope serial reconstruction of the spermatocyte I nuclei at pachytene. *J. Microscopie*, **6,** 557–576. *(136)*

WHITE, J. M., LANG, A., LORKIN, P. A., LEHMANN, H. and REEVE, J. (1972). Synthesis of haemoglobin Lepore. *Nature New Biol.*, **235,** 208–210. *(324)*

WHITEHOUSE, H. L. K. (1963). A theory of crossing-over by means of hybrid deoxyribonucleic acid. *Nature, Lond.*, **199,** 1034–1040. *(342)*

WHITEHOUSE, H. L. K. (1965). A theory of crossing-over and gene conversion involving hybrid DNA. *Proc. 11th int. Congr. Genet.* 1963, **2,** 87–88. Oxford, Pergamon Press. *(359)*

WHITEHOUSE, H. L. K. (1966). An operator model of crossing-over. *Nature, Lond.*, **211,** 708–713. *(367–68)*

WHITEHOUSE, H. L. K. (1967). A cycloid model for the chromosome. *J. Cell Sci.*, **2,** 9–22. *(410)*

WHITEHOUSE, H. L. K. (1970). The mechanism of genetic recombination. *Biol. Rev.*, **45,** 265–315. *(370)*

WHITEHOUSE, H. L. K. (1972). Chromosomes and recombination. *Brookhaven Symp. Biol.*, **23,** 293–325. *(369)*

WHITEHOUSE, H. L. K. and HASTINGS, P. J. (1965). The analysis of genetic recombination on the polaron hybrid DNA model. *Genet. Res.*, **6**, 27–92.
(*345, 359, 367, 447*)

WIGLE, D. T. and DIXON, G. H. (1970). Transient incorporation of methionine at the N-terminus of protamine newly synthesized in trout testis cells. *Nature, Lond.*, **227**, 676–680.
(*293*)

WILCOX, M. S. (1928). The sexuality and arrangement of the spores in the ascus of *Neurospora sitophila. Mycologia*, **20**, 3–16.
(*104*)

WILKIE, D. (1970). Analysis of mitochondrial drug resistance in *Saccharomyces cerevisiae. Symp. Soc. exp. Biol.*, **24**, 71–83.
(*430*)

WILKINS, M. H. F. (1957). Physical studies of the molecular structure of deoxyribose nucleic acid and nucleoprotein. *Cold Spring Harb. Symp. quant. Biol.*, **21**, 75–88.
(*183, 386*)

WILKINS, M. H. F., STOKES, A. R. and WILSON, H. R. (1953). Molecular structure of deoxypentose nucleic acids. *Nature, Lond.*, **171**, 738–740.     (*182, 184*)

WILSON, E. B. (1896). *The Cell in Development and Inheritance*. New York, Macmillan.
(*446*)

WILSON, E. B. (1905). The chromosomes in relation to the determination of sex in insects. *Science, N.Y.*, **22**, 500–502.     (*73–74, 78*)

WILSON, E. B. (1909). Recent researches on the determination and heredity of sex. *Science, N.Y.*, **29**, 53–70.     (*74, 78, 446*)

WIMBER, D. E. and STEFFENSEN, D. M. (1970). Localization of 5S RNA genes on Drosophila chromosomes by RNA–DNA hybridization. *Science, N.Y.*, **170**, 639–641.     (*406*)

WINIWARTER, H. VON. (1900). Recherches sur l'ovogenèse et l'organogenèse de l'ovaire des Mammifères (Lapin et Homme). *Archs Biol., Paris*, **17**, 33–199.
(*63–67, 448, 450, 451*)

WINKLER, H. (1930). *Die Konversion der Gene*. Jena, Gustav Fischer.     (*328, 447*)

WINTERSBERGER, E. and VIEHHAUSER, G. (1968). Function of mitochondrial DNA in yeast. *Nature, Lond.*, **220**, 699–702.     (*430*)

WITKIN, E. M. (1969). The mutability toward ultraviolet light of recombination-deficient strains of *Escherichia coli. Mutation Res.*, **8**, 9–14.     (*356*)

WITTMANN, H. G. (1962). Proteinuntersuchungen an Mutanten des Tabakmosaik-virus als Beitrag zum Problem des genetischen Codes. *Z. VererbLehre*, **93**, 491–530.     (*274*)

WITTMANN, H. G., STÖFFLER, G., HINDENNACH, I., KURLAND, C. G., RANDALL-HAZEL-BAUER, L., BIRGE, E. A., NOMURA, M., KALTSCHMIDT, E., MIZUSHIMA, S., TRAUT, R. R. and BICKLE, T. A. (1971). Correlation of 30S ribosomal proteins of *Escherichia coli* isolated in different laboratories. *Molec. gen. Genet.*, **111**, 327–333.     (*247*)

WITTMANN, H. G. and WITTMANN-LIEBOLD, B. (1967). Protein chemical studies of two RNA viruses and their mutants. *Cold Spring Harb. Symp. quant. Biol.*, **31**, 163–172.
(*275*)

WOLFE, S. L. (1965). The fine structure of isolated chromosomes. *J. Ultrastruct. Res.*, **12**, 104–112.     (*373, 386*)

WOLFE, S. L. and HEWITT, G. M. (1966). The strandedness of meiotic chromosomes from *Oncopeltus*. *J. Cell Biol.*, **31**, 31–42. (*373*)

WOLFF, S. and LUIPPOLD, H. E. (1964). Chromosome splitting as revealed by combined X-ray and labeling experiments. *Expl Cell Res.*, **34**, 548–556. (*376*)

WOLSTENHOLME, D. R. (1971). The size and form of DNA molecules from microsporocytes of *Lilium longiflorum* at different stages of meiosis. *Chromosoma*, **33**, 396–402. (*410*)

WOLSTENHOLME, D. R., KOIKE, K. and RENGER, H. C. (1971). Form and structure of mitochondrial DNA. *Proc. X int. Cancer Cong.*, pp. 627–648. Chicago Year Book Medical Publishers, Inc. (*427–28*)

WOODWARD, D. O. (1959). Enzyme complementation in vitro between adenylosuccinaseless mutants of *Neurospora crassa*. *Proc. natn. Acad. Sci. U.S.A.*, **45**, 846–850. (*224*)

WOODWARD, D. O., EDWARDS, D. L. and FLAVELL, R. B. (1970). Nucleocytoplasmic interactions in the control of mitochondrial structure and function in *Neurospora*. *Symp. Soc. exp. Biol.*, **24**, 55–69. (*429*)

WOODWARD, D. O. and MUNKRES, K. D. (1966). Alterations of a maternally inherited mitochondrial structural protein in respiratory-deficient strains of *Neurospora*. *Proc. natn. Acad. Sci. U.S.A.*, **55**, 872–880. (*424, 429*)

WOODWARD, D. O., PARTRIDGE, C. W. H. and GILES, N. H. (1958). Complementation at the *ad*-4 locus in *Neurospora crassa*. *Proc. natn. Acad. Sci. U.S.A.*, **44**, 1237–1244. (*222–23*)

WOOLLAM, D. H. M. and FORD, E. H. R. (1964). The fine structure of the mammalian chromosome in meiotic prophase with special reference to the synaptinemal complex. *J. Anat.*, **98**, 163–173. (*434*)

WOOLLAM, D. H. M., FORD, E. H. R. and MILLEN, J. W. (1966). The attachment of pachytene chromosomes to the nuclear membrane in mammalian spermatocytes. *Expl Cell Res.*, **42**, 657–661. (*135*)

WRIGHT, M. and BUTTIN, G. (1969). Les mécanismes de dégradation enzymatique du chromsome bactérien et leur régulation. *Bull. Soc. Chim. biol.*, **51**, 1373–1383. (*348*)

WU, R. and TAYLOR, E. (1971). Nucleotide sequence analysis of DNA. II. Complete nucleotide sequence of the cohesive ends of bacteriophage $\lambda$ DNA. *J. molec. Biol.*, **57**, 491–511. (*412*)

WULFF, D. L. and RUPERT, C. S. (1962). Disappearance of thymine photodimer in ultraviolet irradiated DNA upon treatment with a photoreactivating enzyme from baker's yeast. *Biochem. biophys. Res. Commun.*, **7**, 237–240. (*354*)

WYATT, G. R. (1951). The purine and pyrimidine composition of deoxypentose nucleic acids. *Biochem. J.*, **48**, 584–590. (*179*)

WYATT, G. R. and COHEN, S. S. (1952). A new pyrimidine base from bacteriophage nucleic acids. *Nature, Lond.*, **170**, 1072–1073. (*179*)

YANIV, M. and BARRELL, B. G. (1971). Sequence relationship of three valine acceptor tRNAs from *Escherichia coli*. *Nature New Biol.*, **233**, 113–114. (*268*)

YANOFSKY, C. and BONNER, D. M. (1955). Gene interaction in tryptophan synthetase formation. *Genetics, Princeton*, **40**, 761–769. (*170*)

YANOFSKY, C., CARLTON, B. C., GUEST, J. R., HELINSKI, D. R. and HENNING, U. (1964). On the colinearity of gene structure and protein structure. *Proc. natn. Acad. Sci. U.S.A.*, **51**, 266–272.                                                    (*230, 232, 281, 363–64*)

YANOFSKY, C., DRAPEAU, G. R., GUEST, J. R. and CARLTON, B. C. (1967). The complete amino acid sequence of the tryptophan synthetase A protein (α subunit) and its colinear relationship with the genetic map of the A gene. *Proc. natn. Acad. Sci. U.S.A.*, **57**, 296–298.                                                             (*230, 232, 281*)

YANOFSKY, C. and ITO, J. (1966). Nonsense codons and polarity in the tryptophan operon. *J. molec. Biol.*, **21**, 313–334.                                           (*309–10*)

YANOFSKY, C., ITO, J. and HORN, V. (1967). Amino acid replacements and the genetic code. *Cold Spring Harb. Symp. quant. Biol.*, **31**, 151–162.              (*275, 281*)

YATES, F. (1934). Contingency tables involving small numbers and the $\chi^2$ test. *Jl R. statist. Soc.*, Suppl. 1, 217–235.                                               (*30, 34*)

YCAS, M. and VINCENT, W. S. (1960). A ribonucleic acid fraction from yeast related in composition to desoxyribonucleic acid. *Proc. natn. Acad. Sci. U.S.A.*, **46**, 804–811.
                                                                                     (*251*)

YOSHIDA, M., TAKEISHI, K. and UKITA, T. (1970). Anticodon structure of GAA-specific glutamic acid tRNA from yeast. *Biochem. biophys. Res. Commun.*, **39**, 852–7.   (*285*)

YOSHIKAWA, H., O'SULLIVAN, A. and SUEOKA, N. (1964). Sequential replication of the *Bacillus subtilis* chromosome, III. Regulation of initiation. *Proc. natn. Acad. Sci. U.S.A.*, **52**, 973–980.                                                            (*204*)

YOSHIKAWA, H. and SUEOKA, N. (1963). Sequential replication of *Bacillus subtilis* chromosome, I. Comparison of marker frequencies in exponential and stationary growth phases. *Proc. natn. Acad. Sci. U.S.A.*, **49**, 559–566.              (*201*)

YULE, G. U. (1907). On the theory of inheritance of quantitative compound characters on the basis of Mendel's law—a preliminary note. *R. hort. Soc. Rep. 3rd int. Conf. Genet.*, pp. 140–142.                                                     (*31, 37*)

ZACHAU, H. G., DÜTTING, D. and FELDMANN, H. (1966). Nucleotidsequenzen zweier serinspezifischer TransferRibonucleinsäuren. *Angew. Chem.*, **78**, 392–393. Nucleotide sequences of two serine-specific transfer ribonucleic acids. *Angew. Chem., int. Edn*, **5**, 422–423.                                             (*263, 285*)

ZAMENHOF, S., BRAWERMAN, G. and CHARGAFF, E. (1952). On the desoxypentose nucleic acids from several microorganisms. *Biochim. biophys. Acta*, **9**, 402–405.
                                                                                     (*210*)

ZAMENHOF, S. and GRIBOFF, G. (1954). Incorporation of halogenated pyrimidines into the deoxyribonucleic acids of *Bacterium coli* and its bacteriophages. *Nature, Lond.*, **174**, 306–307.                                                       (*196, 235*)

ZELENY, C. (1921). The direction and frequency of mutation in the bar-eye series of multiple allelomorphs of *Drosophila*. *J. exp. Zool.*, **34**, 203–233.        (*158*)

ZEN, S. (1961). Chiasma studies in structural hybrids. VI. Heteromorphic bivalent and reciprocal translocation in *Allium fistulosum*. *Cytologia*, **26**, 67–73.   (*122–23*)

ZICKLER, H. (1934). Genetische Untersuchungen an einen heterothallischen Asko- myzeten (*Bombardia lunata* nov. spec.). *Planta*, **22**, 573–613.              (*329*)

ZIMMERMANN, F. K. (1968). Enzyme studies on the products of mitotic gene conversion in *Saccharomyces cerevisiae*. *Molec. gen. Genet.*, **101**, 171–184.         (*366*)

ZIPSER, D. and NEWTON, W. A. (1967). The influence of deletions on polarity. *J. molec. Biol.*, **25,** 567–569. (*310*)

ZUBAY, G., SCHWARTZ, D. and BECKWITH, J. (1970). Mechanism of activation of catabolite-sensitive genes: a positive control system. *Proc. natn. Acad. Sci. U.S.A.*, **66,** 104–110. (*315*)

# Index

α-helix, 221
α-particles, 150
aberrant tetrads, 328–32, 340–1, 359–63
aberrations, chromosomal. *See* structural change
ABO blood groups, 139–40
abortive transduction, 337
*Abraxas*, 20–22, 74, 76, 441
acceptor RNA, 258
acentric chromosome, 125–9, 343, 414, 445
*Acetabularia*, 42–5, 242, 424, 427, 432, 436
acetate mutants, 56, 430
Achromobacteriaceae, 278, 435
*Acrida*, 121, 123, 441
acridines, 56–7, 150, 236–42, 276–7, 307, 319, 363
acriflavine, 56
Actinomycetales, 278, 301, 436
actinomycin D, 245, 307, 317, 352, 424
activator, 316, 322, 434
acquired characters, 1, 12, 40–1
adaptor hypothesis, 257–61
adenine, 172, 178–80, 183
adenosine derivatives, 266
adenylosuccinase, 218, 222–4
*Aerobacter aerogenes*, 194, 199, 435
*Agrostemma*, 10, 438
alanine, 220
aleurone. See *Zea*
Algae, 278–9, 436–7
alkaline phosphate, 224, 294–5, 297, 306, 316–17
alkaptonuria, 164
alkylating agents, 150
alleles, 9, 137–40, 211–18, 445
  complementation between, 218, 222–30, 319
  heterokaryon test, 166–7
  recombination of, 161–3, 211–18, 227–8, 232–3, 275–6, 281, 327–34, 359–71
*Allium*, 122–3, 381, 382, 437
allosteric inhibition, 396
*Aloe*, 118, 119, 437
*Alosa*, 177, 442
amber mutants, 230, 294–9, 305, 309–21, 364, 366, 425
amber suppressors, 230, 294–8, 309–11, 314, 320, 425
*Ambystoma*, 177, 405, 411, 442
amino-acid code, 273. *See* genetic code
amino-acid site, 270, 288–9
amino-acids, 220
  charged, 277

nucleotide triplets coding for, 273
  sequence in polypeptides, 221–2, 226, 230–5, 274–80, 415–19
  synthesis of, 167–70, 227, 301–3, 319–22
  transport of, 177
amino-acyl transfer RNA, 258, 288–9
amino-acyl transfer RNA synthetase, 258, 268, 290
2-aminopurine, 236–8, 294, 296
*Ammotragus*, 325, 443
Amphibia, 278–9, 393, 442
*Amphidinium*, 386, 436
anaphase, 58–70, 445
  bridge, 125–9, 381
*Anas*, 402, 443
*Anemone*, 382, 438
annelids, 278–9, 440
anthocyanin, 138, 164–5, 319, 388–91, 432–3
anthranilic acid, 167–8
antibody, 139, 163, 325, 415–19
anticodon, 259, 263–9, 282–99, 445
antigen, 139, 163
antiparallel, 200, 282, 376
*Antirrhinum*, 11, 12, 39, 80, 431, 433, 439
*Apis*, 373, 441
arabinose, 315–16, 322
arginine, 166–8, 171, 220
aromatic amino-acids, 320–1
*Arphia*, 72, 441
*Ascaris*. See *Parascaris*
*Ascobolus*, 331–3, 359, 361–3, 367, 371, 439
Ascomycetes, 278–9, 439–40
ascus, 104–8, 327–33
asparagine, 220
aspartic acid, 220
*Aspergillus nidulans*, 439
  adenine (*ad*) mutants, 211–13, 333–4
  arginine mutant, 171
  biotin (*bi*) mutants, 211–12, 214, 333
  competitive inhibition, 171
  cytoplasmic variant, 55
  diploid strains, 211–13, 327
  heterokaryons, 46–7, 55, 211–13
  lysine mutant, 171
  mitotic crossing-over, 212
  negative interference, 333–4
  nitrate, nitrite mutants, 322
  nutritional mutants, 318
  para-aminobenzoic acid (*paba*) mutants, 334
  recombination, 211–15, 327, 333–4, 357, 360

514

white conidia, 46–7, 55
yellow conidia (*y*), 211–12, 333
*Aspergillus niger*, 46–7, 439
assortive mating, 31
*Astasia*, 56, 439
*Aster*, 10, 439
ATP, 194, 254, 258, 271, 348
autoradiography, 64, 177–8, 186–92, 200–6
autosomes, 78–9, 445
auxotrophic mutants, 165–71, 210–13
*Avena*, 33–4, 437
Avogrado's constant, 428
*Azotobacter vinelandii*, 245, 435

β-galactosidase, 296, 302–3, 311, 395
Bacillaceae, 278, 436
*Bacillus cereus*, 317, 436
  subtilis, 201, 204, 255, 346, 365, 366, 402, 436
back-cross, 9, 18, 21, 38–9, 55, 445
back-mutation, 140, 144, 170, 236–41, 363, 367
bacteria, 172, 278–80, 435–6
bacteriophage, 173, 444
*Balaenoptera*, 235, 443
balanced polymorphism, 14
barley. See *Hordeum*
base analogues, 194, 230, 235–9, 251–2
base composition, of DNA. *See* DNA
  of RNA, 247–52
base ratio, 178, 196, 199, 278–80
base tilting, 286, 423
Basidiomycetes, 103, 278–9, 440
*Batrachoseps*, 94–5, 442
B-chromosomes, 392, 446
bead hypothesis, 157–63, 217, 369
bean, broad. See *Vicia*
bean, dwarf. See *Phaseolus*
*Bellevalia*, 189, 385, 437
binomial distribution, 24
biochemical mutants, 165–71, 210
biosynthetic pathways, 167–70, 301–2, 308, 318–22, 325–6, 396, 398
biotin, 166, 168, 211, 333–4
birds, 76, 278–9, 443
bivalent, 64–5, 69, 117–18
blood groups, 139–40, 163
Blue-green Algae, 172, 435–6
*Bombardia*, 329, 440
*Bos buffelus*, 325
*Bos taurus*, 443
  chromatin, 422
  chromosomes, 373
  DNA
    circles, 402
    composition, 178–9, 194, 198, 200
    content of nuclei, 176
    density, 193
    primer, 196, 255
    X-ray crystallography, 182, 386
  histones, 386, 423
  meiotic protein, 345
  nucleohistone, 386
  phosphodiesterase, 196–7
  polypeptides, 235

*Brachystola*, 69, 71, 72, 110, 118, 441
breakage-and-reunion hypothesis, 151–6
break-and-copy hypothesis, 352
bridge, at anaphase, 125–9, 381
broad bean. See *Vicia*
5-bromocytosine, 196
5-bromodeoxyuridine. *See* 5-bromouracil
5-bromouracil, 194, 196, 206, 207, 235–8, 252, 355, 357, 414
Bryophyta, 103, 437
*Bufo*, 383, 442

caesium chloride, 192–3
caffeine, 150
calf. See *Bos taurus*
*Calliptamus*, 119–20, 441
*Cancer*, 406, 440
cancer viruses, 245–6
*Canis*, 31, 443
*Capra*, 325, 443
*Carcharias*, 382, 442
*Carica*, 235, 438
cat. See *Felis*
catabolite repression, 315–16
cattle, domestic. See *Bos taurus*
*Cavia*, 291, 407, 443
centromere, 61, 445
  diffuse, 61
  distance of gene from, 106–10
cephalosporin C, 317
*Cercopithecus*, 204, 207, 443
*Chelidonium*, 10, 438
chemical mutagenesis, 146, 149, 152, 202, 235–9, 295, 319, 362–3
chiasma, 94, 341–2, 408, 445
  classical, 97, 124–5, 130–1, 133–4
  frequencies, 115–17
  reversed, 126–7
  terminal, 131–3, 446
chiasmatype theory, 94–136, 341–2
Chinese hamster. See *Cricetulus*
*Chironomus*, 148, 383–4, 401, 405, 411, 441
*Chlamydomonas*, 436
  cytoplasmic inheritance, 56, 430–1
  DNA composition, 199
  DNA content of chloroplasts, 427–8
  DNA replication, 193–4
  flagellar apparatus, 427
  tetrad analysis, 103
chloral hydrate, 62
chloramphenicol, 387, 428, 430
Chlorophyta, 278, 436
chloroplast, 103, 247, 427–31. *See* plastid
*Chorthippus*, 127, 422, 441
chromatid, 59, 446
  aberrations, 152–7, 374–6, 381–2
  interference, 112–15, 127, 449
  sub-units of, 188–92, 373–4, 381–2, 413–14
chromatin, 58–9, 422–3
chromatography, 178, 200, 219, 253, 256, 305, 309
chromocentre, 148–9, 408
chromomere, 63, 158, 206, 376–88, 393, 399–415, 433–4, 446

*Towards an Understanding of
the Mechanism
of Heredity*

"How odd it is that anyone should not see that all observation must be for or against some view if it is to be of any service!"

*Charles Darwin,* 18 September 1861, in a letter to Henry Fawcett. (*More Letters of Charles Darwin,* edited by Francis Darwin and A. C. Seward (1903). London, John Murray, vol. 1, pp. 194–196.)

Fawcett had just addressed the Botany and Zoology Section at the Manchester Meeting of the British Association for the Advancement of Science 'on the Method of Mr. Darwin in his Treatise on the Origin of Species'. In the discussion that followed, Edwin Lankester expressed the opinion that the facts which Darwin had brought forward in support of the hypothesis of evolution by natural selection were of more value than the hypothesis itself. This gave rise to the remark in Darwin's letter.